Nonlinear Microwave and RF Circuits

Second Edition

For a listing of recent titles in the *Artech House Microwave Library,* turn to the back of this book.

Nonlinear Microwave and RF Circuits

Second Edition

Stephen A. Maas

Artech House
Boston • London
www.artechhouse.com

Library of Congress Cataloging-in-Publication Data
Maas, Stephen A.
Nonlinear microwave and RF circuits / Stephen A. Maas.—2nd ed.
p. cm.
Rev. and updated ed. of: Nonlinear microwave circuits, 1988 and reprinted in 1997.
Includes bibliographical references and index.
ISBN 1-58053-484-8 (alk. paper)
1. Microwave circuits. I. Maas, Stephen A. Nonlinear microwave circuits. II. Title.

TK7876.M284 2003
621.381'32—dc21 2002043664

British Library Cataloguing in Publication Data
Maas, Stephen A.
Nonlinear microwave and RF circuits. — 2nd ed.— (Artech
House microwave library)
1. Microwave circuits 2. Radio circuits 3. Electronic networks,
Nonlinear
I. Titles
621.3'8132

ISBN 1-58053-484-8

Cover design by Gary Ragaglia

685 Canton Street
Norwood, MA 02062

International Standard Book Number: 1-58053-484-8
Library of Congress Catalog Card Number: 2002043664

10 9 8 7 6 5 4 3 2 1

Contents

Preface

Back in the days when I had a lot more energy and a lot less sense, I wrote the first edition of this book. I had just finished writing *Microwave Mixers*, and friends kept asking me, "Well, are you going to write another one?" Sales of *Mixers* were brisk, and the feedback from readers was encouraging, so it was easy to answer, "Sure, why not?" After a year of painful labor, *Nonlinear Microwave Circuits* was born.

The first edition of *Nonlinear Microwave Circuits* was published in 1988. It was well received and continued to sell well, even in a reprint edition, for the next 13 years. Now, it is out of print, and properly so: nonlinear circuit technology has advanced well beyond the material in the first edition of that book. In 1988, general-purpose harmonic-balance simulators had just become available, a workstation computer with an 8-MHz processor and 12 megabytes of memory was the state of the art, cell phones were the size of a shoebox, and the term *microwave bipolar transistor* was an oxymoron. My point isn't that we've come a long way; you know that. My point is that the book was clearly due to be updated.

Nonlinear Microwave Circuits has been almost completely rewritten, mainly to update its specific technical information. The general organization of the book, with the first half presenting theory, and the second design information, is unchanged. A couple of chapters, notably Chapters 4 and 5, are essentially unchanged, for obvious reasons. Chapter 2, on device modeling, is almost twice as long as in the original edition, and I easily could have made it longer. Chapter 3, on harmonic-balance analysis, is likewise much longer. The last seven chapters, which are design oriented, are completely new. In particular, design examples have been modernized, so they show how modern circuit-analysis software can best be exploited to produce first-class components.

Nonlinear Microwave Circuits has become *Nonlinear Microwave and RF Circuits*, a telling change. A large component of the evolution of high-frequency technology, since the first edition, is the importance of RF, wireless, and cellular systems. These depend strongly on heterojunction bipolar transistors, also a technology that has grown to maturity since the publication of the first edition. Similarly, power MOS devices, VHF/UHF transistors in 1988, are extremely important for power applications in the lower end of the microwave region. Finally, while in 1988 the MESFET was the only real option for microwave transistors, now we have high performance HEMT devices for both power and small-signal applications. These new technologies deserve, and have received, a place in this book.

I have many people to thank for their tolerance and assistance in this project. At the top of the list is my wife of 30 years, Julie, who never once has complained about my late nights in my office. My sons, David and Benjamin, also helped enormously, if only by growing up and leaving home. The whole gang at Applied Wave Research also deserve mention and thanks for discussions that clarified many of the dirty little details of making a nonlinear circuit simulator work the way it should. Finally, I am indebted to my colleagues in the nonlinear circuits business, far too many to list, for sharing the benefits of their hard-won experience.

Steve Maas
Long Beach, California
January 2003

Chapter 1

Introduction, Fundamental Concepts, and Definitions

Before we can describe the unique properties of nonlinear microwave circuits and the analytical methods necessary to understand them quantitatively, the author and reader must be certain that they both are speaking the same language. This is no small problem, because many of the terms and concepts inherent in nonlinear circuit theory are completely foreign to linear circuits, and many engineers harbor preconceived ideas about these circuits, ideas that are often not altogether correct. Accordingly, in order to establish a common basis for the following discussions, we begin by folding a few important definitions into an heuristic introduction to microwave nonlinearity.

1.1 LINEARITY AND NONLINEARITY

All electronic circuits are nonlinear: this is a fundamental truth of electronic engineering. The linear assumption that underlies most modern circuit theory is in practice only an approximation. Some circuits, such as small-signal amplifiers, are only very weakly nonlinear, however, and are used in systems as if they were linear. In these circuits, nonlinearities are responsible for phenomena that degrade system performance and must be minimized. Other circuits, such as frequency multipliers, exploit the nonlinearities in their circuit elements; these circuits would not be possible if nonlinearities did not exist. In these, it is often desirable to maximize (in some sense) the effect of the nonlinearities, and even to minimize the effects of annoying linear phenomena. The problem of analyzing and designing such circuits is usually more complicated than for linear circuits; it is the subject of much special concern.

The statement that all circuits are nonlinear is not made lightly. The nonlinearities of solid-state devices are well known, but it is not generally recognized that even passive components such as resistors, capacitors, and inductors, which are expected to be linear under virtually all conditions, are nonlinear in the extremes of their operating ranges. When large voltages or currents are applied to resistors, for example, heating changes their resistances. Capacitors, especially those made of semiconductor materials, exhibit nonlinearity, and the nonlinearity of iron- or ferrite-core inductors and transformers is legendary. Even RF connectors have been found to generate intermodulation distortion at high power levels; the distortion is caused by the nonlinear resistance of the contacts between dissimilar metals in their construction. Thus, the linear circuit concept is an idealization, and a full understanding of electronic circuits, interference, and other aspects of electromagnetic compatibility requires an understanding of nonlinearities and their effects.

Linear circuits are defined as those for which the superposition principle holds. Specifically, if excitations x_1 and x_2 are applied separately to a circuit having responses y_1 and y_2, respectively, the response to the excitation $ax_1 + bx_2$ is $ay_1 + by_2$, where a and b are arbitrary constants, which may be real or complex, time-invariant or time-varying. This criterion can be applied to either circuits or systems.

This definition implies that the response of a linear, time-invariant circuit or system includes only those frequencies present in the excitation waveforms. Thus, linear, time-invariant circuits do not generate new frequencies. (Time-varying circuits generate mixing products between the excitation frequencies and the frequency components of the time waveform; we'll examine this special case later in greater detail.) As nonlinear circuits usually generate a remarkably large number of new frequency components, this criterion provides an important dividing line between linear and nonlinear circuits.

Nonlinear circuits are often characterized as either *strongly nonlinear* or *weakly nonlinear*. Although these terms have no precise definitions, a good working distinction is that a weakly nonlinear circuit can be described with adequate accuracy by a Taylor series expansion of its nonlinear current/voltage (I/V), charge/voltage (Q/V), or flux/current (ϕ/I) characteristic around some bias current or voltage. This definition implies that the characteristic is continuous, has continuous derivatives, and, for most practical purposes, does not require more than a few terms in its Taylor series. (The excitation level, which affects the number of terms required, also must not be too high.) Additionally, we usually assume that the nonlinearities and RF drive are weak enough that the dc operating point is not perturbed. Virtually all transistors and passive components satisfy this definition if the excitation voltages are well within the components' normal operating ranges; that is, well below saturation. Examples of components that do not satisfy this definition are strongly driven transistors and Schottky-barrier diodes, because of their exponential *I*/*V* characteristics; digital logic gates, which have input/output transfer characteristics that vary abruptly with input voltage; and step-recovery diodes, which have very strongly nonlinear capacitance/voltage characteristics under forward bias. If a circuit is weakly nonlinear, relatively straightforward techniques, such as power-series or Volterra-series analysis, can be used. Strongly nonlinear circuits are those that do not fit the definition of weak nonlinearity; they must be analyzed by harmonic balance or time-domain methods. These circuits are not too difficult to handle if they include only single-frequency excitation or comprise only lumped elements. The most difficult case to analyze is a strongly nonlinear circuit that includes a mix of lumped and distributed components, arbitrary impedances, and multiple excitations.

Another useful concept is *quasilinearity.* A quasilinear circuit is one that can be treated for most purposes as a linear circuit, although it may include weak nonlinearities. The nonlinearities are weak enough that their effect on the linear part of the circuit's response is negligible. This does not mean that the nonlinearities themselves are negligible; they may still cause other kinds of trouble. A small-signal transistor amplifier is an example of a quasilinear circuit, as is a varactor-tuned filter.

Two final concepts we will employ from time to time are those of *two-terminal nonlinearities* and *transfer nonlinearities*. A two-terminal nonlinearity is a simple nonlinear resistor, capacitor, or inductor; its value is a function of one independent variable, the voltage or current at its terminals, called a *control voltage* or *control current*. A transfer nonlinearity is a nonlinear controlled source; the control voltage or current is somewhere in the circuit other than at the element's terminals. It is possible for a circuit element to have more than one control, one of which is usually the terminal voltage or current. Thus, many nonlinear elements must be treated as combinations of transfer and two-terminal nonlinearities. An example of a transfer nonlinearity is the nonlinear controlled current source in the equivalent circuit of a field-effect transistor (FET), where the drain current is a function of the gate voltage. Real circuits and circuit elements often include both types of nonlinearities. An example of the latter is the complete FET equivalent circuit described in Section 2.5.4, including nonlinear capacitors with multiple control voltages, transconductance, and drain-to-source resistance.

The need to distinguish between the two types of nonlinearities can be illustrated by an example. Consider a nonlinear resistor, Figure 1.1(a), and

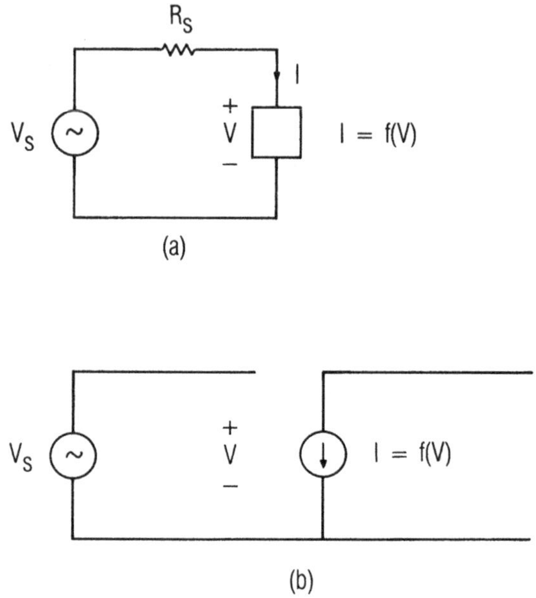

Figure 1.1 (a) Two-terminal nonlinearity; (b) transfer nonlinearity.

a nonlinear but otherwise ideal transconductance amplifier, Figure 1.1(b). Both are excited by a voltage source having some internal impedance R_S. The amplifier's output current is a function of the excitation voltage and the nonlinear transfer function; the current can be found simply by substituting the voltage waveform into the transfer function. In the two-terminal nonlinearity, however, the excitation voltage generates current components in the nonlinear resistor at new frequencies. These components circulate in the rest of the circuit, generating voltages at those new frequencies across R_S and therefore across the nonlinear resistor. These new voltage components generate new current components, and current and voltage components at all possible frequencies are generated.

1.2 FREQUENCY GENERATION

The traditional way of showing how new frequencies are generated in nonlinear circuits is to describe the component's I/V characteristic by a power series, and to assume that the excitation voltage has multiple frequency components. We will repeat this analysis here, as it is a good intuitive introduction to nonlinear circuits. However, our heuristic examination will illustrate some frequency-generating properties of nonlinear circuits that are sometimes ignored in the traditional approach, and will introduce some analytical techniques that complement others we will introduce in later chapters.

Figure 1.2 shows a circuit with excitation V_s and a resulting current I. The circuit consists of a two-terminal nonlinearity, but because there is no source impedance, $V = V_s$, and the current can be found by substituting the source voltage waveform into the power series. Mathematically, the situation is the same as that of the transfer nonlinearity of Figure 1.1(b).

The current is given by the expression

$$I = aV + bV^2 + cV^3 \tag{1.1}$$

where a, b, and c are constant, real coefficients. We assume that V_s is a two-tone excitation of the form

$$V_s = v_s(t) = V_1 \cos(\omega_1 t) + V_2 \cos(\omega_2 t) \tag{1.2}$$

Substituting (1.1) into (1.2) gives, for the first term,

$$i_a(t) = av_s(t) = aV_1 \cos(\omega_1 t) + aV_2 \cos(\omega_2 t) \tag{1.3}$$

After doing the same with the second term, the quadratic, and applying the well-known trigonometric identities for squares and products of cosines, we obtain

$$\begin{aligned} i_b(t) = bv_s^2(t) = \frac{b}{2}\{ & V_1^2 + V_2^2 + V_1^2 \cos(2\omega_1 t) + V_2^2 \cos(2\omega_2 t) \\ & + 2V_1 V_2 [\cos((\omega_1 + \omega_2)t) + \cos((\omega_1 - \omega_2)t)]\} \end{aligned} \tag{1.4}$$

and the third term, the cubic, gives

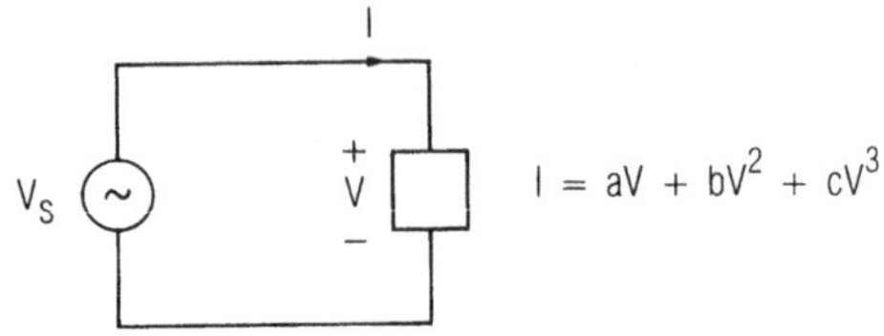

Figure 1.2 Two-terminal nonlinear resistor excited directly by a voltage source.

$$
\begin{aligned}
i_c(t) = cv_s^3(t) = \frac{c}{4}\{ & V_1^3 \cos(3\omega_1 t) + V_2^3 \cos(3\omega_1 t) \\
& + 3V_1^2 V_2 [\cos((2\omega_1 + \omega_2)t) + \cos((2\omega_1 - \omega_2)t)] \\
& + 3V_1 V_2^2 [\cos((\omega_1 + 2\omega_2)t) + \cos((\omega_1 - 2\omega_2)t)] \\
& + 3(V_1^3 + 2V_1 V_2^2) \cos(\omega_1 t) \\
& + 3(V_2^3 + 2V_1^2 V_2) \cos(\omega_2 t)\}
\end{aligned}
\tag{1.5}
$$

The total current in the nonlinear element is the sum of the current components in (1.3) through (1.5). This is the short-circuit current in the element; it consists of a remarkable number of new frequency components, each successive term in (1.1) generating more new frequencies than the previous one; if a fourth- or fifth-degree nonlinearity were included, the number of new frequencies in the current would be even greater. However, in this case, there are only two frequency components of voltage, at ω_1 and ω_2, because the voltage source is in parallel with the nonlinearity. If there were a resistor between the voltage source and the nonlinearity, even more voltage components would be generated via the currents in that resistor, those new voltage components would generate new current components, and the number of frequency components would be, theoretically, infinite. In order to have a tractable analysis, it then would be necessary to ignore all frequency components beyond some point; the number of components retained would depend upon the strength of the nonlinearity, the magnitude of the excitation voltage, and the desired accuracy of the result. The conceptual and analytical complexity of even apparently simple nonlinear circuits is the first lesson of this exercise.

A closer examination of the generated frequencies shows that all occur at a linear combination of the two excitation frequencies; that is, at the frequencies

$$\omega_{m,n} = m\omega_1 + n\omega_2 \tag{1.6}$$

where $m, n = ..., -3, -2, -1, 0, 1, 2, 3, ...$. The term $\omega_{m,n}$ is called a *mixing frequency*, and the current component at that frequency (or voltage component, if there were one) is called a *mixing product*. The sum of the absolute values of m and n is called the *order* of the mixing product. For the m, n to be distinct, ω_1 and ω_2 must be *noncommensurate*; that is, they are not both harmonics of some single fundamental frequency. We will usually assume that the frequencies are noncommensurate when two or more arbitrary excitation frequencies exist.

An examination of (1.3) through (1.5) shows that a kth-degree term in the power series (1.1) produces new mixing frequencies of order k or below; those mixing frequencies are kth-order combinations of the frequencies of the voltage components at the element's terminals. This does not, however, mean that $m + n < k$ in every nonlinear circuit. In the above example, the terminal voltage components were the excitation voltages, so only two frequencies existed. However, if the circuit of Figure 1.2 included a resistor in series with the nonlinear element, the total terminal voltage would have included not only the excitation frequencies, but higher-order mixing products as well. The nonlinear element then would have generated all possible kth-order combinations of those mixing products and the excitation frequencies. Thus, in general, a nonlinear element can generate mixing frequencies involving all possible harmonics of the excitation frequencies, even those where $m + n$ is greater than the highest power in the power series. It does this by generating kth-order mixing products between all the frequency components of its terminal voltage.

Another conclusion one may draw from (1.3) through (1.5) is that the odd-degree terms in the power series generate only odd-order mixing products, and the even-degree terms generate even-order products. This property can be exploited by balanced structures (Chapter 5). Balanced circuits combine nonlinear elements in such a way that either the even- or odd-degree terms in their power series are eliminated, so only even- or odd-order mixing frequencies are generated. These circuits are very useful in rejecting unwanted even- or odd-order mixing frequencies.

The generation of apparently low-order mixing products from the high-degree terms in (1.1) is worth some examination; the terms at ω_1 and ω_2 in (1.5) exemplify this phenomenon. The existence of these terms implies that the fundamental current, for example, is not solely a function of the excitation voltage and the linear term in (1.1); it is dependent on all the odd-degree nonlinearities. Consequently, as V_s is increased, the cubic term becomes progressively more significant, and the fundamental-frequency current components either rise more rapidly or level off, depending on the sign of the coefficient c. A closer inspection of these terms shows that they can be considered to have arisen from the kth-degree term as kth-order mixing products; for example, the ω_1 terms in (1.5) arise as the third-order combinations

$$\omega_1 = \omega_1 + \omega_1 - \omega_1 = \omega_1 + \omega_2 - \omega_2 \tag{1.7}$$

The presence of the negative frequencies might be more convincing if the cosine functions were expressed in their exponential form, $\cos(\omega t) = (\exp(j\omega t) + \exp(-j\omega t))/2$. Thus, when dealing with nonlinear circuits, one must always use a system of analysis that does not exclude the presence of negative frequencies.

It is worthwhile to consider some specific examples, in order to introduce one approach to nonlinear analysis and to gain further insights into the behavior of nonlinear circuits. Figure 1.3 shows a nonlinear circuit consisting of a resistive nonlinearity and a voltage source. The I/V nonlinearity includes only odd-degree terms:

$$I = f(V) = \frac{V}{2} + \frac{V^3}{7} + \frac{V^5}{15} \tag{1.8}$$

The 1Ω resistor complicates things somewhat, but the current can still be found via power-series techniques. First, we use a series reversion to find the voltage as a function of the current:

$$V = f^{-1}(I) = 2.0I - 2.286I^3 + 3.570I^5 + 3.184I^7 + \ldots \tag{1.9}$$

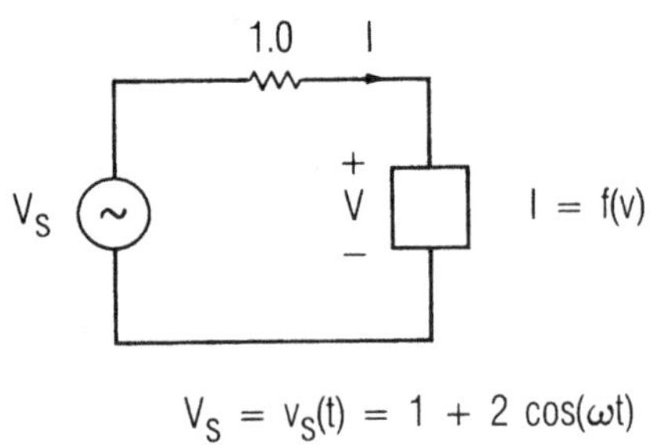

Figure 1.3 A nonlinear resistor, an excitation source, and a linear series resistor.

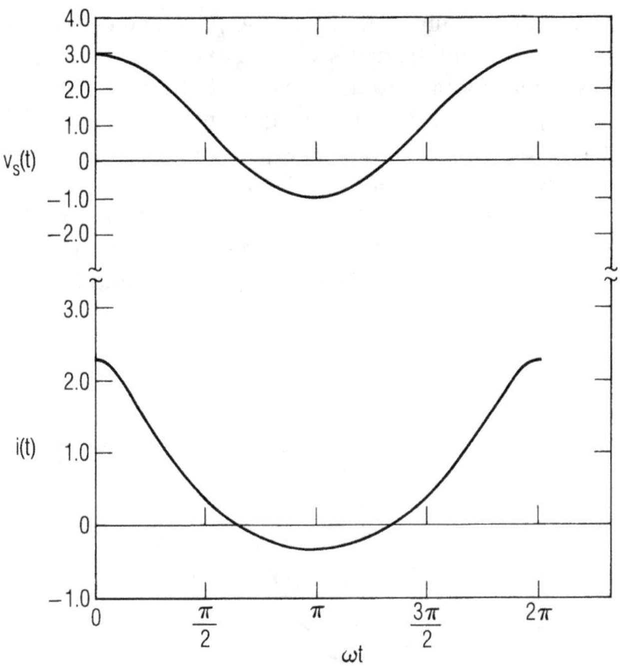

Figure 1.4 Voltage and current waveforms in the circuit of Figure 1.3.

The formula for the series reversion can be found in Abramowitz [1.1, p. 16]. The voltage across the resistor is $1 \cdot I$. Adding this to (1.9) (via Kirchoff's voltage law), we obtain

$$V_s = 3.0I - 2.286I^3 + 3.570I^5 + 3.184I^7 + \dots \tag{1.10}$$

Performing the reversion again gives, for the current,

$$I = 0.333V_s + 0.02822V_s^3 + 0.002271V_s^5 - 0.001375V_s^7 + \dots \tag{1.11}$$

Equation (1.11) expresses I in terms of the known excitation, V_S. It includes only odd terms because all the circuit elements, the nonlinear and linear resistors, have only odd terms in their power series. (We can view the linear resistor as a special case of a nonlinear resistor, having a one-term power "series".) The series in (1.11) is infinite, but it has been truncated after the seventh-degree term; the series does, in fact, include all odd harmonics, thus all odd-order mixing products. To illustrate this point, we assume that $V_S = v_S(t) = 1 + 2\cos(\omega t)$; $v_S(t)$ and the resulting $i(t)$ waveform are shown in Figure 1.4, where the presence of harmonics in the current waveform is evident from its obviously nonsinusoidal shape. The actual harmonics could be found by substituting the expression $v_S(t) = 1 + 2\cos(\omega t)$ into (1.11) and by applying the same algebra as in (1.1) through (1.5). It is also evident at a glance that the dc component of the current is much greater than 0.364A, the current that would be generated by the dc source alone if the ac source were zero. One must not forget that one of the low-order mixing frequencies generated by high-degree nonlinearities is a dc component; thus, the excitation of a nonlinear circuit may offset its dc operating point.

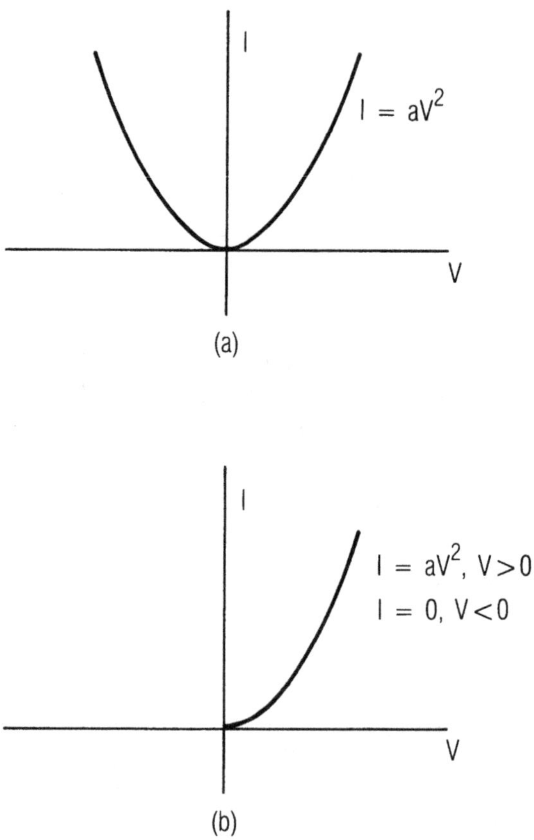

Figure 1.5 (a) *I*/*V* characteristic of the ideal square-law device; (b) *I*/*V* characteristic of a real "square-law" device.

As a second example, consider again the circuit of Figure 1.3 with

$$f(V) = aV^2 \tag{1.12}$$

where a is a constant, as shown in Figure 1.5. Equation (1.12) describes an ideal square-law device. This is a strange situation at the outset, for two reasons: first, the series reversion cannot be applied to (1.12); second, because the squared term generates only even-order mixing products, and the excitation frequency is a first- (i.e., odd-) order mixing product, no excitation-frequency current is possible! It is possible that a true square-law device could be made; however, it would be unstable, because its incremental resistance at some bias voltage V_0, $df(V) / dV$, $V = V_0$, would be negative when $V_0 < 0$. Practical two-terminal "square-law" elements employ solid-state devices and have I/V characteristics like that shown in Figure 1.5(b); the current follows a square law when $V > 0$ but is zero when $V < 0$. This characteristic still presents some analytical problems, because its I/V characteristic has a discontinuous derivative at $V = 0$. The device could, in concept, be operated in such a way that the voltage is always greater than zero, by biasing it at a value V_0 great enough that no negative excitation peaks can drive the terminal voltage to zero. Its power series then becomes

$$f(v + V_0) = a(v + V_0)^2 = a(V_0^2 + 2V_0 v + v^2) \tag{1.13}$$

where a, again, is a constant, and v is the voltage deviation from the bias point. Equation (1.13) includes the linear term $2V_0 v$. Thus, it is rarely possible, in practice, to obtain a true square-law device, or, for that matter, a device having only even-degree terms in its power series; practical devices invariably have at least one odd-order term in their power series. This generalization applies to many devices that are often claimed to be square-law devices, such as FETs.

Now that the pure square-law device has been ignominiously unmasked and shown to be a banal multiterm nonlinearity in disguise, it is interesting to see what happens to the circuit of Figure 1.3 when the nonlinearity includes even-degree terms, plus one odd-degree term, the linear one. By choosing the coefficients carefully, one can define the characteristic over any arbitrary range without generating negative resistances. We assume that

$$I = f(V) = V + 2V^2 + 3V^3 \tag{1.14}$$

After series reversion, and including the 1Ω resistor, we have

$$V_s = 2I - 2I^2 + 8I^3 - 43I^4 + 260I^5 + \ldots \tag{1.15}$$

which has all powers of I. Repeating the reversion again to obtain an expression for I in terms of V_s clearly results in a series having all powers of V_s. Thus, even though the original series contained only one odd-degree term (the linear one), the current contains mixing frequencies of all orders, even and odd, including those orders greater than four, the degree of the original power series.

In summary, the I/V characteristic of a nonlinear circuit or circuit element often can be characterized by a power series. The kth-degree term in the series generates kth-order mixing products of the frequencies in its control voltage or current. Some of these may coincide with lower-order frequencies. Mixing products may also coincide with higher-order frequencies; these are generated as kth-order mixing products between other mixing products. Thus, in general a nonlinear circuit having both even- and odd-degree nonlinearities in its power series generates all possible mixing frequencies, regardless of the maximum degree of its nonlinearities.

A special case of the nonlinear circuit having two-tone excitation occurs where one tone is relatively large, and the other is vanishingly small. This situation is encountered in microwave mixers, where the large tone is the local oscillator (LO), and the small one is the RF excitation. Because the RF excitation is very small, its harmonics are negligibly small, and we can assume that only its fundamental-frequency component exists. The resulting frequencies are

$$\omega = \omega_{\mathrm{RF}} + n\omega_{\mathrm{LO}} \tag{1.16}$$

which can also be expressed by our preferred notation,

$$\omega_n = \omega_0 + n\omega_{\mathrm{LO}} \tag{1.17}$$

where $n = \ldots, -3, -2, -1, 0, 1, 2, 3, \ldots$ and $\omega_0 = |\omega_{\mathrm{RF}} - \omega_{\mathrm{LO}}|$ is the mixing frequency closest to dc; in a mixer, ω_0 is often the intermediate frequency (IF), the output frequency. In (1.16) and (1.17) the mixing frequencies are above and below each LO harmonic, separated by ω_0.

If the total small-signal voltage $v(t)$ is much smaller than the LO voltage $V_L(t)$, the circuit can be assumed to be linear in the RF voltage. The total large-signal and small-signal current $I(t)$ in the nonlinearity of (1.1) is given by

$$I(t) = a(v(t) + V_L(t)) + b(v(t) + V_L(t))^2 + c(v(t) + V_L(t))^3 \tag{1.18}$$

Separating the small-signal part of (1.18), and assuming that $v^2(t) << v(t)$, we find the small-signal current $i(t)$ to be

$$i(t) \approx av(t) + 2bV_L(t)v(t) + 3cV_L^2(t)v(t) + \dots \tag{1.19}$$

This is a linear function of v, even though many of the current components in (1.19) are at frequencies other than the RF. Thus, a microwave mixer, which has an input at RF and output at, for example, ω_0, is a quasilinear component in terms of its input/output characteristics under small-signal excitation.

1.3 NONLINEAR PHENOMENA

The examination of new frequencies generated in nonlinear circuits does not tell the whole story of nonlinear effects, especially the effects of nonlinearities on microwave systems. Many types of nonlinear phenomena have been defined; the foregoing power series techniques can show how these arise from the nonlinearities in individual components or circuit elements. The phenomena described in this section are often considered to be entirely different; we shall see, however, that they are simply manifestations of the same nonlinearities.

1.3.1 Harmonic Generation

One obvious property of a nonlinear system is its generation of harmonics of the excitation frequency or frequencies. These are evident as the terms in (1.3) through (1.5) at $m\omega_1$, $m\omega_2$. The mth harmonic of an excitation frequency is an mth-order mixing frequency. In narrow-band systems, harmonics are not a serious problem because they are far removed in frequency from the signals of interest and inevitably are rejected by filters. In others, such as transmitters, harmonics may interfere with other communications systems and must be reduced by filters or other means.

1.3.2 Intermodulation Distortion

All the mixing frequencies in (1.3) through (1.5) that arise as linear combinations of two or more tones are often called *intermodulation* (IM) *products*. IM products generated in an amplifier or communications receiver often present a serious problem, because they represent spurious signals that interfere with, and can be mistaken for, desired signals. IM products are generally much weaker than the signals that generate them; however, a situation often arises wherein two or more very strong signals, which may be outside the receiver's passband, generate an IM product that is within the receiver's passband and obscures a weak, desired signal. Even-order IM products usually occur at frequencies well above or below the signals that generate them, and consequently are often of little concern. The IM products of greatest concern are usually the third-order ones that occur at $2\omega_1 - \omega_2$ and $2\omega_2 - \omega_1$, because they are the strongest of all odd-order products, are close to the signals that generate them, and often cannot be rejected by filters. Intermodulation is a major concern in microwave systems.

1.3.3 Saturation and Desensitization

The excitation-frequency current component in the nonlinear circuit examined in Section 1.2 was a function of power series terms other than the linear one; recall that (1.5) included components at ω_1 and ω_2 that varied as the cube of signal level. Such components are responsible for gain reduction and desensitization in the presence of strong signals.

In order to describe saturation, we refer to (1.1) to (1.5). From (1.3) and (1.5), and with $V_2 = 0$, we find the current component at ω_1, designated $i_1(t)$, to be

$$i_1(t) = \left(aV_1 + \frac{3}{4}cV_1^3\right)\cos(\omega_1 t) \tag{1.20}$$

If the coefficient c of the cubic term is negative, the response current saturates; that is, it does not increase at a rate proportional to the increase in excitation voltage. Saturation occurs in all circuits because the available output power is finite. If a circuit such as an amplifier is excited by a large and a small signal, and the large signal drives the circuit into saturation, gain is decreased for the weak signal as well. Saturation therefore causes a decrease in system sensitivity, called *desensitization*.

1.3.4 Cross Modulation

Cross modulation is the transfer of modulation from one signal to another in a nonlinear circuit. To understand cross modulation, imagine that the excitation of the circuit in Figure 1.1 is

$$V_s = v_s(t) = V_1 \cos(\omega_1 t) + (1 + m(t))\cos(\omega_2 t) \tag{1.21}$$

where $m(t)$ is a modulating waveform; $|m(t)| < 1$. Equation (1.21) describes a combination of an unmodulated carrier and an amplitude-modulated signal. Substituting (1.21) into (1.1) gives an expression similar to (1.5) for the third-degree term, where the frequency component in $i_c(t)$ at ω_1 is

$$i_c'(t) = \frac{3}{2} c V_1 V_2^2 (1 + 2m(t) + m^2(t))\cos(\omega_1 t) \tag{1.22}$$

where a distorted version of the modulation of the ω_2 signal has been transferred to the ω_1 carrier. This transfer occurs simply because the two signals are simultaneously present in the same circuit, and its seriousness depends most strongly upon the magnitude of the coefficient c and the strength of the interfering signal ω_2. Cross modulation is often encountered on an automobile AM radio when one drives past the transmission antennas of a radio station; the modulation of that station momentarily appears to come in on top of every other received signal.

1.3.5 AM-to-PM Conversion

AM-to-PM conversion is a phenomenon wherein changes in the amplitude of a signal applied to a nonlinear circuit cause a phase shift. This form of distortion can have serious consequences if it occurs in a system in which the signal's phase is important; for example, phase- or frequency-modulated communication systems. The response current at ω_1 in the nonlinear circuit element considered in Section 1.2 is, from (1.3) and (1.5),

$$i_1(t) = \left(aV_1 + \frac{3}{4} c V_1^3\right)\cos(\omega_1 t) \tag{1.23}$$

where $i_1(t)$ is the sum of first- and third-order current components at ω_1. Suppose, however, these components were not in phase. This possibility is not predicted by (1.1) through (1.5) because these equations describe a memoryless nonlinearity. In a circuit having reactive nonlinearities, however, it is possible for a phase difference to exist. The response is then the vector sum of two phasors,

$$I_1(\omega_1) = aV_1 + \frac{3}{4}cV_1^3\exp(j\theta) \tag{1.24}$$

where θ is the phase difference. Even if θ remains constant with amplitude, the phase of I_1 changes with variations in V_1. It is clear from comparing (1.24) to (1.20) that AM-to-PM conversion is most serious as the circuit is driven into saturation.

1.3.6 Spurious Responses

At the end of Section 1.2 we saw that a mixer, with an RF input at ω_{RF} and an LO at ω_{LO}, has currents at the frequencies given by (1.16) or (1.17). It is easy to see that, if the RF is applied at any of those mixing frequencies, currents at all the rest are generated as well. Thus the mixer has some response at a large number of frequencies, not just the one at which it is designed to work. In fact, if the applied signal is very strong, its harmonics are generated and the mixer has spurious responses at any frequency that satisfies the relation

$$\omega_{IF} = m\omega_{RF} + n\omega_{LO} \tag{1.25}$$

where m and n can both be either positive or negative integers. Comparing (1.25) to (1.6) shows that spurious responses are a form of two-tone intermodulation wherein one of the tones is the LO. In microwave technology the concept of spurious responses is used only in reference to mixers.

1.3.7 Adjacent Channel Interference

In many communications systems, especially those used for cellular telephones and other forms of telecommunications, modulated signals are squeezed into narrow, contiguous channels. Nonlinear distortion can generate energy that falls outside the intended channel. This is called *adjacent-channel interference*, *spectral regrowth*, or sometimes *co-channel interference*.

Adjacent-channel interference is fundamentally odd-order intermodulation distortion, and, like most odd-order IM, it is dominated by third-order effects, although higher-order nonlinearities may also contribute. The phenomenon is easy to understand. Volterra analysis (Chapter 4) of a weakly nonlinear, third-order system shows that the output is simply the sum of all possible third-order intermodulation products involving any three-fold combination of excitation frequency components. Like simple third-order intermodulation involving two excitation tones, many of these components fall close to the original excitation spectrum. These components cause adjacent-channel interference. Many components can also fall within the excitation channel as well, distorting the modulated signal.

1.4 APPROACHES TO ANALYSIS

One of the delights of the last decade or two has been the development of a theoretically sound approach to the analysis of nonlinear microwave circuits, and computer software that implements those methods. Previous techniques were questionable attempts to bend linear theory to nonlinear applications, were highly approximate, or were attempts at "black box" characterizations that did not include everything necessary to obtain correct results. Because some of these older methods (and the ideas they are based on) are still in use, it's worthwhile to take a brief look at some of the dominant methods, and to examine their validity.

1.4.1 Load Pull

One straightforward way to characterize a large-signal circuit, such as an amplifier, is to plot on a Smith chart the contours of its load impedances that result in prescribed values of gain and output power. These approximately circular contours can then be used to select an output load impedance that represents the best trade-off of gain against output power. The contours are generated empirically by connecting various loads to the amplifier and by measuring the gain and output power at each value of load impedance. This process, called *load pulling*, has many limitations; the most serious practical one is the difficulty of measuring the load impedances at the device terminals. Load pulling has a major theoretical problem as well: the load impedance at harmonics of the excitation frequency can significantly affect circuit performance, but load pulling is concerned primarily with the load impedance at the fundamental frequency. Furthermore, load pulling is not useful for determining other important properties of nonlinear or quasilinear circuits, for example, harmonic levels or the effects of multitone excitation.

Modern load-pull systems have overcome many of these limitations. Accurate calibration methods have been developed, as have been "harmonic load-pull" systems that account for harmonic tuning as well as fundamental frequency. Such systems can be valuable tools for designing and characterizing power devices. Still, there is need of a design process that does not require, at the outset, the user to make complicated and expensive measurements on his power transistors.

1.4.2 Large-Signal Scattering Parameters

Another approach to the analysis of large-signal, nonlinear circuits is to measure a set of two-port parameters, usually Scattering parameters (called *S parameters*), at the large-signal excitation level. The standard small-signal equations for S-parameter design are then used to predict the performance characteristics of the circuit. This approach may have limited success if the circuit or device is not very strongly nonlinear, and if it is not applied where it is obviously unsuited; for example, to frequency multipliers. Two-port parameters are fundamentally a *linear* concept, however, so the large-signal S-parameter approach represents a futile attempt to force nonlinear circuits to obey linear circuit theory.

In order to see just one example of the problems that arise from bending linear concepts to fit nonlinear problems, consider the meaning of the output reflection coefficient, S_{22}, of a FET or bipolar transistor. For large-signal S-parameter analysis, S_{22} is measured by applying an incident wave to the output port at a power level comparable to that at which the device is used. Now imagine that the device is driven hard at its input, and that the output reflection coefficient is again measured (ignore for a moment the obvious practical difficulties of making such a measurement). If the amplifier is significantly nonlinear, which in all likelihood it will be, one can hardly expect the reflection coefficient to be the same under these conditions, or over the wide range of incident power levels the device is likely to encounter. However, the S-parameter concept is based on the assumption at the that it will be the same.

Nevertheless, it is possible to define a large-signal driving point impedance that is valid for matching a source to the input of a nonlinear circuit. It is defined in the same manner as a linear impedance:

$$Z_{\text{in}}(\omega) = \frac{V(\omega)}{I(\omega)} \tag{1.26}$$

where $V(\omega)$ and $I(\omega)$ are the voltage and current components at the device terminals and at the excitation frequency ω. Other harmonics or mixing products are ignored in determining $Z_{in}(\omega)$. Because the circuit is nonlinear, $Z_{in}(\omega)$ is, in general, a function of the excitation level. We shall use this concept to determine port impedances in the design of many kinds of components described in later chapters.

1.4.3 Time-Domain (Transient) Analysis

An intermediate approach, which is theoretically valid and is frequently used for low-frequency analog and digital design, is to use time-domain techniques. It is a straightforward matter to write time-domain differential equations that describe a nonlinear circuit. Those differential equations are nonlinear, but they can be solved numerically. Although time-domain techniques are most practical for analyzing lumped-element circuits, a limited variety of distributed elements can be used as well. Time-domain analysis is not well suited when components are characterized in the frequency domain. The two major limitations of time-domain analysis are its inability to handle frequency-domain quantities (in particular, S parameters) in any practical way, its difficulty in dealing with transmission lines, and the difficulty of applying it to circuits having multiple noncommensurate excitation frequencies.

1.4.4 Frequency-Domain Methods

Many frequency-domain techniques for analyzing microwave circuits have become popular in recent years. The two most important are called *harmonic-balance analysis* and *Volterra-series analysis*. Harmonic-balance analysis is applicable primarily to strongly nonlinear circuits excited by a single large-signal source; it can be applied to such circuits as transistor power amplifiers, mixers, and frequency multipliers using either diodes or transistors. Volterra-series analysis is applicable to the opposite problem: weakly driven, weakly nonlinear circuits having multiple small-signal excitations at noncommensurate frequencies. As such, it is most useful for evaluating intermodulation characteristics and other nonlinear phenomena in small-signal receiver circuits, especially in amplifiers. With some modifications, the Volterra series can also be used to determine the IM properties of time-varying circuits such as mixers; similarly, harmonic-balance can be extended to certain situations involving noncommensurate signals. Demystifying the theory and practical use of these two techniques is the primary subject of this book.

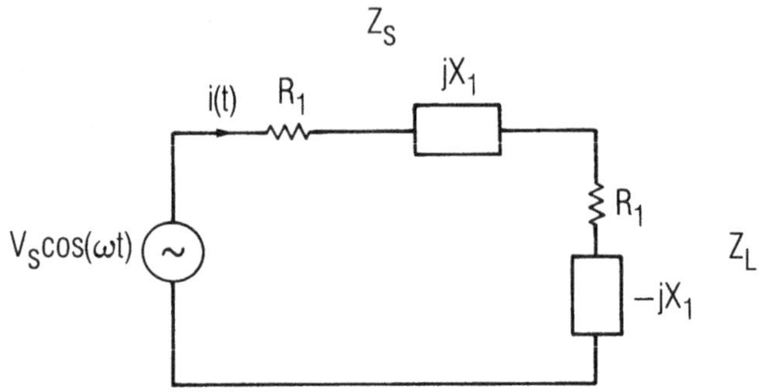

Figure 1.6 Circuit having a matched source and load, illustrating the concept of available power.

1.4.5 The Quasistatic Assumption

All three methods—time-domain analysis, harmonic-balance analysis, and the Volterra series—require a circuit model consisting of lumped components and, for the latter two, impedance elements or multiports. Solid-state device models must consist of linear or nonlinear capacitors, inductors, resistors, and voltage or current sources (nonlinear inductors can be accommodated, although they are rarely encountered in solid-state microwave devices or circuits). Underlying all the nonlinear models described in this book is the *quasistatic assumption*, whereby all nonlinear elements are assumed to change instantaneously with changes in their control voltages. This assumption is also implicit in linear circuit theory; it requires, for example, the charge on a capacitor to be a function solely of the voltage at its terminals. If the capacitor is nonlinear, its incremental capacitance, as well as its charge, must change instantaneously with control voltage. A quasistatic circuit is not necessarily memoryless; a memoryless circuit is one in which no charge or magnetic flux storage elements (no capacitors or inductors) exist, so voltages and currents at any instant do not depend upon previous values of voltage or current. In a quasistatic circuit, the network voltages and currents may depend upon previous values of other voltages or currents, but the capacitances, inductances, resistances, and controlled sources do not depend directly upon their own histories.

The quasistatic assumption is critical to the entire business of both linear and nonlinear circuit analysis. It allows one, for example, to devise equivalent circuits for solid-state devices using only lumped linear and nonlinear elements, and makes many of the techniques of linear circuit theory applicable to at least the linear parts of nonlinear circuits. One of the nicest things about the quasistatic assumption is its range of validity.

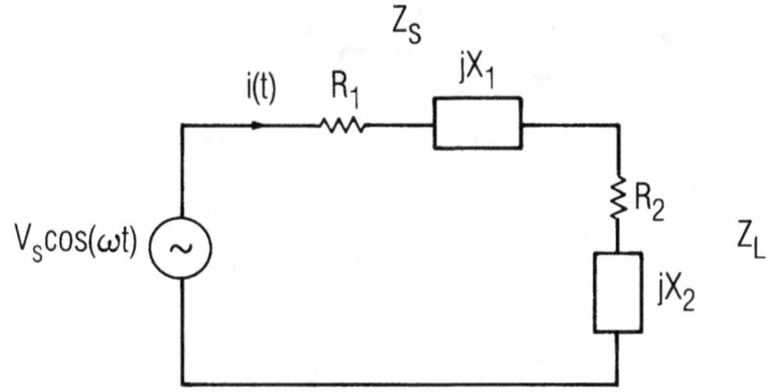

Figure 1.7 Circuit having an unmatched source and load.

Theoretical and experimental studies of silicon and gallium arsenide semiconductors and devices show that time-delay phenomena are usually on the order of picoseconds, or are short compared to the inverse of the highest frequency at which any sensible person would attempt to use the device. Furthermore, the prohibition of time delays is not absolute; in some cases they can still be managed, although with considerably greater difficulty.

1.5 POWER AND GAIN DEFINITIONS

Although it is customary to speak loosely of gain and power in microwave circuits, these quantities can be defined in several different ways. The different definitions of gain are related to the concepts of *available* and *dissipated* power. These concepts are important in both linear and nonlinear circuits, although they are particularly important in nonlinear circuits, where a waveform may have components at many frequencies that may or may not be harmonically related.

Available or *transferable power* is the maximum power that can be obtained from a source. The concept of available power is illustrated in Figure 1.6, in which a sinusoidal voltage source having a peak value V_s has an internal impedance of $R_1 + jX_1$ (unless we state otherwise, all frequency-domain voltages and currents in this book are phasor quantities; thus, their magnitudes are equal to peak sinusoidal quantities, not RMS). The maximum power is obtained from this source if the load impedance equals the conjugate of the source impedance, $Z_L = Z_s^* = R_1 - jX_1$. Under these conditions,

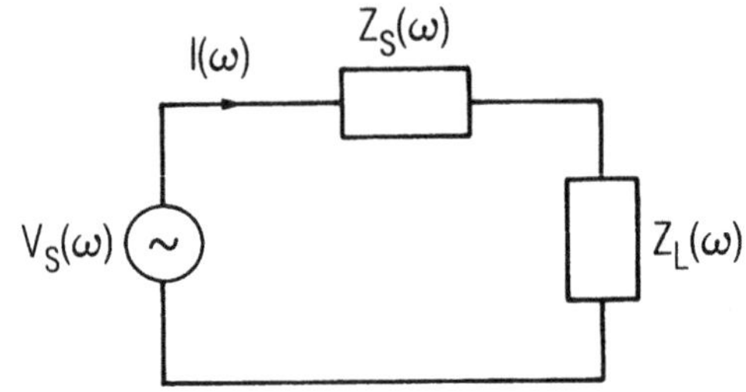

Figure 1.8 Unmatched circuit having a nonsinusoidal voltage-source excitation.

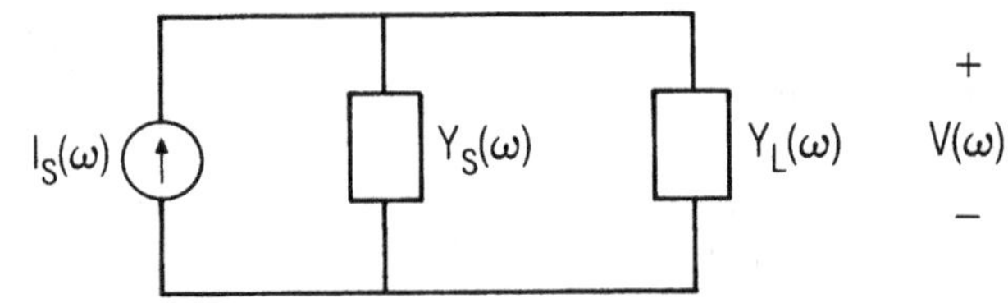

Figure 1.9 Unmatched circuit having a nonsinusoidal current-source excitation.

$$I = \frac{V_s}{2R_1} \tag{1.27}$$

where I is the peak value of the current, $i(t)$. The power dissipated in the load, P_d, is

$$P_d = P_{av} = \frac{1}{2}I^2 R_1 = \frac{1}{2}I^2 \mathrm{Re}\{Z_s\} = \frac{V_s^2}{8\mathrm{Re}\{Z_s\}} \tag{1.28}$$

which is the maximum available from the source, P_{av}. *Dissipated*, or *transferred power* is the power dissipated in a load that may or may not be matched to the source. In Figure 1.7, the load is not conjugate-matched to the source, so the dissipated power is somewhat less than that given in (1.28). In this case,

$$I = \frac{V_s}{((R_1 + R_2)^2 + (X_1 + X_2)^2)^{0.5}} \tag{1.29}$$

and the power dissipated in the load is

$$P_d = \frac{1}{2}I^2 R_2 = \frac{V_s^2 R_2}{2((R_1 + R_2)^2 + (X_1 + X_2)^2)} \tag{1.30}$$

In a nonlinear circuit the voltage source may contain many frequency components, and the source or load impedance may not be the same at each frequency. An example of this situation is the output circuit of a diode frequency multiplier. The multiplier generates many harmonics, all but one of which is undesired, so it has an output filter that allows only the desired harmonic to reach the output port. Thus, the impedance presented to the diode at the desired output frequency is the load impedance, but at all other harmonics it is the out-of-band impedance of the filter. The current in the loop is a function of frequency, as shown in Figure 1.8. Because the load and source are linear, each frequency component can be treated separately without concern for the others. Then the available and transferred power are

$$P_{av}(\omega) = \frac{|V_s(\omega)|^2}{8\text{Re}\{Z_s(\omega)\}} \tag{1.31}$$

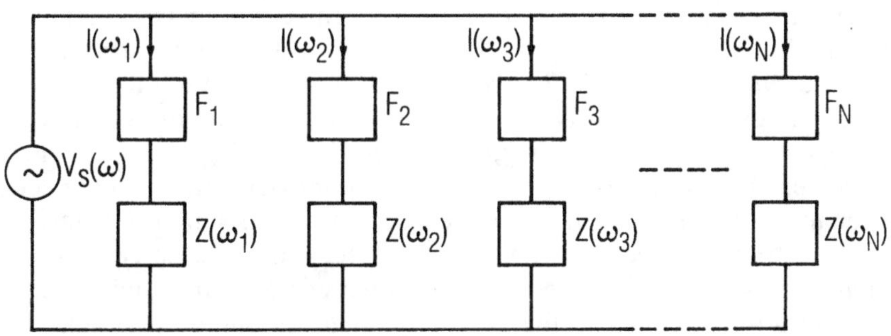

Figure 1.10 Model of a voltage source and load, where the excitation has a number of discrete frequency components.

$$P_d(\omega) = \frac{1}{2}|I(\omega)|^2 \mathrm{Re}\{Z_L(\omega)\} \tag{1.32}$$

An equivalent representation uses a current source and admittances as shown in Figure 1.9. Similarly, the available and dissipated powers are found to be

$$P_{\mathrm{av}}(\omega) = \frac{|I_s(\omega)|^2}{8\mathrm{Re}\{Y_s(\omega)\}} \tag{1.33}$$

$$P_d(\omega) = \frac{1}{2}|V(\omega)|^2 \mathrm{Re}\{Y_L(\omega)\} \tag{1.34}$$

Figure 1.10 shows a model often used when a voltage (or current) source has many discrete frequency components. The load impedance at each frequency is represented by an impedance in series with a filter. The filters $F_1, F_2, ..., F_N$ are ideal series-resonant circuits; that is, they are short circuits at their resonant frequencies and open circuits at all other frequencies. Thus, the current component at only one frequency circulates in each branch. One of these branches is the output circuit; the rest may be arbitrary impedances that represent the combined effects of out-of-band filter or matching circuit terminations, package or other circuit parasitics, or in some cases resonances (called *idlers*) that are purposely introduced to optimize performance. The terminations at intermediate frequencies may have a strong effect upon the circuit's performance, so the design of the output network may have to account for those terminations as well as the one at the output frequency.

The gain of a two-port network can be defined in terms of available and dissipated powers. The two most important gain definitions are *transducer gain* and *maximum available gain.* When a microwave engineer speaks loosely of "gain," he usually means (whether he knows it or not) transducer gain. To see why this is so, imagine a technician using a signal generator and power meter to measure the gain of an amplifier. First, he connects the power meter to the carefully matched output of the signal generator, and notes the power. Because the signal source and the power meter are matched, this is the available power. He then connects the signal generator to the amplifier input, and the power meter to its output, and again notes the output power. The output power is the power dissipated in the load, which is not necessarily conjugate-matched to the amplifier's output port. He calls the ratio of these powers the *gain*, which in this case is the power delivered to the load divided by the power available from the source. This is precisely the definition of transducer gain. Thus,

$$G_t = \frac{P_d \text{ at output}}{P_{\text{av}} \text{ at input}} \tag{1.35}$$

where G_t is the transducer gain.

Transducer gain is a very useful concept because, in microwave systems, it is most important to know how much more or less power a circuit delivers to a standard load (e.g., a 50 coaxial termination), compared to the power that could have been obtained from the source alone. This is precisely what transducer gain tells us. Furthermore, transducer gain is almost always a defined quantity, because it requires only that the source and output powers be finite, and real sources always have finite available power. Thus, the concept is handy in nonlinear circuits where, as our earlier discussion of large-signal S parameters illustrated, it is often impossible to define input and output impedances or reflection coefficients.

Other gain definitions are often useless because they do not tell the engineer what he wants to know, or occasionally result in meaningless or undefined quantities. One such concept is *power gain*, G_p, defined as power delivered to the load divided by power delivered to the two-port's input; thus,

$$G_p = \frac{P_d \text{ at output}}{P_d \text{ at input}} \tag{1.36}$$

We find that the power gain of a low-frequency MESFET amplifier, for example, is meaninglessly high: the FET's output power is modest, but its input impedance is highly reactive, so the input power is close to zero. This result tells nothing about the way the amplifier works in a system. The concept of power gain can give even more bizarre results when applied to other circuits, such as a negative-resistance amplifier without a circulator. The input power of a negative-resistance device is difficult to define, but one could justifiably say that it is negative and equal to the output power. Thus, the power gain of a negative-resistance amplifier is always –1. Even with these strange results, however, the concept of power gain has some limited usefulness; one of these uses the design of linear amplifiers that have prescribed values of transducer gain. This technique is described in Section 8.1.

Available gain, G_a, is defined as the power available from the output divided by the power available from the source; thus,

$$G_a = \frac{P_{av} \text{ at output}}{P_{av} \text{ at input}} \tag{1.37}$$

Available gain is intrinsically not a very useful concept (although it will co-star with power gain in Section 8.1), but its maximum value, called the *maximum available gain*, which occurs when the input of the two-port is conjugate-matched to the source, is very useful. The maximum available gain is, therefore, the highest possible value of the transducer gain, which occurs when both the input and output ports are conjugate-matched. Maximum available gain is defined only if the two-port is unconditionally stable; that is, if the input and output impedances always have positive real parts when any passive load is connected to the opposite port.

1.6 STABILITY

The fundamental definition of a stable electrical network is that its response is bounded when the excitation is bounded. In the case of a linear two-port having a sinusoidal steady-state excitation, this definition leads to a stability criterion: the network's poles must all be in the left half of the complex plane. A stable linear network can be made unstable through an unfortunate choice of source or load impedance; much of the "stability theory" of microwave circuits deals with this possibility, rather than the inherent stability of the circuit itself.

The situation is more complicated in the case of nonlinear circuits. Because the kinds of interactions that can occur in nonlinear circuits are more complex than in linear ones, such circuits often exhibit transient and steady-state phenomena other than sinusoidal oscillation, which, although bounded, are loosely classed as instability. These include parasitic oscillations; spurious outputs that occur only under large-signal excitation; "snap" phenomena, in which the output level or bias conditions change abruptly as input level is varied; chaotic behavior; and the exacerbation of normal noise levels. These may depend on initial conditions; some initial conditions may result in a stable response, others not, so it is strictly correct to speak only of a stable *solution*, not a stable *circuit*. Of course, plain, old-fashioned oscillation is also a possibility. Consequently, it is extremely difficult to devise a meaningful and practical stability criterion for nonlinear circuits.

Even without the academic advantage of a stability criterion, it is usually possible, with care, to design nonlinear or quasilinear circuits that are well-behaved. For example, if a harmonic-balance analysis of a proposed circuit design converges without incident to a solution, one can be confident that it is, by all practical definitions of the term, stable. (It is also stable in theory, because harmonic-balance analysis is a process of perturbing the voltages across the nonlinear elements. If these perturbations do not cause larger perturbations, the circuit must be locally stable. The idea that a circuit is stable if such perturbations do not cause greater perturbations is equivalent to the concept of stability defined earlier.) The converse may not be true, however, because the failure of an iterative technique such as harmonic balance to converge may be caused by numerical problems, not by inherent instability.

In oscillators, we have yet another concept of stability. At start-up, an oscillator is an unstable, linear circuit; it must have poles in the right half plane. However, once the oscillation is established, it must be stable, in the sense that it remains in a steady state and returns to that steady state after any small perturbation. This is a loose description of a concept known as *Liapunov stability.*

Reference

[1.1] M. Abramowitz and I. A. Stegun, *Handbook of Mathematical Functions*, New York: Dover, 1970.

Chapter 2

Solid-State Device Modeling for Quasistatic Analysis

Inherent in nonlinear circuit analysis is the *quasistatic assumption*—the assumption that the current, charge, or flux in a nonlinear element is an algebraic function of one or more control voltages or currents. Thus, when the control voltage changes, the controlled quantity changes instantaneously. As the dominant nonlinearities in a microwave circuit are inevitably those of its solid-state devices, it is important to have quasistatic models for those devices. Models consisting of lumped linear and nonlinear elements are usually most practical. Determining the current-voltage (I/V) or charge-voltage (Q/V) expressions for the nonlinear elements of the equivalent circuit is the key to characterizing the device.

This chapter does not attempt a survey of specific models (which could easily be a book in itself), but instead addresses the theory underlying quasistatic device models and various considerations in their implementation.

2.1 NONLINEAR DEVICE MODELS

Because they are fundamentally linear concepts, impedance and multiport circuit theory cannot describe a nonlinear circuit. Accordingly, the most popular means for characterizing transistors—S, Y, or other multiport parameters—cannot be used to model nonlinear solid-state devices. Instead, the most successful (but by no means the only) method of characterizing such devices is to use a lumped circuit model that includes a mix of linear and nonlinear resistors, capacitors, and controlled sources (nonlinear inductors can also be included, but they are rarely encountered in microwave circuits). The nonlinear elements are invariably assumed to be quasistatic; for microwave FETs and diodes, the quasistatic assumption is valid to at least 100 GHz. The nonlinear elements in transistor and diode models are invariably voltage controlled, usually having one or two control voltages.

Quasistatic modeling is usually not applicable to devices whose operation is dominated by time effects. These include transit-time devices, such as IMPATTs, and Gunn (also called *transferred-electron*) devices. Such devices are so strongly nonlinear that they are rarely used in circuits that have amplitude-modulated or multiple CW excitations, and are usually used as oscillators or amplifiers of CW or constant-amplitude signals. Conversely, the models developed in this chapter are particularly useful in nonlinear and quasilinear circuits commonly employed in communications and radar systems; such components include small-signal amplifiers, linear power amplifiers, and harmonic generators.

An obvious requirement of a good device model is that it be sufficiently accurate and that it maintain its accuracy over a wide frequency range. Solid-state devices, however, are not simply lumped-element circuits, so any such model is necessarily an approximation. Although complex models may be more accurate than simple ones (or may not be; see Section 2.3.12), the natural desire to minimize computational difficulty often dictates that the simplest adequate model be used. Concern for computational difficulty is crucial because many nonlinear analyses require many—perhaps tens or hundreds of thousands—of evaluations of the circuit equations. Thus, the use of unnecessarily complex models may involve excessive computational cost. Unnecessary complexity also may introduce convergence difficulties in harmonic-balance analysis.

Another requirement is that it must be reasonably easy to extract the model's parameters from straightforward measurements. The nonlinear characterization of any solid-state device usually requires a number of measurements. If the number and difficulty of these measurements are excessive, the design cost of the resulting circuit is increased and accuracy may suffer, if only because of the greater chance of error. A nonlinear design technique that requires laborious measurements is not likely to be widely accepted and will always tempt the designer to shortcut the process. The result might be that a more complex technique, which is theoretically very accurate, may be less accurate in practice than a simpler one properly executed. In any case, accuracy is constrained by the device process: there is no sense in creating a model having 1% accuracy if the device process tolerances are 10%.

2.2 NONLINEAR LUMPED CIRCUIT ELEMENTS AND CONTROLLED SOURCES

The nonlinear device models we consider are equivalent circuits, consisting of resistors, capacitors, and controlled sources. In the rare cases where nonlinear inductors occur, they can also be accommodated (Section 2.2.8). The circuit elements can be described by one of two kinds of characteristics: the large-signal, global characteristic, or by an incremental, small-signal characteristic. The former describes the overall *I/V* or *Q/V* relationship and is used for modeling large-signal circuits; the latter describes the deviation of voltage and current or charge in the vicinity of a bias point and is used for modeling small-signal, quasilinear circuits or for Volterra analysis. In the large-signal case, the circuit element is effectively treated as a "black box" having the prescribed *I/V* or *Q/V* characteristic; in the small-signal case, it is a linear or nonlinear small-signal resistor, capacitor, or controlled source having a resistance, capacitance, or small-signal current that is a function of a dc (or occasionally time-varying) control voltage. In this section we examine the relationship between the large-signal and small-signal characterizations, and in particular show how the small-signal characterization can be derived from the large-signal one.

Three concepts critical to the modeling of nonlinear solid-state devices are *voltage control*, *current control*, and *incremental quantities*. A voltage-controlled element is dependent upon a voltage that either may be applied to its terminals or may exist elsewhere in the circuit. The element's value (usually its current, voltage, charge, capacitance, or conductance) must be a single-valued function of the control voltage. For example, it is usually natural to express a diode junction capacitance as a single-valued function of the junction voltage. Conversely, a current-controlled element is one whose value is a single-valued function of a current.

In theory, many elements can be treated as either current- or voltage-controlled. For example, the small-signal junction conductance of a Schottky-barrier diode can be expressed as an exponential function of voltage or as a linear function of current. Either way, the function is single-valued, and a choice of expressing the device as current- or voltage-controlled depends primarily on convenience. In contrast, the current in some types of diodes rises with junction voltage, then drops as voltage is further increased. Such nonlinearities must be treated as voltage-controlled, because the junction voltage is not a single-valued function of current. In practice, most microwave devices are voltage-controlled nonlinearities.

An important question involves the precise definitions of the small-signal resistance and capacitance of nonlinear elements. For example, the global *I/V* characteristic of a linear resistor is given by Ohm's law, $V = RI$.

But suppose a current-controlled nonlinear resistor were used in an application where a small-signal ac current is applied, and a dc control current I_0 exists. The ac component of its voltage should be given by $v(t) = r(I_0)\, i(t)$, where $v(t)$ and $i(t)$ are the small-signal voltage and current, respectively. Figure 2.1 illustrates this case, and it is clear that

$$v(t) = \left.\frac{dV}{dI}\right|_{I = I_0} i(t) \tag{2.1}$$

so

$$r(I_0) = \left.\frac{dV}{dI}\right|_{I = I_0} \tag{2.2}$$

Of course, (2.1) is exact only as the magnitude of $i(t)$ approaches zero. This definition of resistance is called the *incremental resistance* and is valid in small-signal quasilinear analysis. The same idea applies to controlled sources, such as those described by a linear transconductance; thus, in FETs,

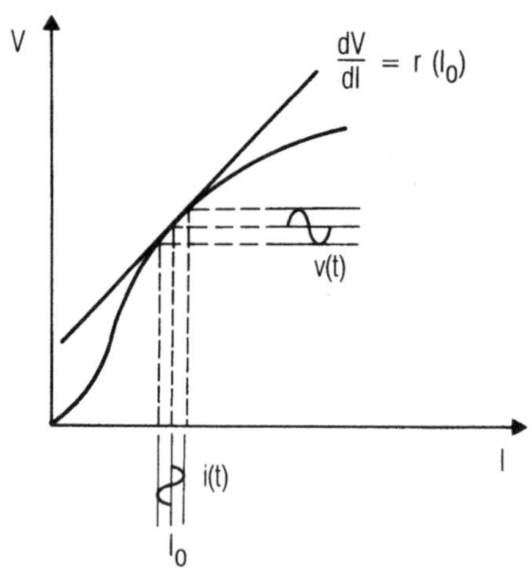

Figure 2.1 Incremental resistance of a nonlinear resistor at the dc bias point, I_0.

$$g_m(V_{g0}) = \left.\frac{dI_d}{dV_g}\right|_{V_g = V_{g0}} \tag{2.3}$$

in which we assumed the drain current I_d to be a function of the gate voltage V_g only. V_{g0} is the gate-bias voltage.

2.2.1 The Substitution Theorem

The expressions (2.1) through (2.3), or the more complete ones that follow, can describe either a single element or a controlled source. In a nonlinear conductance, the control voltage is applied to the element's terminals; in a controlled source, the control voltage is somewhere else in the circuit. This point can be clarified via the substitution theorem, which defines an equivalence between a circuit element and a controlled source.

Figure 2.2(a) shows a linear voltage-controlled current source connected to a network, *N*. Its current is $G\,V$, where V, the control voltage, is the voltage at its terminals. The current is clearly unchanged if a conductance of value G is substituted for the controlled source. The same is true if the I/V relationship of the source is a more complicated nonlinear function of voltage: a conductance having the same I/V characteristic can be substituted, and the representations are equivalent. We now can state the substitution theorem precisely: a linear or nonlinear resistive circuit element having the characteristic $I = f(V)$ is equivalent to a controlled current source having the same characteristic, wherein V is the terminal voltage. Although this definition refers to large-signal V and I, the substitution theorem is equally applicable to a small-signal incremental characteristic. Also, it is applicable by analogy to capacitive elements or current-controlled elements.

One important application of the substitution theorem is shown in Figure 2.2(b), where the I/V characteristic of a nonlinear conductance is described by the power series $I = f(V) = G_1V + G_2V^2 + G_3V^3 + \ldots$. The nonlinear element can be described by an equivalent circuit that includes a linear conductance G_1 and controlled current sources representing the higher-degree terms in the series. Of course, the linear component G_1V could also be represented by a current source if it were more convenient to do so.

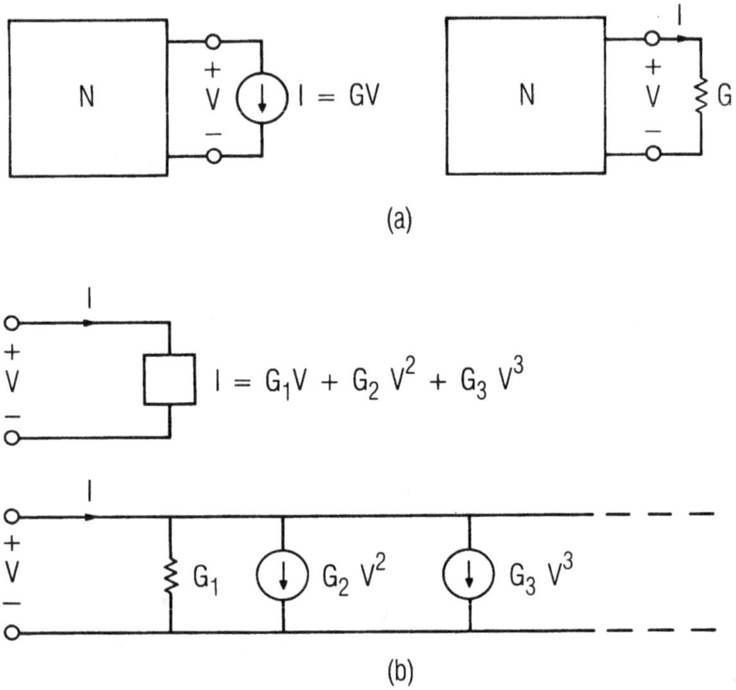

Figure 2.2 The substitution theorem: (a) source-conductance equivalence; (b) nonlinear element equivalence.

2.2.2 Large-Signal Nonlinear Resistive Elements

A nonlinear resistive element, whether a controlled source or two-terminal, can be described either by an I/V function having the form

$$I = f_V(V_1, V_2, \ldots) \tag{2.4}$$

or as a current-controlled element,

$$V = f_I(I_1, I_2, \ldots) \tag{2.5}$$

Most microwave devices are best described as voltage-controlled current sources, so (2.5) is rarely used. Indeed, we shall see in Chapter 3, when we discuss harmonic-balance analysis, that using only voltage-controlled nonlinearities simplifies the description of the linear subcircuit considerably, and a gyrator can be used in the rare cases when a current-controlled element cannot be avoided.

The form of f_V (or f_I) for a particular device requires careful consideration. The obvious requirement is that the function reproduce the measured I/V characteristic of the nonlinearity. However, there are a number of additional considerations, most of which are not obvious. These will be addressed in Section 2.3.

2.2.3 Small-Signal Nonlinear Resistive Elements

An element described by the I/V characteristic $I = f(V)$, as shown in Figure 2.3, is a voltage-controlled conductance. We assume that it has a dc control voltage V_0, which in practice could be a bias voltage, and a small-signal ac voltage $v(t)$. We can expand the current in a Taylor series around V_0 to determine its ac part:

$$f(V_0 + v) = f(V_0) + \frac{d}{dV} f(V)\bigg|_{V = V_0} v + \frac{1}{2}\frac{d^2}{dV^2} f(V)\bigg|_{V = V_0} v^2 + \frac{1}{6}\frac{d^3}{dV^3} f(V)\bigg|_{V = V_0} v^3 + \ldots \tag{2.6}$$

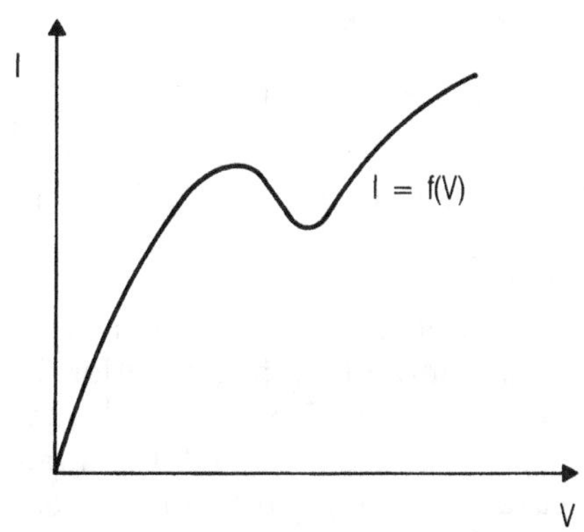

Figure 2.3 A voltage-controlled nonlinear resistive element.

where the indication of time dependence, "(t)", has been deleted from $v(t)$ for simplicity [the "(t)" will be deleted from all the small-signal voltage, current, and charge waveforms in all the equations in this section]. We can assume that $v \ll V_0$ and that the nonlinearity is weak enough so that the series converges. Then the small-signal current i is

$$i = f(V_0 + v) - f(V_0) = \left.\frac{d}{dV} f(V)\right|_{V = V_0} v + \frac{1}{2} \left.\frac{d^2}{dV^2} f(V)\right|_{V = V_0} v^2 + \frac{1}{6} \left.\frac{d^3}{dV^3} f(V)\right|_{V = V_0} v^3 + \ldots \tag{2.7}$$

In (2.7) the current i is the total small-signal current, including dc as well as ac components. Thus, i has dc components even though v has only ac components because the even-degree terms in (2.7) introduce them. Under the stated assumptions, the change in the dc operating point is small compared to $f(V_0)$, so the dc components resulting from v^2, v^4, ... are usually negligible. In the quasilinear case, the terms of degree greater than one are assumed to be negligible, so

$$i = \left.\frac{d}{dV} f(V)\right|_{V = V_0} v = g(V_0)\, v \tag{2.8}$$

where $g(V_0)$ is the incremental conductance at V_0. In the nonlinear case, (2.7) can be expressed as

$$i = g_1 v + g_2 v^2 + g_3 v^3 + \ldots \tag{2.9}$$

and, with the help of the substitution theorem, the nonlinear element can be modeled as shown in Figure 2.4. The linear term, g_1, in (2.9) is the incremental conductance.

Often a nonlinear circuit element is controlled by more than one current or voltage. An example of such a situation is the simplified FET equivalent circuit shown in Figure 2.5, in which the drain current I is a function of both the gate voltage V_1 and the drain voltage V_2; thus, $I = f(V_1, V_2)$. In this case, V_2 is applied to the current source and V_1 is a node voltage elsewhere in the circuit. In most practical nonlinear elements that have multiple control voltages, at least one of the voltages is the

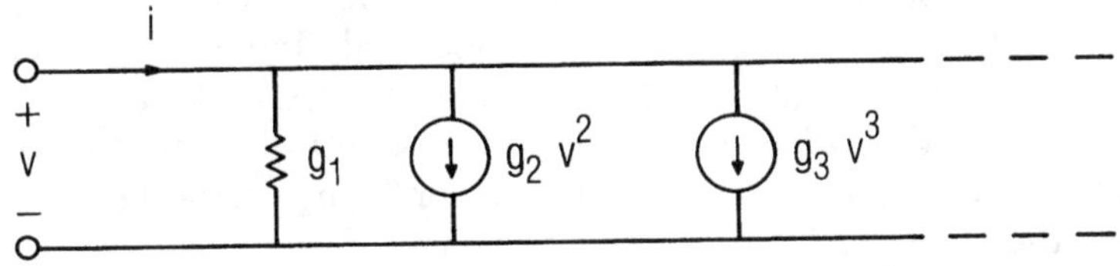

Figure 2.4 Small-signal nonlinear equivalent circuit of the nonlinear conductance.

applied voltage, but the theory makes no such requirement. The function $f(V_1, V_2)$ can be expanded in a two-dimensional Taylor series, and the dc current component subtracted, giving the rather sticky expression

$$\begin{aligned} i = {} & \frac{\partial f}{\partial V_1} v_1 + \frac{\partial f}{\partial V_2} v_2 \\ & + \frac{1}{2}\left(\frac{\partial^2 f}{\partial V_1^2} v_1^2 + 2\frac{\partial^2 f}{\partial V_1 \partial V_2} v_1 v_2 + \frac{\partial^2 f}{\partial V_2^2} v_2^2\right) \\ & + \frac{1}{6}\left(\frac{\partial^3 f}{\partial V_1^3} v_1^3 + 3\frac{\partial^3 f}{\partial V_1 \partial V_2^2} v_1 v_2^2 + 3\frac{\partial^3 f}{\partial V_1^2 \partial V_2} v_1^2 v_2 + \frac{\partial^3 f}{\partial V_2^3} v_2^3\right) + \ldots \end{aligned} \tag{2.10}$$

In (2.10) the notation has been streamlined somewhat, and it is understood that the derivatives are evaluated at the bias points of V_1 and V_2, $V_{1,0}$ and $V_{2,0}$, respectively. In the small-signal, quasilinear case, the high-degree terms are neglected and

$$i = \frac{\partial f}{\partial V_1} v_1 + \frac{\partial f}{\partial V_2} v_2 \tag{2.11}$$

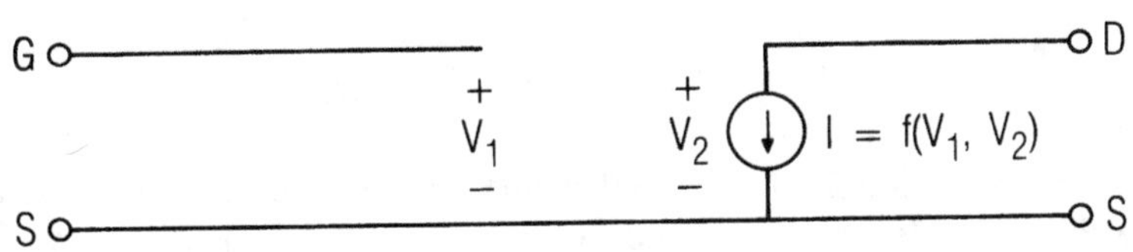

Figure 2.5 A multiply controlled nonlinear element.

The extension of (2.6) through (2.11) to resistances or current-controlled voltage sources is trivial: one need only interchange I and V, and i and v, in (2.6) through (2.11). The same expressions can be used for voltage-controlled voltage sources or current-controlled current sources by substituting the control voltage or current for V, the small-signal excitation for v, and the response for i.

A distressingly common error is to assume that an expression for the small-signal current can be found by expanding the nonlinear conductance in a power series. Specifically, the approach is to find $g(V)$ from (2.8) and to say

$$i = g(V_0 + v)v \tag{2.12}$$

with $g(V_0)$ given by (2.8),

$$i = \frac{d}{dV}f(V)\bigg|_{V=V_0} v + \frac{d^2}{dV^2}f(V)\bigg|_{V=V_0} v^2 + \frac{1}{2}\frac{d^3}{dV^3}f(V)\bigg|_{V=V_0} v^3 + \ldots \tag{2.13}$$

which is clearly not the same as (2.7). There is no reason why (2.12) should be equivalent to (2.7); $g(V_0 + v)$ is just the linear conductance at a slightly different control voltage. This error is particularly insidious, because the linear terms in (2.13) and (2.7) are fortuitously the same. The correct incremental I/V characteristic can be obtained from the $g(V)$ characteristic; the method is described in Section 2.2.6.

2.2.4 Large-Signal Nonlinear Capacitance

A nonlinear capacitor's large-signal charge, Q_c, is described by

$$Q_c = f_Q(V_1, V_2, \ldots) \tag{2.14}$$

The considerations for the functional form of f_Q are identical to those of the resistive element, described in Section 2.2.2. The current in the nonlinear capacitor is

$$I = \frac{dQ_c}{dt} = \frac{\partial f_Q}{\partial V_1}\frac{dV_1}{dt} + \frac{\partial f_Q}{\partial V_2}\frac{dV_2}{dt} + \ldots \tag{2.15}$$

For a simple nonlinear capacitor having one control voltage, we have

$$I = C(V)\frac{dV}{dt} \tag{2.16}$$

where we define

$$C(V) = \frac{\partial f_Q}{\partial V} \tag{2.17}$$

$C(V)$ is the incremental capacitance, which we shall study in more detail in Section 2.2.5. $C(V)$ is the capacitance that would be measured if the nonlinear element were biased at dc voltage V, and a small ac voltage were applied to it. Equation (2.17) implies that f_Q can be measured indirectly, by first measuring $C(V)$ and then integrating it to obtain charge. Another method is to differentiate f_Q to obtain $C(V)$, and to fit the parameters of $C(V)$ to the measured capacitance. One of these approaches is almost always necessary, as it is usually impossible to measure charge directly.

2.2.5 Small-Signal Nonlinear Capacitance

Initially, we assume that the voltage $V = V_0$ is the sole dc control voltage and that it is applied to the capacitor's terminals. By expanding the charge function in a Taylor series, as with the conductance, and subtracting the dc component of the charge, we obtain the small-signal component of the charge:

$$
\begin{aligned}
q &= f_Q(V_0 + v) - f_Q(V_0) \\
&= \frac{d}{dV} f_Q(V)\bigg|_{V = V_0} v + \frac{1}{2}\frac{d^2}{dV^2} f_Q(V)\bigg|_{V = V_0} v^2 \\
&\qquad + \frac{1}{6}\frac{d^3}{dV^3} f_Q(V)\bigg|_{V = V_0} v^3 + \ldots
\end{aligned}
\tag{2.18}
$$

Again, for simplicity, in (2.18) the small-signal voltage $v(t)$ is written as v and $q(t)$ as q. The small-signal current is the time derivative of the charge:

$$
\begin{aligned}
i &= \frac{dq}{dt} \\
&= \frac{d}{dV} f_Q(V)\bigg|_{V = V_0} \frac{dv}{dt} + \frac{d^2}{dV^2} f_Q(V)\bigg|_{V = V_0} v \frac{dv}{dt} \\
&\qquad + \frac{1}{2}\frac{d^3}{dV^3} f_Q(V)\bigg|_{V = V_0} v^2 \frac{dv}{dt}
\end{aligned}
\tag{2.19}
$$

Equation (2.19) can be expressed as

$$
i = (C_1(V_0) + C_2(V_0)v + C_3(V_0)v^2 + \ldots)\frac{dv}{dt} \tag{2.20}
$$

which is the series form of the incremental capacitance. For the quasilinear case, this expression reduces to a simple linear capacitance,

$$
i = C_1(V_0)\frac{dv}{dt} \tag{2.21}
$$

where

$$
C_1(V_0) = \frac{d}{dV} f_Q(V)\bigg|_{V = V_0} \tag{2.22}
$$

Equation (2.22) is the standard definition of capacitance of a Schottky-barrier or *pn* junction device. It is the same as the incremental capacitance given in (2.17).

In both the small-signal capacitance and conductances, we assumed that the element is biased at some dc voltage and that a much smaller ac voltage is superimposed. This situation is common in many nonlinear microwave problems; for example, in calculating intermodulation distortion in small-signal amplifiers. However, in many circuits a large ac signal may also exist, such as the LO waveform in a mixer or a saturating signal in a small-signal amplifier. If the nonlinearity is strong, or if the signal is very large (or, as in a diode mixer, both), a large number of terms must be used in the series expansion to give adequate computational accuracy. Carrying expressions like (2.10) to a high number of terms is difficult enough, but, as we shall see in Chapter 4, the task of analyzing even a relatively simple nonlinear circuit by such a long series is nearly impossible. It is possible, however, to circumvent these difficulties by expanding the Q/V or I/V characteristic in a Taylor series and using the large ac voltage (plus any dc bias voltage, of course) as the central "point." This expansion allows the small-signal voltage at any instant to be treated as a small deviation from the central value, so the minimum number of Taylor series terms can be used. The trade-off in this approach is that the Taylor series coefficients are time-varying, and thus must be differentiated along with the small-signal voltage in such expressions as (2.19). This approach, which is examined further in Chapter 3, allows an accurate and tractable analysis of such phenomena as intermodulation distortion in mixers, which appears at first to be an extraordinarily difficult problem.

2.2.6 Relationship Between *I/V*, *Q/V* and *G/V*, *C/V* Expansions

The series expansions developed in the previous section, describing the incremental conductances and capacitances, were derived from static I/V and Q/V characteristics. Sometimes, however, it is more convenient to begin with incremental C/V or G/V data (i.e., the linear capacitance or conductance as a function of bias voltage). This situation arises often in the modeling of solid-state devices, in which C/V or G/V characteristics are often the easier ones to measure. In this case, we must find the Taylor-series expansions of the I/V or Q/V characteristics from a series expansion of the C/V or G/V characteristic. The Taylor-series expansion of the characteristic $I = f(V)$ is, from (2.6),

$$f(V_0+v) = f(V_0) + \frac{d}{dV}f(V)\Big|_{V=V_0} v + \frac{1}{2}\frac{d^2}{dV^2}f(V)\Big|_{V=V_0} v^2$$
$$+\frac{1}{6}\frac{d^3}{dV^3}f(V)\Big|_{V=V_0} v^3 + \ldots \tag{2.23}$$
$$= f(V_0) + g_1 v + g_2 v^2 + g_3 v^3 + \ldots$$

and the expansion of $G(V)$ is

$$G(V_0+v) = G(V_0) + \frac{d}{dV}G(V)\Big|_{V=V_0} v + \frac{1}{2}\frac{d^2}{dV^2}G(V)\Big|_{V=V_0} v^2$$
$$+\frac{1}{6}\frac{d^3}{dV^3}G(V)\Big|_{V=V_0} v^3 + \ldots \tag{2.24}$$
$$= f(V_0) + \zeta_1 v + \zeta_2 v^2 + \zeta_3 v^3 + \ldots$$

We note that

$$G(V) = \frac{df}{dV} \tag{2.25}$$

and after substituting (2.25) into (2.24) and comparing the result to (2.23), we see immediately that

$$\begin{aligned} g_1 &= \zeta_0 \\ g_2 &= \zeta_1/2 \\ g_3 &= \zeta_2/3 \\ \ldots &\quad \ldots \\ g_n &= \zeta_{n-1}/n \end{aligned} \tag{2.26}$$

We can do the same with the Taylor-series expansion of the Q/V characteristic and the expansion of the C/V characteristic. The Q/V expansion has the form

$$Q(V_0 + v) = f_Q(V_0) + C_1 v + C_2 v^2 + C_3 v^3 + \ldots \tag{2.27}$$

and the C/V characteristic has the expansion

$$C(V_0 + v) = \gamma_1 v + \gamma_2 v^2 + \gamma_3 v^3 + \ldots \tag{2.28}$$

Comparing their Taylor-series terms as in (2.23) through (2.26) gives the identical result,

$$C_n = \gamma_{n-1} / n \tag{2.29}$$

Having said all this, we should note that determining the g_n or C_n coefficients from an expansion of the $G(V)$ or $C(V)$ function is not always practical. Many devices are very linear so these coefficients are small, and small variations in the measured $G(V)$ or $C(V)$ function can cause large errors in the high-degree terms. In this case, it is better to extract these values from indirect measurements; for example, from measurements of harmonics generated by the device.

2.2.7 Multiply Controlled Nonlinear Capacitors

Capacitors, like conductances, can be controlled by more than one voltage. Capacitors having multiple control voltages are found in many solid-state device models. Modeling capacitances in such devices is a tricky business; done incorrectly, it can result in nonconservation of charge, or in such bizarre phenomena as dc currents in capacitors. We first consider the easier problem of weakly nonlinear capacitances, and then address the greater problem of large-signal nonlinear capacitances.

2.2.7.1 Small-Signal Case

In this case the large-signal Q/V characteristic is

$$Q_c = f_Q(V_1, V_2, \ldots) \tag{2.30}$$

It is rarely necessary to consider more than two control voltages, so we can limit our discussion to the expression $Q_c = f_Q(V_1, V_2)$. As before, we expand this function in a two-dimensional Taylor series about the bias points $V_{1,0}$ and $V_{2,0}$. After subtracting the dc charge components to obtain the small-signal charge, we have

$$\begin{aligned} q = {} & \frac{\partial f_Q}{\partial V_1} v_1 + \frac{\partial f_Q}{\partial V_2} v_2 \\ & + \frac{1}{2}\left(\frac{\partial^2 f_Q}{\partial V_1^2} v_1^2 + 2\frac{\partial^2 f_Q}{\partial V_1 \partial V_2} v_1 v_2 + \frac{\partial^2 f_Q}{\partial V_2^2} v_2^2\right) \\ & + \frac{1}{6}\left(\frac{\partial^3 f_Q}{\partial V_1^3} v_1^3 + 3\frac{\partial^3 f_Q}{\partial V_2^2 \partial V_1} v_1 v_2^2 + 3\frac{\partial^3 f_Q}{\partial V_1^2 \partial V_2} v_1^2 v_2 + \frac{\partial^3 f_Q}{\partial V_2^3} v_2^3\right) \end{aligned} \tag{2.31}$$

where the partial derivatives are evaluated at the dc bias voltages $V_{1,0}$ and $V_{2,0}$. The current is obtained by taking the derivative with respect to time.

Fortunately, the dependence of Q_c on one voltage is often less strong than on the other, and in those cases (2.31) can be simplified considerably. However, before deleting terms wildly, one should be careful not to throw out the baby with the bathwater. The terms in (2.31) generate different frequency components under sinusoidal steady-state conditions, so deleting certain terms, even if they are very small, may delete the intermodulation component of interest. One of the advantages of working in the frequency domain with capacitive nonlinearities is that much of the complexity evident in (2.31) is circumvented; in Chapter 3 we shall show that the process of taking the derivative in the time domain can be performed in the frequency domain merely by multiplying a matrix by a diagonal matrix.

2.2.7.2 Large-Signal Case

Equations (2.14) and (2.15) give expressions for the current in a multiply controlled nonlinear capacitor under large-signal excitation. We saw that the current in the capacitor is

$$I = \frac{dQ_c}{dt} = \frac{\partial f_Q}{\partial V_1}\frac{dV_1}{dt} + \frac{\partial f_Q}{\partial V_2}\frac{dV_2}{dt} + \ldots \tag{2.32}$$

In the usual case, (2.32) describes a capacitor whose charge is a function of its terminal voltage and one or more voltages elsewhere in the circuit. Terms involving voltages other than the capacitor's terminal voltages are sometimes called *transcapacitances*, a term that was first used in [2.1]. For example, if V_1 is the terminal voltage, $\partial f_Q / \partial V_2$ is a transcapacitance. The concept is analogous to transconductance; the transcapacitance represents a dependence of charge upon a remote voltage.

Determining the charge function from small-signal measurements can be difficult. For example, consider (2.31) and (2.32) with only two control voltages, V_1 and V_2. The small-signal current is given by

$$i = C_1 \frac{dV_1}{dt} + C_2 \frac{dV_2}{dt} \tag{2.33}$$

where

$$C_1 = \frac{\partial f_Q}{\partial V_1} \qquad C_2 = \frac{\partial f_Q}{\partial V_2} \tag{2.34}$$

That is, we need to find one capacitance, C_1, and one transcapacitance, C_2. If these can be found, it is a simple matter to integrate them to obtain f_Q, or even easier to differentiate a given expression for f_Q and to fit its parameters to these capacitances. Unfortunately, it is usually difficult to separate these two terms. For this reason, other approaches to modeling multiply controlled capacitances are often used. This situation arises most frequently in modeling microwave FETs, and is discussed further in Section 2.5.7.

2.2.7.3 Multiterminal Capacitance

Although the capacitors considered in this section may be controlled by multiple voltages, the charge itself resides on a two-terminal element (or, if you wish, a single branch of a circuit). In many cases, especially FET gate capacitances, the capacitor may have more than two terminals. Such capacitors are often approximated as a set of two-terminal, multiply controlled capacitances, but this simplification invariably leads to problematical behavior [2.2, 2.3].

Multiterminal capacitors appear frequently in linear circuits. For example, in the analysis of coupled strip transmission lines, we define a capacitance matrix

$$\begin{bmatrix} Q_1 \\ Q_2 \\ \cdots \\ Q_K \end{bmatrix} = \begin{bmatrix} C_{11} & C_{12} & \cdots & C_{1K} \\ C_{21} & C_{22} & \cdots & C_{2K} \\ \cdots & \cdots & \cdots & \cdots \\ C_{K1} & C_{K2} & \cdots & C_{KK} \end{bmatrix} \begin{bmatrix} V_1 \\ V_2 \\ \cdots \\ V_K \end{bmatrix} \tag{2.35}$$

where Q_i are the charges on the K strips and V_j are the voltages. The C_{ij} have the predictable definition,

$$C_{ij} = \left. \frac{Q_i}{V_j} \right|_{V_k = 0,\, k \neq j} \tag{2.36}$$

If there were only $K - 1$ strips and one of the Q_i represented the charge on the ground plane (which usually is not included in the charge vector), we would have

$$\sum_{i=1}^{K} Q_i = 0 \tag{2.37}$$

that is, charge neutrality would apply, as it does when we consider the charges on both plates of a simple parallel-plate capacitor.

In many nonlinear models, especially those describing modern metal oxide-semiconductor (MOS) devices, we follow a similar approach. The Q_i are called *pin* (or *terminal*) *charges*, which are functions of the terminal voltages. We then use a vector of functions to describe those charges:

$$\begin{aligned} Q_1 &= f_1(V_1, V_2, \ldots, V_K) \\ Q_2 &= f_2(V_1, V_2, \ldots, V_K) \\ &\cdots \\ Q_K &= f_K(V_1, V_2, \ldots, V_K) \end{aligned} \tag{2.38}$$

In most solid-state devices, charge neutrality applies, so

$$\sum_{i=1}^{K} Q_i = 0 \tag{2.39}$$

for all combinations of voltages. The current in each terminal is

$$I_i = \frac{dQ_i}{dt} = \frac{d}{dt} f_i(V_1, V_2, \ldots, V_K) \tag{2.40}$$

and, clearly,

$$\sum_{i=1}^{K} I_i = 0 \tag{2.41}$$

When the device is dc biased, (2.38) can be converted to a small-signal, incremental capacitance matrix. Its elements are

$$C_{ij} = \frac{\partial}{\partial V_j} f_i(V_1, V_2, \ldots, V_K) \tag{2.42}$$

which is evaluated at the dc values of all the controlling voltages, V_1 through V_K.

2.2.8 Nonlinear Inductance

A nonlinear inductance is described by its flux-current characteristic,

$$\Phi = F_\Phi(I) \tag{2.43}$$

where Φ is its magnetic flux. We shall see in later chapters that a nodal formulation is most convenient to describe the linear part of a circuit containing nonlinear elements. The nodal formulation, however, cannot accommodate current as an independent variable. Other methods, such as the modified nodal formulation, can do so, but they involve additional complexity.

A simple solution is to use a gyrator. A gyrator is a two-port element that has the admittance matrix

$$Y = \begin{bmatrix} 0 & \frac{1}{R} \\ -\frac{1}{R} & 0 \end{bmatrix} \tag{2.44}$$

where R is called the *gyrational resistance*. If we set $R = 1$, the gyrator converts current at either port to voltage at the other, and therefore converts a capacitance at one port to an inductance at the other. To realize the nonlinear inductor, we simply connect a gyrator to the circuit and terminate it with a nonlinear capacitor having the charge characteristic,

$$Q(V) = F_{\Phi}(V) \tag{2.45}$$

The Φ/V characteristic at its input port is then given by (2.43). Gyrators can also be used to realize controlled voltage sources, current-controlled sources, circulators, and transformers.

2.3 NUMERICAL AND HUMAN REQUIREMENTS FOR DEVICE MODELS

Solid-state device models are used in circuit simulators, operated by human beings. As such, models must satisfy requirements imposed by the limitations of both of these entities. The methods used in circuit simulators are well known and their requirements can be clearly enumerated; the methods used by human beings are less easily categorized. Still, a model that does not conform to those methods, however arbitrary, is not particularly useful.

The dominant method of nonlinear circuit simulation is harmonic balance analysis (Chapter 3). Because the dominant implementations of both harmonic-balance and transient analysis use Newton iteration in their solutions, the requirements imposed by both methods are similar. We consider some of the necessary requirements in this section.

2.3.1 Continuous Derivatives in I/V or Q/V Expressions

Convergence of both harmonic-balance analysis and transient analysis requires continuous first and second derivatives of the I/V or Q/V expression. If this requirement is not satisfied, convergence robustness is degraded. Certain kinds of analysis may require more derivatives to be reproduced accurately by the model.

Newton-based harmonic-balance analysis is an iterative method. It estimates a solution and uses the derivatives of the *I*/*V* expressions at each iteration to improve the estimate. If a derivative has a "kink" in it (i.e., the second derivative is discontinuous), it may not point to an improved solution. In some cases, satisfying this requirement may actually make the nonlinearity stronger, and convergence is more reliable than with a weaker nonlinearity. The diode junction *I*/*V* characteristic described in Section 2.4.2.4 is an example.

Discontinuities in derivatives are likely to occur when different expressions are used for different ranges of control voltage. When this practice is followed, it is essential that derivatives be matched at the boundaries of the ranges.

2.3.2 Accuracy of Derivatives

For accurate *n*th-order IM simulations, the function must reproduce accurately not only the *I*/*V* characteristic, but also its first *n* derivatives. The reason for this requirement can be clarified by Volterra-series theory, but previous discussions hint at the reason. We saw in Chapter 1 that the *n*th power of a polynomial dominated in generating *n*th-order mixing products, and in Section 2.2.3 we saw that the coefficient of the *n*th degree term is the *n*th derivative multiplied by a constant. Other types of analysis place accuracy requirements on particular derivatives; for example, even-order derivatives are necessary for dc quantities, which are necessary for accurate calculations of efficiency.

2.3.3 Range of Expressions

The *I*/*V* function must be well-behaved far outside of the range of voltages or currents that the device experiences in practice. Harmonic-balance analysis is an iterative method, and it is common, during intermediate iterations, for extraordinarily large or small voltages to exist.

A wide variety of numerical difficulties can be introduced simply by the form of the equations. For example, it is extraordinarily easy to generate numerical underflow or overflow in the computation of logarithmic or exponential functions, especially in diode *I*/*V* characteristics. In C or C++ compilers, the range limits of standard functions can be found in the header file *float.h*. Limits on integers are given in *limits.h*.

2.3.4 Transient-Analysis Models in Harmonic-Balance Analysis

Many common harmonic-balance models have been copied directly from transient-analysis programs, mainly SPICE [2.4]. Transient-analysis models are sometimes not well-suited for use in harmonic-balance analysis. Transient analysis uses iterative methods to solve the circuit equations at each time point in the transient response; these time points are closely spaced, so the circuit voltages and currents change little between solutions. This is not the case in harmonic-balance analysis, where huge changes are not only possible, but very likely. Many models used in transient simulators take advantage of the fact that changes, from iteration to iteration, are usually small, so little regard is given for their numerical performance far outside of the voltage and current ranges they are expected to experience. When such models are used in harmonic-balance analysis, their deficiencies quickly become apparent.

2.3.5 Matrix Conditioning

As with most circuit-simulation methods, harmonic-balance analysis requires the solution of large systems of linear equations. Most importantly, at each iteration a large Jacobian matrix must be factored. A large admittance matrix for the linear subcircuit must also be created, a process that requires considerable matrix manipulation.

The Jacobian is used to estimate an improved solution at each harmonic-balance iteration. In theory, it is possible for the Jacobian to be singular, so it has no solution. In practice, however, it is more likely for the Jacobian to be technically nonsingular, but so close to singular that the solution is inaccurate. We say that such matrices are *ill conditioned*. The main effect of ill conditioning is to lose numerical precision in the solutions. In extreme cases, virtually all precision is lost, and the result is best described by the well-known computer-science term, *garbage*.

Ill conditioning can be caused in many ways. One frequent cause is unusually large or small entries in the Jacobian. For example, when a linear differentiator is used in a model having a division-by-capacitance scheme (Section 2.5.7.1), the Jacobian has large, off-diagonal terms. Because they include terms of the form $jk\omega_0$, where k is a large harmonic and ω_0 is the fundamental excitation frequency, strongly nonlinear capacitances of any type can create an ill conditioned Jacobian. Disconnected or unilateral circuits are also causes of ill conditioning.

Perhaps the easiest way to have an inaccurate (although not necessarily ill conditioned) Jacobian is to have an ill conditioned admittance matrix of the linear subcircuit. These occur when a node in the linear subcircuit is disconnected, often by partitioning the circuit into the linear and nonlinear parts, or by connecting two nodes by a low impedance. See Sections 3.3.7.4 and 3.3.9.6 for further information.

2.3.6 Limiting the Range of Control Voltages

Occasionally it is necessary to limit the range of control voltages. Such limits must be applied in a numerically acceptable way. Perhaps the worst way to limit a variable's range is simply by truncating it to a maximum or minimum of the allowable range, because the truncation introduces a discontinuity that is difficult for harmonic-balance analysis to handle. The result is poor convergence.

A number of functions can be used to limit voltage range without creating discontinuities. For example,

$$V_{lim} = V_{\min} + \ln(\exp(V - V_{\min}) + 1) \tag{2.46}$$

returns

$$\begin{aligned} V_{lim} &= V & V &\gg V_{\min} \\ V_{lim} &= V_{\min} & V &\ll V_{\min} \end{aligned} \tag{2.47}$$

with a smooth but rather gradual transition around $V_{lim} = V_{\min}$. Unfortunately, this method is subject to numerical overflow or underflow in the exp function. A better formula is

$$V_{lim} = V_{\min} + 0.5[V - V_{\min} + \sqrt{(V - V_{\min})^2 + \delta}] \tag{2.48}$$

The parameter δ controls the shape of the $V_{lim}(V)$ curve near $V_{\min}$. To limit the maximum excursion, use

$$V_{lim} = V_{\max} - 0.5[V_{\max} - V + \sqrt{(V_{\max} - V)^2 + \delta}] \tag{2.49}$$

Even with these functions, one must be careful. Equation (2.49), for example, could be written

$$V_{lim} = V - 0.5[V - V_{\max} + \sqrt{(V - V_{\max})^2 + \delta}] \tag{2.50}$$

but this evaluates to zero, not V_{max}, when V is so large that limited numerical precision causes $V - V_{max}$ to be evaluated as V. Similarly, (2.49) evaluates as zero, not V, when V_{max} is very large and V is small.

2.3.7 Use of Polynomials

The idea of using polynomials to model difficult I/V or Q/V expressions is seductive. After all, well-known numerical techniques are available for fitting a polynomial to an arbitrary function, and, if the process fails, simply increasing the degree of the polynomial usually does the trick. The derivatives of polynomials are also devoid of discontinuities, so they satisfy this important requirement for device modeling, explained in Section 2.3.1.

Unfortunately, polynomials have significant disadvantages, some of which we list below:

1. High-degree polynomial functions often have small-scale ripple that may not be visible in a plot of the I/V or Q/V characteristic, but becomes quite clearly evident in the derivatives.
2. The *normal equation*, used to fit polynomials to measured data, is notoriously ill conditioned, so small variations in the data can result in large changes in the polynomial coefficients and in the quality of the fit. (Singular-value decomposition can sometimes minimize this problem.)
3. Outside the range of the data used to generate the polynomial approximation, the polynomial can have undesirable behavior; for example, there may be regions of negative incremental resistance, which can prevent convergence. It is also possible to obtain spurious solutions, which may be nonphysical.
4. Finally, polynomials restrict numerical range. For example, double-precision arithmetic limits positive real numbers to the range (10^{-307}, 10^{+308}). If a tenth-degree polynomial is used, the range of the independent variable is limited to approximately (10^{-30}, 10^{+30}) or numerical overflow or underflow results. This may seem like a minor point, but, in fact, numbers outside the latter range occur frequently in nonlinear circuit analysis.

In spite of these caveats, occasionally it may be best to express a quantity by a polynomial. In such cases, it is most efficient to calculate the polynomial

$$f(V) = a_0 + a_1 v + a_2 v^2 + a_3 v^3 + \ldots \tag{2.51}$$

as

$$f(V) = a_0 + v(a_1 + v(a_2 + v(a_3 + \ldots))) \tag{2.52}$$

Also, it is a poor practice to use the C or C++ pow() function for raising a real number to an integer power; (2.52) is far more efficient.

2.3.8 Loops of Control Voltages

In both harmonic-balance and transient analysis, the circuit simulator attempts to determine the values of a set of control voltages or, occasionally, currents. This process can be successful only if those quantities are (1) independent, and (2) adequate to define the state of the system. In circuit-theory terminology, they must be *state variables*. If, however, a loop of three control voltages exists, only two of those voltages are independent; the third is linearly dependent on the other two. If that third quantity is treated in the simulator as an independent variable, successful convergence is unlikely.

One solution is to break the loop with a low-value resistor or some other component that does not affect the simulation results. This expedient sometimes works, but it usually creates an ill-conditioned Jacobian matrix (Sections 2.3.5 and 3.3.7.4). A better solution is to reformulate the problem to use only independent quantities. Figure 2.6 illustrates how this can be accomplished. This process generally works well, but occasionally it can create additional solutions to the nonlinear circuit equations.

2.3.9 Default Parameters

Inconsistencies in default parameters frequently cause errors in porting models between simulators. In netlist versions of SPICE, for example, not all model parameters need be provided; if they are not, a default value is used, or some other behavior ensues. Clearly, if such models are ported from SPICE to another simulator, the default behavior must be the same, or all parameters must be listed.

In some cases, a parameter depends on whether another one was "provided," that is, entered in the MODEL statement of the netlist. This type of behavior can be difficult to port to a schematic-capture simulator, where all parameters are listed in the parameter-entry dialog box and,

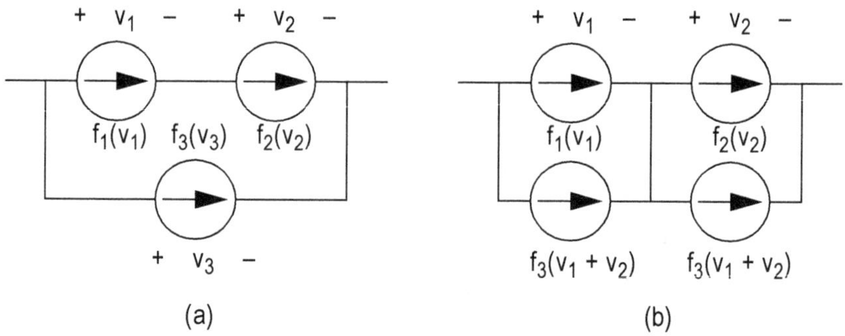

Figure 2.6 Conversion of three linearly dependent control voltages to two independent ones: (a) v_3 depends on v_1 and v_2; (b) the troublesome branch, f_3, has been converted to two elements in parallel with the f_1 and f_2 branches.

therefore, provided. The user should be aware of the way the schematic-capture simulator handles this situation.

In SPICE, a number of parameters are interpreted as infinity (in practice, a very high value) if zero is entered. Clearly, the simulator receiving such models must behave identically, or an appropriate nonzero parameter value must be provided.

2.3.10 Error Trapping

Models are used in circuit simulators; circuit simulators are used by humans. Human imperfections cause many errors in circuit simulation, and these imperfections are often exacerbated by poor model design. Unfortunately, it is difficult to anticipate and trap all possible errors in parameter entry, but still, an attempt to include comprehensive error trapping must be made. Often, it is easy to design models in such a way that errors are unlikely.

As a simple example, consider an I/V function with terms of the form,

$$I(V) = \ldots + \frac{V}{k_1} + \ldots \tag{2.53}$$

where k_1 is a user-supplied model parameter. If a naive user enters $k_1 = 0$, an error occurs. It is a simple matter to use instead

$$I(V) = \ldots + c_1 V + \ldots \tag{2.54}$$

where $c_1 = 1 / k_1$. The problem is solved, and no special error trapping is necessary.

Many users of nonlinear circuit simulators attempt to copy parameter sets between different simulators. Unless the models, parameter names, and default behavior are identical in the two simulators, errors can result. For example, in many simulators illegal zero entries are automatically changed to a reasonable value. If one implementation of a model performs this modification, but another doesn't, an error is certain to occur.

It is almost impossible to list the number of ways in which models can ambush an unsuspecting user. It is essential, however, for model designers to be aware of this problem and to anticipate it as best they can.

2.3.11 Lucidity of Models and Parameters

The underlying logic of a model, and the effect of its parameters, must be clear. A model that confuses the user is unlikely to be used properly, and it is unlikely that sensible parameters will be found for it.

For example, it is logical to formulate a model for FET channel current as

$$I_d(V_g, V_d) = f_G((V_g) \cdot f_D(V_d)) \tag{2.55}$$

where f_G is a function of the gate voltage, V_g, and f_D is a function of the drain voltage, V_d. Even if f_G is modified to allow some degree of dependence on V_d, and f_D on V_g, this type of formulation is consistent with users' understanding of FET I/V characteristics and therefore is comprehensible. It is unlikely that such a model will be misused. On the other hand, a complex expression mixing V_g and V_d in an incomprehensible way is much more likely to create problems.

2.3.12 Does Complexity Improve a Model?

Most designers intuitively accept the idea that a complex model is more likely to be accurate than a simple one. This idea is correct, within limits. At some point, however, additional complexity does not improve a model, because the increased likelihood of error, which comes with complexity, tends to cancel progressively more minor improvements in accuracy. Unfortunately, the response to this situation is often to increase the model's complexity further, in the hope that the additional complexity will solve the problem.

Excessive complexity often results from an attempt to make a device model that gives highly accurate results for all kinds of analyses. Another cause is an attempt to model every possible phenomenon, without considering whether it has any significant effect on the results of the analysis. For example, the requirements for a model used in intermodulation analysis are very different from those for calculating the conversion loss of a mixer or the output power of an amplifier. By focusing on the important characteristics and ignoring minor ones, one can create models that are both simple and accurate.

2.4 SCHOTTKY-BARRIER AND JUNCTION DIODES

Virtually all microwave mixer diodes and many varactors use Schottky (metal-to-semiconductor) junctions instead of *pn* junctions or point contacts. *pn* junction diodes are never used in microwave circuits as resistive diodes, although they are often used as varactors. A Schottky-barrier diode consists of a metal contact deposited on a semiconductor; such contacts can be made with far better uniformity than point contacts, and they do not have the recombination-time limitations of *pn* junctions. Inexpensive silicon Schottky-barrier diodes are capable of good performance as mixers at frequencies well into the millimeter-wave region. Gallium arsenide diodes, which are somewhat more expensive, can realize mixers at terahertz frequencies. Gallium arsenide Schottky-barrier varactors, which generally have higher Q factors than silicon varactors, are commonly used in millimeter-wave frequency multipliers.

The Schottky-barrier diode is perhaps the simplest modern solid-state microwave device in existence and the easiest to characterize accurately. The junction I/V and capacitance characteristics can be expressed by simple closed-form equations that are accurate for almost all purposes; there is little need to make a trade-off in the diode model between accuracy and simplicity. Furthermore, the diode model developed in this section is accurate to frequencies of at least a few hundred gigahertz and, with minor modifications, to even higher frequencies. As a result, circuit modeling of mixers and frequency multipliers, including noise, intermodulation, and conversion efficiency, has been highly successful and can now be considered a mature practice.

2.4.1 Structure and Fabrication

Figure 2.7 shows the general structure of a Schottky-barrier diode; most Schottky devices are similar. The diode is fabricated on a high-conductivity *n*-type (*n*+) substrate; because the electron mobilities of all practical *n* dopants are much greater than those of *p* materials, *n* materials are used almost exclusively in microwave Schottky devices. A very pure, high-conductivity *n*+ buffer layer is grown on top of the substrate to assure low series resistance and to prevent impurities in the substrate from diffusing into the epitaxial layer during processing. The buffer is usually a few microns thick, and the buffer and substrate are doped as heavily as possible, usually on the order of 10^{18} atoms/cm^3 for GaAs, somewhat higher for silicon. An *n* epitaxial layer (sometimes called the *epilayer* or, simply, the *epi*) is grown on top of the buffer. In GaAs mixer diodes, the epilayer is doped to $1 \cdot 10^{17}$ to $2 \cdot 10^{17}$ cm^{-3} and is usually 1,000Å to 1,500Å thick.

The contact of the metal anode to the epitaxial layer forms the rectifying junction. Platinum and titanium are the most common anode materials for GaAs diodes. A gold layer is usually plated onto the metal anode to prevent corrosion and to facilitate a bond wire, ribbon, air bridge, or whisker connection. The anode metal rarely covers the entire top surface of the chip; the size and shape of the anode are selected to give the appropriate combination of junction capacitance and series resistance for the intended application. The circular anodes of microwave diodes vary in

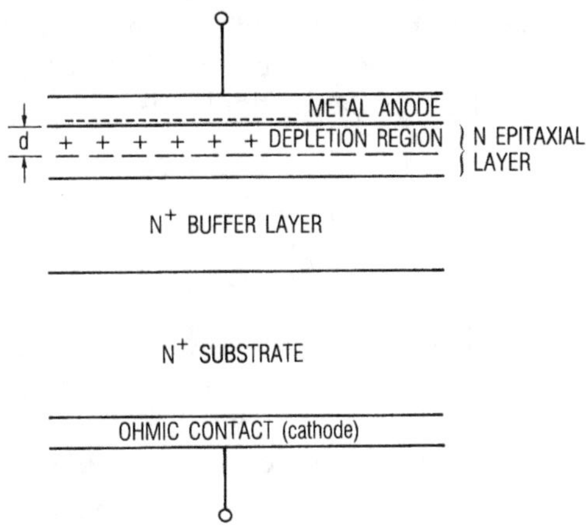

Figure 2.7 Cross section of a Schottky-barrier diode.

diameter from 1.5 microns for millimeter-wave devices to 10 to 20 microns for use at lower frequencies. For practical reasons, in many diodes a large number of anodes are defined on the top surface of a single chip and are isolated from each other by an oxide (SiO_2) layer. An ohmic contact to the substrate must be made; alloyed gold-germanium is commonly used on GaAs. The ohmic contact is usually formed on the bottom of the substrate, but it can be formed on the top of the diode (e.g., for beam-lead devices) if appropriate means are used to isolate the anode from the cathode and to minimize the parasitic capacitance that arises from their proximity.

2.4.2 The Schottky-Barrier Diode Model

2.4.2.1 Junction Capacitance

The physics of conduction and capacitance in Schottky barriers will not be covered here; the interested reader should consult [2.5 – 2.7]. For present purposes it is enough to note that the contact of the metal to the semiconductor allows some of the free electrons in the semiconductor to collect on the surface of the metal. The semiconductor immediately under the anode (imaginatively called the *depletion region*) is depleted of electrons and contains only positively charged donor ions. Because of these ions, an electric field, which opposes further movement of electrons, is set up between the anode and the semiconductor, and a state of equilibrium is reached. Also because of this electric field, a potential difference, called the *diffusion potential* or *built-in voltage*, exists between the neutral semiconductor and the anode.

The width of the depletion region can be found from the doping density and material parameters of the semiconductor. The depletion width d of an ideal junction having uniform epitaxial doping is

$$d = \sqrt{\frac{2\phi\varepsilon_s}{qN_d}} \tag{2.56}$$

where ϕ is the diffusion potential; N_d is the doping density, assumed to be uniform throughout the epilayer; ε_s is the electric permittivity of the semiconductor; and q is the electron charge, $1.6 \cdot 10^{-19}$ coul. If a dc voltage V is applied to the junction, the depletion width changes. The width of the biased depletion region becomes

$$d = \sqrt{\frac{2(\phi - V)\varepsilon_s}{qN_d}} \tag{2.57}$$

If the junction is reverse-biased, the depletion region becomes wider and more electrons move to the anode, leaving behind more positive charge in the form of ionized donor atoms. Conversely, if the diode is forward-biased, the depletion region narrows and less charge is stored. Thus, a negative voltage stores more negative charge on the anode, and a positive voltage reduces it. The junction, therefore, operates as a nonlinear capacitor.

As the forward bias is increased, the electric field in the junction becomes weaker and presents less of a barrier to electrons. More electrons have sufficient thermal energy to cross the barrier, and forward conduction occurs. The current is proportional to the number of electrons having energy greater than the barrier energy; that number is an exponential function of barrier height. Thus, the *I*/*V* characteristic is an exponential function, one of the strongest nonlinear functions found in solid-state devices. Because conduction occurs almost entirely as the result of thermal emission of electrons—majority carriers—over a barrier, the Schottky-barrier diode is often called a *majority carrier device*.

In conventional Schottky diodes, the epitaxial layer is never fully depleted of charge in normal operation, even at the highest reverse voltages. Consequently there is always some undepleted epitaxial material between the depletion region and the buffer layer, especially under forward bias, when the depletion region is narrow. Because this material has a relatively high resistivity, especially compared to the substrate, it represents a parasitic resistance in series with the diode junction. In mixers and frequency multipliers, series resistance is an important loss mechanism.

Figure 2.8 shows the equivalent circuit of a Schottky-barrier diode. The diode consists of three elements, two of which, the junction capacitance and conductance, are nonlinear. The third element, the parasitic series resistance R_s, is also nonlinear, but because it varies only slightly under forward bias, it is usually treated as a linear resistance. The series resistance of a varactor diode, which is operated with reverse bias and rarely experiences forward conduction, varies somewhat more with junction voltage. However, even in that case R_s is usually approximated as a linear element.

A remarkably accurate junction charge function can be derived from a simple analysis. It is

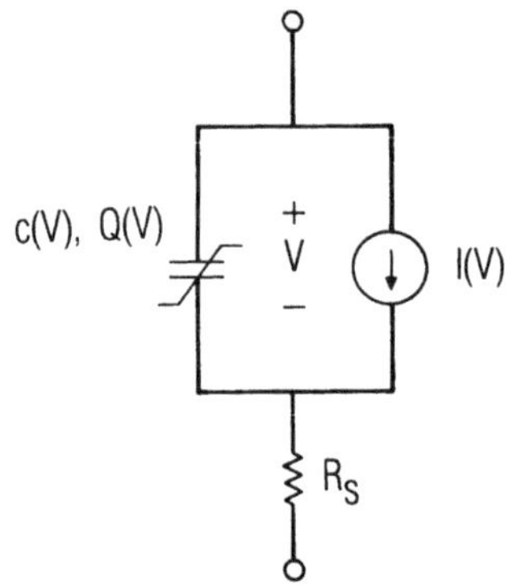

Figure 2.8 Equivalent circuit of a Schottky-barrier diode.

$$Q(V) = \frac{-C_{j0}\phi}{1-\gamma}\left(1-\frac{V}{\phi}\right)^{1-\gamma} \tag{2.58}$$

The small-signal incremental junction capacitance is

$$C(V) = \frac{dQ}{dV} = \frac{C_{j0}}{\left(1-\frac{V}{\phi}\right)^{\gamma}} \tag{2.59}$$

where ϕ is the diffusion potential and C_{j0} is the zero-voltage junction capacitance. If the junction is uniformly doped, γ is 0.5. V is the junction voltage shown in Figure 2.8; that is, excluding the voltage dropped across the series resistance. It is defined as positive if the junction is forward biased. Equations (2.58) and (2.59) are strictly valid only if the epilayer is never completely depleted. It is interesting to note that the reverse-biased junction has the same capacitance as a parallel-plate capacitor whose plate spacing equals the depletion width, and whose dielectric constant equals that of the semiconductor.

If the doping is nonuniform, (2.59) may not describe the capacitance adequately over a wide voltage range. In this case (2.59) may be used piecewise, with different γ parameters for different voltage ranges, or an entirely empirical expression may be used (see Section 2.3.2 for warnings about the pitfalls of this practice). Occasionally a diode's doping profile is purposely designed to maximize or to minimize its capacitive nonlinearity. In varactor diodes, the capacitive nonlinearity is made as strong as possible, to increase their usefulness as voltage-controlled tuning elements or as efficient frequency multipliers. One of the most extreme cases is that of the hyperabrupt varactor, in which the doping concentration actually decreases with distance from the junction. Hyperabrupt varactors can have $\gamma = 1.5$ or even $\gamma = 2.0$ over at least part of their reverse voltage ranges. These varactors usually have relatively high series resistance, because the undepleted part of the epitaxial layer is very lightly doped, and are consequently unsuited for use in frequency multipliers. They are most useful in tuning applications, especially in voltage-controlled oscillators, where the strong, controlled nonlinearity can be used to achieve a wide and nearly linear frequency/voltage characteristic.

2.4.2.2 Harmonic-Balance Capacitance Model

Equation (2.58) has an obvious problem as $V \to \phi$: its derivative, (2.59), becomes infinite. This characteristic in real devices is not particularly important, because virtually all diodes conduct strongly at $V << \phi$, so the junction voltage is clamped to a value well below ϕ. In the circuit simulator, however, there is no such limitation, and $V \geq \phi$ can easily occur.

One solution, first used in SPICE, is to define a quantity F_c and to approximate the charge function as a quadratic at voltages above $F_c\phi$. As long as the derivatives are matched at $V = F_c\phi$, the first and second derivatives are continuous at $V \geq \phi$. $F_c\phi$ should be set to a value larger than the maximum junction voltage, of course, which is determined by the circuit and the I/V characteristic. Note that a linear extension of the charge function is not adequate, as it results in a discontinuous second derivative (see Section 2.3); the extension must be quadratic.

The charge function, thus modified, is

$$Q(V) = C_{j0}F_1 + \frac{C_{j0}}{F_2}\left(F_3 V + 0.5\gamma\frac{V^2}{\phi} + F_4\right) \tag{2.60}$$

where

$$\begin{aligned} F_1 &= \frac{\phi}{1-\gamma}(1-(1-F_c)^{1-\gamma}) \\ F_2 &= (1-F_c)^{1+\gamma} \\ F_3 &= 1-F_c(1+\gamma) \\ F_4 &= F_c\phi(F_c(1+0.5\gamma)-1) \end{aligned} \tag{2.61}$$

Equations (2.60) and (2.61) apply at $V > F_c\phi$; at $V < F_c\phi$, (2.59) applies. A problem still occurs when $\gamma = 1$. This problem is easy to trap, but model developers should be mindful of it.

This modification improves the performance of the model for more subtle reasons. In computer circuit analysis, we work with finite increments of voltage, and use the derivative to estimate the change in charge over each increment. The derivative of (2.58), Equation (2.59), is accurate near ϕ only as those increments approach zero, but for the finite increments used in simulation, the derivative is often a poor estimate of the change in charge. Thus, the derivative of the modified charge function may actually do a better job of finding the solution than the correct one. Similarly, when the derivative is large, a numerical estimate of the derivative ($\Delta I / \Delta V$) may work better than an analytical one.

2.4.2.3 *I*/*V* Characteristic

The *I*/*V* characteristic of a Schottky diode can be expressed by a simple relation, which is derived under the assumption that conduction occurs primarily via the thermionic emission of electrons over a barrier. Other mechanisms, such as tunneling, occur as well, but for Schottky diodes of moderate doping densities, operated close to room temperature, the thermionic-emission assumption is valid and agrees remarkably well with measurements. The *I*/*V* characteristic of the junction of a Schottky-barrier diode (i.e., not including the voltage drop across the series resistance) has the same general form as that of a *pn* junction diode,

$$I(V) = I_{sat}\left(\exp\left(\frac{qV}{\eta KT}\right) - 1\right) \tag{2.62}$$

where q is the electron charge, K is Boltzmann's constant, $1.37 \cdot 10^{-23}$ J/K, and T is absolute temperature. The ideality factor η accounts for unavoidable imperfections in the junction and for other secondary phenomena that thermionic emission theory can not predict. η is always greater than 1.0 and, in a well-made diode, should be less than 1.20. I_{sat}, a proportionality constant, is called the *current parameter*, or, because (2.62) implies $I(V) = I_{sat}$ as $V \rightarrow -\infty$, the *reverse-saturation current*. An expression for I_{sat} is

$$I_{sat} = A^{**} T^2 W_j \exp\left(\frac{q\phi_b}{KT}\right) \tag{2.63}$$

where A^{**} is the modified Richardson constant; W_j is the junction area; and ϕ_b is the barrier height in volts, a constant usually approximately 0.1 V greater than the diffusion potential. A^{**} is approximately 96 A cm^{-2} K^{-2} for silicon and 4.4 A cm^{-2} K^{-2} for GaAs.[1] One should be careful about taking (2.63) too seriously; because of such secondary effects as charge generation and surface imperfections in the junction, I_{sat} can differ significantly from the value given by (2.63). Equation (2.63) can be used, however, to draw some general conclusions. For example, the value of the Richardson constant in GaAs is lower than in silicon, which implies that the knee of the *I*/*V* characteristic occurs at a higher voltage for GaAs diodes than for silicon diodes. It also implies that the device is highly sensitive to temperature.

Figure 2.9(a) shows the *I*/*V* characteristic of a Schottky diode in Cartesian coordinates, and Figure 2.9(b) shows the same characteristic graphed on semilog axes. The semilog graph is a straight line having a slope of one decade of current per 58.5η mV of junction-voltage change at low current levels and at 295K. Imperfections in the diode design or fabrication can be identified readily by deviations from that straight line. For example, excessive tunneling current at low voltages reduces the slope to nearly half the usual value, as does junction damage due to electrical overstress. The curve deviates from a straight line at the high current end because of the voltage dropped across the parasitic series resistance, R_s.

At high reverse voltages, junction breakdown results from avalanching. Avalanche breakdown voltage increases as doping density is reduced, but series resistance also increases. Thus, there is a trade-off in diode design between low R_s and high reverse-breakdown voltage. GaAs diodes generally have greater reverse-breakdown voltages than silicon, partly because the higher electron mobility in GaAs allows lower series resistance to be achieved with lighter doping. In many types of balanced mixers, breakdown voltage is irrelevant, because the diodes are connected in

---

1. It is impossible to provide a precise value for this parameter. See [2.7] for more information.

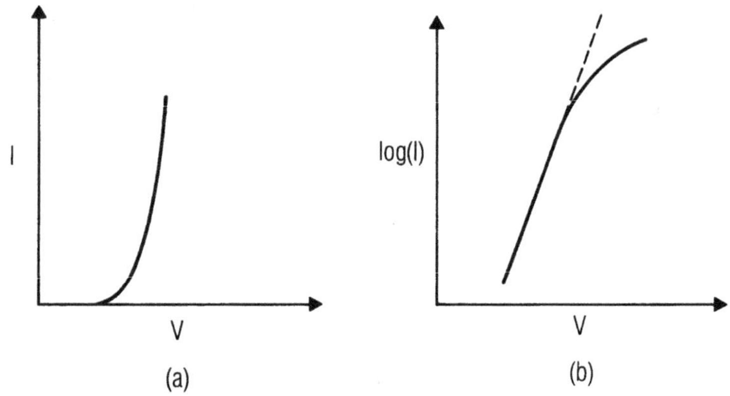

Figure 2.9 *I/V* characteristic of a Schottky-barrier diode: (a) in Cartesian coordinates; (b) on semilogarithmic axes.

parallel but reversed (a so-called *antiparallel* connection). In such an arrangement the reverse voltage on any diode never exceeds the diode's forward voltage drop.

2.4.2.4 Harmonic-Balance *I/V* Model

Equation (2.62) is subject to numerical overflow during computation when V is large. In many compilers, for example, the maximum argument of the *exp* function in IEEE standard double-precision arithmetic is 709; since $q/\eta KT \sim 40$, V is limited to a little less than 18V.

The solution to this problem, as with the capacitance, is a quadratic extension of the exp function above some threshold value, V_t. The first and second derivatives must be matched at the threshold, and the threshold should be much greater than the expected maximum junction voltage.

The new expression, at $V > V_t$, is

$$I(V) = I_{sat}\exp(\delta V_t)\left(1 + \delta(V - V_t) + \frac{\delta}{2}(V - V_t)^2\right) \tag{2.64}$$

where $\delta = qV/\eta KT$. As with capacitance, the function must be quadratic. A linear extension avoids numerical overflow but introduces a discontinuity in the second derivative.

One must also be careful of numerical underflow at large negative values of V. In diode or BJT *I/V* characteristics, we often calculate terms of the form $\exp(\delta V) - 1$. Double-precision arithmetic has at most 15 digits of precision, so any value of $\exp(\delta V) < 10^{-15}$ is indistinguishable from zero in that expression. Thus, it is usually acceptable simply to approximate $\exp(\delta V) - 1 \approx -1$ when $\delta V < \ln(10^{-15})$ or $V < -35/\delta$.

2.4.3 Mixer Diodes

Because they have virtually no minority-carrier effects, Schottky-barrier diodes are very fast-switching devices. As such, they are ideal for use in a diode mixer, which is often idealized as a high-frequency switch. Very high-quality silicon Schottky diodes are available at low cost, and for applications requiring the best possible conversion loss and noise figure, GaAs diodes can be obtained at only slightly greater expense. Diode technology today is sufficiently mature to allow mixers at frequencies above 1,000 GHz to be fabricated.

Figure 2.10 shows the cross section of a mixer diode chip. The vertical structure of the diode is identical to that shown in Figure 2.7, but the area of the junction is defined precisely; the anode is formed as a circular dot. Chips usually have a number of these, to facilitate connection of the anode wire, or to allow for the selection of an anode of the desired size.

In operation, the mixer diode operates as a variable-resistance diode or as a switch, which in many respects is the same thing. The incremental

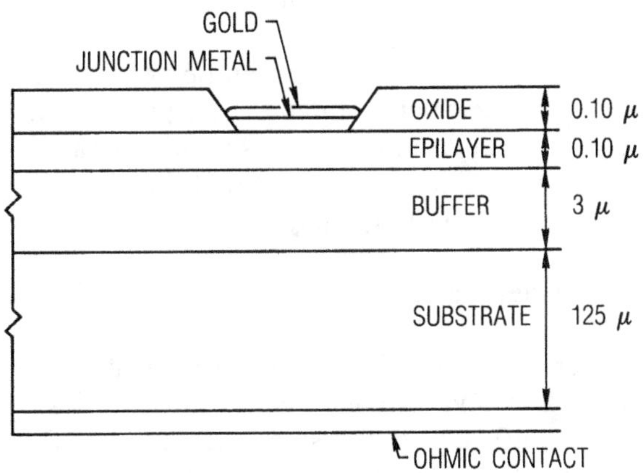

Figure 2.10 Cross section of a chip diode. The dimensions are typical for high-performance mixers.

small-signal conductance of the junction can be found by differentiating (2.62):

$$g(V) = \frac{d}{dV}I(V) = \frac{q}{\eta\, KT}I_{sat}\exp\left(\frac{qV}{\eta\, KT}\right) \approx \frac{q}{\eta\, KT}I(V) \qquad (2.65)$$

The junction conductance is proportional to the large-signal junction current. Virtually all high-frequency Schottky-barrier mixer diodes are uniformly doped, so (2.59), with $\gamma = 0.5$, describes the junction capacitance accurately. However, it is often not valid to assume that the dc series resistance, which can be found from Figure 2.9(b), represents the series resistance at millimeter-wave frequencies. Skin effect causes the high-frequency series resistance to be greater than the dc value because the high-frequency current forms a thin sheet at the surface of the chip and is nearly zero in the bulk substrate. The increased path length and reduced cross-sectional area of this thin current sheet increase the series resistance of the diode.

The cutoff frequency f_c is a figure of merit for a mixer diode. The cutoff frequency is traditionally calculated from dc quantities (thus the common misnomer *dc cutoff frequency*), without regard to skin-effect enhancement of the series resistance. The cutoff frequency is defined as

$$f_c = \frac{1}{2\pi R_s C_{j0}} \qquad (2.66)$$

Cutoff frequencies of mixer diodes often can be very high, on the order of several thousand gigahertz. Such high cutoff frequencies do not imply that a diode can be used in terahertz mixers; f_c is valid only as a figure of merit. For good performance, a mixer diode's cutoff frequency should be at least 10 times the mixer's operating frequency.

2.4.4 Schottky-Barrier Varactors

Frequency-multiplier varactors are often realized as $p+$ structures on n substrates in both GaAs and silicon. Because the diodes' $p+$ regions are difficult to fabricate uniformly in the small anode sizes necessary for very high-frequency operation, these diodes are limited to the lower microwave and perhaps millimeter-wave regions. Furthermore, at high frequencies the importance of minimizing series resistance and maximizing capacitance variation becomes progressively greater, and Schottky-barrier varactors are generally superior in these respects. Frequency multipliers having input frequencies above approximately 50 GHz usually employ Schottky-barrier varactors; often such multipliers generate output power at frequencies of several hundred gigahertz. High-performance Schottky-barrier varactors for such applications are invariably realized in GaAs.

The structure of a Schottky-barrier varactor is qualitatively the same as that of a mixer diode, shown in Figure 2.10. In order to achieve both good efficiency and high output power, varactors require higher breakdown voltages than mixer diodes; accordingly, the doping density in a varactor's epilayer is very low (typically 10^{16} to 10^{17} atoms/cm^3) and its junction area is relatively large.

The large junction area provides greater capacitance than would be tolerable in a mixer diode, usually approximately 0.1 pF for operation near 50 GHz. The large area also facilitates heat dissipation, an important consideration; most of the multiplier's input power is dissipated in the diode. Because of the low doping level, the series resistance of the Schottky varactor is greater than that of a mixer diode of the same size, and the cutoff frequency is significantly lower. The mixer-diode equivalent circuit, shown in Figure 2.8 and described by (2.58) through (2.62), is generally valid for Schottky-barrier varactors, as long as the parameters, especially C_{j0} and γ, are appropriately modified. The diffusion potential is usually around 1V, higher than that of a mixer diode, and because of second-order effects, γ is often somewhat lower (approximately 0.45). I_{sat} also differs; however, in normal operation, the diode is usually not driven into forward conduction, so the forward *I/V* characteristic is of secondary concern.

Several figures of merit can be defined for varactors. One of the most important is the *dynamic cutoff frequency*, f_{cd}:

$$f_{cd} = \frac{S_{\max} - S_{\min}}{2\pi R_s} \tag{2.67}$$

where S is elastance, or inverse capacitance. $S_{\min}$ is the minimum elastance, which occurs as the junction voltage approaches ϕ. $S_{\min}$ is often negligible, so (2.67) becomes

$$f_{cd} = \frac{S_{\max}}{2\pi R_s} \tag{2.68}$$

where S_{max} is the elastance at reverse breakdown or at some other standard reverse voltage, often –6V. Infrequently $S_{max} = 1/C_{j0}$. Clearly, one should always determine precisely how the f_{cd} of a particular varactor is defined.

Dynamic cutoff frequency is an important quantity. It is possible to create varactors having very high static cutoff frequencies, as defined by (2.66), but poor nonlinearity. Such devices are inefficient and have low f_{cd}. Another figure of merit is the dynamic Q, Q_δ:

$$Q_\delta = \frac{S_{max}}{2\pi f_0 R_s} = \frac{f_{cd}}{f_0} \tag{2.69}$$

where f_0 is the frequency at which Q_δ is evaluated.

Schottky-barrier varactors have very high Q_δ, allowing good efficiency to be achieved at high frequencies. However, Schottky varactors are limited in power handling capability; they can be driven only to the point at which the junction begins to conduct. If the input level is increased beyond this point, efficiency suffers, and output power saturates; this phenomenon is illustrated in Chapter 7. Although limited to lower frequencies, p^+n junction varactors largely circumvent this problem.

2.4.5 p^+n Junction Varactors

At microwave frequencies, silicon or GaAs p^+n junction varactors are preferred. The dc I/V characteristic of a p^+n junction has the same general form as that of a Schottky barrier (2.62), and the depletion capacitance expression (2.59) is also generally applicable, although $\gamma \neq 0.5$. p^+n diodes have greater capacitance variation, and thus provide greater efficiency, at high drive levels. These properties are the result of the long minority-carrier lifetimes that obviate the use of *pn* junction diodes in mixers. When the junction is forward-biased during the positive part of a high-frequency RF cycle, charge is injected into the junction region. Most of that charge (which consists of holes from the p^+ region injected into the n region) does not have time to recombine with electrons, so it is stored momentarily and removed when the RF current swings negative. This injected charge is stored, not conducted, and thus increases the capacitance variation of the diode. This phenomenon is called *diffusion charge storage*.

The amount of stored diffusion charge can be very great, so the forward-bias capacitance is substantial. When the varactor is driven so that the voltage peaks at ϕ and its reverse-breakdown voltage, the varactor is said to be *nominally driven*. If it is driven harder, the junction conducts, causing diffusion charge storage, and the varactor is said to be *overdriven*.

This charge-storage phenomenon is also used in the step-recovery diode (also called the *SRD* or *snap diode*), described in Section 2.4.7. The main functional difference between the varactor and step-recovery diode is that the SRD obtains its capacitance variation almost entirely by diffusion charge storage, while the p^+n varactor's operation depends less on high diffusion capacitance than on a gradual capacitance variation over its entire forward and reverse-voltage range.

A disadvantage of the p^+n structure is the p-diffusion step required in its fabrication. The diffusion process limits the minimum size of the p^+ region, so the minimum capacitance of the diode is limited as well. The p^+ region also has higher series resistance than the metal anode of a Schottky diode, so p^+n varactors have lower f_{cd} than Schottky varactors. These properties limit p^+n varactors to frequencies below approximately 50 GHz.

A cross section of a p^+n junction varactor is shown in Figure 2.11. The initial part of the varactor's fabrication is much like that of a mixer diode: an n epitaxial layer is grown on an n^+ substrate. A p^+ region is then diffused into the epitaxial layer, and ohmic contacts are formed on the p^+ and n^+ regions for the anode and cathode, respectively. An oxide-isolated anode like the structure used in Schottky-barrier diodes is not optimum for the p^+n varactor, because in oxide-isolated diodes the junction's electric field is stronger near the edge of the anode. The nonuniform electric field would cause avalanche breakdown to occur near the anode's edge, at a relatively low voltage, much lower than if the field were uniform. In the p^+n varactor, the anode area is formed by etching a mesa, making the electric field more uniform over the junction, thereby increasing the breakdown voltage. Diodes fabricated in this manner are called *diffused epitaxial varactors*.

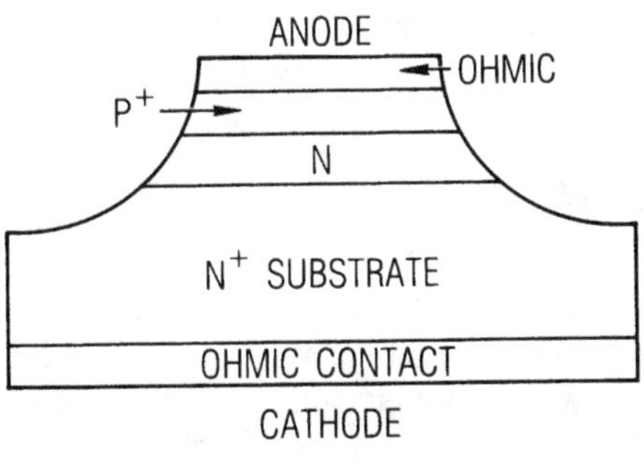

Figure 2.11 Cross section of a p^+n varactor. The mesa structure provides a higher breakdown voltage than a planar design.

A variant of the diffused epitaxial varactor is the *punch-through varactor*. The epitaxial layer of this device is so thin that it is completely depleted at a modest reverse voltage, usually a little more than half the breakdown voltage. The varactor has the reverse-bias capacitance characteristic of (2.59) at low reverse voltage; at higher voltage, the epi is fully depleted, or *punched through*, and, like the Mott diode, the *C/V* characteristic is nearly flat. The advantage of this structure is reduced sensitivity to changes in input power level, compared to a multiplier using a conventional varactor. The disadvantage is the reduced capacitance range, which results in a lower dynamic Q and therefore lower efficiency.

Most microwave-frequency p^+n varactors are realized in silicon. The minority-carrier lifetime in silicon is greater than in GaAs, so for lower-frequency operation (i.e., at output frequencies below about 20 GHz), charge-storage properties of silicon diodes are better than those of GaAs devices. At higher frequencies, GaAs has the advantage of lower series resistance and consequently higher Q_δ. Because of the additional series resistance of the p^+ region and its ohmic contact, both silicon and GaAs p^+n diodes have lower Q than comparable Schottky diodes.

2.4.6 Varactor Modeling

2.4.6.1 Capacitance

The above discussion indicates that the simple capacitance expression of (2.58) does not hold well for most types of varactor diodes. Devices having p^+n structure may have decidedly nonuniform doping, and thus very different *C/V* characteristics from the ideal. Even Schottky-barrier varactors may not follow (2.58) well, as the capacitance variation decreases rapidly at the point where the reverse voltage depletes the epilayer. Because the structure of such devices can differ dramatically, precise modeling must be largely ad hoc, and capacitance functions must be designed for the particular device.

2.4.6.2 Series Resistance

The greater capacitance variation of varactor diodes implies that their depletion widths vary considerably over their ranges of junction voltage. Since the series resistance consists largely of the undepleted epilayer, the series resistance likewise varies significantly. The assumption that series resistance is linear, in the model shown in Figure 2.8, may not be valid for such devices.

In lightly doped varactors, electrons may approach saturated drift velocity in the resistive epilayer. This phenomenon increases the incremental resistance in a nonlinear manner. One approach to modeling saturation in the series resistance is by the function

$$I(V) = I_s \tanh\left(\frac{V}{I_s R_{s0}}\right) \tag{2.70}$$

where R_{s0} is the low-current value of the series resistance and I_s is the saturation current. Another expression [2.8] is

$$V(I) = R_{s0}(I + \alpha I^7) \tag{2.71}$$

This expression provides a softer saturation characteristic. Unfortunately, it is more difficult to invert to obtain the $I(V)$ form, which usually is required by circuit simulators.

2.4.6.3 Substrate Impedance

Diodes used in submillimeter mixers and frequency multipliers are subject to additional phenomena that can affect their performance. At high frequencies, the inertia of the electrons in the substrate cannot be neglected, and it gives rise to an inductive impedance component. Similarly, the lossy substrate is subject to dielectric relaxation effects, which create a capacitive reactance. These combined effects create a parallel resonance in the terahertz range, adding a high impedance in series with the diode. A detailed treatment of these effects is beyond the scope of this book; they are discussed more completely in [2.8, 2.9].

2.4.7 Step-Recovery Diodes

Like a varactor, a step-recovery diode (*SRD*; also called a *snap diode*) uses capacitance variation to generate harmonics. However, it does so by storing charge under forward bias and by switching very rapidly to a high-impedance state when the diode is discharged. The multiplier is adjusted so that the diode switches at the instant the reverse current is maximum, thus generating a large and very short-lived voltage pulse during each excitation cycle. The resulting pulse train is rich in harmonic content, so it need only be filtered to obtain harmonic output. An SRD multiplier is used primarily for high-harmonic multiplication at high power levels. A typical application of an SRD is to multiply an input frequency of a few hundred megahertz to an output of several gigahertz. Step-recovery diodes are also used as pulse generators anywhere that short pulses (on the order of tens of picoseconds) are needed. Examples of such applications are fast sampling gates (e.g., for sampling oscilloscopes), time-domain reflectometers, and low-cost pulse-radar sensors.

An SRD must have high charge storage in the forward direction, low capacitance in the reverse direction, low series resistance, and, for power applications, high reverse-breakdown voltage. Its switching time must also be short, because switching speed establishes its high-frequency limit of operation. To meet these requirements, an SRD must have a relatively long charge storage time (long recombination time), and the charge that is injected into the junction while it is forward-biased must not travel so far that it cannot be removed during the reverse-bias interval. Finally, the depletion region must not be too wide, or transit-time effects reduce the multiplier's efficiency at high frequencies.

SRDs have the *pin* structure shown in Figure 2.12, in which the *i* region is a layer of undoped (intrinsic) or lightly doped semiconductor. The *i* region is formed by the overlap of the *p* and *n* regions, both of which have steep doping profiles. Such profiles create a narrow depletion region and a strong built-in electric field, which opposes the diffusion of charge into the junction. During forward conduction, holes and electrons are injected into the *i* region, where they recombine very slowly; the *i* layer thus becomes a region in which charge is stored. When the SRD is reverse-biased, the *i* layer is fully depleted; because of the wide depletion width, which includes the entire *i* layer, reverse capacitance is very low. The *i* region also provides a high reverse-breakdown voltage.

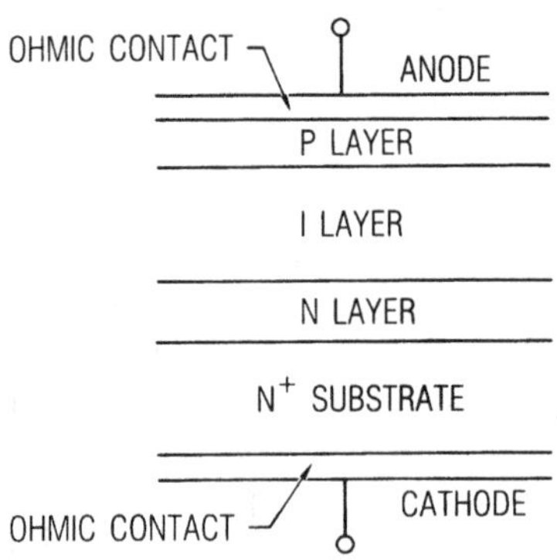

Figure 2.12 A step-recovery diode uses a *pin* structure on an n^+ substrate.

The forward *I*/*V* characteristic of the SRD obeys (2.62) under dc bias. Because the depleted region includes the *i* region, the reverse-capacitance characteristic can usually be treated as a constant. Under forward bias, the diode can be modeled as a *pn* junction in parallel with the diffusion capacitance. The stored diffusion charge is

$$Q_s = \tau I \tag{2.72}$$

where τ is the recombination time, or minority-carrier lifetime, of the material. Although a depletion charge exists in forward conduction, it is invariably negligible in comparison to Q_S. The reverse-bias junction capacitance of the SRD is

$$C_s = \frac{\varepsilon_s A}{d} \tag{2.73}$$

where *A* is the area of the junction and *d* is the depletion width. The largest part of *d* is the width of the *i* region, which is large and independent of voltage. Consequently, when the SRD is reverse-biased, its capacitance is very low and nearly constant. In an SRD frequency multiplier or pulse generator, it is important that all the charge injected into the junction during the positive excursion of the excitation cycle be removed during the negative excursion. Recombination of charge during that time reduces efficiency, because the recombined charge is lost as conduction current. Minimizing charge recombination requires that the minority-carrier lifetime be long compared to the period of an excitation cycle; because of its longer minority-carrier lifetime, silicon is invariably used for SRDs instead of GaAs. Like other diodes, the SRD has parasitic series resistance. This resistance arises in the ohmic contacts to the *p* and *n* regions, and in the undepleted parts of those regions. Because series resistance introduces loss and reduces multiplier efficiency, it is as important to minimize series resistance in an SRD as in any other type of diode.

2.5 FET DEVICES

It is no overstatement to say that the GaAs MESFET and its variants, including the high electron-mobility transistor (HEMT) have revolutionized low-noise microwave electronics and microwave systems. FETs also make excellent mixers, having low noise figures, broad bandwidths, and conversion gains, and as frequency multipliers they exhibit high efficiency, gain, and output power. FETs are commonly used in quasilinear applications, especially as small-signal and medium-power amplifiers, where an understanding of their nonlinearities is critical in minimizing the less attractive aspects of their performance, primarily intermodulation distortion and saturation.

Silicon metal oxide-semiconductor field-effect transistor (MOSFET) technology has progressed to the point where such devices can be used at microwave frequencies. New technologies are making MOSFETs attractive for use in a wide variety of RF applications. Laterally diffused MOSFETs (LDMOS) are attractive for high-power amplifiers at frequencies up to a few gigahertz, and submicron lithography has produced silicon MOSFETS with cutoff frequencies of tens of gigahertz. Interestingly, in spite of their maturity, these devices continue to improve.

Virtually all types of FET devices are highly symmetrical; they can be operated with negative drain-to-source voltage and current. This allows them to be used as resistive elements in switches, attenuators, and mixers.

2.5.1 MESFET Operation

Figure 2.13 shows a cross section of a GaAs metal epitaxial-semiconductor field effect transistor (MESFET). The MESFET is fabricated by first growing a very pure, semi-insulating buffer layer on a semi-insulating GaAs substrate, then growing an *n*-doped epitaxial layer that is used to realize the FET's active channel. Three connections are made to the channel: the source and drain ohmic contacts and, between them, the

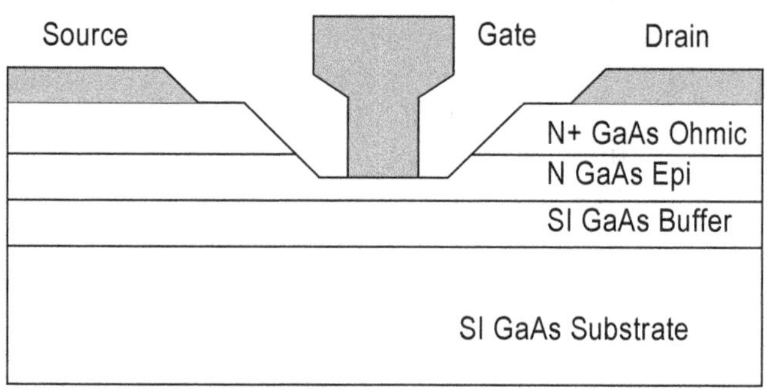

Figure 2.13 Cross section of a GaAs MESFET. Modern FETs all use the recessed channel *T*-shaped gate. The *T* gate minimizes gate resistance while retaining a short gate length.

Schottky-barrier gate. The epilayer is made thicker than necessary for the channel and is etched to the correct channel thickness in the gate region. This *recessed gate* structure allows the layer of epitaxial material under the source and drain ohmic contacts to be quite thick, much thicker than the channel, minimizing the parasitic source and drain resistances. Reducing the source resistance is especially important for low-noise devices; it is also important for achieving good conversion efficiency in FET mixers, frequency multipliers, and power amplifiers.

The MESFET is biased by the two sources: V_{ds}, the drain-to-source voltage, and V_{gs}, the gate-to-source voltage. These voltages control the channel current by varying the width of the gate-depletion region and the longitudinal electric field. In order to develop a qualitative understanding of MESFET operation, imagine first that $V_{gs} = 0$ and V_{ds} is raised from zero to some low value, as shown in Figure 2.14(a). When $V_{gs} = 0$, the depletion region under the Schottky-barrier gate is relatively narrow, and as V_{ds} is increased, a longitudinal electric field and current are established in the channel. Because of V_{ds}, the voltage across the depletion region is greater at the drain end than at the source end, so the depletion region becomes wider at the drain end. The narrowing of the channel and the increased V_{ds} increase the electric field near the drain, causing the electrons to move faster; although the channel's conductive cross section is reduced, the net effect is increased current. When V_{ds} is low, the current is approximately proportional to V_{ds}. If, however, the gate reverse bias is increased while the drain bias is held constant, the depletion region widens and the conductive channel becomes narrower, reducing the current. When $V_{gs} = V_t$, the turn-on (or *threshold*) voltage, the channel is fully depleted and the drain current is zero, regardless of the value of V_{ds}.[2] Thus, both V_{gs} and V_{ds} control the drain current. When the FET is operated in this manner (i.e., when both V_{gs} and V_{ds} have a strong effect on the drain current), it is said to be in its *linear*, or *voltage-controlled resistor* region.

If V_{ds} is increased further, as in Figure 2.14(b), the channel current increases, the depletion region becomes even wider at the drain end, and the conductive channel becomes narrower. The current clearly must be constant throughout the channel, so as the conductive channel near the drain becomes narrower, the electrons must move faster. However, the electron velocity cannot increase indefinitely; the average velocity of the

---

2. In fact, the current does not turn off abruptly, in part because the conductivity of the buffer layer is not zero and the edge of the depletion region is not distinct. Thus, the threshold voltage is somewhat indistinct as well. It can be defined, for example, as the point where the drain current decreases to some particular fraction of its zero-voltage value.

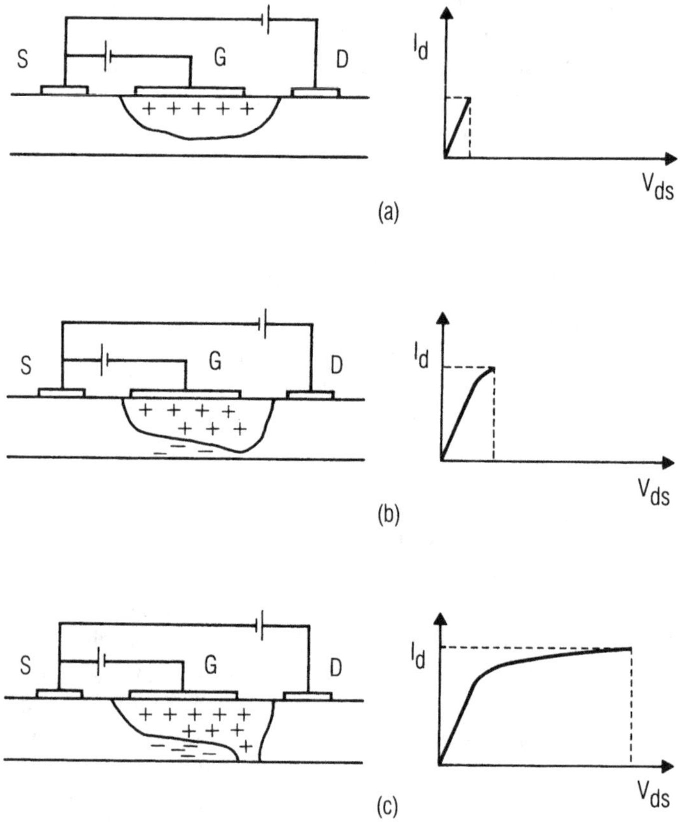

Figure 2.14 GaAs MESFET operation: (a) very low V_{ds} (i.e., a few tenths of a volt); (b) V_{ds} at the saturation point; (c) current saturation.

electrons in GaAs can not exceed a velocity called their *saturated drift velocity*, approximately $1.3 \cdot 10^7$ cm/s. If V_{ds} is increased beyond the value that causes velocity saturation (usually only a few tenths of a volt), the electron concentration rather than velocity must increase to maintain current continuity throughout the channel. Accordingly, a region of electron accumulation forms near the drain end of the gate. Conversely, after the electrons transit the channel and move at saturated velocity into the wide area between the gate and drain, an electron depletion region is formed. That depletion region is positively charged because of the positive donor ions remaining in the crystal. As V_{ds} continues to increase, Figure

2.14(c), progressively more of the voltage increase is dropped across this region, called a *dipole layer*, and less is dropped across the unsaturated part of the channel. Eventually a point is reached where further increases in V_{ds} are dropped entirely across the charge domain and do not substantially increase the drain current; at this point the electrons move at saturated drift velocity over a large part of the channel length. When the FET is operated in this manner, which is the normal mode of operation for small-signal devices, it is said to be in its *saturation region*, or in *saturated operation.* All FET amplifiers and most FET mixers and frequency multipliers are biased into saturation. One notable exception is the FET resistive mixer, which we shall examine in Chapter 11.

It is important to recognize that the charge domain begins to form at drain-to-source voltages well below those corresponding to the horizontal portion of the drain *I/V* characteristic, so the charge domain affects the *I/V* characteristic throughout almost the entire range of V_{ds}.

The terms *linear region* and *saturation region* are unfortunate, because they seem to indicate exactly the opposite of their true meaning: small-signal, quasilinear operation takes place in the FET's saturation region, not in its linear region. Further confusion arises because the same terms are used, with opposite meaning, to describe the operating regions of bipolar transistors: a bipolar transistor is said to be in saturation when the collector/emitter voltage is very low. For better or worse, this terminology is widely accepted, so even with some misgivings we will use it throughout the rest of this book.

As in the Schottky-barrier diode, the Schottky-barrier gate depletion region represents a capacitance. At low drain voltages, the gate-to-channel capacitance has nearly the ideal Schottky-barrier voltage dependence of (2.58), but as V_{ds} increases, the situation becomes more complex. At $V_{ds} \approx 0$ (and notwithstanding the arguments made in Section 2.2.7.3), the gate capacitance is distributed along the channel, but it frequently is modeled approximately as two equal capacitors, one between the gate and source, and the other between the gate and drain. These capacitances are related to the change in gate-depletion charge with changes in gate-to-source voltage V_{gs} and gate-to-drain voltage V_{gd}, respectively. As V_{ds} is increased and the FET begins saturated operation, however, drain-voltage changes are shielded from the gate depletion region by the dipole layer. Further changes in V_{ds} no longer increase the charge in the depletion region, so the gate-to-drain capacitance drops to a point where it consists of little more than stray capacitance between metallizations. In saturation the gate-to-source capacitance represents the full gate-depletion capacitance, so the gate-to-source capacitance increases to approximately twice the value it had in linear operation.

2.5.2 HEMT Operation

A HEMT differs from a conventional MESFET in that the channel is formed by a heterojunction instead of a simple epitaxial layer. Because the channel is not doped, impurity scattering is minimized and high electron mobilities result. The mobility increases as temperature decreases, so substantial improvement in gain and noise figure can be achieved at low, even cryogenic, temperatures.

Figure 2.15 shows a simple HEMT. Instead of a doped epilayer, the device has an n^+ AlGaAs layer and a very thin undoped InGaAs layer immediately underneath it. (Not shown in the figure is an extremely thin, undoped AlGaAs spacer layer between the AlGaAs and InGaAs layers. This spacer is on the order of 50Å thick and prevents scattering by ions in the AlGaAs layer.) Because of the band structure of the semiconductors, electrons from the AlGaAs layer accumulate in the InGaAs layer near the interface; the charge density of this electron layer is controlled by the gate voltage. The charge density is generally very low, making such devices difficult to use as power amplifiers and, to some degree, frequency multipliers. The high transconductance, however, provides a high cutoff frequency and very low noise figure. These characteristics makes HEMTs ideal for low-noise amplifiers and active mixers at frequencies well into the millimeter wave range.

The AlGaAs-InGaAs device is usually called a *pseudomorphic HEMT*, or *pHEMT*,[3] because of the lattice mismatch between the AlGaAs and InGaAs layers. Other types of pHEMTs are possible, as well as devices with multiple heterojunctions. The latter provide greater channel charge density, and thus are useful as power amplifiers. The wide variety of materials, layer thicknesses, and device geometries in modern HEMT technology provides many degrees of freedom for optimizing the device's channel; in contrast, the only degrees of freedom in MESFET channel design are thickness and doping density.

Models of HEMTs are not very different from those of MESFETs. One of the greater differences between MESFETs and HEMTs is in the shape of the transconductance curve, as a function of gate voltage. In MESFETs, the transconductance usually increases monotonically with gate voltage, possibly with a peak at positive V_{gs}; in HEMTs, it often has a pronounced peak.

---

3. Pronounced "pee-hemt." An affected acronym, to be sure, but such things are beloved of engineers. It may have been written this way to prevent people from pronouncing it as "femt."

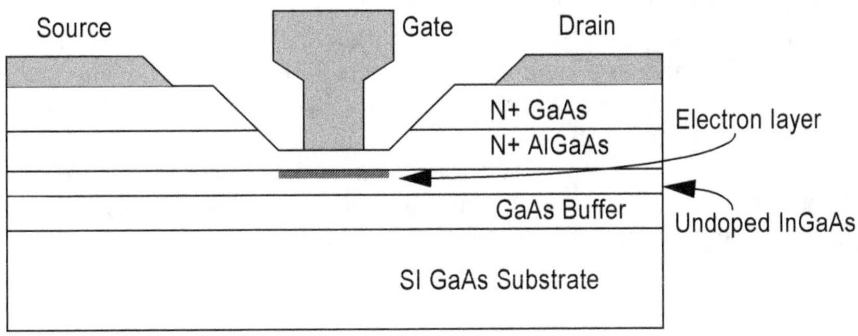

Figure 2.15 Cross section of a simple AlGaAs-InGaAs-GaAs HEMT.

2.5.3 MOSFET Operation

The operation of MOSFETs has been so thoroughly described in previous books that we review it only briefly here. It is important to note, however, that advances in semiconductor technology and submicron lithography have resulted in MOSFETs that are useful at RF and microwave frequencies. MOS technologies, especially complementary MOS (CMOS) can be very useful, especially for low-power, low-cost RF ICs.

All RF and microwave devices are enhancement mode, *n* channel silicon devices. They consist of a lightly doped *p* substrate and a gate, which can be either metal or semiconductor, insulated from the substrate by a very thin oxide (SiO_2) layer. At low gate voltages, no channel exists, so no conduction is possible. When the gate voltage exceeds a positive threshold voltage, V_t, an inversion layer of electrons is formed under the gate, and that layer acts as a channel. (This is similar in some ways to a HEMT, and, in fact, HEMTs have been compared in their operation to MOSFETs.) Simple analysis gives an expression for the charge density in the channel when $V_d = 0$ and $V_g \geq V_t$:

$$Q_{ch} = W_g L_g C_{ox}(V_g - V_t) \tag{2.74}$$

where L_g is the gate length, W_g is the gate width, and C_{ox} is the oxide capacitance, the parallel-place capacitance, per area, between the gate and channel. We use V_d and V_g instead of V_{ds} and V_{gs}, to represent the internal voltages, which do not include voltage drop across the source and drain contact resistances, R_s and R_d, respectively.

A number of effects can complicate (2.74). One of the most important is called *backgating*, the effect of the voltage between the substrate and the channel, which acts as a kind of second gate. Others are oxide and interface charges, short- and narrow-channel effects, weak inversion (or subthreshold effects), and nonuniform substrate doping.

As in other types of FETs, application of drain bias causes the voltage between the gate and channel to be lower (i.e, more negative) at the drain end. The charge disappears at the drain end when

$$V_g - V_d \leq V_t \tag{2.75}$$

and this condition represents the onset of current saturation, much as the completely depleted channel, from a combination of velocity saturation and pinch-off, causes saturation in MESFETs. Velocity saturation, however, plays only a minor role in the operation of silicon devices.

One of the more interesting developments is the laterally diffused MOSFET, or LDMOS, device. These are used primarily for power amplifiers at frequencies from VHF to a few gigahertz. Figure 2.16 shows a cross section of an LDMOS device. An advantage of this structure is the direct electrical connection of the source to the mounting surface; in contrast, other power FETs have the drain connected to the substrate. This eliminates the need for wire bonds or insulators between the chip and mounting surface, thus minimizing source inductance and resistance. It

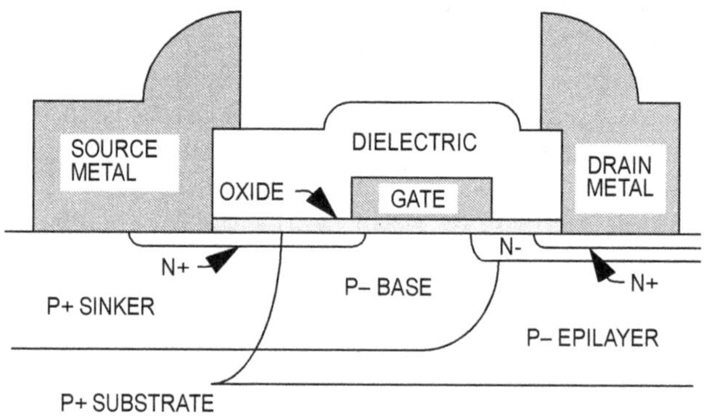

Figure 2.16 Cross section of an LDMOS device.

also provides better cooling. Other advantages are a low-resistance gate and a long, lightly doped area between the channel and drain contacts, which minimizes gate-to-drain capacitance and provides a high breakdown voltage.

2.5.4 MESFET Modeling

Figure 2.17 shows a lumped-element equivalent circuit of the MESFET that can be used either in a small-signal or a large-signal analysis. R_g is the ohmic resistance of the gate, and R_s and R_d are the source and drain ohmic contact resistances, respectively. R_1 is the resistance of the semiconductor region under the gate (called the *intrinsic resistance*, R_i, in some texts) between the source and channel; R_2 is a similar resistance, which is negligible in ordinary, current-saturated operation. It may be significant when the FET is operated in its linear region or in inverse mode. C_{ds} is the drain-to-source capacitance, which is dominated by metallization capacitance, and is therefore often treated as a constant. C_{gs} and C_{gd} are the gate-to-channel capacitances; by expressing these as capacitances instead of charges we imply the use of a division-by-capacitance model (see Section 2.5.7.1), although many MESFET models use a division-by-charge characterization. I_d is the nonlinear channel-current source. The diodes in parallel with C_{gs} and C_{gd} account for forward or reverse (avalanche) gate conduction. I_d, C_{gs}, and C_{gd} are functions of the gate voltage V_g and either

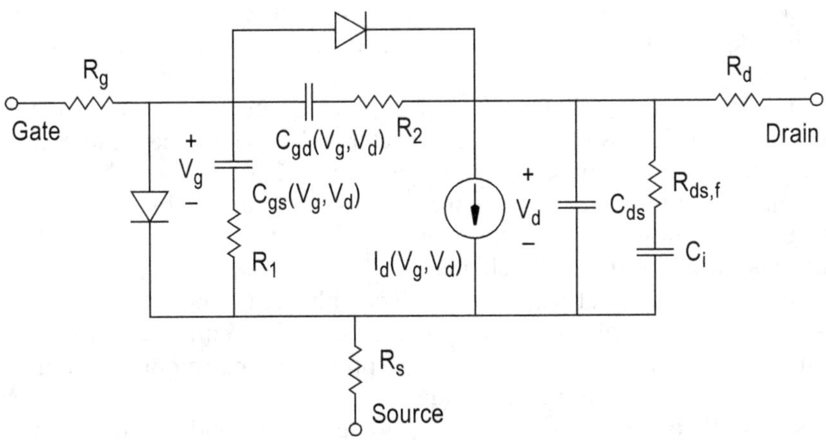

Figure 2.17 Equivalent circuit of a GaAs MESFET. Essentially the same circuit can be used for HEMTs.

the drain voltage V_d or the gate-to-drain voltage, V_{gd}. V_g and V_d are called the *internal* gate and drain voltages, to distinguish them from the voltages at the FET's terminals, V_{gs} and V_{ds}, called the *external* voltages. V_{gd} represents the internal quantity; we do not use the external gate-to-drain voltage. V_g and V_d are related to V_{gs} and V_{ds} as follows:

$$V_g = V_{gs} - R_s I_d \tag{2.76}$$

and

$$V_d = V_{ds} - (R_s + R_d) I_d \tag{2.77}$$

$R_{ds,f}$ and C_i require explanation. In silicon devices, the drain-to-source conductance is accurately represented by the partial derivative of I_d w.r.t. V_d. In III-V devices, that derivative is valid only at very low frequencies, at most a few megahertz. At higher frequencies, the resistance is a factor of three to ten lower than at dc; this phenomenon is sometimes called *drain dispersion*. The combination of $R_{ds,f}$ and C_i models this effect. Because of the low transition frequency, C_i is often remarkably large, on the order of microfarads.

Modeling drain dispersion is a difficult task, complicated by the nonlinearity of the drain-to-source resistance. The use of $R_{ds,f}$ and C_i is actually a rather poor approach to the problem as it has several undesirable characteristics; for example, a linear $R_{ds,f}$ does not pinch off properly at $V_g = V_t$. It is important, in this formulation, that $R_{ds,f}$ be a linear element; making $R_{ds,f}$ nonlinear invariably produces dc currents that are open-circuited by C_i.

In virtually all models, C_{gs} and C_{gd} are treated as distinct, nonlinear capacitors. In fact, the FET's gate-to-channel capacitance should best be treated as a multiterminal capacitor (Section 2.2.7.3). The subject of FET capacitance models is subtle; we address it further in Section 2.5.7. However, for now, we will limit ourselves to a discussion of customary practices for modeling these elements.

In spite of many attempts to produce "physical" models, ones that are based on the physical operation of the FET device, empirical models have been by far the most successful. The expressions that model the nonlinear circuit elements in empirical models are chosen only to reproduce the measured *I*/*V* or *Q*/*V* characteristics of the device. Indeed, most physical models include significant empirical elements, turning them, fundamentally, into empirical models. For these reasons this book is concerned exclusively with empirical, equivalent-circuit models of all the devices it describes.

2.5.4.1 MESFET Channel Current

The current in the drain-current source is the "heart" of a MESFET model. Many widely used models are, primarily, drain-current models. To follow tradition, we designate the channel current as I_d, although I_d is equal to the drain-terminal current only at dc.

The drain current is a function of internal gate and drain voltages, V_g and V_d. It can be expressed satisfactorily via an empirical expression. The advantage of an empirical expression is that the expression and its derivatives (in particular, the transconductance, $\partial I_d / \partial V_g$) usually can be evaluated with less computation—hence less computer time—than a physical model. The greatest advantage of a physical model (i.e., one in which the current is calculated from the physical parameters and dimensions of the device) is in its use to relate the device structure and physical characteristics directly to the performance of the circuit. Although it is sometimes assumed that physical models are inherently more accurate than empirical ones, this has not been the case in practice.

The considerations listed in Section 2.3 are particularly important for modeling MESFET channel current. Additional caveats are presented below.

Multiquadrant Operation

In the past, it was often considered adequate for a model to allow only "one quadrant" operation; that is, one quadrant of the I_d / V_d plane, $I_d > 0$ and $V_d > 0$. In fact, in many large-signal circuits, V_d not only drops into the linear region, but can momentarily become negative. In other circuits, especially such passive circuits as FET resistive mixers, switches, and attenuators, the FET is biased at $V_d = 0$. In these cases, the model must operate properly near zero drain bias.

Many models do not do well under these conditions. For example, many use a hyperbolic tangent to model the dependence on V_d; that is, they have the form

$$I_d(V_g, V_d) = f_G(V_g)\tanh(\alpha V_d) \tag{2.78}$$

where $f_G(V_g)$ is a function describing the gate-voltage dependence. This expression causes the current to be an odd function of drain voltage and creates an inflection point (zero second derivative) at $V_d = 0$. The current of the real device, however, does not behave this way; it has a finite second derivative at $V_d = 0$ and $-I_d$ increases monotonically with $-V_d$.

Because a MESFET is a highly symmetrical device, it is a common practice to use a one-quadrant model and to reverse its voltages, when V_d goes negative, so the drain voltage in the model is always positive. The calculated I_d is then reversed at exit; some SPICE MOSFET models, for example, do this. Unfortunately, it is easy, in such models, to have discontinuous derivatives at $V_d = 0$. These can lead to poor convergence in analyses of active circuits and to erroneous analyses of passive circuits.

Pinch-off Considerations

Making the drain current pinch off at $V_g = V_t$ is not enough; the transconductance *and its first derivative* must also be zero at $V_g = V_t$. In an expression having the form of (2.78), $f_G(V_g)$ must satisfy these requirements. For example, it is common to use

$$f_G(V_g) = a_0 + a_1 V_g + a_2 V_g^2 + a_3 V_g^3 \tag{2.79}$$

Imposing the obvious requirements that (1) $I_d = 0$ at $V_g = V_t$, (2) $G_m = 0$ at $V_g = V_t$, and (3) $I_d = I_{dss}$ at $V_g = 0$ defines three of the four a_n coefficients, even without imposing the derivative constraint on $f_G(V_t)$. Thus, we really have at most only one coefficient for adjusting the shape of $f_G(V_g)$. Similar problems exist in other types of functions.

External and Internal Voltages

Both the physical and empirical I/V models describe only the I/V dependence on the internal voltages V_g and V_d. Usually we wish to know the I/V dependence upon the external voltages V_{gs} and V_{ds}, because these are observable. The dc values of these quantities differ because the voltage drops across R_s and R_d; there is no dc voltage drop across R_g, so it need not be considered.

It is desirable, for easiest fitting of an I/V function to measured data, to use points on a rectangular grid; that is, where V_g is held constant while V_d is varied, and V_d is held constant while V_g is varied. When the voltages across the parasitic resistances are subtracted, however, those points are no longer on the desired grid. Two-dimensional interpolation is then needed to return the points to a rectangular grid. Such methods are standard material in books on numerical methods.

Drain Dispersion

In silicon junction FETs, it is accurate to assume that the transconductance, G_m, and the drain-to-source conductance, G_{ds}, are given by the expressions

$$G_m = \frac{\partial I_d}{\partial V_g} \tag{2.80}$$

$$G_{ds} = \frac{\partial I_d}{\partial V_d} \tag{2.81}$$

In FETs, (2.80) is reasonably accurate, but (2.81) is not accurate above, at most, a few megahertz. The increase in G_{ds} at high frequencies is called *drain dispersion*. Drain dispersion is a difficult phenomenon to model. The traditional method, using the combination of $R_{ds,f}$ and C_i, shown in Figure 2.17, is often inadequate; it allows drain current when the device is pinched off and does not account for bias dependence of G_{ds}. Because of the theoretical difficulty of modeling this phenomenon, it is usually treated heuristically.

2.5.4.2 Modeling C_{gs} and C_{gd}

The nonlinear capacitances C_{gs} and C_{gd} account for the displacement current through the gate depletion region; they are large-signal capacitances.[4] These capacitances are logically functions of V_g and V_{gd}; however, for simplicity we would like them to have the same control voltages as I_d. Thus, they are usually treated as functions of V_g and V_d.

If the FET remains in saturation, C_{gs} is usually modeled successfully as a Schottky-barrier capacitance, as long as $V_g > V_t$. Below V_t, the depletion region cannot expand further, so the capacitance decreases rapidly. In saturation, C_{gs} is usually not very sensitive to V_d. When V_d drops so low that the FET enters its linear region, C_{gs} must be reduced to approximately half its saturation value.

---

4. We view the capacitances in the division-by-capacitance sense; see Section 2.5.7.

It is almost always valid to assume C_{gd} to be constant in saturation. However, in large-signal operation, the FET's drain voltage waveform may reach low values, so the FET drops periodically into linear operation. At this point C_{gd} increases significantly and depends on both V_g and V_d; at $V_d = 0$, $C_{gs} = C_{gd}$. This phenomenon has important implications for the design and analysis of passive FET components.

2.5.4.3 Extrinsic Capacitances

In both discrete and integrated devices, the capacitances of contact pads and interconnection metal are small but not negligible; in fact, in most devices C_{gd} consists almost entirely of intermetallic capacitance. The extrinsic parts of C_{gs} and C_{ds} are usually treated as capacitances between their respective terminals and the FET's source. In fact, they are capacitances between the pads and mounting surface, which may or may not be connected to the source. Both users and designers of models should be mindful of such details.

2.5.5 HEMT Modeling

Although the operation of HEMTs is qualitatively similar to that of MESFETs, they are sufficiently different in detail that most MESFET models do not work well for HEMTs. For this reason, specific HEMT models have been developed. As HEMTs continue to supplant MESFETs in most microwave and even RF applications, these models become progressively more important.

From a modeling perspective, the main differences between MESFETs and HEMTs are the following:

1. The transconductance of a HEMT shows a pronounced peak, usually well short of the maximum gate voltage. In extreme cases, the transconductance can decrease, at high gate voltages, to half its maximum value.
2. As gate voltage increases from threshold, the device turns on much more abruptly than a MESFET. This is, to a large degree, a consequence of its higher transconductance.
3. The threshold voltage is often much higher (i.e., more positive) than in MESFETs. It is possible for it to be close to or even greater than zero, creating enhancement mode devices.

4. The knee voltage of the gate-to-channel junction is usually greater. Consequently, the maximum gate-to-source voltage is greater, and the I_{sat} value of the gate-to-channel diodes is greater.
5. The drain-to-source resistance of HEMTs is generally lower than that of MESFETs, and this tends to mask dispersion effects. (Additionally, HEMTs use a high-quality buffer layer that introduces fewer traps and thus less dispersion.) The low resistance is probably more a consequence of the short gate lengths used in modern devices, not so much an inherent characteristic of the device.
6. Although the capacitance behaves qualitatively as described in Section 2.5.4, it differs in detail. For example, it is not unusual for C_{gs}, as a function of V_g, to exhibit a peak, while in MESFETs it usually varies monotonically. Also, because of the disappearance of the channel charge layer, HEMT capacitances decrease more rapidly than MESFETs at low (i.e., more negative) gate voltages.

The MESFET equivalent circuit in Figure 2.17 is valid for HEMTs. Similarly, considerations related to FET capacitances in Section 2.5.7 apply fully to HEMTs.

The peaked transconductance can be surprisingly difficult to model. One elegant approach is that of Angelov [2.10, 2.11] who uses the expression

$$I_d(V_g, V_d) = I_{pk}(1 + \tanh(\psi))\tanh(\alpha V_d)(1 + \lambda V_d) \tag{2.82}$$

where

$$\psi(V_g, V_d) = \sum_{i=1}^{3} p_i(V_g - V_{pk})^i \qquad V_{pk} = V_{pk0} - \gamma V_d \tag{2.83}$$

and p_i, α, γ, and λ are empirically determined parameters of the model. V_{pk0} and I_{pk} are the gate voltage and drain current, respectively, at peak transconductance. This model does not require clipping of I_d below some user-specified pinch-off voltage; pinch-off is implicit in the model, and it is much better behaved near pinch-off than, for example, (2.79).

2.5.6 MOSFET Modeling

The earliest MOSFET models, such as the SPICE Level 1 model, were based on a primitive, one-dimensional analysis. These were simple square-law *I*/*V* models whose deficiencies became clear almost immediately. The result was a continual stream of "improvements," resulting in a cottage industry devoted to the development of ever newer models. At this writing, one popular simulator actually includes more than 50 MOSFET models!

Some of the problems addressed by modern MOSFET models are the following:

1. *Effective gate width and length.* The length of a FET's gate has a strong effect on its performance. Because of processing limitations, the length of a MOSFET's gate is never precisely what was intended. There is always some overlap with the source and drain diffusions, and especially in short-gate devices, a number of physical effects make the gate behave as if it were longer than it is. The same is roughly true of the width, but the gate width is much less critical.
2. *Short-channel effects.* Much MOSFET theory is based on an assumption that the gate is long compared to the channel dimensions. This is clearly not the case in modern MOSFETs, where gate length may be only a small fraction of one micrometer.
3. *Subthreshold effects.* As with other types of FETs, MOSFETs do not pinch off precisely. The indistinct threshold voltage is caused by weak inversion in the channel.
4. *Mobility variation and velocity saturation.* Simple models treat electron mobility as a constant quantity. Electron mobility is constant only in relatively weak electric fields. In high fields, mobility decreases and electrons eventually reach a limiting velocity, called the *saturation velocity.*
5. *Drain effects.* Early models considered only pinch-off at the drain end of the channel. This is clearly an oversimplification.
6. *Substrate current.* Silicon substrates generally do not have high resistivity, and substrate current can also be generated by impact ionization.

Much of the complication in MOSFET models arises from a perceived need to have "physical models" of the devices, so the models attempt to reproduce as many physical phenomena as possible. This situation contrasts strongly with other devices, especially microwave MESFET and HEMT models, which use empirical equations almost exclusively. The need for physical MOSFET models comes largely from digital electronics, where development speed is critical and circuits are often designed in parallel with process development. In such an environment, devices are not available for measurement and empirical modeling, so physical models become necessary. In RF design, however, there is far less commercial pressure to improve chip performance rapidly, and adequate time is available to generate models from measured devices. Empirical MOSFET models, for RF devices, probably would be simpler and much more practical.

Early in the development of MOSFET models, the problem of charge (or capacitance) partitioning became visible. It is clear that the gate-to-channel capacitance of a MOSFET is largely the parallel-plate capacitance of the gate; however, the best way to divide it between the gate-to-source capacitance, C_{gs}, and the gate-to-drain capacitance, C_{gd}, is not clear. In normal, saturated operation, the entire capacitance probably should be assigned to C_{gs}, but in linear operation, it must be divided in some manner between C_{gs} and C_{gd}.

One of the earliest MOSFET capacitance models that attempted to deal with this problem came from Meyer [2.12]. The Meyer model is implemented in the Level 1 Berkeley SPICE model but was eventually recognized to violate charge conservation. This characteristic is particularly troublesome in a transient simulator, as charge nonconservation can result in numerical overflow; in harmonic balance analysis, charge nonconservation affects accuracy but rarely causes numerical difficulties. Interestingly, the Ward-Dutton model [2.13], a far superior model, was implemented in the SPICE Level 2 model, but Meyer was again used in the Level 3. Nevertheless, the Ward-Dutton model is included in many other simulators' implementations of the Level 3 model. Whatever its deficiencies, the SPICE Level 3 MOSFET model has been a de facto standard for many years.

More recently the BSIM model was developed at the Berkeley campus of the University of California to be an industrywide standard MOSFET model. BSIM itself has undergone considerable evolution; at this writing, the "standard" version is BSIM3, and BSIM4 is under development. BSIM3 is a very complex model. It offers a number of options for the mobility models, charge models, and charge partitioning (using a division-by-charge approach; see Section 2.5.7.2).

See [2.14] for a good description of the Meyer model and the problems of charge nonconservation. Section 2.5.7 provides for further information on the capacitance problem in FET devices. The author also suggests that users of highly complex MOSFET models, especially BSIM3, consider the points raised in Section 2.3.12.

2.5.7 FET Capacitances

FET devices have both a gate-to-drain and gate-to-source capacitance. These are usually controlled by at least two voltages. As such, it is tempting to treat them simply as a pair of multiply controlled capacitances. Traditionally, this is exactly what is done.

As long as the FET remains in normal, forward, current-saturated conduction, this type of model usually presents few problems. However, in many circuits, the FET is forced into its linear region and sometimes even into inverse operation. In such cases, the two-capacitance models are rarely satisfactory. It is well known, for example, that such models can create an impulse of current when the FET switches from forward to inverse operation [2.3].

There are two ways to convert the single gate depletion charge into two individual capacitances: one is to divide the depletion charge into two parts, the gate-to-drain and gate-to-source charges, and the other is to divide the capacitance in two. These two approaches are, surprisingly, quite different. We call these, respectively, *division by charge* and *division by capacitance*.

2.5.7.1 Division by Capacitance

The FET's reactive gate current I_g is

$$I_g = \frac{dQ_g}{dt} = \frac{\partial Q_g}{\partial V_g}\frac{dV_g}{dt} + \frac{\partial Q_g}{\partial V_{gd}}\frac{dV_{gd}}{dt} \tag{2.84}$$

where Q_g is the total gate depletion charge, which we assume to be controlled by both V_g and V_{gd}. It seems reasonable to assume that

$$\begin{aligned} I_s &= \frac{\partial Q_g}{\partial V_g}\frac{dV_g}{dt} \\ I_d &= \frac{\partial Q_g}{\partial V_{gd}}\frac{dV_{gd}}{dt} \end{aligned} \tag{2.85}$$

where I_s and I_d are the *reactive* parts of the source and drain current, respectively, and clearly $I_g = I_s + I_d$. (To avoid confusion with multiple minus signs, we define the reference direction for I_s and I_d pointing out of the device terminals.) This is a convenient treatment, since it defines two capacitances,

$$C_{gs} = \frac{\partial Q_g}{\partial V_g}$$
$$C_{gd} = \frac{\partial Q_g}{\partial V_{gd}} \tag{2.86}$$

whose currents depend only on the time derivatives of their own terminal voltages, and not on changes in any remote voltage. The resulting small-signal equivalent circuit consists, simply, of these capacitances, evaluated at the dc bias voltages. The small-signal circuit is completely consistent with the large-signal, and it requires no transcapacitances.

This approach is not convenient for harmonic-balance analysis. To obtain I_s and I_d, either (2.85) must be evaluated, which requires time-domain differentiation, or charge increments of Q_g must be accumulated, which requires storage of previous values. (In harmonic-balance analysis, it is much more convenient to calculate charge waveforms and to differentiate them in the frequency domain by multiplying by $j\omega$.) In transient analysis, time derivatives are readily available, so implementing this type of model involves no special difficulties.

It is important to note that there is no strong theoretical justification for the division by capacitance. Equation (2.85) is largely a conjecture, justified by its intuitive reasonableness and analytical convenience.

2.5.7.2 Division by Charge

Another option is to divide Q_g into two independent charges. Then

$$Q_g = Q_{gs} + Q_{gd} \tag{2.87}$$

In general, both Q_{gs} and Q_{gd} are functions of V_g and V_{gd}. Differentiating (2.87) with respect to time confirms that

$$I_g = I_s + I_d \tag{2.88}$$

As with (2.85), we have

$$I_s = \frac{dQ_{gs}}{dt} = \frac{\partial Q_{gs}}{\partial V_g}\frac{dV_g}{dt} + \frac{\partial Q_{gs}}{\partial V_{gd}}\frac{dV_{gd}}{dt}$$
$$I_d = \frac{dQ_{gd}}{dt} = \frac{\partial Q_{gd}}{\partial V_g}\frac{dV_g}{dt} + \frac{\partial Q_{gd}}{\partial V_{gd}}\frac{dV_{gd}}{dt} \quad (2.89)$$

and in this case the reactive source and drain currents result from both capacitances and transcapacitances. Substituting (2.89) into (2.88) gives

$$I_g = \left(\frac{\partial Q_{gs}}{\partial V_g} + \frac{\partial Q_{gd}}{\partial V_g}\right)\frac{dV_g}{dt} + \left(\frac{\partial Q_{gd}}{\partial V_{gd}} + \frac{\partial Q_{gs}}{\partial V_{gd}}\right)\frac{dV_{gd}}{dt} \quad (2.90)$$

Clearly,

$$\frac{\partial Q_g}{\partial V_g} = \frac{\partial Q_{gs}}{\partial V_g} + \frac{\partial Q_{gd}}{\partial V_g}$$
$$\frac{\partial Q_g}{\partial V_{gd}} = \frac{\partial Q_{gd}}{\partial V_{gd}} + \frac{\partial Q_{gs}}{\partial V_{gd}} \quad (2.91)$$

and (2.90) has the same form as (2.84).

Now, consider the first term of (2.90). In (2.85), this term represented I_s. However, it comprises terms that, in (2.89), represent parts of both I_s and I_d. Thus, the division by capacitance and the division by charge are not equivalent and, in fact, are contradictory. Indeed, (2.87) does not solve the problems of fictitious source or drain currents. It also introduces another serious problem, charge nonconservation, because it is possible for a periodic excitation that conserves Q_g to result in nonperiodic Q_{gs} and Q_{gd} [2.3].

A consequence of (2.87) is that, if V_g and V_{gd} change in such a way that Q_g does not change, there may still be a reactive drain-to-source current. This is not entirely unreasonable, since any change in V_g and V_{gd} involves a change in the shape of the gate depletion region, which may in turn create a displacement current in I_s and I_d. However, the fictitious reactive current is not simply a displacement current; it is caused by the artificial transfer of "ownership" of charge from Q_{gs} to Q_{gd} or from Q_{gd} to Q_{gs}.

It is difficult to decide which of these representations is preferable. Division by capacitance seems intuitively to be more consistent with the behavior of real devices, but those intuitive assumptions are difficult to justify rigorously. Division by charge is much more suitable for harmonic-balance analysis, as well as for other frequency-domain methods, but predicts phenomena that do not occur in real devices and requires the disturbing use of quantities that are not state variables. In practice, division by capacitance is frequently used for transient analysis and division by charge for harmonic-balance and other frequency-domain methods. This results in inconsistencies between transient and harmonic-balance analyses of the same circuit.

A serious problem in the division-by-charge approach is the difficulty in determining the transcapacitances. Often these are ignored, and the resulting small-signal equivalent circuit is inconsistent with the large-signal.

The lack of transcapacitances is an advantage of the division-by-capacitance formulation. Theoretically, the capacitances and transcapacitances could be separated by repeated measurements with $V_{gd} = 0$ and $V_{gd} = 0$, but because of the parasitic resistances R_g, R_s, and R_d, these conditions are almost impossible to create. Measurement of transcapacitances is a subject of great research interest.

2.5.7.3 Harmonic-Balance Simulation of FET Capacitances

To determine the current in a nonlinear capacitor, a harmonic-balance simulator first calculates the large-signal charge waveform, $Q(t_n)$, where t_n are time increments. It then Fourier transforms $Q(t_n)$ and finally multiplies each harmonic in the frequency domain by $j\omega$. In effect, for the gate-to-source reactive current, it calculates

$$I_s(t) = \frac{Q_{gs}(V_g + \Delta V_g, V_d + \Delta V_d) - Q_{gs}(V_g, V_d)}{\Delta t} \tag{2.92}$$

where Δt is the difference between two time points. This formulation is clearly valid only when Q_{gs} represents a division-by-charge model. For division by capacitance, the simulator must calculate

$$I_s(t) = \frac{Q_g(V_g + \Delta V_g, V_d) - Q_g(V_g, V_d)}{\Delta t} \tag{2.93}$$

where Q_g is the gate charge. The author has never seen this formulation included in a harmonic-balance simulator, although it would not be impossible to do so. A simpler approach is to generate the time derivative of V_g, a linear operation, and to form

$$I_g(t) = C_{gs}(V_g, V_d)\frac{dV_g}{dt} \tag{2.94}$$

where C_{gs} is defined by (2.86).

It is sometimes assumed, incorrectly, that a division-by-charge model can be generated by extracting C_{gs} from a linear, small-signal FET equivalent circuit and integrating as

$$Q_{gs}(V_g, V_d) = \int_{V_g} C_{gs}(V_g, V_d)\, dV_g \tag{2.95}$$

The charge expression generated in this manner corresponds to (2.93), not (2.92), and is thus invalid for a division-by-charge model. It happens, however, that (2.95) is valid for a division-by-charge model if C_{gs} is the equivalent circuit's gate-to-source capacitance, extracted from an equivalent circuit that includes transcapacitances.

2.5.7.4 MOSFET Capacitance and Terminal Charges

Models using a terminal capacitance formulation, discussed in Section 2.2.7.3, are rare in HEMTs and MESFETs, but in MOSFETs such models are largely the standard. They are used, for example, in some implementations of the SPICE Level 3 and BSIM3 models [2.15]. The Ward-Dutton capacitance model, included in many MOSFET models, serves as an example [2.13]. The BSIM3 implementation is far too complex to be described here; the reader should see instead [2.14] or [2.15].

The charge in the active region of a MOSFET consists of the gate charge, Q_{GATE}; the inversion charge in the channel, Q_{INV}; and the depletion charge in the substrate under the inversion layer, Q_{DEP}. (Capacitances associated with the source and drain diffusions, and gate overlay, are treated separately.) For charge neutrality, we have

$$Q_{GATE} + Q_{DEP} + Q_{INV} = 0 \tag{2.96}$$

Clearly, we need determine only two of these quantities; the third can be found from the neutrality condition.

The Ward-Dutton model provides expressions for Q_{DEP} and Q_{GATE}; Q_{INV} is found from (2.96). The form of these expressions is not important for our purposes; it is available, in any case, in [2.14]. The inversion charge is connected to the drain and source terminals, so it realizes those terminal charges; however, it is not entirely clear how it should be divided between those terminals. In the linear region, it is assumed that the inversion charge is divided equally between the drain and source:

$$Q_S = Q_D = \frac{1}{2}Q_I \tag{2.97}$$

where Q_D and Q_S are the drain and source pin charges, respectively. In saturation, the model leaves the charge-partitioning problem on the user's doorstep and steals away silently into the night. It uses the expressions,

$$\begin{aligned} Q_D &= XQC \cdot Q_{INV} \\ Q_S &= (1 - XQC) \cdot Q_{INV} \end{aligned} \tag{2.98}$$

where XQC is a user-selected constant, between zero and one, that defines the charge partitioning between the terminals.

The advantage of this model, as in any properly formulated terminal-charge model, is its assurance of charge conservation. A difficulty is the obviously arbitrary division of charge between the source and drain. Even the more modern BSIM3 model does not solve this problem, and indeed offers the user a selection of three charge partitions: 0%/100%; 40%/60%; and 50%/50%. It would be useful to have a theoretically sound criterion for making this division.

2.6 BIPOLAR DEVICES

Two types of bipolar transistors are used in microwave and RF circuits: bipolar junction transistors (BJTs) and heterojunction bipolar transistors (HBTs). BJTs are sometimes called *homojunction transistors*, to distinguish them from heterojunction devices. Bipolar devices have higher gain than FETs at low frequencies and lower levels of low-frequency noise. Unlike FETs, they require only a single bias polarity and can operate at very low supply voltages; these are significant advantages in battery-powered circuits. They are often preferred for RF integrated circuits (RFICs), and especially for low-noise oscillators.

2.6.1 BJT Operation

As both BJTs and HBTs operate in a similar manner, we begin by describing BJTs and then address the differences between them and HBTs.

Figure 2.18 shows a schematic cross section of a BJT, and Figure 2.19 shows how they are implemented in ICs. Discrete devices are similar; however, they are built on an n^+ conductive substrate, instead of a *p* substrate, so the collector connection is made directly to the substrate's bottom surface. This structure reduces collector resistance, compared to the IC, and facilitates heat removal. Microwave BJTs are exclusively *npn* devices, although *pnp* devices are occasionally used in low-frequency parts of RFICs.

When the transistor's base-to-emitter (BE) junction is forward-biased, electrons are injected into the base. The base, however, is very thin and lightly doped, so the probability of an electron recombining with a hole in the base is small. Instead, the electrons pass into the collector. In this way, a voltage applied to the BE junction controls a large current in the collector.

To achieve high current gain, base current must be minimized. Base current consists primarily of hole injection from the base into the emitter; this process is minimized by light base doping and heavy emitter doping. The base must also be kept thin, to minimize transit time and charge-storage capacitance. The resulting high base resistance presents a fundamental limitation to the high-frequency performance of a BJT.

Since the BJT current is essentially that of the base-to-emitter *pn* junction, it should be no surprise that the current in a BJT is given by an exponential function:

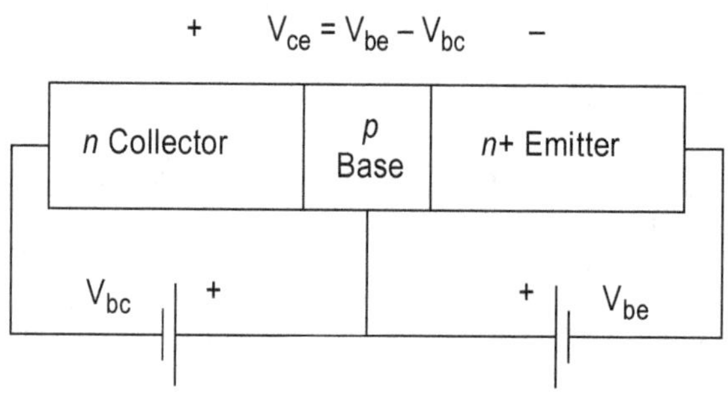

Figure 2.18 Structure and biasing of a BJT.

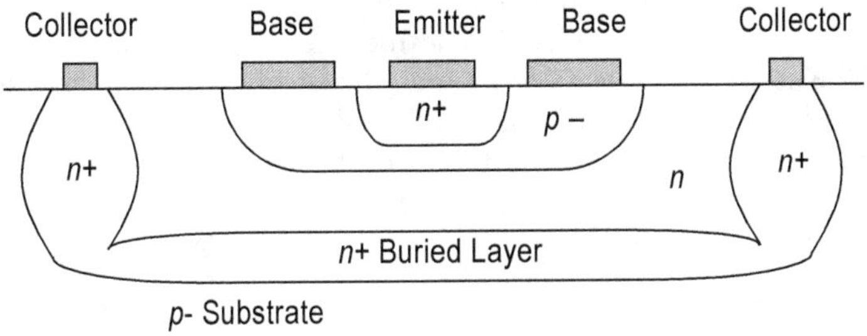

Figure 2.19 Cross section of a BJT used in an integrated circuit.

$$I_{cf} = \alpha I_e = I_s\left(\exp\left(\frac{qV_{be}}{\eta_f KT}\right) - 1\right) \tag{2.99}$$

where I_{cf} is the forward collector current, I_e is the emitter current, and V_{be} is the base-to-emitter voltage. α is a coefficient close to 1.0, which accounts for base current. More frequently, the forward current gain β_f is used, where

$$\beta_f = \frac{I_{cf}}{I_{be}} = \frac{\alpha}{1-\alpha} \tag{2.100}$$

and I_{be} is base-to-emitter current. The remaining quantities are the same as those in a Schottky-barrier diode (Section 2.4.2).

Figure 2.20(a) shows a model that describes this behavior. The BE diode determines the emitter current, and the controlled source provides the collector current, which is on the order of 1% less. The base provides the remaining current. This circuit, however, does not account for the base-to-collector (BC) junction, which creates a similar reverse current. Its current, I_{cr}, is

$$I_{cr} = I_s\left(\exp\left(\frac{qV_{bc}}{\eta_r KT}\right) - 1\right) \tag{2.101}$$

where V_{bc} is the base-to-collector voltage. A reverse current gain, β_r, analogous to (2.100), can also be defined. In Figure 2.20(b)[5] we have modified the equivalent circuit to include the reverse current. This structure is the core of virtually all BJT and HBT models.

The reverse current is negligible at collector voltages used in normal operation, but not at low voltages. The total collector current, I_{ct}, is the difference between (2.101) and (2.99):

$$I_{ct} = I_{cf} - I_{cr} = I_s\left(\exp\left(\frac{qV_{be}}{\eta_f KT}\right) - \exp\left(\frac{qV_{bc}}{\eta_r KT}\right)\right) \tag{2.102}$$

If we assume that $\eta_f \sim \eta_r \sim 1$, and note that the collector-to-emitter voltage $V_{ce} = V_{be} - V_{bc}$, we obtain

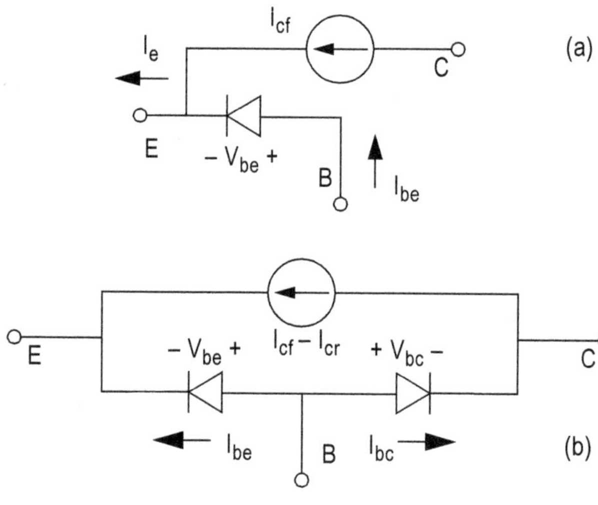

Figure 2.20 (a) An equivalent circuit describing (2.99); (b) the complete equivalent circuit including both forward and reverse conduction.

---

5. Some references and texts show a different circuit that uses two current sources. The circuit in Figure 2.20 is equivalent to those. The configuration in our figure is used in most circuit simulators because it avoids spurious incorrect solutions to the circuit equations.

$$I_{ct} = I_s \exp\left(\frac{qV_{be}}{KT}\right)\left(1 - \exp\left(\frac{-qV_{ce}}{KT}\right)\right) \tag{2.103}$$

The inclusion of reverse conductance thus causes the collector *I*/*V* characteristic to have a familiar $1 - \exp(-x)$ shape, with amplitude controlled by an exponential function of V_{be}.

Differentiating (2.103) with large V_{ce} gives the transconductance:

$$G_m = \frac{\partial I_{ct}}{\partial V_{be}} = \frac{q}{\eta KT} I_s \exp\left(\frac{qV_{be}}{\eta KT}\right) \approx \frac{q}{\eta KT} I_{cf} \tag{2.104}$$

Since $q/\eta KT \approx 40$ at room temperature, the transconductance of a BJT is quite high. This gives a BJT very high low-frequency gain, compared to most FETs. However, because the BE capacitance is also high, the cutoff frequencies of conventional homojunction BJTs are considerably lower than those of microwave FETs. Although a few advanced BJTs can be used at frequencies of tens of gigahertz, it is unusual to see BJTs used above approximately 10 GHz. Heterojunction bipolar transistors (Section 2.6.2) can operate at much higher frequencies.

The largest capacitance in a BJT comes from the combination of depletion capacitance in the BE junction and charge storage capacitance, sometimes called *diffusion capacitance*. The depletion component is modeled by the same expression as for a Schottky diode, (2.58) and (2.59). The stored charge, $Q_{s,\,be}$, is

$$Q_{s,\,be} = \tau_f I_{cf} \tag{2.105}$$

where τ_f is called the *forward base transit time*. τ_f actually includes several other delay terms, especially the time required for electrons to transit the depletion regions on both sides of the base, and for this reason it varies somewhat with V_{be} and V_{bc}. The current gain-bandwidth product, f_t, can be approximated by

$$f_t = \frac{1}{2\pi\tau_f} \tag{2.106}$$

The base-transit time is a fundamental property of the device; a narrow base is necessary for high-speed operation. The small-signal diffusion capacitance is found by differentiating (2.105):

$$C_{s,be} = \frac{dQ_{s,be}}{dV_{be}} = \frac{q\tau_f}{\eta_f KT} I_s \exp\left(\frac{qV_{be}}{\eta_f KT}\right) \approx \frac{q\tau_f}{\eta_f KT} I_{cf} \tag{2.107}$$

In normal forward operation, the BC capacitance is a simple depletion capacitance given by (2.59) and (2.60). Because the BC junction is strongly reverse-biased and both sides are lightly doped, this capacitance is relatively small.

2.6.2 HBT Operation

The operation of an HBT is fundamentally the same as that of a BJT. An HBT has the same *npn* structure as a BJT, although its implementation is very different. An HBT uses a BE heterojunction instead of a simple *pn* junction. The heterojunction employs dissimilar semiconductor materials to provide a barrier between the emitter and base, allowing heavy base doping, which minimizes base resistance and maximizes cutoff frequency. Advanced fabrication techniques used for HBTs, which would make no economic sense for conventional BJTs, contribute to improved performance as well. Unlike BJTs, HBTs are rarely available as discrete devices; almost all are used in IC technologies.

While conventional BJTs are invariably silicon devices, HBTs are realized in many III-V technologies. Silicon HBTs are also possible; silicon-germanium HBTs provide high performance at lower cost than III-V devices.

Figure 2.21 shows the structure of a simple HBT. In contrast to the planar BJT of Figure 2.19, its mesa structure is decidedly nonplanar. Although more complicated (and, of course, more expensive) to fabricate than the planar BJT, the structure provides better definition of the emitter, lower parasitic resistances, and lower fringing capacitance. As with BJTs, power HBTs can be fabricated by paralleling a number of devices having long, narrow emitters.

Equations (2.99) through (2.107) are generally applicable to HBTs as well as BJTs. The differences in the devices are largely details and occur mainly at the extremes of their operation. For example, the heavily doped base makes HBTs, unlike BJTs, largely immune to high-level injection effects. We consider these matters further when we examine HBT modeling in Section 2.6.4.

2.6.3 BJT Modeling

Ebers and Moll [2.16] created the first practical large-signal BJT model, and Gummel and Poon [2.17] extended it to include phenomena that the Ebers-Moll model did not. The Gummel-Poon model, in turn, was extended somewhat and included in SPICE, and that resulting model has been dominant for at least 35 years. More recently, advanced BJT models have been proposed, but, because of its historical dominance and its availability in virtually all circuit simulators, the SPICE Gummel-Poon (SGP) model remains in wide use. It is important to examine the SGP model, as it addresses the most important characteristics of BJT operation, and more advanced models are arguably variations on the theme it has established.

Figure 2.22 shows the complete large-signal equivalent circuit of a BJT. As well as the nonlinear elements described above, it includes contact resistances R_e, R_c, and R_b, and a parasitic collector-to-substrate capacitance, C_{js}. C_{je} and C_{jc} are the depletion components of the BE and BC capacitances, respectively, and C_{de} and C_{dc} are the diffusion capacitances. The model also shows an extra pair of diodes, marked with the currents $I_{be,l}$ and $I_{bc,l}$, which model BE and BC leakage. Although it is not part of the original model [2.17], the SPICE implementation accounts for nonlinearity in the base resistance, R_b. The circuit is designed to model the device in both forward and reverse operation; reverse operation is common in digital circuits, but occurs only rarely in microwave ones. Thus, in many cases the parameters describing reverse conduction can be ignored.

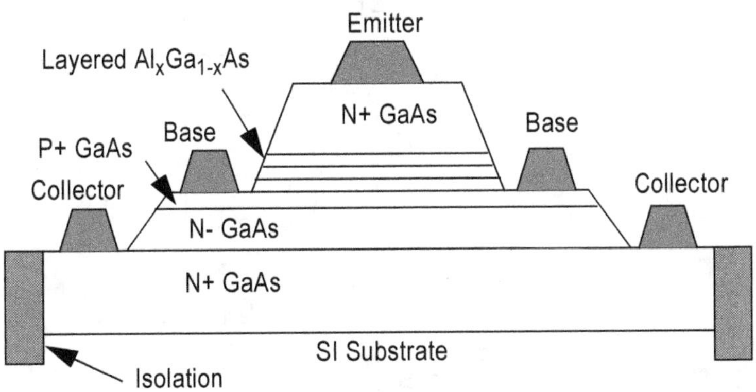

Figure 2.21 The mesa structure of a simple AlGaAs/GaAs HBT. Layered AlGaAs provides a heterojunction.

Two of the most important phenomena included in the SGP model are high-level injection and the Early effect. At high BE currents, the charge injected into the base is no longer small relative to the doping concentration, and the increased base charge prevents the collector current from increasing as fast as (2.103) implies. Although (2.103) implies that the collector-to-emitter resistance is infinite, this is clearly not the case. As V_{ce} increases, the base depletion region widens on the collector side, and the base width decreases. The resulting increased current gain creates a slope in the I/V characteristic; this phenomenon is called the *Early effect*. The SGP model accounts for both high-level injection and the Early effect by including a term, Q_b, the normalized majority base charge:

$$I_{ct} = \frac{I_s}{Q_b}\left(\exp\left(\frac{qV_{be}}{\eta_f KT}\right) - \exp\left(\frac{qV_{bc}}{\eta_r KT}\right)\right) \tag{2.108}$$

At moderate V_{ce} and I_{ct}, $Q_b \to 1$, and (2.108) reduces to (2.102). Q_b is a complicated expression involving the Early voltages and BE / BC voltages:

$$Q_b = \frac{q_1}{2}(1 + \sqrt{1 + 4q_2}) \tag{2.109}$$

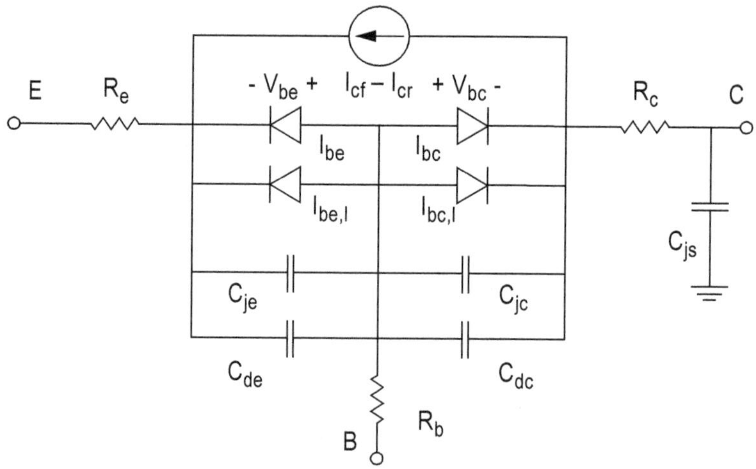

Figure 2.22 Large-signal BJT equivalent circuit used in the Gummel-Poon model. The equivalent circuit is generally applicable to HBTs as well.

where

$$
\begin{aligned}
q_1 &= 1 + \frac{V_{be}}{V_B} + \frac{V_{bc}}{V_A} \\
q_2 &= \frac{I_s}{I_{KF}}\left(\exp\left(\frac{qV_{be}}{\eta_f KT}\right) - 1\right) + \frac{I_s}{I_{KR}}\left(\exp\left(\frac{qV_{bc}}{\eta_r KT}\right) - 1\right)
\end{aligned}
\tag{2.110}
$$

V_A, V_B are the Early voltages and I_{KF}, I_{KR} are parameters describing high-level injection effects. It is important to note that (2.109) is an asymptotic approximation of a much more complex set of expressions, so it may not be valid in many cases; for example, when the Early voltages are low. Unfortunately, in silicon RF BJTs, Early voltages are usually quite low, and may violate this condition. For information on the details of this formulation, see [2.6].

The forward and reverse base currents are modeled by the diodes I_{be} and I_{bc}. Their currents are given by

$$
\begin{aligned}
I_{be} &= \frac{I_{cf}}{\beta_f} \\
I_{bc} &= \frac{I_{cr}}{\beta_r}
\end{aligned}
\tag{2.111}
$$

where β_f and β_r are the forward and reverse current gains, respectively. The currents in the leakage diodes, $I_{be,l}$ and $I_{bc,l}$, are significant only at low base-to-emitter voltages.

The forward and reverse capacitances consist of both depletion and diffusion components. The depletion component is modeled as in (2.60) and (2.61); the diffusion component is treated as in (2.105). In microwave devices, the reverse diffusion component is rarely significant, but the reverse depletion component is critical. The collector-to-base depletion capacitance, C_{jc}, is a distributed capacitance; to model it as such, it is usually split between the internal and external base nodes.

The SGP model includes additional effects. One is the change in τ_f at high V_{ce} and I_{ct}; another is the scaling of the equations with temperature. SGP does not account for many other important phenomena; for example, base push-out, or *Kirk effect*, avalanche breakdown, and self-heating. Furthermore, its handling of voltage and current dependence of τ_f do not always work well for advanced devices, especially HBTs. τ_f is especially important; if it is inaccurate, the BE capacitance also is inaccurate. Finally, the lack of self-heating is a serious deficiency of the SGP model. These phenomena are addressed in more modern BJT models, such as VBIC [2.18], MEXTRAM [2.19], and HICUM [2.20].

2.6.4 HBT Modeling

HBTs offer dramatically improved high-frequency performance compared to BJTs. Models for HBTs, however, are not very different. Indeed, most of the dominant BJT models, including the SGP model, are adequate for HBTs in most ordinary types of analysis, although some redefinition of the parameters may be necessary.

Equation (2.111) implies that there is a broad range of collector current, in a BJT, over which the current gain is constant. In HBTs this is not the case; the current gain generally increases monotonically with collector current. This effect can be modeled by setting β_f and β_r to large values to turn off the base current, and using the leakage diodes to model the complete base current.

HBTs have much higher Early voltages than BJTs; indeed, it is usually adequate to set V_A and V_B in (2.110) to large values to turn off the Early effect entirely. The low values of high-frequency |S22| often exhibited by HBTs is caused by feedback effects, not low V_A. Similarly, because of their heavily doped bases, HBTs do not exhibit high-level injection effects, so I_{KF} and I_{KR} in (2.110) are likewise very large. The scaling of τ_f with collector current, in the SGP model, is inaccurate for HBTs, as are some of its thermal scaling equations. Finally, in HBTs the current gain (β_f) decreases with temperature, while in silicon BJTs it increases.

One possible approach to an HBT model would be to retain the core of the SGP model while adding new phenomena, such as self-heating, and correcting things that are not well modeled for HBTs, such as τ_f and β_f. The model of Anholt [2.21] does precisely that. More extensive models, designed particularly for HBTs, are the Angelov [2.22] and the UCSD [2.23] models. All these models, as with the advanced BJT models, preserve the core of the Ebers-Moll model shown in Figure 2.20(b). There is also evidence that the advanced BJT models listed in Section 2.6.3 are versatile enough to be used for some kinds of HBTs as well.

2.7 THERMAL MODELING

Solid-state devices are temperature sensitive, and their temperature sensitivity is a nonlinear phenomenon. FETs are only moderately temperature-sensitive, while bipolar devices and diodes are much more so. Early device models included thermal scaling equations, in which parameters were functions of temperature, and the user was responsible for estimating the temperature of the device. More modern device models include self-heating, in which the temperature and its effects are calculated as part of the nonlinear analysis. We consider the latter in this section.

To account for temperature, an *I*/*V* expression like (2.4) must be modified as

$$I = f_V(V_1, V_2, \ldots, T_j) \tag{2.112}$$

where T_j is the temperature of the device at the location in the chip where its effects are significant, such as a FET channel or diode junction, so we can loosely call it the *junction temperature*. T_j is usually given by

$$T_j = \theta_{jc} P_d + T_c \tag{2.113}$$

where θ_{jc} is the thermal resistance between the junction and mounting surface, P_d is the power dissipated in the device, and T_c is the temperature of the mounting surface.

A few details should be considered. First, P_d includes both dc and RF power dissipation, so it is a function of time. That time function varies instantaneously with the sinusoidal RF carrier waveform, and also with the much longer time scale of the modulating waveform. The thermal mass of the chip and mounting surface filter out the RF frequency temperature variations, but often do not remove all the modulation-frequency variations. The periodic temperature variation gives rise to so-called *memory effects* in power devices.

A second consideration is that θ_{jc} is dominated by the thermal conductivity of the semiconductor device, which is invariably nonlinear. Thus, we should have

$$T_j = f_\theta(P_d) \tag{2.114}$$

Unfortunately, the thermal resistance must be described as a function of temperature. It is common to write $\theta_{jc}(T_j) = T_j/P_d$, so the thermal nonlinearity is expressed as

$$T_j = \theta_{jc}(T_j) P_d \tag{2.115}$$

Equation (2.115) is a pretty scary formulation. Furthermore, the temperature of the device is not constant throughout the semiconductor, so (2.115) is, in any case, invalid. A better approach is to view the problem incrementally; thus,

$$dT = P_d\,\theta(T_s)\frac{dl}{A} \tag{2.116}$$

where T_s is the temperature of the semiconductor at some point, dl is an increment of length through the material, and dT is the temperature change in that increment. Then $\theta(T_s)$ is the thermal resistivity at temperature T_s, and A is the cross-sectional area of the thermally conductive region. Equation (2.116) is best integrated numerically, by dividing the conductive region into a number of length increments Δl. We begin with T_0 as the baseplate temperature. Then, the temperature change from the nth to the $(n+1)$th point is

$$T_{n+1} = P_d\frac{\Delta l}{A}\left[\frac{1}{2}(\theta(T_n) + \theta(T_{n+1}))\right] + T_n \tag{2.117}$$

where we have approximated the thermal resistivity in the interval as the average of the resistivities at the two end points. At each interval, Newton's method (or any one of several other numerical methods) can be used to solve (2.117) to obtain T_{n+1}. The process continues interval to interval until the complete temperature profile is obtained. This method is still imperfect, as it is a one-dimensional integration, while the heat flow in solid-state devices clearly has a three-dimensional structure. Still, it illustrates the considerations necessary for correctly calculating temperature increase in thermally nonlinear media.

Most device models approximate θ_{jc} as a linear quantity. In this case, the temperature increase can be modeled by the thermoelectric equivalent circuit shown in Figure 2.23. In this circuit, P_d is analogous to current, T_j is analogous to voltage, and θ_{jc} is analogous to resistance. The capacitor, C_θ, models the thermal storage of the structure, and the thermal time constant is $\theta_{jc} \cdot C_\theta$. When an electrothermal equivalent circuit is used, T_j in (2.112) can be treated like any other control voltage.

A third problem is that the simple thermal equivalent circuit of Figure 2.23 does not adequately describe many devices. In particular, power transistors usually consist of multiple cells that are thermally coupled to each other as well as to the mounting surface. In this case, a thermal

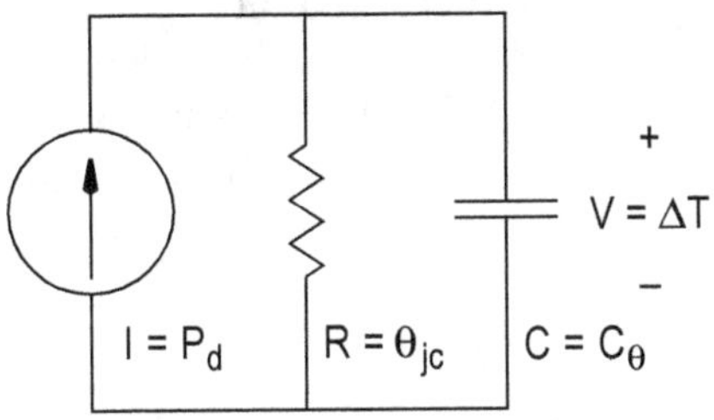

Figure 2.23 Electrothermal equivalent circuit of a single device.

resistance matrix may be used, where the temperature increase (over the mounting-surface temperature) ΔT_n at each of N cells is

$$\begin{bmatrix} \Delta T_1 \\ \Delta T_2 \\ \cdots \\ \Delta T_N \end{bmatrix} = \begin{bmatrix} \theta_{11} & \theta_{12} & \cdots & \theta_{1N} \\ \theta_{21} & \theta_{22} & \cdots & \theta_{2N} \\ \cdots & \cdots & \cdots & \cdots \\ \theta_{N1} & \theta_{N2} & \cdots & \theta_{NN} \end{bmatrix} \begin{bmatrix} P_1 \\ P_2 \\ \cdots \\ P_N \end{bmatrix} \tag{2.118}$$

where P_n is the power dissipated in the nth cell. The thermal resistance matrix in (2.118) is analogous to an impedance matrix; that is,

$$\theta_{ij} = \left.\frac{\Delta T_i}{P_j}\right|_{P_n = 0,\, n \neq j} \tag{2.119}$$

In this case, using capacitors to represent the thermal mass is tricky. It is probably best simply to use a single capacitor at each node for this purpose. This method cannot account for nonlinearity in the thermal resistance.

Modeling the effects of temperature on nonlinear elements is the other half of the task, and this depends on the type of element. In diodes, for example, the temperature dependence in (2.62) and (2.63) is clear (although the SPICE diode model uses a somewhat different expression for I_{sat}). Contact resistances are often treated as linear functions of temperature; for example,

$$R(T) = R_0(1 + K_R(T - T_0)) \tag{2.120}$$

where R_0 is the resistance measured at T_0 and K_R is a temperature coefficient. Such expressions are usually adequate over the range of temperatures that a device experiences. Depending on the magnitude of K_R, the function can return a negative value for the resistance. For this reason, $R(T)$ must be limited in some numerically acceptable manner (see Section 2.3) to positive values.

2.8 PARAMETER EXTRACTION

A solid-state device that is to be used in small-signal, linear applications can usually be characterized adequately by S or Y parameters. In order to model a nonlinear device, however, it is necessary to measure the circuit element values and to determine their dependence upon one or more control voltages or currents (usually voltages) within the circuit. Invariably, the *C/V* and *I/V* characteristics of the nonlinear elements are needed. The process for determining the model parameters, from measurements of the device, is called *parameter extraction.*

The methods used to determine the model parameters depend, understandably, upon the type of device. In general, however, the *I/V* characteristics and sometimes certain resistances can be determined, with good accuracy, from dc measurements. *C/V* characteristics naturally require RF measurements, although occasionally measurements with a capacitance meter are adequate.

In the past, transistors were modeled by first measuring dc *I/V* characteristics and then "fitting" the small-signal linearized equivalent circuit to measured S parameters. S parameters were measured over a wide range of frequencies, and the resistance and capacitance values were adjusted by numerical optimization until the calculated S parameters of the equivalent circuit agreed with those measured. The process was repeated with a large set of bias voltages, and eventually a table of *C/V* (occasionally *I/V)* characteristics was generated. Finally, the *C/V* or *I/V* expression was fit to the tabulated data numerically.

This process has a number of deficiencies. The most serious is that the equivalent circuit's set of element values, representing a particular set of S parameters, is not unique. Thus, the element values have large variability. Even linear elements appear variable as well, and the process often returns such nonphysical results as negative resistances. A second problem is that the linear equivalent circuit is sometimes not precisely equivalent to the linearized large-signal equivalent circuit. This problem, known as the *consistency problem*, occurs most often in FETs, when transcapacitances are ignored and it is assumed that the gate-to-source and gate-to-drain capacitances, in the small-signal equivalent circuit, can be treated as simple capacitors. This problem has been addressed in Section 2.5.7. A final problem is the huge amount of data that must be taken, and the amount of human intervention and personal skill needed to develop the model.

Because of these problems, more recent techniques involve *direct extraction* methods. In these, values of the equivalent circuit elements are calculated directly from measurements, usually Y parameters that are obtained by conversion from S parameters. The Y parameters may be measured at a number of bias voltages, including *cold device* measurements; that is, an unbiased device. Optimization is not used, although occasionally statistical methods are used to select the most meaningful data or to perform a curve fit. In some cases, quantities that are difficult to measure, such as source and drain contact resistances in FETs, are determined by measurement of test cells on fabrication wafers. Finally, planar electromagnetic analysis can be used to determine intermetallic capacitances and to model cell interconnections in large devices. The resulting models have considerably less variability than in methods based on optimization, and if performed carefully, are accurate even at frequencies higher than those used for the original measurements.

2.8.1 Diode Parameter Extraction

The important dc parameters of a diode junction, I_{sat}, η, and R_s, can be found very easily from a direct measurement of the I/V characteristic. Figure 2.24 shows the measured I/V characteristic of a Schottky-barrier diode, plotted on semilog axes. The I/V characteristic is nearly a straight line, deviating noticeably at currents above approximately 1 mA because of the voltage drop across the series resistance R_s. Simple manipulations of (2.62) show that the room-temperature slope of the straight-line portion of the curve is 58.5 mV per decade of current at 295K (22C). The diode's slope parameter η can be found by measuring the slope of the closest-fit straight line, in mV/decade of current, and dividing by 58.5. The deviation of the curve from the extrapolated straight line at its high-current end is the voltage dropped across the series resistance. Thus,

$$R_s = \frac{\Delta V}{\Delta I} \tag{2.121}$$

where V is the voltage deviation and I is the current at which V is determined. Finally, I_{sat} is found from any pair of points (I, V) along the straight-line portion of the curve:

$$I_{sat} = I(V)\exp\left(\frac{-qV}{\eta KT}\right) \tag{2.122}$$

Calculating the series resistance of small-diameter GaAs diodes from dc I/V measurements requires care because junction heating at even modest current can affect the accuracy of R_S. The thermal resistance of a chip diode is approximately 4 C/mW, enough to shift the I/V curve slightly toward lower voltages and thus make the series resistance appear lower than it is. As an alternative, resistance can be extracted from on-wafer S-parameter measurements, much as is done for transistors.

Direct measurement of the C/V characteristic presents some practical difficulties because the junction capacitance of many types of diodes, especially millimeter-wave mixer diodes, is on the order of femtofarads. One solution is to measure the capacitance of a large diode and to scale it according to area. This process is not highly accurate because the fringing capacitance at the edge of the anode, which does not scale in proportion to area, is significant. On-wafer RF measurements are usually adequate for all but the smallest millimeter-wave devices. For these, some of the old-fashioned measurement techniques may still be best. See [2.5] for further information.

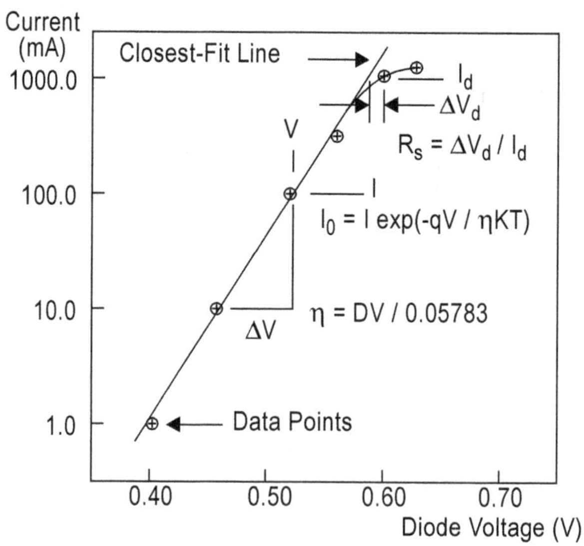

Figure 2.24 Diode I/V characteristic, plotted on semilog axes, summarizing the determination of its I/V parameters.

2.8.2 FET Parameter Extraction

2.8.2.1 Linear Elements

One of the most difficult problems in FET parameter extraction is to separate the effects of source resistance, R_s, and transconductance, G_m. These quantities compensate each other to a large degree; that is, one can obtain much the same results with either high G_m and R_s or low. It is surprisingly difficult to measure these uniquely.

A number of approaches can be used. One is to determine R_s from dc measurements of the forward-biased gate-to-channel junction. One of the earliest methods, from Fukui [2.24], is still used occasionally. A better approach is that of Yang and Long [2.25], somewhat improved by Holstrom et al. [2.26]. This method determines both R_s and the drain parasitic resistance, R_d. Other direct extraction methods [2.27–2.29] include methods to separate the effects of these quantities.

If one of the series resistances R_s, R_d, or R_g can be found or estimated (e.g., from a test cell), the rest can be determined easily from a cold FET with a forward-biased gate. The low-frequency equivalent circuit is shown in Figure 2.25. A little analysis shows that

$$\begin{aligned} R_s &= Z_{12} - 0.5R_{ch} \\ R_g &= Z_{11} - Z_{12} - R_j \\ R_d &= Z_{22} - 0.5R_{ch} - Z_{12} \end{aligned} \tag{2.123}$$

In the above equations, R_j is the junction resistance and R_{ch} is the channel resistance, which must be determined in some other way.

Capacitances C_{gs}, C_{ds}, and C_{gd}, and some of the remaining resistances, are most easily found from low-frequency Y parameters. The parameters must be measured at a frequency low enough so the resistances in series with them are negligible, but high enough so they can be measured accurately in a standard 50Ω microwave measurement system. The parasitic resistances R_s, R_d, and R_g first must be determined and their effects removed from the Y matrices. This can be accomplished with a circuit simulator by connecting resistances of value $-R_s$, $-R_d$, and $-R_g$ in series with their respective terminals and recalculating the Y matrices. Other parasitics, such as bond-wire inductance, can be removed in a similar manner.

The capacitances can then be found from simple circuit analysis. We first find C_{gd} from Y_{12}:

$$Y_{12} = -j\omega C_{gd} \tag{2.124}$$

Then, C_{gs} and the resistance R_1 are found from Y_{11}:

$$Y_{11} = j\omega C_{gd} + \frac{j\omega C_{gs}}{1 + j\omega C_{gs} R_1} \tag{2.125}$$

C_{ds} and the drain-to-source resistance R_{ds} are found from Y_{22}:

$$Y_{22} = \frac{1}{R_{ds}} + j\omega C_{ds} \tag{2.126}$$

and G_m, if desired, can be found from Y_{21}:

$$Y_{21} = \frac{G_m}{1 + j\omega C_{gs} R_i} - j\omega C_{gd} \tag{2.127}$$

These capacitances should be measured over a wide range of frequencies, and data in the frequency range showing the least variability should be selected and averaged.

It is important to note that this method does not determine the transcapacitances, and therefore is valid only for a division-by-capacitance model. It is not possible to obtain parameters of division-by-charge models without assuming the existence of transcapacitances and evaluating them.

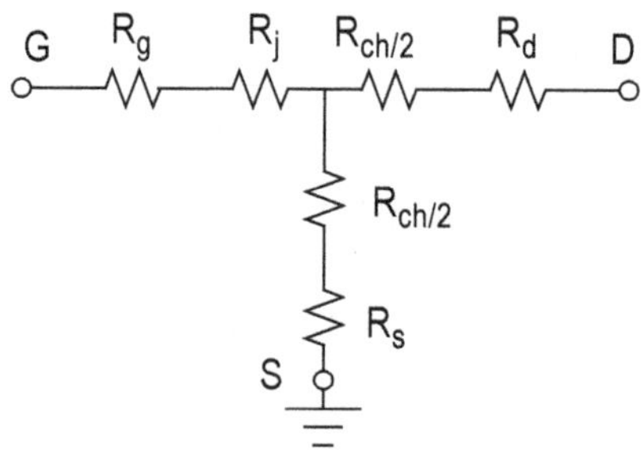

Figure 2.25 Low-frequency equivalent circuit of a FET with forward gate bias.

2.8.2.2 Nonlinear Elements

Large-signal models can be created by curve-fitting the measured *I*/*V* data to the model's *I*/*V* function. The parameters of charge functions are readily determined from measured capacitances.

It is difficult to determine the small-signal nonlinearities for Volterra analysis by differentiating a measured *I*/*V* characteristic. In any weakly nonlinear device, the Taylor-series coefficients are, by definition, relatively small. (If they are not small, the device is not weakly nonlinear!) Repeated differentiation introduces numerical noise, which quickly becomes large relative to the nonlinearities.

A better approach is to extract the Taylor coefficients of the gate *I*/*V* characteristic from RF measurements. A workable method involves exciting the device with a weak RF signal, measuring the harmonics, and using a Volterra analysis to determine the coefficients [2.30]. Figure 2.26 shows the test setup. A 50-MHz signal, well filtered and at a level of approximately –30 dBm, is applied to the device, and the levels of the harmonics at 100 MHz and 150 MHz are measured by a spectrum analyzer. Because the 50-MHz output is much greater than the harmonics, a filter is needed to reject it, or distortion generated in the spectrum analyzer may interfere with the measurement of harmonics generated by the device. The Taylor-series coefficients are found from the following:

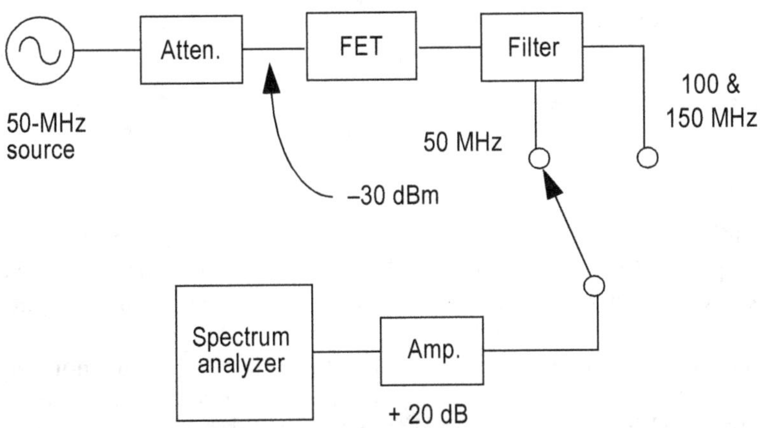

Figure 2.26 Measurement system for characterizing weak nonlinearities in FETs.

$$g_1 = \frac{|y_{21}| g_{ds}\left(R_s + R_d + \dfrac{1}{g_{ds}}\right)}{1 - |y_{21}| R_s} \tag{2.128}$$

$$g_2 = g_1(1 + g_1 C_R R_s)^2 \sqrt{\frac{\mathrm{IM}_2}{2R_{\mathrm{in}}P_s}} \tag{2.129}$$

$$g_3 = \frac{2g_2^2 C_R R_s}{1 + g_1 C_R R_s} \pm \frac{g_1(1 + g_1 C_R R_s)^3 \sqrt{\mathrm{IM}_3}}{2R_{\mathrm{in}}P_s} \tag{2.130}$$

where g_n are the Taylor-series coefficients, R_{in} is the source resistance (invariably 50Ω), g_{ds} is the drain-to-source conductance, P_s is the available source power at the device, and IM_n are the ratios of output harmonic power to linear power,

$$\begin{aligned} \mathrm{IM}_2 &= \frac{P_o(100\mathrm{MHz})}{P_o(50\mathrm{MHz})} \\ \mathrm{IM}_3 &= \frac{P_o(150\mathrm{MHz})}{P_o(50\mathrm{MHz})} \end{aligned} \tag{2.131}$$

The coefficient C_R is

$$C_R = \frac{1}{g_{ds}(R_L + R_d + R_s + \dfrac{1}{g_{ds}})} \tag{2.132}$$

The double root in (2.130) arises from the spectrum analyzer's inability to measure phase. One can determine the correct root by comparing the two roots to the derivative of g_2. Since $g_1 = G_m$, (2.127) can be used in place of (2.128).

This method characterizes only the gate-to-drain nonlinearity, sometimes loosely called the *nonlinear transconductance*. An extension of this method, which characterizes all the I/V nonlinearities, can be found in [2.31].

2.8.3 Parameter Extraction for Bipolar Devices

We saw that measuring R_s in FETs was a particularly difficult problem. Measuring the analogous quantity in bipolar devices, the emitter resistance, R_e, is actually quite easy. This is true of both HBTs and BJTs. To a good approximation, R_e is given by

$$R_e = \text{Re}\{1/Z_{12}\} \tag{2.133}$$

and this relation is valid over a wide frequency range. In principle, the base resistance, R_b, could also be measured by performing a common-emitter to common-base transformation and again calculating (2.133), but this process is considerably less accurate. The same is true of measuring the collector resistance, R_c.

The *I*/*V* parameters for a Gummel-Poon model can be obtained from so-called *Gummel plots*. A Gummel plot is an *I*/*V* plot of the transistor's collector current, I_c, as a function of base-to-emitter voltage, V_{be}, when the device is configured as shown in Figure 2.27. This measurement produces an *I*/*V* curve much like that of a diode (Figure 2.24), and the same methods can be used to find the parameters of I_s, η_f, and I_{KF} of (2.110). Reverse Gummel plots, in which the collector is treated as the emitter and the emitter as the collector, are used to find the reverse parameters, η_r and I_{KR}. The assumption is made that I_s is the same for both forward and reverse characteristics (called the *reciprocity condition*); if it is not, the collector current, as predicted by (2.110), may not be zero when $V_{ce} = 0$.

In evaluating Gummel plots, it is important to subtract the voltage dropped across the parasitic resistances, primarily R_e, from V_{be}. If this is not done, it is difficult to separate the effects of high-level injection from simple resistive voltage drop.

The depletion component of the base-to-emitter capacitance, C_{be}, is a function of V_{be} only, while the diffusion component, because of its depen-

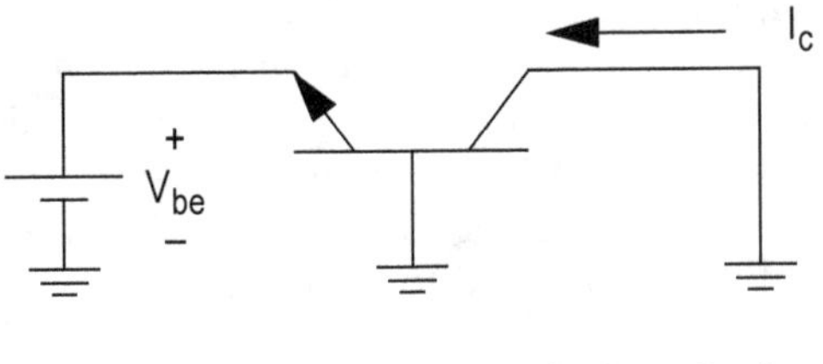

Figure 2.27 Circuit used for producing Gummel plots. Setting $V_{bc} = 0$ reduces the second term in (2.102) to 1.0, giving it the same form as a diode *I*/*V* equation.

dence on transit time, is a weak function of V_{bc} as well. For purposes of parameter extraction, the dependence on V_{bc} can usually be ignored, so no transcapacitances are needed. Separation of depletion and diffusion capacitance is tricky, but it is essential for proper modeling. The simplest, and probably the best method, is to determine the depletion component by measurements at values of V_{be} low enough so that no significant collector current results. To determine the diffusion component, the depletion component can be calculated from (2.59) and subtracted from the capacitance measured at higher V_{be}. If this is done correctly, the diffusion capacitance should follow (2.105) accurately, at least at frequencies that are low compared to $1/\tau_f$.

Another intriguing possibility is to calculate the model parameters from the device geometry and the measured characteristics of the substrate. This process is one step removed from a "physical model," in that no attempt is made to analyze the device in the way that a solid-state device simulator does. Instead, a lumped-element model is used for the intrinsic transistor, and its parameters, as well as the parasitics, are calculated from the substrate characteristics and the device's geometry. One such example, described in [2.32], reputedly is quite successful.

2.8.4 Final Notes on Parameter Extraction

Parameter extraction depends on measurements, but measurements are always imperfect. Thus, it makes no sense to force a model to agree with measurements more closely than the measurement accuracy. This point seems obvious, but it is frequently missed or ignored.

A second concern is that RF measurements may be more accurate in some frequency ranges than in others. For example, network-analyzer measurements of gate-to-source capacitance in a FET are unlikely to be accurate at low frequencies, where the capacitive reactance is much higher than the analyzer's system impedance. The same is true of gate-to-drain capacitance, but measurement accuracy may be better in a higher frequency range. The extraction process must select the data that are most reliable, for example by selecting the region where the variation is minimal.

Finally, one must be careful in using the original data to validate a model. A model should be validated by showing that it successfully reproduces the phenomenon it is intended to model. It proves little to show that it reproduces only the data used to generate it.

References

[2.1] D. E. Root and B. Hughes, "Principles of Nonlinear Active Device Modeling for Circuit Simulation," *Proc. 32nd IEEE MTT ARFTG Conference*, 1988, p. 3.

[2.2] H. Statz et al., "GaAs FET Device and Circuit Simulation in SPICE," *IEEE Trans. Electron Devices*, Vol. ED-34, No. 2, Feb. 1987, p. 160.

[2.3] I. W. Smith et al., "On Charge Nonconservation in FETs," *IEEE Trans. Electron Devices*, Vol. ED-34, No. 12, Dec. 1987, p. 2565.

[2.4] L. W. Nagel, "SPICE2: A Computer Program to Simulate Semiconductor Circuits," Electronics Research Laboratory Report No. ERL-M520, University of California, Berkeley, 1975.

[2.5] S. Maas, *Microwave Mixers*, Norwood, MA: Artech House, 1993.

[2.6] S. M. Sze, *Physics of Semiconductor Devices*, New York: John Wiley & Sons, 1981.

[2.7] E. H. Rhoderick, "Metal-Semiconductor Contacts," *IEE Proc.*, Part I, Vol. 129, 1982, p. 1.

[2.8] E. L. Kollberg et al., "Current Saturation in Submillimeter-Wave Varactors," *IEEE Trans. MTT*, Vol. MTT-40, 1992, p. 831.

[2.9] M. T. Faber, J. Chramiec, and M. E. Adamski, *Microwave and Millimeter Wave Diode Frequency Multipliers*, Norwood, MA: Artech House, 1995.

[2.10] I. Angelov, H. Zirath, and N. Rorsman, "A New Empirical Nonlinear Model for HEMT and MESFET Devices," *IEEE Trans. Electron Devices*, Vol. ED-40, 1992, p. 2258.

[2.11] I. Angelov, L. Bengtsson, and M. Garcia. "Extensions of the Chalmers Nonlinear HEMT and MESFET Model," *IEEE Trans. Microwave Theory Tech.*, Part I, Vol. MTT-44, 1996, p. 1664.

[2.12] J. Meyer, "MOS Models and Circuit Simulation," *RCA Review*, Vol. 32, 1971, p. 42.

[2.13] D. Ward and R. Dutton, "A Charge Oriented Model for MOS Transistor Capacitances," *IEEE J. Solid State Circuits*, Vol. SC-13, 1978, p. 703.

[2.14] D. Foty, *MOSFET Modeling with SPICE*, Upper Saddle River, NJ: Prentice Hall, 1977.

[2.15] Y. Cheng and C. Hu, *MOSFET Modeling and BSIM3 User's Guide*, Boston: Kluwer, 1999.

[2.16] J. J. Ebers and J. L. Moll, "Large-Signal Behavior of Junction Transistors," *Proc. IRE*, Vol. 42, 1954, p. 1761.

[2.17] H. K. Gummel and H. C. Poon, "An Integral Charge Control Model of Bipolar Transistors," *Bell Sys. Tech. J.*, Vol. 49, May/June 1970, p. 827.

[2.18] C. McAndrew et al., "VBIC95, The Vertical Bipolar Inter-Company Model," *IEEE J. Solid-State Circuits*, Vol. 31, 1996, p. 1476.

[2.19] H. C. de Graaf and F. M. Klaasen, *Compact Transistor Modelling for Circuit Design*, New York: Springer-Verlag, 1990.

[2.20] H.-M. Rein et al., "A Semi-Physical Bipolar Transistor Model for the Design of Very High-Frequency Analog ICs," *Proc. IEEE Bipolar and BiCMOS Circuits and Technology Meeting*, 1992, p. 217.

[2.21] R. Anholt, *Electrical and Thermal Characterization of MESFETs, HEMTs, and HBTs*, Norwood, MA: Artech House, 1995.

[2.22] I. Angelov, "A Simple HBT Large-Signal Model for CAD," *IEEE MTT Int. Microwave Symp. Dig.*, 2002.

[2.23] UCSD Electrical Engineering Dept., High-Speed Devices Group, *HBT Modeling, rev. 9.001A*, http://hbt.ucsd.edu, March 2000.

[2.24] H. Fukui, "Determination of Basic Device Parameters of a GaAs MESFET," *Bell Syst. Tech. J.*, Vol. 58, 1979, p. 771.

[2.25] L. Yang and S. Long, "New Method to Measure the Source and Drain Resistance of the GaAs MESFET," *IEEE Electron Dev. Ltrs.*, Vol. ED-7,1986, p. 75.

[2.26] R. P. Holstrom, W. L. Bloss, and J. Y. Chi, "A Gate Probe Method of Determining Parasitic Resistance in MESFETs," *IEEE Electron Dev. Ltrs.*, Vol. ED-7, 1986, p. 410.

[2.27] G. Dambrine et al., "A New Method for Determining the FET Small-Signal Equivalent Circuit," *IEEE Trans. MTT*, Vol. 36, 1988, p. 1151.

[2.28] M. Berroth and R. Bosch, "Broad-Band Determination of the FET Small-Signal Equivalent Circuit," *IEEE Trans. Microwave Theory Tech.*, Vol. 38, 1990, p. 891.

[2.29] N. Rorsman et al., "Accurate Small-Signal Modeling of HFETs for Millimeter-Wave Applications," *IEEE Trans. Microwave Theory Tech.*, Vol. 44, 1996, p. 432.

[2.30] S. Maas and A. M. Crosmun, "Modeling the Gate I/V Characteristic of a GaAs MESFET for Volterra-Series Analysis," *IEEE Trans. Microwave Theory Tech.*, Vol. MTT-37, 1989, p. 1134.

[2.31] J. C. Pedro and J. Perez, "Accurate Simulation of GaAs MESFET's Intermodulation Distortion Using a New Drain-Source Current Model," *IEEE Trans. Microwave Theory Tech.* Vol. 42, 1994, p. 25.

[2.32] D. J. Walkey, M. Schröter, and S. Voinigescu, "Predictive Modelling of Lateral Scaling in Bipolar Transistors", *Proc. IEEE Bipolar/BiCMOS Circuits and Technology Meeting*, 1995, p. 74.

Chapter 3

Harmonic-Balance Analysis and Related Methods

This chapter is concerned with two of the most important techniques for analyzing nonlinear circuits. The first, called *harmonic-balance analysis*, is most useful for strongly or weakly nonlinear circuits that have single or multitone excitation. Harmonic balance analysis is applicable to a wide variety of problems in such microwave circuits as power amplifiers, frequency multipliers, and mixers. Harmonic-balance calculates a circuit's steady-state response. It works particularly well when a circuit has a mix of long and short time constants and, in fact, was originally proposed to solve the problems inherent in analyzing such circuits [3.1].

The second technique, *large-signal/small signal analysis*, is used for nonlinear circuits that are excited by two tones, one of which is very large and the other is vanishingly small. This situation is encountered most frequently in mixers, in which a diode or transistor is excited by a large-signal local oscillator and a much smaller RF signal. The circuit is first analyzed via harmonic balance, under LO excitation alone, and is converted into a small-signal linear, time-varying equivalent. The time-varying circuit is then analyzed as a quasilinear circuit under small-signal RF excitation. The quasilinear assumption is not always necessary, and the small-signal analysis can be extended to include nonlinear effects such as intermodulation.

3.1 WHY USE HARMONIC-BALANCE ANALYSIS?

Transient analysis methods predate harmonic balance methods. Thus, the existence of harmonic-balance analysis implies that transient methods are not adequate for many kinds of circuits. In fact, the methods are pleasantly complementary: harmonic balance works well where transient analysis does not, and transient analysis usually outperforms harmonic balance in the kinds of problems where it is applicable.

Three problems can make time-domain techniques impractical. First, matching circuits may contain such elements as dispersive transmission lines, transmission-line discontinuities, and multiport subnetworks described by S or Y parameters. These are difficult to analyze in the time domain. Second, the circuit's time constants may be large compared to the period of the fundamental excitation frequency. When long time constants exist, it becomes necessary to continue the numerical integration of the equations through many—perhaps thousands—of excitation cycles, until the transient part of the response has decayed and only the steady-state part remains. This long integration is an extravagant use of both computer time and the engineer's patience; furthermore, numerical truncation errors in the long integration may become large and reduce the accuracy of the solution. Although algorithms exist to ameliorate this difficulty [3.2, 3.3], implementing them is an extra complication. Third, each linear or nonlinear reactive element in the circuit adds a differential equation to the set of equations that describes the circuit. A large circuit can have many reactive elements, so the set of equations that must be solved may be very large. For this reason, time-domain analysis is notoriously slow.

The greatest advantage of time-domain analysis is its ability to handle very strong nonlinearities in large circuits. Its robustness results in part from the fact that small time steps can be used in the time-domain integration. As long as the nonlinearities are continuous, the time steps can always be made short enough so that the circuit voltages and currents change very little between steps.

3.2 AN HEURISTIC INTRODUCTION TO HARMONIC-BALANCE ANALYSIS

Figure 3.1 shows a simple dc diode circuit, which we wish to analyze. Knowing that the diode's I/V characteristic is given by (2.62), we can easily write an equation for the circuit,

$$I = I_{sat}(\exp(\delta(V_s - IR)) - 1) \tag{3.1}$$

where $\delta = q / \eta KT$ (see Section 2.4.2). This equation cannot be solved algebraically. It must be solved numerically or, if only moderate accuracy is adequate, graphically. The usual method is to estimate *I*, substitute it into (3.1), and see if it satisfies the equation. If it does not, *I* is modified and the process repeated until the equation is solved. A variety of numerical methods can be used for this purpose. We do not need a method that solves the problem completely; all we need is to improve an estimated solution. Then, we need only repeat the process a number of times, using the result of each iteration as the starting estimate for the next one. Eventually, the error is reduced to the point where it is deemed negligible.

Thus, we need four things:

1. An initial estimate of the solution;
2. A numerical method for improving an estimated solution;
3. A criterion for determining whether the process has indeed improved the solution at any particular iteration step;
4. A way to decide when the solution is adequate.

These needs are easily satisfied for the circuit in Figure 3.1, but they might not be so clear in more complex circuits. Let's look at a slightly more complicated problem, shown in Figure 3.2(a), which consists of an RF source, which may include a dc component, a diode, and a complex impedance, $Z(\omega)$. We excite our diode, with the RF source, at the frequency ω_p. We know from Chapter 1 that the diode generates harmonics of both current and voltage, and $Z(\omega)$ can be expected to vary with harmonic frequency; thus, we could write it $Z(k\omega_p)$, where *k* is the harmonic number.

Although still simple, this circuit illustrates the most significant difficulties in analyzing a nonlinear RF or microwave circuit. Since the impedance is represented in the frequency domain, it is impossible to analyze this circuit in precisely the same manner as the previous one. However, with a few changes, we can use a similar approach. First, we assume that we know the diode voltage (consisting of its complex

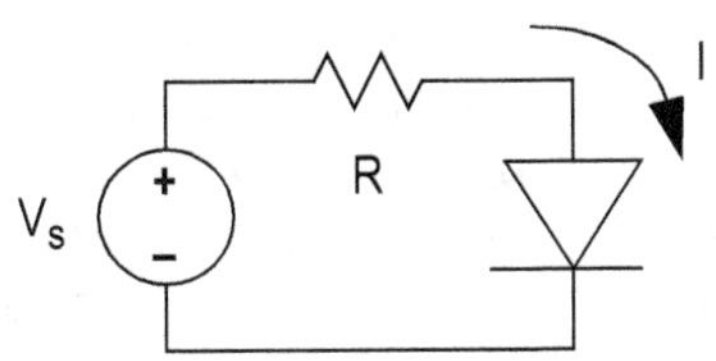

Figure 3.1 A simple dc-biased diode.

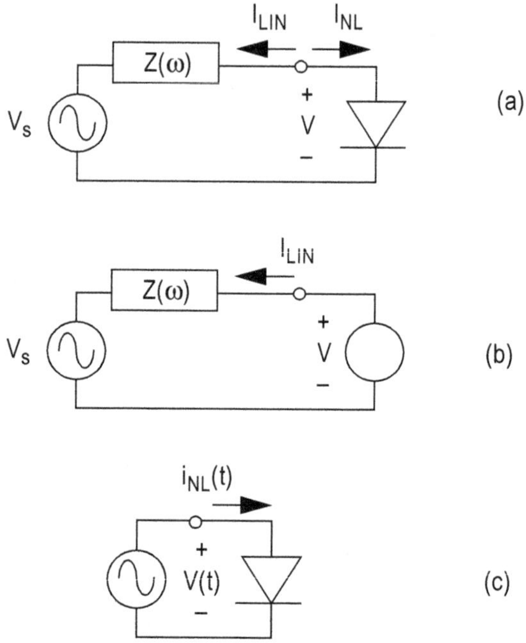

Figure 3.2 A diode excited by an RF circuit (a) can be divided into a pair of equivalent circuits, one describing the linear part (b), and another, the nonlinear part (c).

components at all harmonic frequencies, $k\omega_p$). We then create the equivalent circuit in Figure 3.2(b), which can be analyzed easily in the frequency domain, giving

$$I_{LIN}(k\omega_p) = \frac{V(k\omega_p) - V_s(k\omega_p)}{Z(k\omega_p)} \tag{3.2}$$

Of course, if V_s consists of a dc and a sinusoidal component, only two components of V_s, $V_s(0)$ and $V_s(\omega_p)$, are nonzero. V_s need not be sinusoidal, but for our present purposes, it must be periodic.

Using Fourier theory, we convert $V(k\omega_p)$ into a time waveform, $V(t)$. We then create the circuit in Figure 3.2(c) and find the current in the diode junction algebraically from (2.62):

$$I_{NL}(t) = I_{sat}(\exp(\delta V(t)) - 1) \tag{3.3}$$

If necessary, we can find $I_{NL}(k\omega_p)$ by Fourier transformation. The only remaining problem is that we really don't know $V(k\omega_p)$. However, we do know how to tell whether a particular $V(k\omega_p)$ is a solution: substitute it ($V(t)$ or $V(k\omega_p)$, as appropriate) into (3.2) and (3.3), and see if Kirchoff's current law is satisfied at all the harmonics:

$$I_{LIN}(k\omega_p) + I_{NL}(k\omega_p) = 0 \tag{3.4}$$

If (3.4) is satisfied, we have a solution.

We now can summarize the solution process as follows:

1. Create an initial estimate of $V(k\omega_p)$, $k = 0, 1, \ldots, K$, where K is the maximum harmonic with which we need be concerned. This estimate may be extremely crude; for example, $V(k\omega_p) = 0$ for all k.
2. Use (3.2) to obtain $I_{LIN}(k\omega_p)$.
3. Inverse-Fourier transform $V(k\omega_p)$ to obtain $V(t)$.
4. Use (3.3) to determine $I_{NL}(t)$.
5. Fourier transform $I_{NL}(t)$ to obtain $I_{NL}(k\omega_p)$.
6. Substitute $I_{LIN}(k\omega_p)$ and $I_{NL}(k\omega_p)$ into (3.4). Of course, (3.4) probably will not be satisfied. Define an error function at each harmonic, f_k, where

$$f_k = I_{LIN}(k\omega_p) + I_{NL}(k\omega_p) \qquad k = 0, 1, \ldots, K \tag{3.5}$$

   Note that *each* f_k is implicitly a function of *all* voltage components $V(k\omega_p)$.

7. Modify $V(k\omega_p)$ and repeat the process from step 2. Use some appropriate numerical method that can be trusted to decrease the $|f_k|$.
8. Continue until all $K + 1$ errors f_k are negligibly small.

In step 7, we have assumed the existence of some "appropriate numerical method." This assumption is not unreasonable because, fortunately, the mathematicians have been here ahead of us. There exists a large body of mathematical theory addressing the problem of finding zeros of multiple sets of nonlinear equations (see, for example, [3.4]); this is one application of that theory.

Looking at all this a little more closely, we encounter some dilemmas. For example, if we improve f_k at several harmonics, but it increases at a few others, is this an improvement? Are some harmonics more important than others? What about termination criteria? If the error is small at the harmonic of interest (say, the second harmonic in a frequency doubler), is that good enough, or must all harmonic errors be reduced? And, to what degree? Answering these questions is, obviously, essential; we address them throughout the rest of this chapter.

3.3 SINGLE-TONE HARMONIC-BALANCE ANALYSIS

Having introduced a method for solving simple nonlinear-circuit problems, we now must generalize it to larger circuits. Although earlier work involved the application of harmonic-balance analysis to simple circuits, more recent work has enabled it to be used more generally, often in circuits having large numbers of circuit elements.

We begin by examining single-tone circuits, ones having periodic excitations at a single fundamental frequency. This includes periodic, nonsinusoidal excitations, as long as they can be expressed by a one-dimensional Fourier series. In later sections, we show how harmonic-balance analysis can be applied to circuits having more complex excitations.

3.3.1 Circuit Partitioning

In general, microwave and RF circuits have a large number of both linear and nonlinear circuit elements. These can be grouped as shown in Figure 3.3 to form two subcircuits, one linear and the other nonlinear. The linear subcircuit can be treated as a multiport and described by its Y parameters, S parameters, or by some other multiport matrix. The nonlinear elements are modeled by their global *I*/*V* or *Q*/*V* characteristics, described in Chapter 2, and must be analyzed in the time domain. Thus, the circuit is reduced to an $(N + 2)$-port network, with nonlinear elements connected to N of the ports and voltage sources connected to the other two ports. [The $(N + 1)$th and $(N + 2)$th ports represent, of course, the input and output ports in a two-port network. Usually, a sinusoidal source is connected to only one of those ports; however, sources are shown at both ports in Figure 3.3 for generality.] $Z_s(\omega)$ and $Z_L(\omega)$, the source and load impedances, respectively, are "absorbed" into the linear subcircuit; they are still in series with the input and output ports, and for some purposes it may be necessary to resurrect them as separate entities. The voltages and currents at each port can be expressed in the time or the frequency domain; because of the nonlinear elements, however, the port voltages and currents have frequency components at harmonics of the excitation. Although in theory an infinite number of harmonics exist at each port, we shall assume throughout this chapter that the dc component and the first K harmonics (i.e., $k = 0 \ldots K$) describe all the voltages and currents adequately. Consequently, all higher harmonics can be ignored. Ignoring the higher harmonics is equivalent to setting the embedding impedances to zero at those frequencies; see Section 3.3.6.

The circuit in Figure 3.3 is successfully analyzed when either the steady-state voltage or current waveforms at each port are known. Alternatively, knowledge of the frequency components at all ports constitutes a solution, because the frequency components and time waveforms are related by the Fourier series. If, for example, we know the frequency-domain port voltages, we can use the Y-parameter matrix of the linear subcircuit to find the port currents. The port currents can also be found by inverse-Fourier transforming the voltages to obtain their time-domain waveforms and calculating the current waveforms from the nonlinear

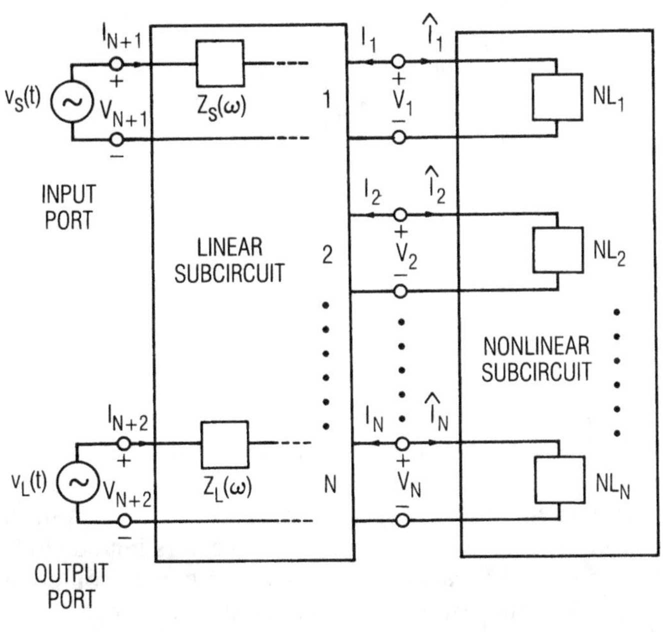

Figure 3.3 A nonlinear microwave circuit can be divided into linear and nonlinear subcircuits with the source and load impedances $Z_s(\omega)$ and $Z_L(\omega)$ absorbed into the linear subcircuit.

elements' I/V equations. The idea of harmonic balance is to find a set of port voltage waveforms (or, alternatively, the harmonic voltage components) that give the same currents in both the linear-network equations and the nonlinear-network equations; that is, the currents satisfy Kirchoff's current law. When that set is found, it must be a solution.

(Note that we were careful to say *a* solution, not *the* solution. Nonlinear circuits, in general, have multiple solutions. Fortunately, in practical circuits, a single solution usually dominates. Nevertheless, we should remain aware of the possibility of multiple solutions in any nonlinear circuit.)

If we express the frequency components of the port currents as vectors, Kirchoff's current law requires that

$$\begin{bmatrix} I_{1,0} \\ I_{1,1} \\ I_{1,2} \\ \cdots \\ I_{1,K} \\ I_{2,0} \\ I_{2,1} \\ \cdots \\ I_{2,K} \\ \cdots \\ I_{N,K} \end{bmatrix} + \begin{bmatrix} \hat{I}_{1,0} \\ \hat{I}_{1,1} \\ \hat{I}_{1,2} \\ \cdots \\ \hat{I}_{1,K} \\ \hat{I}_{2,0} \\ \hat{I}_{2,1} \\ \cdots \\ \hat{I}_{2,K} \\ \cdots \\ \hat{I}_{N,K} \end{bmatrix} = \begin{bmatrix} 0 \\ 0 \\ 0 \\ \cdots \\ 0 \\ 0 \\ 0 \\ \cdots \\ 0 \\ \cdots \\ 0 \end{bmatrix} \tag{3.6}$$

where $I_{n,k}$ is a phasor quantity, the kth harmonic component of the current at port n, in the linear subcircuit; $\hat{I}_{n,k}$, with the circumflex accent, is the port current in the nonlinear subcircuit. Equation (3.6) shows the general form of the voltage, current, and charge vectors; all such vectors in this chapter use this form unless indicated otherwise. The vectors include only positive-frequency components, because the negative-frequency components, being the complex conjugates of the positive-frequency ones, can be found immediately if needed. Eliminating the negative-frequency components from (3.6) reduces its complexity considerably.

First we consider the linear subcircuit. The admittance equations are

$$\begin{bmatrix} \mathbf{I}_1 \\ \mathbf{I}_2 \\ \mathbf{I}_3 \\ \cdots \\ \mathbf{I}_N \\ \mathbf{I}_{N+1} \\ \mathbf{I}_{N+2} \end{bmatrix} = \begin{bmatrix} \mathbf{Y}_{1,1} & \mathbf{Y}_{1,2} & \cdots & \mathbf{Y}_{1,N} & \mathbf{Y}_{1,N+1} & \mathbf{Y}_{1,N+2} \\ \mathbf{Y}_{2,1} & \mathbf{Y}_{2,2} & \cdots & \mathbf{Y}_{2,N} & \mathbf{Y}_{2,N+1} & \mathbf{Y}_{2,N+2} \\ \mathbf{Y}_{3,1} & \mathbf{Y}_{3,2} & \cdots & \mathbf{Y}_{3,N} & \mathbf{Y}_{3,N+1} & \mathbf{Y}_{3,N+2} \\ \cdots & \cdots & \cdots & \cdots & \cdots & \cdots \\ \mathbf{Y}_{N,1} & \mathbf{Y}_{N,2} & \cdots & \mathbf{Y}_{N,N} & \mathbf{Y}_{N,N+1} & \mathbf{Y}_{N,N+2} \\ \mathbf{Y}_{N+1,1} & \mathbf{Y}_{N+1,2} & \cdots & \mathbf{Y}_{N+1,N} & \mathbf{Y}_{N+1,N+1} & \mathbf{Y}_{N+1,N+2} \\ \mathbf{Y}_{N+2,1} & \mathbf{Y}_{N+2,2} & \cdots & \mathbf{Y}_{N+2,N} & \mathbf{Y}_{N+2,N+1} & \mathbf{Y}_{N+2,N+2} \end{bmatrix} \begin{bmatrix} \mathbf{V}_1 \\ \mathbf{V}_2 \\ \cdots \\ \cdots \\ \mathbf{V}_N \\ \mathbf{V}_{N+1} \\ \mathbf{V}_{N+2} \end{bmatrix} \tag{3.7}$$

The current vector **I**, from (3.6), has been written as a set of subvectors, where

$$\mathbf{I}_n = \begin{bmatrix} I_{n,0} \\ I_{n,1} \\ \cdots \\ I_{n,K} \end{bmatrix} \tag{3.8}$$

that is, $\mathbf{I}_n$ is the vector of harmonic currents at the *n*th port. Similarly,

$$\mathbf{V}_n = \begin{bmatrix} V_{n,0} \\ V_{n,1} \\ \cdots \\ V_{n,K} \end{bmatrix} \tag{3.9}$$

The elements of the admittance matrix $\mathbf{Y}_{m,n}$ in (3.7) are all submatrices; each submatrix is a diagonal, whose elements are the values $Y_{m,n}$ at each harmonic of the fundamental excitation frequency, $k\omega_p$, $k = 0 \ldots K$:

$$Y_{m,n} = diag[Y_{m,n}(k\omega_p)] \qquad k = 0, 1, 2, \ldots, K \tag{3.10}$$

that is,

$$\mathbf{Y}_{m,n} = \begin{bmatrix} Y_{m,n}(0) & 0 & 0 & \dots & 0 \\ 0 & Y_{m,n}(\omega_p) & 0 & \dots & 0 \\ 0 & 0 & Y_{m,n}(2\omega_p) & \dots & 0 \\ \dots & \dots & \dots & \dots & 0 \\ 0 & 0 & 0 & \dots & Y_{m,n}(K\omega_p) \end{bmatrix} \tag{3.11}$$

$\mathbf{V}_{N+1}$ and $\mathbf{V}_{N+2}$, the excitation vectors, have the form,

$$\begin{bmatrix} \mathbf{V}_{N+1} \\ \mathbf{V}_{N+2} \end{bmatrix} = \begin{bmatrix} V_{b1} \\ V_s \\ 0 \\ 0 \\ \dots \\ V_{b2} \\ 0 \\ \dots \\ 0 \end{bmatrix} \tag{3.12}$$

where V_{b1} and V_{b2} are the dc voltages at ports $N+1$ and $N+2$, respectively, and V_s is the excitation voltage at port $N+1$. Equation (3.12) implies that the port $N+1$ excitation includes a dc and a fundamental frequency source, while the $N+2$ port includes only dc. This is the usual situation; it corresponds, for example, to a FET amplifier that has gate and drain bias and gate excitation. A two-terminal device would normally have only one bias source, and in this case the $N+2$ port might not exist. On the other hand, a very complex IC might have a number of dc sources, and perhaps many RF sources as well. Finally, if the excitation were periodic but not sinusoidal, the vector on the right in (3.12) would include its harmonic components instead of zeros. The extension to these cases is straightforward.

Partitioning the Y matrix in (3.7) gives an expression for **I**, the vector of currents in ports 1 to N:

$$\begin{bmatrix} \mathbf{I}_1 \\ \mathbf{I}_2 \\ \cdots \\ \mathbf{I}_N \end{bmatrix} = \begin{bmatrix} \mathbf{Y}_{1,\,N+1} & \mathbf{Y}_{1,\,N+2} \\ \mathbf{Y}_{2,\,N+1} & \mathbf{Y}_{2,\,N+2} \\ \cdots & \cdots \\ \mathbf{Y}_{N,\,N+1} & \mathbf{Y}_{N,\,N+2} \end{bmatrix} \begin{bmatrix} \mathbf{V}_{N+1} \\ \mathbf{V}_{N+2} \end{bmatrix} + \begin{bmatrix} \mathbf{Y}_{1,\,1} & \mathbf{Y}_{1,\,2} & \cdots & \mathbf{Y}_{1,\,N} \\ \mathbf{Y}_{2,\,1} & \mathbf{Y}_{2,\,2} & \cdots & \mathbf{Y}_{2,\,N} \\ \cdots & \cdots & \cdots & \cdots \\ \mathbf{Y}_{N,\,1} & \mathbf{Y}_{N,\,2} & \cdots & \mathbf{Y}_{N,\,N} \end{bmatrix} \begin{bmatrix} \mathbf{V}_1 \\ \mathbf{V}_2 \\ \cdots \\ \mathbf{V}_N \end{bmatrix} \tag{3.13}$$

or

$$\mathbf{I} = \mathbf{I}_s + \mathbf{Y}_{N \times N}\mathbf{V} \tag{3.14}$$

where $\mathbf{Y}_{N \times N}$ is the $N \times N$ submatrix of $\mathbf{Y}$ corresponding to its first N rows and columns. $\mathbf{I}_s$ represents a set of current sources in parallel with the first N ports; the first matrix term in (3.13) transforms the input- and output-port excitations into this set of current sources, so the $(N + 1)$th and $(N + 2)$th ports need not be considered further. The equivalent representation is shown in Figure 3.4. This transformation allows us to express the harmonic-balance equations as functions of currents at only the first through Nth ports, the ones connected to nonlinear elements.

3.3.2 The Nonlinear Subcircuit

The nonlinear-element currents, represented by the current vector on the right in (3.6), can result from nonlinear capacitors, resistors, controlled sources, or occasionally inductors. Because nonlinear inductors occur rarely in RF and microwave circuits, we need not consider them at this point. Furthermore, we assume that the nonlinear elements are all voltage controlled. These assumptions do not limit us severely; simple methods, such as the use of a gyrator, can be employed to circumvent them. Inverse Fourier transforming the voltages at each port gives the time-domain voltage waveforms at each port:

$$\mathcal{F}^{-1}\{\mathbf{V}_n\} \rightarrow v_n(t) \tag{3.15}$$

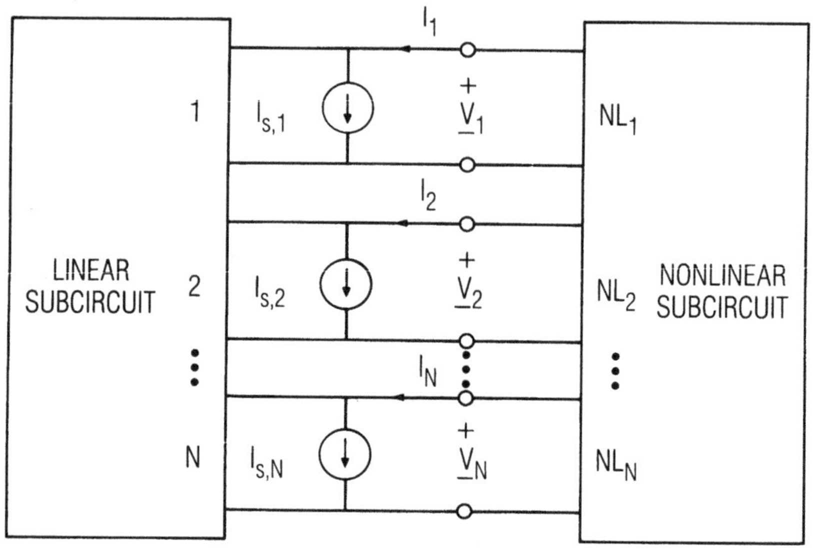

Figure 3.4 The circuit of Figure 3.3, in which the excitation voltage sources at ports $N+1$ and $N+2$ have been transformed into current sources at ports 1 to N.

We first examine nonlinear capacitors. Because the port voltages uniquely determine all voltages in the network, a capacitor's charge waveform can be expressed as a function of those voltages:

$$q_n(t) = f_{qn}(v_1(t), v_2(t), \ldots, v_N(t)) \tag{3.16}$$

Fourier transforming the charge waveform at each port gives the charge vectors for the capacitors at each port:

$$\mathcal{F}\{q_n(t)\} \rightarrow \mathbf{Q}_n \tag{3.17}$$

and the charge vector, $\mathbf{Q}$, is

$$\mathbf{Q} = \begin{bmatrix} \mathbf{Q}_1 \\ \mathbf{Q}_2 \\ \dots \\ \mathbf{Q}_N \end{bmatrix} = \begin{bmatrix} Q_{1,0} \\ Q_{1,1} \\ Q_{1,2} \\ \dots \\ Q_{1,K} \\ Q_{2,0} \\ \dots \\ Q_{2,K} \\ \dots \\ Q_{N,K} \end{bmatrix} \quad (3.18)$$

The nonlinear-capacitor current is the time derivative of the charge waveform. Taking the time derivative corresponds to multiplying by $j\omega$ in the frequency domain, so

$$i_{c,n}(t) = \frac{dq_n(t)}{dt} \leftrightarrow jk\omega_p Q_{n,k} \qquad k = 0, 1, \dots, K \quad (3.19)$$

Equation (3.19) can be written as

$$\mathbf{I}_c = j\Omega\mathbf{Q} \quad (3.20)$$

where Ω is the diagonal matrix

$$\Omega = \begin{bmatrix} 0 & 0 & 0 & \cdots & \cdots & \cdots & \cdots & \cdots & 0 \\ 0 & \omega_p & 0 & \cdots & \cdots & \cdots & \cdots & \cdots & 0 \\ 0 & 0 & 2\omega_p & \cdots & \cdots & \cdots & \cdots & \cdots & 0 \\ \cdots & \cdots & \cdots & \cdots & \cdots & \cdots & \cdots & \cdots & \cdots \\ 0 & \cdots & \cdots & \cdots & K\omega_p & 0 & 0 & \cdots & 0 \\ 0 & \cdots & \cdots & \cdots & 0 & 0 & 0 & \cdots & 0 \\ 0 & \cdots & \cdots & \cdots & 0 & 0 & \omega_p & \cdots & 0 \\ \cdots & \cdots & \cdots & \cdots & \cdots & \cdots & \cdots & \cdots & \cdots \\ 0 & \cdots & \cdots & \cdots & \cdots & \cdots & \cdots & \cdots & K\omega_p \end{bmatrix} \tag{3.21}$$

This matrix has N cycles of $(0, ..., K)\omega_p$ along the main diagonal.

Similarly, the current in a nonlinear conductance or a controlled current source is

$$i_{g,n}(t) = f_n(v_1(t), v_2(t), \ldots, v_N(t)) \tag{3.22}$$

Fourier transforming these gives

$$\mathcal{F}\{i_{g,n}(t)\} \rightarrow \mathbf{I}_{G,n} \tag{3.23}$$

and the vector

$$\mathbf{I}_G = \begin{bmatrix} \mathbf{I}_{G,1} \\ \mathbf{I}_{G,2} \\ \cdots \\ \mathbf{I}_{G,N} \end{bmatrix} \tag{3.24}$$

Substituting (3.14), (3.20), and (3.24) into (3.6) gives the expression

$$\mathbf{F}(\mathbf{V}) = \mathbf{I}_s + \mathbf{Y}_{N\times N}\mathbf{V} + j\Omega\mathbf{Q} + \mathbf{I}_G = \mathbf{0} \tag{3.25}$$

Equation (3.25) represents a test to determine whether a trial set of port voltage components is the correct one; that is, if $\mathbf{F}(\mathbf{V}) = \mathbf{0}$, then $\mathbf{V}$ is a valid solution. It also represents an equation that can be solved to obtain the port-voltage vector, **V**. **F**(**V**), called the *current-error vector*, represents the difference between the current calculated from the linear and nonlinear subnetworks, at each port and at each harmonic, for a trial-solution vector **V**.

3.3.2.1 Example: Formulation of the Current-Error Vector

We shall derive the current-error vector of the circuit in Figure 3.5, which consists of an ideal diode (one having no series resistance or junction capacitance) and a linear embedding network described by an admittance matrix. As before, the source impedance $Z_s(\omega)$ is absorbed into the linear network. Figure 3.5 might represent, for example, the local-oscillator circuit in a diode mixer. Because only one nonlinear element exists, $N = 1$ and the vector $\mathbf{V} = \mathbf{V}_1$. The admittance matrix of the embedding network, $\mathbf{Y}_m$, can be written as

$$\mathbf{Y}_m = \begin{bmatrix} \mathbf{Y}_{1,1} & \mathbf{Y}_{1,2} \\ \mathbf{Y}_{2,1} & \mathbf{Y}_{2,2} \end{bmatrix} \tag{3.26}$$

$\mathbf{Y}_{1,1}$ is a submatrix that corresponds to $\mathbf{Y}_{N\times N}$ in (3.13) and (3.14), and $\mathbf{Y}_{1,2}$ corresponds to the leftmost submatrix in (3.13). Then

$$\mathbf{I}_s = \mathbf{I}_{s,1} = \mathbf{Y}_{1,2}\mathbf{V}_2 \tag{3.27}$$

When $\mathbf{V}_2$ is transformed through the Y network, the equivalent circuit of Figure 3.6 results. The linear-circuit equations then depend only upon $\mathbf{I}_{s,1}$ and the admittances seen by the diode at each harmonic, the elements of

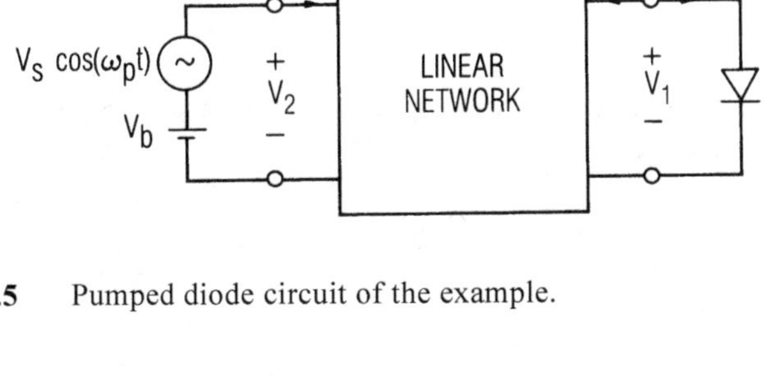

Figure 3.5 Pumped diode circuit of the example.

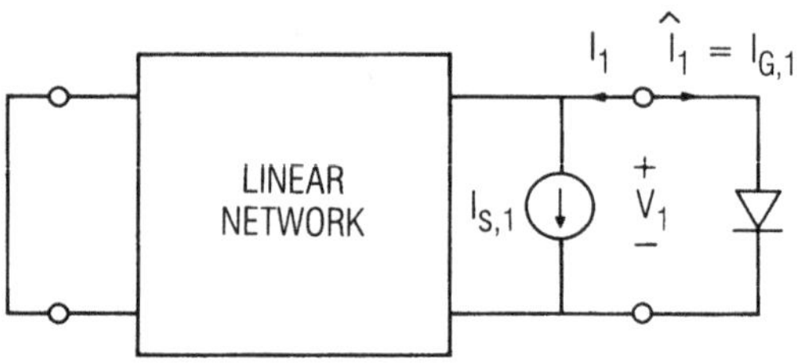

Figure 3.6 Simplified circuit for the example, with the two-port linear subcircuit reduced to a one-port.

$\mathbf{Y}_{1,1}$, are often called the *embedding admittances*. $\mathbf{V}_2$ in (3.27) consists of only the fundamental source $V_s\cos(\omega_p t)$ and the dc bias source V_b, so

$$\mathbf{V}_2 = \begin{bmatrix} V_b \\ V_s \\ 0 \\ \cdots \\ 0 \end{bmatrix} \tag{3.28}$$

We must generate an initial estimate of either $v_1(t)$ or $\mathbf{V}_1$; because this is a simple circuit, that estimate is easy to produce. Previous experience suggests that a sinusoidal waveform clipped at approximately 0.6V should be a good initial estimate of $v_1(t)$. Fourier transforming $v_1(t)$ gives the components of $\mathbf{V}_1$. The diode current $i_1(t)$ is found from the I/V equation of an ideal junction, given by (2.62). Because there is no capacitance, $q_1(t) = 0$. Fourier transforming $i_1(t)$ gives the components of the vector $\mathbf{I}_{G,1}$, so the current-error vector is

$$\mathbf{F}(\mathbf{V}) = \mathbf{Y}_{1,2}\mathbf{V}_2 + \mathbf{Y}_{1,1}\mathbf{V}_1 + \mathbf{I}_{G,1} \tag{3.29}$$

One might wonder if the other Y parameters, $\mathbf{Y}_{2,1}$ and $\mathbf{Y}_{2,2}$, have any use. Indeed they do. Once $\mathbf{V}_1$ is known, they can be used to find the input current from the source, $\mathbf{I}_2$:

$$\mathbf{I}_2 = \mathbf{Y}_{2,1}\mathbf{V}_1 + \mathbf{Y}_{2,2}\mathbf{V}_2 \tag{3.30}$$

Knowing $\mathbf{I}_2$, the vector of input currents at all the harmonics, we can calculate many useful quantities. The RF input power is

$$P_{in} = \frac{1}{2}\mathrm{Re}\{ V_{2,1} I_{2,1}^{*}\} \tag{3.31}$$

where the raised asterisk indicates the complex conjugate. The power dissipated in the source impedance at $k\omega_p$ is

$$P_k = \frac{1}{2}|I_{2,k}|^2\mathrm{Re}\{ Z_s(k\omega_p)\} \tag{3.32}$$

This quantity is more important than it might seem at first, because a nonlinear circuit using a two-terminal device is often modeled in such a way that the linear subcircuit consists only of $Z_s(\omega)$; in this case, $Z_s(\omega)$ is the impedance seen by the diode at each harmonic, including the source and output impedances. For example, this approach is often used to model a diode frequency multiplier, wherein the source impedance is $Z_s(\omega_p)$ and the load at the kth harmonic is $Z_s(k\omega_p)$. We can also find the fundamental-frequency input impedance,

$$Z_{\mathrm{in}} = \frac{V_s}{I_{2,1}} - Z_s(\omega_p) \tag{3.33}$$

As explained in Chapter 1, one cannot define a true input impedance of a nonlinear circuit because an impedance implies a V/I relationship that is independent of voltage or current magnitude. However, the "quasi-impedance" given by (3.33) can be used in much the same way as a linear-circuit impedance. Specifically, Z_{in} is the input impedance to which the source should be matched in order to optimize power transfer at the specific value of V_s.

3.3.3 The Linear Subcircuit

Many methods exist for generating the N-port admittance matrix of a linear circuit. Perhaps the simplest is to generate an indefinite admittance matrix and convert it into a port matrix. The process is straightforward and can be implemented readily on a computer. Of course, a matrix must be produced for each harmonic frequency in the analysis.

The indefinite admittance matrix is created by "stamping" the matrix with a pattern of admittances for each element. For example, if we have a simple two-terminal element connected from node $n1$ to $n2$, whose admittance is Y, we add Y to the (n_1, n_1) and (n_2, n_2) positions and add $-Y$ to the (n_1, n_2) and (n_2, n_1) positions. Similar procedures are used for more complex elements, such as controlled sources and multiports. An N-node circuit results in an $N \times N$ matrix.

To convert the indefinite admittance matrix to a port admittance matrix, we must create a port impedance matrix and invert it. To obtain the impedance matrix, we first select the nodes corresponding to port 1, excite them with unity current, and measure the voltage between the nodes representing each of the ports. This produces the first column of the impedance matrix. Moving the excitation to port 2 produces the second column, and proceeding in this manner to the last port produces the entire matrix.

Specifically, suppose node 1 is the positive terminal of port 1 and node 3 is the negative. The matrix equation is

$$\begin{bmatrix} Y_{1,1} & Y_{1,2} & Y_{1,3} & \cdots & Y_{1,N} \\ Y_{2,1} & Y_{2,2} & Y_{2,3} & \cdots & Y_{2,N} \\ Y_{3,1} & Y_{3,2} & Y_{3,3} & \cdots & Y_{3,N} \\ \cdots & \cdots & \cdots & \cdots & \cdots \\ Y_{N,1} & Y_{N,2} & Y_{N,3} & \cdots & Y_{N,N} \end{bmatrix} \begin{bmatrix} V_1 \\ V_2 \\ V_3 \\ \cdots \\ V_N \end{bmatrix} = \begin{bmatrix} 1 \\ 0 \\ -1 \\ \cdots \\ 0 \end{bmatrix} \tag{3.34}$$

Solving (3.34) gives us all the V_n. Since port 1 is excited by a unit current, $Z_{1,1}$ in the port impedance matrix is simply the port voltage: $Z_{1,1} = V_1 - V_3$. Similarly, if port 2 were connected to nodes 4 and 7, $Z_{2,1} = V_4 - V_7$. When all $Z_{n,1}$ have been found, we move the excitation to port 2 ($I_4 = 1$ and $I_7 = -1$, to continue the example) and repeat the process. Finally, the impedance matrix must be inverted. This is rarely a lengthy process, because the number of ports is usually far less than the number of nodes.

A few notes are in order. First, the solution of (3.34) is best obtained by factoring the matrix. This allows the matrix equation to be solved repeatedly, for each new right-hand (current) vector, by only a simple back-substitution operation instead of a lengthy matrix reduction. As the indefinite admittance matrix is very sparse, sparse-matrix techniques can dramatically reduce computation time compared to full-matrix methods. Second, it often happens that certain elements, such as transformers and controlled voltage sources, do not have an admittance representation and cannot be used directly to form the admittance matrix. One common solution is to approximate the unrealizable element by realizable ones; for example, a voltage-controlled voltage source can be approximated as a voltage-controlled current source with a low-value resistor in shunt with the current source. Another solution, which provides better matrix conditioning, is to cascade the current source with a gyrator. Similarly, transformers can be realized as interconnections of gyrators. Finally, a persistent problem in formulating the matrix is that breaking the connections with the nonlinear elements often results in disconnected nodes. If nothing were done to accommodate them, the indefinite admittance matrix would be singular. The conventional solution is to interconnect all nodes by high-value resistors, but an ill-conditioned matrix usually results. A better solution is to shunt each port with a moderate-value resistor, typically 100Ω, when the indefinite matrix is formulated. To remove the resistors, simply subtract their admittances from the main diagonal of the port admittance matrix.

3.3.4 Solution Algorithms

The one remaining problem, and the nastiest part of the whole business, is to solve (3.25) to obtain **V**. Each of the $K + 1$ frequency components of **V** at each port is a variable, and each component has a real and imaginary part. Thus, there are $2N(K + 1)$ variables to be determined (we concede that the dc components do not have imaginary parts; however, it is usually easier in the analysis to carry the dc terms' imaginary parts than to try to circumvent them). For example, a FET frequency-multiplier analysis might include nonlinear elements at three ports, and have eight significant harmonics plus dc at each port. Thus, $N = 3$, $K = 8$, and there are 54 variables in (3.25)! Solving a set of equations having so many variables, especially in view of the circuit's nonlinear nature, is no small task.

A number of algorithms have been proposed for solving (3.25). Some of these are obvious applications of existing numerical techniques, but others show great ingenuity. Today, there is a strong consensus that Newton's method is preferred for harmonic-balance simulation, and virtually all harmonic-balance simulators use it. We describe Newton's method in Section 3.3.4.3.

3.3.4.1 Optimization

At first glance, solving (3.25) looks a lot like an optimization problem. Therefore, we might be able to solve it by minimizing the magnitude squared of the current-error function; that is, minimize ε where

$$\varepsilon = \mathbf{F}^{*T}(\mathbf{V})\mathbf{F}(\mathbf{V}) \tag{3.35}$$

and *T represents the complex conjugate transpose of the vector. Libraries of scientific subroutines often include a general-purpose functional optimization routine, so, with this method, a large and difficult part of the computer programming is prepackaged for us. However, the error function in (3.35) destroys a lot of information about the individual contribution of each variable to the error, so optimization routines may have convergence problems, especially when a large number of variables must be optimized simultaneously. Because of these limitations, optimization is a reasonable approach only for relatively simple problems, in which the ease of programming outweighs the inefficiency.

3.3.4.2 Relaxation Methods

A number of relaxation methods, which are both simple to implement and intuitively satisfying, have been proposed. As the name implies, relaxation methods use simple algorithms that encourage the voltages to move gradually (or *relax*) toward the solution. Often the methods are largely heuristic. For example, the bisection method, used to find the zero of a nonlinear function and described in virtually all basic numerical methods texts, is a kind of relaxation method. An advantage of these methods is that they are simple to implement, often not requiring the generation of *I*/*V* derivatives or even an initial estimate of the solution.

Two of the most popular relaxation methods are those of Hicks and Khan [3.5] and Kerr [3.6]. Although seemingly quite different, it is possible to show that these methods are equivalent, and that Kerr's method is a reflection form of Hicks and Khan.

Relaxation methods are largely obsolete. The worst problems are (1) unpredictable (and often disappointing) convergence characteristics, and (2) inapplicability to large systems. Their importance today is largely historical, as they represent an important step in the development of nonlinear circuit simulation technology. As a more practical matter, they may yet be useful for a "quick and dirty" solution of a special problem.

3.3.4.3 Newton's Method

Newton's method is a powerful algorithm for finding the zeros of a set of multivariate nonlinear functions. Because the harmonic-balance method involves finding the zeros of $\mathbf{F}(\mathbf{V})$, Newton's method is an obvious choice as a solution algorithm. Newton's method is an iterative technique; it estimates the zero of a function by using its first derivative to extrapolate to the axis of the independent variable. Its power comes from its use of all the derivatives of $\mathbf{F}(\mathbf{V})$, with respect to the voltage components of $\mathbf{V}$, in each iteration. Newton's method is used in virtually all modern harmonic-balance software.

This iterative process is most easily illustrated by applying it to a one-dimensional problem. Figure 3.7(a) shows a function of one variable, $f(x)$, and a Newton estimate of its zero. One can write, for the linear extrapolation,

$$f(x_0) - \left.\frac{df}{dx}\right|_{x = x_0} \Delta x = 0 \tag{3.36}$$

$f(x_0)$ and its derivative are known, (3.36) can be solved easily to obtain Δx, and a new estimate of the zero is found as $x_0 - \Delta x$. The function and its derivative are again evaluated, at the estimated zero, and the process is repeated until the zero is determined with the required accuracy.

It is important to realize that Newton's method can fail. For example, Figure 3.7(b) shows what can happen when the process is started near a relative minimum of the function: the second estimate of the zero returns the process to a point near the original one, and it oscillates within a limited region. The process can easily get caught in that region and never find the zero. It could also land close to the minimum, where df/dx is nearly zero, so the next estimate of the zero would either be hopelessly far from the zero's real location, or would cause a numerical exception. Newton's method can be trusted to converge only when it is started sufficiently close to the zero. "Sufficiently close" can be hard to determine, but it generally means that (1) the function is well behaved between the zero and starting point, and (2) it is not too strongly nonlinear. Thus, Newton's method usually requires approximate knowledge of the location of the zero before it begins.

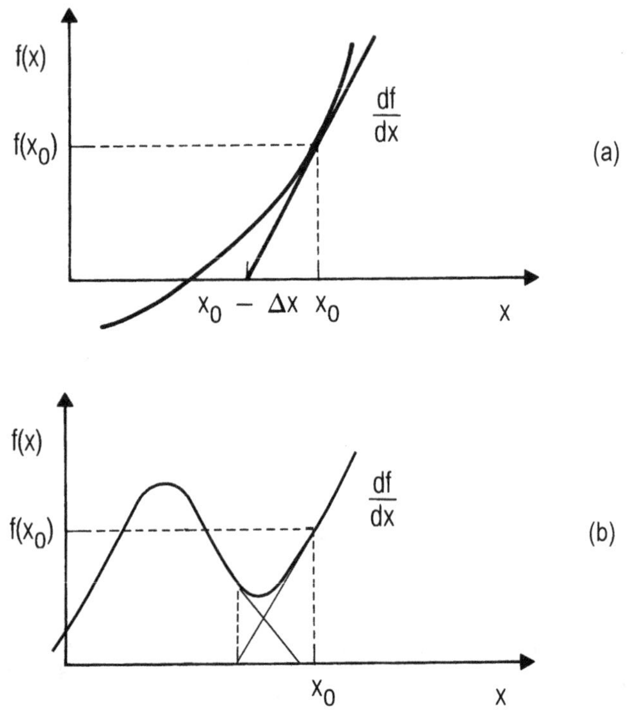

Figure 3.7 Newton's method in one dimension. In (a) the process finds the zero easily by making repeated linear estimates of its location. When a relative minimum exists, as in (b), the process can become trapped near the minimum.

3.3.5 Newton Solution of the Harmonic-Balance Equation

3.3.5.1 Iterative Process and Jacobian Formulation

Our error function, $\mathbf{F}(\mathbf{V})$, is in fact a set of multidimensional functions, and we need to find all zeros simultaneously. The analog of (3.36) applied to a set of multidimensional functions is

$$\mathbf{F}(\mathbf{V}^p) - \left.\frac{d\mathbf{F}(\mathbf{V})}{d\mathbf{V}}\right|_{\mathbf{V}=\mathbf{V}^p} \Delta\mathbf{V} = \mathbf{0} \tag{3.37}$$

where $\mathbf{V}^p$ is the pth estimate of the solution vector. With

$$\mathbf{V}^p - \mathbf{V}^{p+1} = \Delta\mathbf{V} \tag{3.38}$$

the updated vector, $\mathbf{V}^{p+1}$, is

$$\mathbf{V}^{p+1} = \mathbf{V}^p - \left(\frac{d\mathbf{F}(\mathbf{V})}{d\mathbf{V}}\right)^{-1} \mathbf{F}(\mathbf{V}^p) \tag{3.39}$$

Equation (3.39) involves the derivative of a vector, **F**, with respect to another vector, **V**. The result is a matrix, called the *Jacobian* of **F**, designated $\mathbf{J}_\mathbf{F}$:

$$\mathbf{J}_\mathbf{F} = \left.\frac{d\mathbf{F}(\mathbf{V})}{d\mathbf{V}}\right|_{\mathbf{V} = \mathbf{V}^p} \tag{3.40}$$

The Jacobian contains the derivatives of all the components of the error vector with respect to the components of **V**. Thus, it contains information on the sensitivity of changes in every component of **F** resulting from changes in any component of **V**. This amount of information is the maximum possible from a linearized system of equations.

The form of the Jacobian is

$$\mathbf{J_F} = \begin{bmatrix} \frac{\partial F_{1,0}}{\partial V_{1,0}} & \frac{\partial F_{1,0}}{\partial V_{1,1}} & \frac{\partial F_{1,0}}{\partial V_{1,2}} & \cdots & \frac{\partial F_{1,0}}{\partial V_{1,K}} & \frac{\partial F_{1,0}}{\partial V_{2,0}} & \cdots & \frac{\partial F_{1,0}}{\partial V_{N,K}} \\ \frac{\partial F_{1,1}}{\partial V_{1,0}} & \frac{\partial F_{1,1}}{\partial V_{1,1}} & \frac{\partial F_{1,1}}{\partial V_{1,2}} & \cdots & \frac{\partial F_{1,1}}{\partial V_{1,K}} & \frac{\partial F_{1,1}}{\partial V_{2,0}} & \cdots & \frac{\partial F_{1,1}}{\partial V_{N,K}} \\ \cdots & \cdots & \cdots & \cdots & \cdots & \cdots & \cdots & \cdots \\ \frac{\partial F_{1,K}}{\partial V_{1,0}} & \frac{\partial F_{1,K}}{\partial V_{1,1}} & \frac{\partial F_{1,K}}{\partial V_{1,2}} & \cdots & \frac{\partial F_{1,K}}{\partial V_{1,K}} & \frac{\partial F_{1,K}}{\partial V_{2,0}} & \cdots & \frac{\partial F_{1,K}}{\partial V_{N,K}} \\ \frac{\partial F_{2,0}}{\partial V_{1,0}} & \frac{\partial F_{2,0}}{\partial V_{1,1}} & \frac{\partial F_{2,0}}{\partial V_{1,2}} & \cdots & \frac{\partial F_{2,0}}{\partial V_{1,K}} & \frac{\partial F_{2,0}}{\partial V_{2,0}} & \cdots & \frac{\partial F_{2,0}}{\partial V_{N,K}} \\ \cdots & \cdots & \cdots & \cdots & \cdots & \cdots & \cdots & \cdots \\ \frac{\partial F_{2,K}}{\partial V_{1,0}} & \frac{\partial F_{2,K}}{\partial V_{1,1}} & \cdots & \cdots & \cdots & \cdots & \cdots & \frac{\partial F_{2,K}}{\partial V_{N,K}} \\ \cdots & \cdots & \cdots & \cdots & \cdots & \cdots & \cdots & \cdots \\ \frac{\partial F_{N,K}}{\partial V_{1,0}} & \cdots & \cdots & \cdots & \cdots & \cdots & \cdots & \frac{\partial F_{N,K}}{\partial V_{N,K}} \end{bmatrix} \quad (3.41)$$

The elements of the Jacobian are the derivatives

$$\frac{\partial F_{n,k}}{\partial V_{m,l}} \quad (3.42)$$

where n and m are the port indices $(1, N)$, and k and l are the harmonic indices $(0, ..., K)$. Determining these quantities requires some effort. We begin by taking the derivative of (3.25):

$$\mathbf{J_F} = \frac{d\mathbf{F}(\mathbf{V})}{d\mathbf{V}} = \mathbf{Y}_{N \times N} + \frac{\partial \mathbf{I}_G}{\partial \mathbf{V}} + j\Omega \frac{\partial \mathbf{Q}}{\partial \mathbf{V}} \quad (3.43)$$

Thus, we must find $\partial \mathbf{I}_G / \partial \mathbf{V}$ and $\partial \mathbf{Q} / \partial \mathbf{V}$. We begin with the former. $\partial \mathbf{I}_G / \partial \mathbf{V}$, a matrix, has the same form as $\partial \mathbf{F} / \partial \mathbf{V}$; that is, its elements are $\partial \mathbf{I}_{n,k} / \partial \mathbf{V}_{m,l}$.

First, we note that a Fourier series can be expressed a number of different ways; we express it here as follows:

$$i_n(t) = \sum_{k=-K}^{K} I_{n,k} \exp(jk\omega_p t) \tag{3.44}$$

where $k \neq 0$. The components are

$$I_{n,k} = \frac{1}{T} \int_0^T i_n(t) \exp(-jk\omega_p t) \tag{3.45}$$

Thus, $I_{n,k}$, $k > 0$, is half the phasor value of the current component at $k\omega_p$. To determine its derivative w.r.t. $V_{m,l}$, we must consider both positive and negative frequency components of $V_{m,l}$. Then,

$$dI_{n,k} = \frac{\partial I_{n,k}}{\partial V_{m,l}} dV_l + \frac{\partial I_{n,k}}{\partial V_{m,-l}} dV_{-l} \tag{3.46}$$

where

$$dV_{-l} = dV_l^* \tag{3.47}$$

From (3.45), the Fourier component $\partial I_k / \partial V_l$ is

$$\frac{\partial I_{n,k}}{\partial V_{m,l}} = \frac{1}{T} \int_0^T \frac{\partial i_n}{\partial v_m} \frac{\partial v_m}{\partial V_{m,l}} \exp(-jk\omega_p t) \tag{3.48}$$

with

$$v_m(t) = \sum_{l=-K}^{K} V_{m,l} \exp(jl\omega_p t) \tag{3.49}$$

We see immediately that

$$\frac{\partial v_m}{\partial V_{m,l}} = \exp(jl\omega_p t) \tag{3.50}$$

Substituting (3.50) into (3.48) gives

$$\frac{\partial I_{n,k}}{\partial V_{m,l}} = \frac{1}{T}\int_0^T \frac{\partial i_n}{\partial v_m} \exp(-j(k-l)\omega_p t) \tag{3.51}$$

For the negative-frequency component

$$\frac{\partial v_m}{\partial V_{m,-l}} = \exp(-jl\omega_p t) \tag{3.52}$$

so

$$\frac{\partial I_{n,k}}{\partial V_{m,-l}} = \frac{1}{T}\int_0^T \frac{\partial i_n}{\partial v_m} \exp(-j(k+l)\omega_p t) \tag{3.53}$$

We find that the derivatives are components of the Fourier expansion of the derivative waveform.

The terms (3.53) tend to occur at higher frequencies than those in (3.51) and thus have less effect on the convergence process. In many simple situations, (3.53) can be neglected. However, including it noticeably improves convergence in strongly nonlinear circuits.

We still need to put all this into the same form as (3.8). First, we note that the derivative terms are complex, so we can write

$$\begin{aligned} \frac{\partial I_{n,k}}{\partial V_{m,l}} &= G^R_{k-l} + jG^I_{k-l} \\ \frac{\partial I_{n,k}}{\partial V_{m,-l}} &= G^R_{k+l} + jG^I_{k+l} \end{aligned} \tag{3.54}$$

where $G_p = G^R_p + jG^I_p$ is the pth Fourier-series component of $g(t) = \partial i_n / \partial v_m$, evaluated at $v_m(t)$, and

$$\begin{aligned} dV_{m,\,l} &= dV^R_{m,\,l} + j dV^I_{m,\,l} \\ dV_{m,\,-l} &= dV^R_{m,\,l} - j dV^I_{m,\,l} \end{aligned} \tag{3.55}$$

Substituting into (3.46) gives

$$\begin{aligned} dI^R_{n,\,k} &= (G^R_{k-l} + G^R_{k+l})dV^R_{m,\,l} + (-G^I_{k-l} + G^I_{k+l})dV^I_{m,\,l} \\ dI^I_{n,\,k} &= (G^I_{k-l} + G^I_{k+l})dV^R_{m,\,l} + (G^R_{k-l} - G^R_{k+l})dV^I_{m,\,l} \end{aligned} \tag{3.56}$$

Thus, each term in (3.41) must be treated as a 2×2 submatrix,

$$\begin{bmatrix} dI^R_{n,\,k} \\ dI^I_{n,\,k} \end{bmatrix} = \begin{bmatrix} G^R_{k-l} + G^R_{k+l} & -G^I_{k-l} + G^I_{k+l} \\ G^I_{k-l} + G^I_{k+l} & G^R_{k-l} - G^R_{k+l} \end{bmatrix} \begin{bmatrix} dV^R_{m,\,l} \\ dV^I_{m,\,l} \end{bmatrix} \tag{3.57}$$

Unfortunately, it is not possible to write (3.56) as a simple, complex equation. Worse, when $k = l = 0$, the matrix in (3.57) becomes

$$\begin{bmatrix} G_0 & 0 \\ 0 & 0 \end{bmatrix} \tag{3.58}$$

The second row and second column of the Jacobian, for each port, are both zero. The Jacobian is, therefore, singular and (3.37) cannot be solved! The problem arises from the fact that dc components must have zero imaginary parts. To circumvent this difficulty, we can set the (2, 2) position of (3.58) to some arbitrary value; for example,

$$\begin{bmatrix} G_0 & 0 \\ 0 & G_0 \end{bmatrix} \tag{3.59}$$

Then, as long as $G_0 \neq 0$ and the imaginary parts of the dc voltage and current components are consistently set to zero, all should be well. Of course, another solution is simply to delete the row and column. Yet another solution is to derive the Jacobian as described in Section 3.6.8.

The third term of (3.43) is handled in the same manner. To obtain $\partial Q_{n,k}/\partial V_{m,l}$, we use the small-signal capacitance waveform, $\partial q_n/\partial v_m$, instead of the conductance waveform, $\partial i_n/v_m$, in (3.51), and proceed identically. The result can be written,

$$\begin{bmatrix} dI^R_{n,k} \\ dI^I_{n,k} \end{bmatrix} = \begin{bmatrix} 0 & -k\omega_p \\ k\omega_p & 0 \end{bmatrix} \begin{bmatrix} C^R_{k-l} + C^R_{k+l} & -C^I_{k-l} + C^I_{k+l} \\ C^I_{k-l} + C^I_{k+l} & C^R_{k-l} - C^R_{k+l} \end{bmatrix} \begin{bmatrix} dV^R_{m,l} \\ dV^I_{m,l} \end{bmatrix} \tag{3.60}$$

where $C_p = C^R_p + jC^I_p$ represents the pth Fourier-series component of $c(t) = \partial q_n/\partial v_m$. The matrix containing the $k\omega_p$ terms represents a single entry of the Ω matrix (3.21).

Since we have separated the real and imaginary parts of $dI_{n,k}$ and $dV_{m,l}$, we need to treat the Y matrix similarly. The 2×2 submatrix representing a Y parameter has the form

$$Y_{n,m}(k\omega_p) \rightarrow \begin{bmatrix} Y^R{}_{n,m}(k\omega_p) & -Y^I{}_{n,m}(k\omega_p) \\ Y^I{}_{n,m}(k\omega_p) & Y^R{}_{n,m}(k\omega_p) \end{bmatrix} \tag{3.61}$$

For dc, we require only the matrix component in the (1, 1) position.

3.3.5.2 Jacobian Structure

The Jacobian consists of an $N \times N$ matrix of square submatrices, each of which has dimension $K + 1$. Each submatrix represents the harmonic components for a particular nonlinear-element port; that is, if the current or charge at port n depends upon the voltage at port m, the (n, m) submatrix is filled with Fourier terms. Added to each (n, m) submatrix is a diagonal matrix of $Y_{n,m}$ at each harmonic, $0 \ldots K\omega_p$. Thus, some submatrices are filled and some are diagonal. It is also possible for some to be empty.

Equation (3.62) shows a possible form of the matrix when $N = 3$ and $K = 3$. Filled blocks along the main diagonal occur when a port has a two-terminal nonlinear element connected to it. Off-diagonal filled blocks result from controlled nonlinear current sources. The diagonal matrix in the (3, 3) position implies that the voltage at this port is a control voltage for one of the other nonlinearities, but there is no nonlinear element connected to it.

It is interesting to note that this matrix is rather sparse, so sparse-matrix methods may be useful in solving it. Sparse-matrix methods, unfortunately, usually work well only when the matrix is extremely sparse, and the Jacobian, in the harmonic-balance problem, is usually not sparse enough to benefit more than modestly from such methods. It also may be possible to exploit the special structure of this matrix in other, more elegant ways to speed its factorization.

$$\begin{bmatrix}
\begin{bmatrix} x & x & x & x \\ x & x & x & x \\ x & x & x & x \\ x & x & x & x \end{bmatrix} &
\begin{bmatrix} x & x & x & x \\ x & x & x & x \\ x & x & x & x \\ x & x & x & x \end{bmatrix} &
\begin{bmatrix} x & & & \\ & x & & \\ & & x & \\ & & & x \end{bmatrix} \\
\begin{bmatrix} x & & & \\ & x & & \\ & & x & \\ & & & x \end{bmatrix} &
\begin{bmatrix} x & x & x & x \\ x & x & x & x \\ x & x & x & x \\ x & x & x & x \end{bmatrix} &
\begin{bmatrix} x & & & \\ & x & & \\ & & x & \\ & & & x \end{bmatrix} \\
\begin{bmatrix} x & & & \\ & x & & \\ & & x & \\ & & & x \end{bmatrix} &
\begin{bmatrix} x & & & \\ & x & & \\ & & x & \\ & & & x \end{bmatrix} &
\begin{bmatrix} x & & & \\ & x & & \\ & & x & \\ & & & x \end{bmatrix}
\end{bmatrix} \tag{3.62}$$

3.3.5.3 Example: Jacobian Formulation

The circuit of the previous example and Figures 3.5 and 3.6 will be solved by means of Newton's method. **F(V)** is given by (3.29); differentiating gives the terms of the Jacobian:

$$\mathbf{J}_{\mathbf{F}} = \frac{d\mathbf{F}(\mathbf{V})}{d\mathbf{V}_1} \tag{3.63}$$

or

$$\frac{\partial F_{1,k}}{\partial V_{1,l}} = Y_{1,k}(k = l) + \frac{\partial I_{G;1,k}}{\partial V_{1,l}} \tag{3.64}$$

where

$$Y_{1,k}(k=l) = \begin{cases} Y_{1,1}(k\omega_p) & k = l \\ 0 & k \neq l \end{cases} \tag{3.65}$$

The partial derivative within the integral sign can be interpreted as the incremental junction conductance of the diode:

$$\begin{aligned} g(t) = \frac{\partial i_{g,1}}{\partial v} &= \frac{d}{dv}[I_{sat}(\exp(\delta V) - 1)] \\ &= \delta I_{sat}\exp(\delta v) \\ &\approx \delta i_{g,1}(t) \end{aligned} \tag{3.66}$$

Fourier-transforming $g(t)$ gives the frequency components G_k, $k = -K \ldots 0 \ldots K$. The Jacobian $\mathbf{J_F}$ is

$$\mathbf{J_F} =$$

$$\begin{bmatrix} \begin{bmatrix} Y_{11}(0) + G_0 & 0 \\ 0 & 0 \end{bmatrix} & \begin{bmatrix} 2G^R{}_1 & 2G^I{}_1 \\ 0 & 0 \end{bmatrix} & \begin{bmatrix} 2G^R{}_2 & 2G^I{}_2 \\ 0 & 0 \end{bmatrix} & \cdots \\ \begin{bmatrix} 2G_1^R & 0 \\ 2G_1^I & 0 \end{bmatrix} & \begin{bmatrix} Y_{11}^R(\omega_p) + G_0 + G_2^R & -Y_{11}^I(\omega_p) + G_2^I \\ Y_{11}^I(\omega_p) + G_2^I & Y_{11}^R(\omega_p) + G_0 - G_2^R \end{bmatrix} & \begin{bmatrix} G_{-1}^R + G_3^R & -G_{-1}^I + G_3^I \\ G_{-1}^I + G_3^I & G_{-1}^R - G_3^R \end{bmatrix} & \cdots \\ \cdots & \cdots & \cdots & \cdots \\ \cdots & \cdots & \cdots & \cdots \\ \cdots & \cdots & \cdots & \cdots \end{bmatrix} \tag{3.67}$$

Note that the terms arising from the negative frequencies (3.53) involve rather high harmonics, so they decrease to insignificance rapidly.

The solution is found by the following process:

1. Form an initial estimate of the waveform $v_1(t)$. As in the previous example, a clipped sinusoid is a good initial estimate.

2. Fourier-transform $v_1(t)$ to obtain $\mathbf{V}_1^0$, the initial estimate in the frequency domain. The superscript represents the iteration number.
3. Find the conductance waveform $g(t)$ from (3.66) and Fourier transform it.
4. Form $\mathbf{J}_\mathbf{F}$ and $\mathbf{F}(\mathbf{V}^0)$ from (3.67)and (3.29).
5. Solve (3.39) to obtain a new estimate of the voltage vector, $\mathbf{V}_1{}^1$.
6. Fourier-transform the diode current, which was found in step 3, and form the vector $\mathbf{I}_{G,1}$.
7. Use (3.29) to determine $\mathbf{F}(\mathbf{V}^1)$.
8. If the magnitudes of the components of $\mathbf{F}(\mathbf{V}^1)$ are small enough, the solution has been found. Otherwise, inverse Fourier transform to obtain $v_1(t)$ and repeat from step 3 to obtain $\mathbf{V}_1{}^2$.

3.3.6 Selecting the Number of Harmonics and Time Samples

In theory, the waveforms generated in nonlinear analysis have an infinite number of harmonics, so a complete description of the operation of a nonlinear circuit would appear to require current and voltage vectors of infinite dimension. Fortunately, the magnitudes of frequency components invariably decrease with frequency; otherwise the time waveforms would represent infinite power. Accordingly, it is always possible to ignore all harmonics above some maximum number, which we have designated K. An important consideration in implementing a harmonic-balance analysis is the selection of K. Selecting K too small results in poor accuracy, and often poor convergence; conversely, selecting K too large slows the solution process, which under the best circumstances is time-consuming, and increases the use of computer memory.

Perhaps the simplest criterion for selecting K is to consider the magnitudes of the capacitances in the device's equivalent circuit. Above some frequency the capacitive susceptances are greater than the circuit's conductances, so they effectively are short circuits, and their voltage components are negligibly small. This criterion can be applied easily to a diode, for example, where the junction capacitance short-circuits all voltage components across the only other nonlinearity, the resistive junction.

Another important consideration in the selection of K is the strength of the dominant nonlinearity and the magnitude of the excitation. It is often possible to generate a simplified equivalent circuit for the nonlinear device and to approximate the voltages and currents in it well enough to form a rough estimate of the frequency-component magnitudes. For example, in a strongly driven FET, one can often approximate the gate voltage as a sinusoid and the drain current as a rectangular pulse train. The length of each pulse is equal to the length of time that the gate voltage is above V_t, the *threshold voltage*. A pulse train's Fourier series is found easily, and because the actual drain-current pulse is invariably softer than a rectangular pulse, this series establishes an upper bound to the relative magnitudes of the drain current's harmonic components. In Chapter 1 we saw that an nth-degree nonlinearity generates only n harmonics directly, although higher harmonics are possible as mixing products between these frequencies. These higher harmonics are usually much weaker than those generated directly, so it rarely makes sense to pick K much larger than the highest degree nonlinearity in the circuit. Conversely, if we wish to determine the levels of high harmonics, we must be careful to model the circuit nonlinearities by using polynomials (or other functions having polynomial expansions) of a degree great enough to generate those harmonics.

The nature of the problem to be solved often places some constraints on K. If the current or voltage at some harmonic k are to be found, $K > k$ is an obvious requirement. It is perhaps less obvious that the errors introduced by harmonic truncation are usually greater at higher harmonics than at the lower ones, so we really must choose K considerably larger. Calculating the magnitudes of high harmonics accurately also requires that convergence be more complete, so that the errors in all the high-harmonic components are small.

The properties of the fast Fourier-transform algorithm (FFT), used to obtain the frequency components from the time waveforms, also places constraints on K. One requirement of the FFT is that the number of harmonics must always be an integer power of two. (Forms of the FFT have been devised that do not have this requirement, but they are not used much in harmonic-balance simulators.) The second, a consequence of the sampling theorem, is that the number of time samples must be twice the number of frequency components. It is not necessary to include all these harmonics in the harmonic-balance equations; it is possible to use, for example, only 10 harmonics in the equations but to calculate 16 via the FFT. It is essential, however, to use all the time samples required by the FFT. Furthermore, there are good reasons to use even more time samples. Using the minimum number of samples required by the sampling theorem may result in aliasing errors, where the neglected high harmonics affect the accuracy of lower-harmonic components. The simplest way to minimize aliasing errors is to *oversample*, that is, to use a sampling rate 25% to 30% greater than the minimum, or 2.5 to 2.6 times the minimum number of required samples. In the above example, 10 harmonics require a sample rate of 25 or 26 time samples per cycle. The next highest power of two is 32, so 32 samples should be used, with 16 harmonics in the FFT. The higher six harmonics are simply discarded in formulating the current-error vector.

The intended use of the analysis also affects the number of harmonics that must be considered. In Section 3.4 we shall see that a conversion-matrix analysis involving mixing products around the kth local-oscillator harmonic requires $K = 2k$ harmonics, plus the dc component, in the large-signal analysis. Obtaining good accuracy in the IM analysis of mixers or other time-varying circuits often requires even more harmonics, however, and it is often very difficult to estimate K beforehand. In these problems one must determine K empirically by increasing it until consistent results, independent of K, are obtained.

What is the effect of discarding the harmonics $k > K$? It implies that the voltage across the nonlinear elements at those frequencies is zero, so the impedance looking into the embedding network from the element terminals is a short circuit. It is sometimes possible, although rarely practical, to formulate a dual case to the one we have described, wherein the element currents, not the voltages, are the independent variables. In this case, harmonic truncation would set the currents to zero, which implies open circuits at the higher frequencies.

3.3.7 Matrix Methods for Solving (3.37)

Solving (3.37) involves solving a set of linear equations. Certainly, there is no shortage of literature describing methods for solving linear equations; however, certain methods have been found especially useful for harmonic-balance analysis, so we examine them here.

3.3.7.1 Direct Solvers

Direct or "full" solvers are those described in most basic linear-algebra texts. Especially when norm reduction methods are used (Section 3.3.8), the most practical is LU decomposition, as solutions can be obtained for multiple right-hand sides with a single factorization.

The principle behind LU decomposition is very simple. Suppose we must solve the matrix equation,

$$\mathbf{A}\mathbf{x} = \mathbf{b} \tag{3.68}$$

for **x**, where **A** is a matrix and **x**, **b** are vectors. We factor the matrix **A** into a *lower triangular matrix*, **L**, in which the entries above the diagonal are zero, and an *upper triangular matrix*, **U**, in which the entries below the diagonal are zero. Then, we have

$$\mathbf{LUx} = \mathbf{b} \tag{3.69}$$

Let

$$\mathbf{Ux} = \mathbf{m} \tag{3.70}$$

where **m** is a vector, and solve, in two steps,

$$\begin{aligned} \mathbf{Lm} &= \mathbf{b} \\ \mathbf{Ux} &= \mathbf{m} \end{aligned} \tag{3.71}$$

The two steps in (3.71) can be solved by back-substitution operations, which are computationally inexpensive; virtually all the work is in factoring the matrix. Once the matrix is factored, it can be used repeatedly to solve (3.68) at very low cost. This property is especially valuable in harmonic-balance analysis. Another nice property is that LU factorization can be performed "in place": that is, without using more memory than what is required to hold the original matrix, **A**. **A** is destroyed in the factorization and is replaced by **L** and **U**.

Direct solvers scale poorly for harmonic-balance analysis. The time required to factor the matrix varies approximately as the cube of its dimension; thus, doubling the size of the matrix increases computation time by a factor of eight. This characteristic clearly makes direct solvers impractical for analyses of large circuits.

3.3.7.2 Sparse Solvers

A sparse matrix is one that contains mostly zero entries. Conventional sparse solvers use LU decomposition to factor the matrix but exploit the sparsity of certain kinds of matrices to improve efficiency. The improvement comes from avoiding the need to multiply and add large numbers of zero entries. In most such methods, the zero entries are not stored, so a saving of memory results as well.

As a sparse matrix is reduced, it tends to *fill in*; that is, entries that originally were zero are converted to nonzero numbers. Avoiding such "fill-ins" is important to the success of a sparse-matrix method. Usually, there is a trade-off between fill-ins and optimal pivoting, so sparse matrix methods may not be as robust as direct, full solvers.

If the matrix is very sparse, the time required to factor it may vary by as little as the 1.5 power of its dimension. This is a considerable improvement over direct methods. However, it is rarely possible to achieve adequate sparsity in the Jacobian to achieve this kind of performance.

3.3.7.3 Krylov-Subspace Techniques and Inexact Newton Iteration

Krylov subspace techniques are a class of iterative methods for solving sparse linear systems of equations. There is now a general consensus that a technique called the *generalized minimum residual*, or *GMRES*, is the preferred one, of many available, for harmonic-balance analysis. Although some of the material in this section may be valid for other methods, it should be considered specific to GMRES.

Iterative methods minimize the residual, $\mathbf{r}$, of (3.68):

$$\mathbf{r} = \mathbf{b} - \mathbf{A}\hat{\mathbf{x}} \tag{3.72}$$

where $\hat{\mathbf{x}}$ is an estimate of the solution. This can be done efficiently only when $\hat{\mathbf{x}}$ can be estimated with at least moderate accuracy, so $\mathbf{r}$ is not too large. To obtain such conditions, we must *precondition* the matrix; that is, multiply it by an estimate of the inverse. Thus,

$$\mathbf{PAx} = \mathbf{Pb} \tag{3.73}$$

where $\mathbf{P}$, the *preconditioner*, is an estimate of $\mathbf{A}^{-1}$. Other Krylov methods require different kinds of preconditioning; in GMRES, it is also possible to perform *right preconditioning*:

$$\mathbf{AP}^{-1}\mathbf{y} = \mathbf{b} \tag{3.74}$$

to obtain $\mathbf{y}$, and then solve

$$\mathbf{y} = \mathbf{Px} \tag{3.75}$$

In this case, it is essential that (3.75) be solvable at low computational cost. In harmonic-balance analysis, a suitable preconditioner is the inverse of the admittance matrix of the linear subcircuit, which is generated in the process of creating the port Y matrix, and need be inverted only once in the solution process. Another option for the preconditioner is the inverse of a severely pruned (Section 3.3.9.1) version of the Jacobian, but this must be regenerated periodically as the Jacobian changes.

An advantage of Krylov techniques is that (3.37) need not be fully solved in each iteration; the iterative process need only proceed until $\Delta\mathbf{V}$ decreases the error function. This approach to the solution, called *inexact Newton*, can provide significantly improved efficiency. In the early harmonic-balance iterations, a Newton step is, at best, a poor estimate of the zero, so an accurate solution of (3.37) has little value. At each step, the matrix need only be solved until some condition on the improved solution is found; the usual criterion [3.7, 3.8] is

$$\|\mathbf{F}(\mathbf{V}) - \mathbf{J}(\mathbf{V})\Delta\mathbf{V}\| < \alpha\|\mathbf{F}(\mathbf{V})\| \tag{3.76}$$

where α is selected at the beginning of the pth harmonic-balance iteration to be

$$\alpha = \frac{\|\mathbf{F}(\mathbf{V}^p) - \mathbf{F}(\mathbf{V}^{p-1}) + \mathbf{J}(\mathbf{V}^{p-1})\Delta\mathbf{V}^{p-1}\|}{\|\mathbf{F}(\mathbf{V}^{p-1})\|} \tag{3.77}$$

Setting $\alpha = 0$ in (3.76) corresponds to ordinary, exact Newton iterations.

The author has observed that Krylov solvers are distinctly inferior to direct solvers in handling poorly conditioned Jacobian matrices (Section 3.3.7.4). For further information on Krylov-subspace methods, see [3.9–3.11].

3.3.7.4 Matrix Conditioning

It is well known that, if $\mathbf{A}$ is singular and $\mathbf{b}$ is nonzero, (3.68) has no unique solution. In many cases, however, $\mathbf{A}$ is nonsingular, but it is so close to being singular that the solution is indistinct. In this case, we say that the matrix is *ill conditioned*, and the result is an inaccurate solution, $\mathbf{x}$.

The accuracy of the solution is controlled by the *condition number*, $\kappa(\mathbf{A})$. Then,

$$\frac{\delta\|\mathbf{x}\|}{\|\mathbf{x}\|} \le \kappa(\mathbf{A})\frac{\delta\|\mathbf{b}\|}{\|\mathbf{b}\|} \tag{3.78}$$

where $\|\mathbf{x}\|$ is the *maximum norm* of the N-dimensional vector, $\mathbf{x}$,

$$\|\mathbf{x}\| = \max|x_i| \qquad 1 \le i \le N \tag{3.79}$$

and

$$\kappa(\mathbf{A}) = \|\mathbf{A}\|\,\|\mathbf{A}^{-1}\| \tag{3.80}$$

where

$$\|\mathbf{A}\| = \max\frac{\|\mathbf{A}\mathbf{x}\|}{\|\mathbf{x}\|} \tag{3.81}$$

over all nonzero **x**.

A loose interpretation of (3.78) is that a certain fractional error in **b**, which may come from a loss of numerical precision or limiting the degree of the harmonic series in the FFT, results in a proportionately larger error in **x**, when $\kappa(\mathbf{A})$ is large. An error in a particular component of **b**, b_i, does not simply affect x_i, but can create errors in any or, more commonly, all the components of **x**. If $\kappa(\mathbf{A})$ is very large, as is the case when **A** is nearly singular, the error can be greater than **x** itself, rendering the solution useless. Convergence failure in Newton-based harmonic-balance analysis is often caused by an ill-conditioned Jacobian.

It is disturbingly easy, in harmonic-balance analysis, to encounter an ill-conditioned Jacobian or Y matrix. Some of the causes are discussed in Section 3.3.9.6.

3.3.8 Norm Reduction

The need for norm-reduction methods can be illustrated by reviewing the operation of Newton's method in one dimension. For example, consider the problem of finding the zero of the function shown in Figure 3.8. Although the nonlinearities are weak, and we have no relative minima near the zero, the process can still become trapped near the zero, or even diverge. However, suppose that, instead of the full Newton step in (3.36), we take a partial step; precisely,

$$\Delta x = \beta f(x_0)\left.\frac{df}{dx}\right|_{x = x_0}^{-1} \tag{3.82}$$

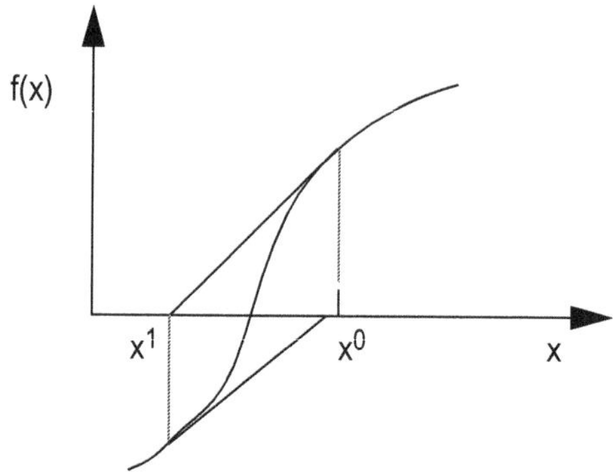

Figure 3.8 A situation where norm-reduction methods prevent failure of Newton's method. Newton's method can fail when the function has an inflection point near the zero. In this case, reducing the step size can provide success.

where β is a constant, between zero and one, that we can adjust as needed. Now, we adjust β to obtain the step size that minimizes $f(x)$, and simply continue the Newton steps. Unless our algorithm for selecting β is extraordinarily inept, this process always finds the zero in the situation illustrated in Figure 3.8.

In the multidimensional case, (3.39) is modified to form

$$\mathbf{V}^{p+1} = \mathbf{V}^{p} - \beta\left(\frac{d\mathbf{F}(\mathbf{V})}{d\mathbf{V}}\right)^{-1}\mathbf{F}(\mathbf{V}^{p}) \tag{3.83}$$

where β, as in the one-dimensional case, is a real constant. The usual process for adjusting β is to begin with a full Newton step (β = 1). If that step reduces the current error, it is retained; if not, β is reduced by some factor, and the process is repeated until the error decreases. The process devolves to a direct, linear search over a single variable, β. Note that it is not necessary to solve (3.37) every time that β is modified; β simply multiplies $\Delta\mathbf{V}$ and $\mathbf{F}(\mathbf{V})$ is recalculated. Therefore, the process is computationally much less expensive than doing a full Newton step.

3.3.9 Optimizing Convergence and Efficiency

Even under the best circumstances, convergence problems are sometimes encountered in the algorithm, especially in circuits that are complex, strongly nonlinear, or strongly excited. Various methods have been developed to improve convergence and to make the process more efficient.

The effectiveness of Newton's method comes from its use of all the derivatives of the error function with respect to each frequency component at each port. In principle, this allows it to select a $\Delta\mathbf{V}$ vector that decreases *all* the components of the error function at any step. As a result, Newton's method is capable of achieving convergence with a very large number of variables, as long as the nonlinearity is not too strong. The disadvantage of this algorithm is in the large amount of computer memory and computation time required to generate the Jacobian and to solve the matrix equation (3.37). The Jacobian is a square matrix of dimension $2N(K + 1)$; in our earlier example of a FET circuit having three nonlinear elements and eight harmonics plus dc, the Jacobian is 54×54. Because the Jacobian is complex, solving (3.37) for this simple case involves solving a 54×54 set of real linear equations. It is not unusual for a single RF IC to have several hundred transistors and, with multitone excitation, tens or hundreds of frequency components. Analyzing such a circuit is a computationally expensive proposition.

In most cases, the entire matrix, the solution vector, and the update vector must remain in memory simultaneously; thus, Newton's method requires a large amount of computer memory. (Many of the matrix entries are zero in large problems, so the use of sparse-matrix techniques can ameliorate this situation somewhat, as can the use of Krylov methods.) Finally, generating the Jacobian requires taking a large number of derivatives. Many solid-state device models are very complex, and expressions for the derivatives require several times the computation of the static I/V and Q/V functions. Evaluating such functions may be a significant part of the time required for the entire analysis.

3.3.9.1 Pruning the Matrix: Removal of Small Values from the Jacobian

Because the method scales poorly with matrix size, direct factorization of the Jacobian is rarely used. Instead, some type of sparse-matrix method, whether conventional or iterative, is preferred. Such methods are most efficient when the matrix is very sparse; they can be worse than direct methods if the matrix's sparsity is inadequate.

Often, many of the elements of the Jacobian are very small and have little effect on the Newton update, so it makes sense to increase the sparsity simply by eliminating all elements whose magnitudes are below some threshold. These are invariably the elements farthest from the diagonals of the Jacobian's blocks. Eliminating them converts each block into a banded matrix.

The matrix can often be pruned rather severely without affecting convergence. As we shall see, a certain degree of inaccuracy in the Jacobian is often tolerable. A Newton step is, after all, only an approximation of the zero, and is expected only to decrease the error function. The Jacobian need only be accurate enough so that the update vector, $\Delta\mathbf{V}$, decreases the error function.

3.3.9.2 Samanskii Iteration

It is an empirical observation that, after the first few iterations, and especially close to convergence, the Jacobian does not change much between iterations. We can exploit this fact by simply reusing the Jacobian for several consecutive Newton iterations. This practice works well; the key is to have an intelligent method for deciding when the process has become so inefficient that it is best to reformulate the Jacobian. If a Jacobian is used too long, the improvement in $\mathbf{F}(\mathbf{V})$ becomes gradually smaller; if it is reformulated too often, efficiency suffers. Generally, the Jacobian is reformulated when the norm of $\mathbf{F}(\mathbf{V})$ fails to be reduced by some predetermined amount.

Samanskii iteration should be used with care in a norm-reduction process involving the NU norm (Section 3.3.9.4). If the Jacobian is inaccurate, the accuracy of the norm suffers accordingly, and it can become difficult to tell whether a norm-reduction step results in an improvement in the error.

3.3.9.3 Continuation Methods

Continuation methods circumvent convergence problems at the cost of increased computation time. In a continuation method, some parameter of the circuit is varied gradually, so convergence can be achieved at each step. At the first step, the parameter is adjusted to make the circuit nearly linear, and convergence is achieved easily. The solution of that step is used as the initial estimate for the next step, the parameter is adjusted to make the circuit somewhat more strongly nonlinear, and the process is repeated. The process continues in this manner until convergence is achieved with the full value of the circuit parameter.

The most commonly used continuation method is called *source stepping*. In that process, the magnitude of an RF source (or occasionally one or more dc sources) is the continuation parameter. In the continuation process, the excitation is varied stepwise from a low level to the desired excitation level, in such a way that convergence is achieved at each step.

The amount of increase, per step, depends on the strength of the circuit's nonlinearity.

Continuation works best when (1) it is adaptive (i.e., if convergence fails, the step size is reduced and the process repeated), and (2) an estimate of the new solution, rather than simply the solution of the previous step, is used as the step's starting value.

The new solution is estimated as follows. From (3.25) we have

$$\mathbf{F}(\mathbf{V}) = \mathbf{I}_s + \mathbf{Y}_{N \times N}\mathbf{V} + j\Omega\mathbf{Q} + \mathbf{I}_G = \mathbf{0} \tag{3.84}$$

When a step has converged,

$$\mathbf{I}_s = -(\mathbf{Y}_{N \times N}\mathbf{V} + j\Omega\mathbf{Q} + \mathbf{I}_G) \tag{3.85}$$

Differentiating, we obtain

$$\frac{\partial \mathbf{I}_s}{\partial \mathbf{V}} = -\frac{\partial}{\partial \mathbf{V}}(\mathbf{Y}_{N \times N}\mathbf{V} + j\Omega\mathbf{Q} + \mathbf{I}_G) = -\mathbf{J}_\mathbf{F} \tag{3.86}$$

or

$$\frac{\partial \mathbf{V}}{\partial \mathbf{I}_s} = -\mathbf{J}_\mathbf{F}^{-1} \tag{3.87}$$

Thus, the inverted Jacobian can be used at the end of each continuation step to estimate the port voltages at the next step.

3.3.9.4 Yeager and Dutton's NU Norm

In a linear or nearly linear circuit, each Newton step should reduce all components of $\mathbf{F}(\mathbf{V})$. In reality, however, some components of $\mathbf{F}(\mathbf{V})$ decrease, while others may change very little or even increase. How do we determine whether a Newton step is a good one? Such a determination is essential when norm reduction methods are used.

One possible method is to calculate the Euclidean norm of $\mathbf{F}(\mathbf{V})$, designated $|\mathbf{F}(\mathbf{V})|$. It happens, however, that $|\mathbf{F}(\mathbf{V})|$ is a poor choice, because the direction of the Newton step, in multidimensional space, does not necessarily minimize $|\mathbf{F}(\mathbf{V})|$. The step direction that minimizes $|\mathbf{F}(\mathbf{V})|$ is its gradient; however, our Newton step is the gradient of $\mathbf{F}(\mathbf{V})$, not of $|\mathbf{F}(\mathbf{V})|$.

In a classic paper, Yeager and Dutton [3.12] showed that a good Newton step does not, in general, coincide with the gradient of $|\mathbf{F}(\mathbf{V})|$, and can even be perpendicular[1] to it. They propose, instead, a new norm, called the *NU* (*Newton Update*) *norm*, that is weighted by the Jacobian and therefore has a gradient that coincides with the Newton step. The use of this norm improves the performance of a harmonic-balance simulator significantly.

The NU norm, at iteration p, is defined as

$$N_{NU} = |\mathbf{J}^{-1}(\mathbf{V}^p)\mathbf{F}(\mathbf{V})| \tag{3.88}$$

where $|*|$ indicates the L_2 (Euclidean) norm, $\mathbf{V}^p$ is the voltage vector at the pth iteration, and $\mathbf{V} = \mathbf{V}^p - \beta\Delta\mathbf{V}$; that is, $\mathbf{V}$ is evaluated at the norm-reduction step. The steepest descent in this norm always coincides with the direction of the step $\Delta\mathbf{V}$.

3.3.9.5 Parametric Models

If the multidimensional error surface can be "flattened" (i.e., the nonlinearity reduced), the convergence characteristics of harmonic-balance analysis can be improved. Rizzoli [3.13] has shown that this can be done by making both the current and voltage functions of an abstract parameter; if that parameter is x, we form $i = f_i(x)$ and $v = f_v(x)$. As an example, consider a diode having the I/V relation,

$$I = I_{sat}(\exp(\delta V) - 1) \tag{3.89}$$

The I/V equation can be written in the parametric form

1. In $2N(K+1)$-dimensional space. Don't try to visualize this.

$$v(t) = \begin{cases} V_1 + \frac{1}{\delta}\ln(1 + \delta(x(t) - V_1)) & x(t) > V_1 \\ x(t) & x(t) \le V_1 \end{cases} \tag{3.90}$$

$$i(t) = \begin{cases} I_s \exp(\delta V_1)(1 + \delta(x(t) - V_1)) - I_s & x(t) > V_1 \\ I_s(\exp(\delta x) - 1) & x(t) \le V_1 \end{cases}$$

V_1 can be selected arbitrarily, but it is best if it is a small number, typically $1 / \delta$. When $x(t) < V_1$, $x(t) = v(t)$ and (3.90) is identical to (3.89). At higher voltages, however, $i(t)$ becomes a linear function of $x(t)$, and the diode voltage, $v(t)$, becomes a logarithmic function of $x(t)$. This preserves the exponential relationship of (3.89), while dividing the strong nonlinearity of (3.89) between $v(t)$ and $i(t)$, making it effectively much weaker.

In this process, we have replaced one dependent variable, $i(t)$, with two, $v(t)$ and $i(t)$, both of which are functions of $x(t)$, the independent variable. This makes the problem larger, although the improvement in convergence should compensate for the increase in problem size. The greatest limitation of this method is that existing, industry-standard models are not formulated in this manner, and in many cases it is difficult to translate models into this form. The simulator then must be formulated to work optimally with both parametric and nonparametric models, an additional complication.

3.3.9.6 Nodal Formulation and Ill Conditioning in the Y matrix

In Sections 3.3.1 and 3.3.3, we assumed that the Y matrix and its inverse both exist. In many cases, however, partitioning the circuit results in a disconnected subcircuit, which has no admittance or impedance matrix. If nothing is done about this situation, the analysis clearly must fail.

One simple solution is to replace each nonlinear branch with a moderate-value resistor; a resistance of 100Ω usually works well. These resistors prevent the Y matrix from being disconnected, so a singular Y matrix is much less likely to occur. The resistors can be removed from the N-port Y matrix by simply subtracting their conductances from the main diagonal.

Another solution is to use a nodal formulation instead of the port formulation that we have described in this chapter [3.14]; in this case, the node voltages, not the branch voltages of the nonlinear elements, become the independent quantities. A nodal formulation may be more tolerant of circuit disconnections, although isolated nodes still result in a singular matrix.

In a nodal formulation, each node voltage in the circuit becomes an independent variable. The number of variables remains $2N(K + 1)$, but N is the number of nodes, not the number of control voltages. In microwave and RF circuits, the number of nodes is likely to be greater than the number of nonlinear elements, so the nodal formulation increases the size of the Jacobian. In analog ICs, however, the number of nonlinearities may be on the same order as the number of nodes, so the disadvantage may be minor or even nonexistent. When a nodal formulation is used in the analysis of microwave or RF ICs, large parts of the linear subcircuit can often be reduced to smaller nodal blocks, reducing the size of the problem.

3.3.9.7 Termination Criteria

In Newton-based harmonic-balance analysis, we decrease the error vector $\mathbf{F}(\mathbf{V})$ until the errors are negligible. Defining, precisely, what we mean by *negligible* is something of a dilemma. The problems arise from the fact that the current components at various ports and harmonics may be vastly different in magnitude; a factor of 10^6 or even 10^8 difference between the largest and smallest components is not unusual.

To illustrate the difficulties, we examine a few possibilities:

Limit the Euclidean Norm

One possibility is to require that the Euclidean norm of the error function be less than some maximum value. Mathematically, we require that

$$|\mathbf{F}(\mathbf{V})| < \varepsilon \tag{3.91}$$

where ε is a scalar value. In this case, the larger current components, which normally have the larger errors, dominate in establishing $|\mathbf{F}(\mathbf{V})|$; the smaller errors contribute little. Thus, even when $|\mathbf{F}(\mathbf{V})|$ is small, the errors in smaller current components may be quite large. The errors in the smaller currents could be controlled by requiring that $|\mathbf{F}(\mathbf{V})|$ be smaller, but this may put unrealistic demands on the errors in large components. Then, the maximum allowable errors for large current components may be so small that convergence is impossible.

Limit the Absolute Magnitudes of the Current Components

Another possibility is to require that the absolute magnitudes of all components of **F**(**V**) be less than some maximum value; specifically,

$$F_{n,\,k} < \varepsilon \qquad \text{for all } n,\, k \tag{3.92}$$

As with the previous criterion, a value of ε that is adequate to guarantee the accuracy of small current components may be too stringent for large ones.

Limit the Relative Magnitudes of the Current Components

A possible solution is to limit the relative magnitudes of the current-error components instead of the absolute ones. Specifically, we require that

$$2\left|\frac{(I_{n,\,k} + \hat{I}_{n,\,k})}{I_{n,\,k} - \hat{I}_{n,\,k}}\right| < \varepsilon \qquad \text{for all } n,\, k \tag{3.93}$$

That is, we compare the current error, $|I_{n,\,k} + \hat{I}_{n,\,k}|$, to the absolute current, defined as the average of the current in the linear and nonlinear subcircuits, $|I_{n,\,k} - \hat{I}_{n,\,k}|/2$. Although an improvement over the previous criteria, this criterion has the opposite problem: a reasonable value of ε for large current components makes unreasonably severe demands on the convergence of small components. For example, in a power amplifier, we might well want the fundamental-frequency error to be less than 1%, but a 1% error is too severe for weak intermodulation components, whose accuracy is on the order of a few decibels at best.

Another problem is that (3.93) is meaningful only near convergence. When the iterative process is far from convergence, $I_{n,\,k} + \hat{I}_{n,\,k} \sim I_{n,\,k} - \hat{I}_{n,\,k}$ so the relative error sits stoically at a value of 2. This does not affect the Newton iterations, but it gives the user no information about the progress of the convergence, so he cannot tell if the problem is proceeding normally toward convergence.

Combined Relative and Absolute Criteria

A final solution is to combine relative and absolute criteria. In this scheme, each current-error component, $|I_{n,\,k} + \hat{I}_{n,\,k}|$, is observed. If it satisfies either the relative or absolute error criterion, the component is considered converged. The analysis terminates when all components are converged in this sense. Of course, the entire error vector must be examined on each convergence test.

This scheme naturally accommodates both large and small current components. Large components usually converge on the basis of the relative error, and small ones on the fractional error, as the relative criterion is a weaker one for large components and the absolute criterion is weaker for small components. It is a simple matter to select the criteria so they ensure that all errors are sufficiently small at termination.

3.3.9.8 Initial Estimate

One important property of Newton's method is that its speed and reliability of convergence depend strongly upon the initial estimate of the solution vector. Formulating the initial estimate may not be difficult in analyzing a specific type of circuit, but it may be difficult to conceive of a way to form initial estimates in a general-purpose circuit-analysis program, which must accommodate a wide variety of circuits that have a concomitant variety of possible responses.

For nearly linear circuits, such as class-A power amplifiers, the linear response is a good initial estimate. The response can be found by setting the excitation level to a small value and the harmonic number, K, equal to one, so the size of the problem is relatively small. When the solution has completed, the results are scaled to the correct excitation level and K is reset to the desired value for the large-signal analysis.

In strongly nonlinear circuits, such as class-B or -C amplifiers, frequency multipliers, and mixers, an initial estimate is more difficult to generate. Occasionally the nature of the circuit allows a good estimate; for example, in diode mixers, the diode-voltage waveform invariably is a clipped sinusoid. In difficult cases, it may be best first to do a dc analysis, then to apply the RF signal and increase it using a continuation method.

3.4 LARGE-SIGNAL/SMALL-SIGNAL ANALYSIS USING CONVERSION MATRICES

Large-signal/small-signal analysis, or *conversion matrix analysis*, is useful for a large class of problems wherein a nonlinear device is driven, or "pumped," by a single large sinusoidal signal; another signal, much smaller, is applied; and we seek only the linear response to the small signal. The most common application of this technique is in the design of mixers and in nonlinear noise analysis. The process involves first analyzing the nonlinear device under large-signal excitation only, usually by the harmonic-balance method. The nonlinear elements in the device's equivalent circuit are then linearized to create small-signal, linear, time-varying elements, and finally a small-signal analysis is performed. The method is much more efficient than multitone harmonic-balance analysis but provides only the linear response of the circuit. It cannot be used for determining saturation or intermodulation distortion in mixers, but it is a good method for calculating a mixer's conversion efficiency and its RF and IF port impedances. The results of the harmonic-balance analysis can be used for finding LO voltage and current waveforms, and LO port impedance.

3.4.1 Conversion Matrix Formulation

Figure 3.9 shows a nonlinear resistive element driven by a large-signal voltage, V, generating a current I. The nonlinear element has the I/V relationship $I = f(V)$. Following the process outlined in Chapter 2, we can find the incremental small-signal current by assuming that V consists of the sum of a large-signal component V_0 and a small-signal component v. The current resulting from this excitation can be found by expanding $f(V_0 + v)$ in a Taylor series,

$$f(V_0 + v) = f(V_0) + \left.\frac{d}{dV}f(V)\right|_{V=V_0} v + \left.\frac{1}{2}\frac{d^2}{dV^2}f(V)\right|_{V=V_0} v^2 + \left.\frac{1}{6}\frac{d^3}{dV^3}f(V)\right|_{V=V_0} v^3 + \dots \tag{3.94}$$

The small-signal, incremental current is found by subtracting the large-signal component of the current,

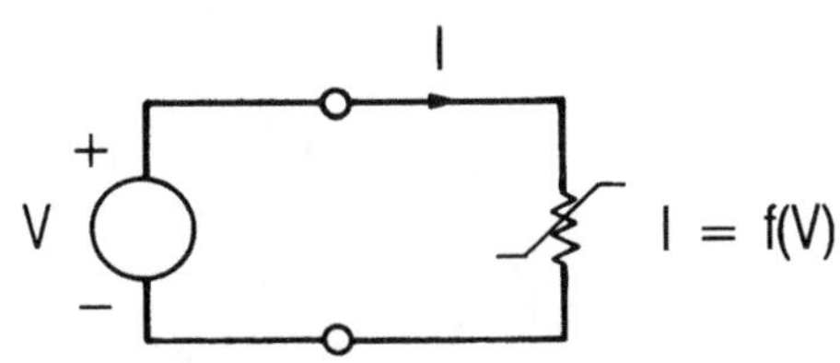

Figure 3.9 Nonlinear resistive element driven by a large excitation.

$$i(v) = I(V_0 + v) - I(V_0) \tag{3.95}$$

If $v << V_0$, v^2, v^3, ... are negligible (and, in any event, are nonlinear, so they do not contribute to the linear response). Then,

$$i(v) = \left.\frac{d}{dV}f(V)\right|_{V = V_0} v \tag{3.96}$$

V_0 need not be a dc quantity; it can be a time-varying large-signal voltage $V_L(t)$ (in fact, V_0 and V_L are control voltages). We assume that this is the case, and also that $v = v(t)$, a function of time. Then

$$i(t) = \left.\frac{d}{dV}f(V)\right|_{V = V_L(t)} v(t) \tag{3.97}$$

Equation (3.97) can be expressed as

$$i(t) = g(t)v(t) \tag{3.98}$$

The time-varying conductance in (3.98), $g(t)$, is the derivative of the element's I/V characteristic at the large-signal voltage. This is the usual definition of small-signal conductance for static elements. By an analogous derivation, one could have a current-controlled resistor with the V/I characteristic

$$V = f_R(I) \tag{3.99}$$

and obtain the small-signal v/i relation

$$v(t) = r(t)i(t) \tag{3.100}$$

where

$$r(t) = \left.\frac{d}{dI}f_R(I)\right|_{I = I_L(t)} \tag{3.101}$$

Often, the nonlinear element is a function of more than one control voltage. A conductance controlled by two voltages has $I = f_2(V_1, V_2)$. $f_2(V_1, V_2)$ can be expanded in a two-dimensional Taylor series, and after subtracting the large-signal current component and retaining only the linear terms,

$$i(t) = g_1(t)v_1(t) + g_2(t)v_2(t) \tag{3.102}$$

where

$$\begin{aligned} g_1(t) &= \left.\frac{\partial}{\partial V_1} f_2(V_1, V_2)\right|_{\substack{V_1 = V_{L,1}(t) \\ V_2 = V_{L,2}(t)}} \\ g_2(t) &= \left.\frac{\partial}{\partial V_2} f_2(V_1, V_2)\right|_{\substack{V_1 = V_{L,1}(t) \\ V_2 = V_{L,2}(t)}} \end{aligned} \tag{3.103}$$

Equation (3.102) shows that a nonlinear conductance having two control voltages is equivalent to two conductances in parallel. One must be a controlled current source, and the other may be either a controlled source or a time-varying two-terminal conductance. When the I/V characteristic is a function of more than two voltages, (3.102) can be extended in the manner one would expect:

$$i(t) = g_1(t)v_1(t) + g_2(t)v_2(t) + g_3(t)v_3(t) + \ldots \tag{3.104}$$

It is unusual, however, to encounter a nonlinear element having more than two control voltages.

The same process can be followed with a capacitor. A nonlinear capacitor has the Q/V characteristic $Q = f_Q(V)$, and by a similar derivation, the incremental, small-signal charge is

$$q(t) = \left.\frac{d}{dV} f_Q(V)\right|_{V = V_L(t)} v(t) \tag{3.105}$$

or

$$q(t) = c(t)v(t) \tag{3.106}$$

The capacitor's current is the time derivative of the charge:

$$i(t) = \frac{d}{dt}q(t) = c(t)\frac{d}{dt}v(t) + v(t)\frac{d}{dt}c(t) \tag{3.107}$$

Like a conductance, a capacitance can have multiple control voltages. In a manner analogous to (3.102) to (3.104), the small-signal charge is

$$q(t) = c_1(t)v_1(t) + c_2(t)v_2(t) + c_3(t)v_3(t) + \dots \tag{3.108}$$

and the current is found by differentiating with respect to time:

$$\begin{aligned} i(t) = \frac{d}{dt}q(t) &= c_1(t)\frac{d}{dt}v_1(t) + v_1(t)\frac{d}{dt}c_1(t) \\ &+ c_2(t)\frac{d}{dt}v_2(t) + v_2(t)\frac{d}{dt}c_2(t) + \dots \end{aligned} \tag{3.109}$$

A nonlinear element excited by two tones supports currents and voltages at the mixing frequencies $m\,\omega_1 + n\,\omega_2$, where m and n are integers. If we assume that one of those tones, ω_1, has such a low level that it does not generate harmonics, and the other is a large-signal sinusoid at ω_p, the mixing frequencies are $\omega = \pm\omega_1 + n\,\omega_p$. This equation represents the set of frequency components shown in Figure 3.10, which consists of two tones on either side of each large-signal harmonic frequency, separated by $\omega_0 = |\omega_1 - \omega_p|$. A more compact representation of the mixing frequencies is

$$\omega_n = \omega_0 + n\omega_p \tag{3.110}$$

which is shown in Figure 3.11 and includes only half of the mixing frequencies: the negative components of the lower sidebands and the positive components of the upper sidebands. This set of frequencies is adequate for two reasons: first, the small-signal analysis is linear, so by the superposition principle, the results for positive and negative components can be separated; and second, positive- and negative-frequency components are complex conjugate pairs, so knowledge of only one is

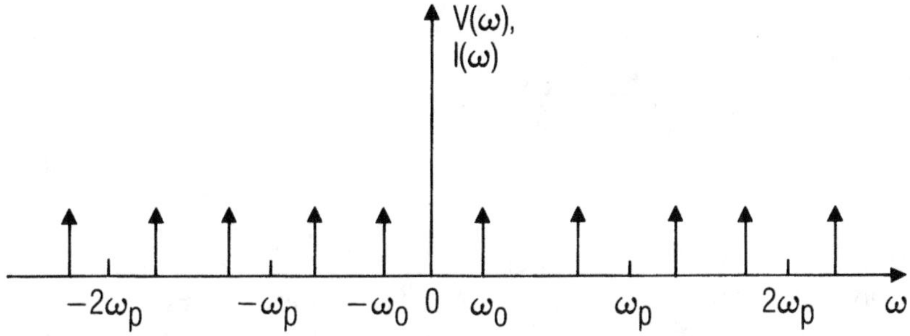

Figure 3.10 Spectrum of small-signal mixing frequencies in the pumped nonlinear element.

necessary. We will carry only the components in (3.110) in the following analysis, with confidence that the others can be generated when necessary.

The frequency-domain currents and voltages in a time-varying circuit element are related by a *conversion matrix*. We begin by deriving the conversion matrix that represents a time-varying conductance. The small-signal voltage and current can be expressed in the frequency notation of (3.110) as

$$v'(t) = \sum_{n=-\infty}^{\infty} V_n \exp(j\omega_n t) \tag{3.111}$$

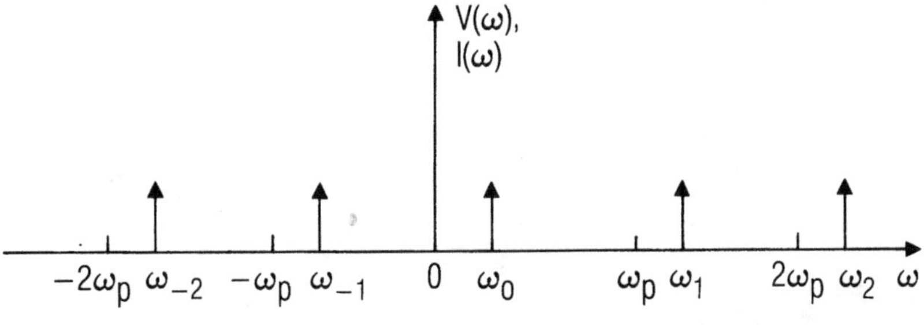

Figure 3.11 Spectrum of small-signal mixing frequencies illustrating the frequency notation of (3.110).

and

$$i'(t) = \sum_{n=-\infty}^{\infty} I_n \exp(j\omega_n t) \qquad (3.112)$$

where the primes indicate that $v'(t)$ and $i'(t)$ are sums of the positive- and negative-frequency phasor components in (3.110) and are not the complete time waveforms. Above all, (3.111) and (3.112) are not Fourier series, in spite of their superficial resemblance. The conductance waveform $g(t)$ can be expressed by its Fourier series,

$$g(t) = \sum_{n=-\infty}^{\infty} G_n \exp(jn\omega_p t) \qquad (3.113)$$

and the voltage and current are related by Ohm's law,

$$i'(t) = g(t)v'(t) \qquad (3.114)$$

Substituting (3.111) through (3.113) into (3.114) gives the relation,

$$\sum_{k=-\infty}^{\infty} I_k \exp(j\omega_k t) = \sum_{n=-\infty}^{\infty} \sum_{m=-\infty}^{\infty} G_n V_m \exp(j\omega_{m+n} t) \qquad (3.115)$$

Equating terms on both sides of the equation in (3.115) results in a set of equations that can be expressed in matrix form:

$$
\begin{bmatrix} I^*_{-N} \\ I^*_{-N+1} \\ I^*_{-N+2} \\ \cdots \\ \cdots \\ I^*_{-1} \\ I_0 \\ I_1 \\ \cdots \\ \cdots \\ I_N \end{bmatrix} = \begin{bmatrix} G_0 & G_{-1} & G_{-2} & \cdots & G_{-2N} \\ G_1 & G_0 & G_{-1} & \cdots & G_{-2N+1} \\ G_2 & G_1 & G_0 & \cdots & G_{-2N+2} \\ \cdots & \cdots & \cdots & \cdots & \cdots \\ \cdots & \cdots & \cdots & \cdots & \cdots \\ G_{N-1} & G_{N-2} & G_{N-3} & \cdots & G_{-N-1} \\ G_N & G_{N-1} & G_{N-2} & \cdots & G_{-N} \\ G_{N+1} & G_N & G_{N-1} & \cdots & G_{-N+1} \\ \cdots & \cdots & \cdots & \cdots & \\ \cdots & \cdots & \cdots & \cdots & \\ G_{2N} & G_{2N-1} & G_{2N-2} & \cdots & G_0 \end{bmatrix} \begin{bmatrix} V^*_{-N} \\ V^*_{-N+1} \\ V^*_{-N+2} \\ \cdots \\ \cdots \\ V^*_{-1} \\ V_0 \\ V_1 \\ \cdots \\ \cdots \\ V_N \end{bmatrix} \tag{3.116}
$$

Two details in (3.116) must be clarified. First, the vectors in (3.116) have been truncated to a limit of $n = N$ for I_n and V_n, and $n = 2N$ for G_n. We assume that V_n, I_n, and G_n are negligible beyond these limits. The second detail is that the negative-frequency components (V_n, I_n where $n < 0$) are shown as conjugate. The conjugates are caused by a change of definition; according to (3.110), ω_n is negative when $n < 0$, so the I_n and V_n are negative-frequency components when $n < 0$. We would rather define them as phasors, which are always positive-frequency components. Positive- and negative-frequency components are related as $V_{-n} = V_n^*$ and $I_{-n} = I_n^*$, so if we wish V_n, I_n to represent positive-frequency components, they must be V_n^*, I_n^*. Thus the conversion matrix relates ordinary phasor voltages to currents at each mixing frequency. The main advantage of making this change is that the conversion matrix is now completely compatible with conventional linear, sinusoidal steady-state analysis.

The dual case, a time-varying resistor, has an unsurprising result. The conversion matrix is

$$\begin{bmatrix} V^*_{-N} \\ V^*_{-N+1} \\ V^*_{-N+2} \\ \cdots \\ \cdots \\ V^*_{-1} \\ V_0 \\ V_1 \\ \cdots \\ \cdots \\ V_N \end{bmatrix} = \begin{bmatrix} R_0 & R_{-1} & R_{-2} & \cdots & R_{-2N} \\ R_1 & R_0 & R_{-1} & \cdots & R_{-2N+1} \\ R_2 & R_1 & R_0 & \cdots & R_{-2N+2} \\ \cdots & \cdots & \cdots & \cdots & \cdots \\ \cdots & \cdots & \cdots & \cdots & \cdots \\ R_{N-1} & R_{N-2} & R_{N-3} & \cdots & R_{-N-1} \\ R_N & R_{N-1} & R_{N-2} & \cdots & R_{-N} \\ R_{N+1} & R_N & R_{N-1} & \cdots & R_{-N+1} \\ \cdots & \cdots & \cdots & \cdots & \\ \cdots & \cdots & \cdots & \cdots & \\ R_{2N} & R_{2N-1} & R_{2N-2} & \cdots & R_0 \end{bmatrix} \begin{bmatrix} I^*_{-N} \\ I^*_{-N+1} \\ I^*_{-N+2} \\ \cdots \\ \cdots \\ I^*_{-1} \\ I_0 \\ I_1 \\ \cdots \\ \cdots \\ I_N \end{bmatrix} \tag{3.117}$$

where the R_n are the Fourier components of the resistance waveform. As one might expect, the resistance-form conversion matrix of any element is the inverse of its conductance-form matrix, as long as the element can be defined either as a time-varying conductance or resistance.

The conversion matrix of a capacitor is only slightly more complicated. The capacitor's charge is given by

$$q'(t) = c(t)v'(t) \tag{3.118}$$

and $c(t)$ has the Fourier series

$$c(t) = \sum_{n=-\infty}^{\infty} C_n \exp(jn\omega_p t) \tag{3.119}$$

The current is

$$i'(t) = \frac{d}{dt}q'(t) \tag{3.120}$$

and $q'(t)$ has the form

$$q'(t) = \sum_{n=-\infty}^{\infty} Q_n \exp(j\omega_n t) \tag{3.121}$$

Substituting (3.111), (3.119), and (3.121) into (3.118) gives

$$\sum_{k=-\infty}^{\infty} Q_k \exp(j\omega_k t) = \sum_{n=-\infty}^{\infty} \sum_{m=-\infty}^{\infty} C_n V_m \exp(j\omega_{m+n} t) \tag{3.122}$$

The current can be found by differentiating. In the frequency domain, differentiation corresponds to multiplying by $j\omega$, so

$$\sum_{k=-\infty}^{\infty} I_k \exp(j\omega_k t) = \sum_{n=-\infty}^{\infty} \sum_{m=-\infty}^{\infty} j\omega_{m+n} C_n V_m \exp(j\omega_{m+n} t) \tag{3.123}$$

Equating terms at the same frequency gives the matrix equation

$$\mathbf{I} = j\Omega\mathbf{C}\mathbf{V} \tag{3.124}$$

where **I** and **V** represent the frequency-component current and voltage vectors and C represents the conversion matrix for the capacitance. **I** and **V** are identical to the vectors in (3.116) and (3.117), and **C** has the same form as the conductance and resistance matrices in those equations. The matrix Ω is a diagonal matrix; its elements are $j\omega_{-N}$ to $j\omega_N$:

$$\Omega = \begin{bmatrix} j\omega_{-N} & 0 & \dots & 0 \\ 0 & j\omega_{-N+1} & \dots & 0 \\ \dots & \dots & \dots & \dots \\ 0 & 0 & \dots & j\omega_N \end{bmatrix} \tag{3.125}$$

3.4.1.1 Example: Conversion Matrix of a Time-Varying Element

We form the conversion matrix of the circuit shown in Figure 3.12(a). It consists of a conductance in series with a switch; the switch is opened and closed with a duty cycle of 0.5, so the combination has the waveform shown in Figure 3.12(b). Its Fourier series, when $t_0 = 0.5T$, is

$$\begin{aligned} g(t) = G_p[0.5 &+ 0.318\exp(j\omega_p t) + 0.318\exp(-j\omega_p t) \\ &- 0.106\exp(j3\omega_p t) - 0.106\exp(-j3\omega_p t) \\ &+ 0.064\exp(j5\omega_p t) + 0.064\exp(-j5\omega_p t) + \ldots \end{aligned} \tag{3.126}$$

The conversion matrix when $2N = 6$ is

$$\mathbf{G} = G_p \begin{bmatrix} 0.5 & 0.318 & 0 & -0.106 & 0 & 0.064 & 0 \\ 0.318 & 0.5 & 0.318 & 0 & -0.106 & 0 & 0.064 \\ 0 & 0.318 & 0.5 & 0.318 & 0 & -0.106 & 0 \\ -0.106 & 0 & 0.318 & 0.5 & 0.318 & 0 & -0.106 \\ 0 & -0.106 & 0 & 0.318 & 0.5 & 0.318 & 0 \\ 0.064 & 0 & -0.106 & 0 & 0.318 & 0.5 & 0.318 \\ 0 & 0.064 & 0 & -0.106 & 0 & 0.318 & 0.5 \end{bmatrix} \tag{3.127}$$

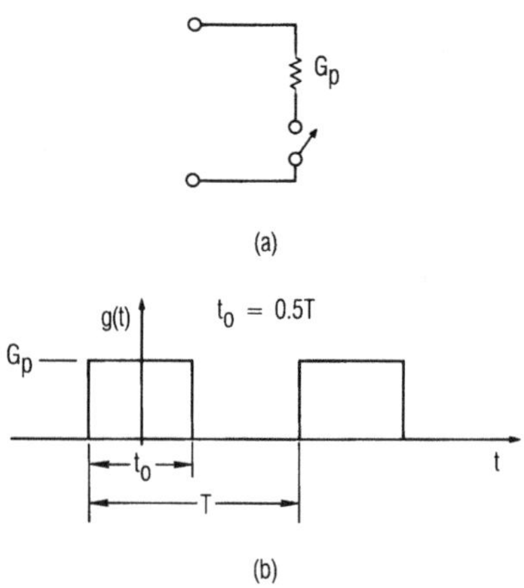

Figure 3.12 (a) Time-varying conductance; (b) conductance waveform, $g(t)$.

which relates the mixing products up to ω_3, those close to the third harmonic of the large-signal excitation.

3.4.2 Applying Conversion Matrices to Time-Varying Circuits

In order to mix ordinary, constant-value, and time-varying components in the same equations, the constant-value elements must have a conversion matrix form. This form is a diagonal matrix, and the element value must occupy all the locations on the main diagonal. The conversion matrix of a frequency-sensitive time-invariant element, such as a fixed impedance or admittance, is also a diagonal; however, the matrix elements are the impedance or admittance at the frequency corresponding to the location in the matrix. For example, the impedance-form conversion matrix of a static, lumped impedance is

$$\mathbf{Z} = \begin{bmatrix} Z^*(-\omega_{-N}) & 0 & \cdots & 0 & 0 & 0 & \cdots & 0 \\ 0 & Z^*(-\omega_{-N+1}) & \cdots & 0 & 0 & 0 & \cdots & 0 \\ \cdots & \cdots & \cdots & \cdots & \cdots & \cdots & \cdots & \cdots \\ 0 & 0 & \cdots & Z^*(-\omega_{-1}) & 0 & 0 & \cdots & 0 \\ 0 & 0 & \cdots & 0 & Z^*(\omega_0) & 0 & \cdots & 0 \\ 0 & 0 & \cdots & 0 & 0 & Z(-\omega_1) & \cdots & 0 \\ \cdots & \cdots & \cdots & \cdots & \cdots & \cdots & \cdots & \cdots \\ 0 & 0 & \cdots & 0 & 0 & 0 & \cdots & Z(\omega_N) \end{bmatrix} \tag{3.128}$$

When $n < 0$ ω_n is negative, so the impedance or admittance in the ω_n position is $V_n^*/I_n^* = Z^*(-\omega_n)$; thus the entry must be the conjugate of the positive-frequency impedance or admittance at that frequency.

Equations (3.116) and (3.117) can be expressed, like (3.124), as

$$\mathbf{I} = \mathbf{GV} \tag{3.129}$$

$$\mathbf{V} = \mathbf{RI} \tag{3.130}$$

These relations have the same form as those that define the *I*/*V* relations of linear, time-invariant resistance, conductance, and capacitance in the sinusoidal steady state. The only difference is that these are matrix equations, and the latter are scalar. The individual current and voltage components in the **V** and **I** vectors must satisfy Kirchoff's current and voltage laws in any linear circuit using time-varying elements, just as in time-invariant circuits. Therefore, the matrix equations can be used in exactly the same way as the scalar ones, as long as the requirements of matrix arithmetic are met: the order of multiplication must be preserved, and one must invert and multiply instead of dividing.

This realization allows all the tools of conventional sinusoidal, steady-state analysis to be applied to time-varying circuits. For example, the conversion matrix for two elements in parallel is the sum of their individual admittance-form matrices, and for two elements in series, it is the sum of their impedance-form matrices. One can also generate transfer functions and input/output impedances or admittances in terms of conversion matrices.

A second property of the conversion matrices is that they can be treated in all ways like multiport admittance or impedance matrices; the "ports" in the conversion matrix are currents and voltages at different frequencies, not physically separate ports. In theory, one could separate the frequency components by filters and create a physically separate port for each, without changing any of the circuit's properties. Indeed, in designing components that include time-varying elements, such as mixers, one tries to separate at least a few of the frequency components in this manner, in order to realize input and output ports, and to terminate other mixing products optimally. This property allows multiport-circuit concepts to be employed in interconnecting time-varying circuits, interfacing them with matching networks, and determining their gain, impedances, and stability. One can even convert the admittance- or impedance-form conversion matrix to an S-parameter form. These points are illustrated by the following examples.

3.4.2.1 Example: Conversion Matrix of a Simple Circuit

We derive the conversion matrix that represents the circuit shown in Figure 3.13. This circuit consists of a time-varying conductance and capacitance in parallel and a resistor in series (this is a common model of a pumped mixer diode). We assume that a large-signal analysis has been performed, and that the time waveforms and conversion matrices of each circuit element have been determined. $\mathbf{C}_j$ and $\mathbf{G}_j$ are the conversion matrices representing $c_j(t)$ and $g_j(t)$, respectively.

Because the capacitor and conductance are in parallel, the conversion matrix is the sum of the admittance-form conversion matrices of each component:

$$\mathbf{Y}_j = \mathbf{G}_j + j\Omega\mathbf{C}_j \tag{3.131}$$

and their impedance-form conversion matrix is the inverse:

$$\mathbf{Z}_j = \mathbf{Y}_j^{-1} = (\mathbf{G}_j + j\Omega\mathbf{C}_j)^{-1} \tag{3.132}$$

The conversion matrix for the resistor is $\mathbf{R1}$, where $\mathbf{1}$ is the $2N + 1 \times N + 1$ identity matrix. $\mathbf{R1}$ is in series with $\mathbf{Z}_j$, so the impedance-form conversion matrix of the entire circuit is the sum of $\mathbf{R1}$ and $\mathbf{Z}_j$:

$$\mathbf{Z}_c = \mathbf{R1} + \mathbf{Z}_j = \mathbf{R1} + (\mathbf{G}_j + j\Omega\mathbf{C}_j)^{-1} \tag{3.133}$$

The admittance-form matrix, if needed, is just the inverse of the impedance-form matrix.

3.4.2.2 Example: Two-Port Conversion Matrix

We calculate the conversion matrix that represents the simplified FET equivalent circuit shown in Figure 3.14(a), which could represent a FET mixer. It has two nonlinear circuit elements, $I_d(V_g, V_d)$ and $C_g(V_g)$, and all the remaining elements are linear. The circuit is treated as a two-port, so a two-port admittance-form matrix is needed. It has the form

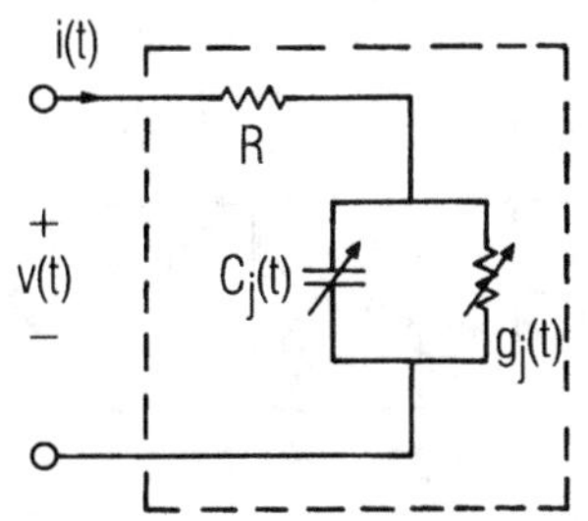

Figure 3.13 Pumped diode equivalent circuit of the example.

$$\begin{bmatrix} \mathbf{I}_1 \\ \mathbf{I}_2 \end{bmatrix} = \begin{bmatrix} \mathbf{Y}_{1,1} & \mathbf{Y}_{1,2} \\ \mathbf{Y}_{2,1} & \mathbf{Y}_{2,2} \end{bmatrix} \begin{bmatrix} \mathbf{V}_1 \\ \mathbf{V}_2 \end{bmatrix} \tag{3.134}$$

where $\mathbf{I}_1$, $\mathbf{I}_2$, $\mathbf{V}_1$, and $\mathbf{V}_2$ are current and voltage vectors as shown in (3.116) and (3.117), and the $\mathbf{Y}_{m,n}$ submatrices are each complete conversion matrices. Thus, (3.134) relates not only the currents and voltages at the mixing frequencies and at each port, but also includes transfer terms between ports.

Again, we assume that a large-signal analysis has been performed and that the nonlinear elements have been converted to their incremental, time-varying forms. The drain current source can be split into two elements, $g_m(t)$ and $g_d(t)$, according to (3.102) and (3.103); the former element is a controlled source, representing the time-varying transconductance, and the latter is a time-varying drain-to-source conductance. The resulting circuit is shown in Figure 3.14(b).

The submatrices are defined in a manner entirely analogous to static admittance matrices:

$$\mathbf{I}_1 = \mathbf{Y}_{1,1}\mathbf{V}_1 \qquad \mathbf{V}_2 = \mathbf{0} \tag{3.135}$$

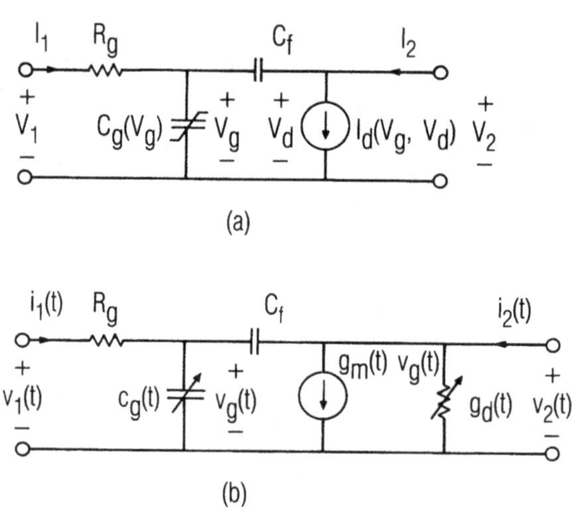

Figure 3.14 (a) FET nonlinear equivalent circuit for the example; (b) time-varying linear equivalent circuit.

and so on, where **0** is the zero vector. The time-varying quantities $c_g(t)$, $g_m(t)$, and $g_d(t)$ have conversion matrices designated $\mathbf{C}_g$, $\mathbf{G}_m$, and $\mathbf{G}_d$, respectively.

We begin by finding $\mathbf{Y}_{1,1}$. When port 2 is shorted at all harmonics, $c_g(t)$ and C_f are in parallel, so we can immediately write

$$\mathbf{I}_1 = \{[j\Omega(\mathbf{C}_g + \mathbf{C}_f\mathbf{1})]^{-1} + \mathbf{R}_g\mathbf{1}\}^{-1}\mathbf{V}_1 \tag{3.136}$$

and $\mathbf{Y}_{1,1}$ is found by comparing (3.136) to (3.135). $\mathbf{Y}_{2,1}$ is just a little more trouble. When the output is shorted,

$$\mathbf{V}_g = [j\Omega(\mathbf{C}_g + \mathbf{C}_f\mathbf{1})]^{-1}\mathbf{I}_1 \tag{3.137}$$

and

$$\mathbf{I}_2 = (\mathbf{G}_m - j\Omega\mathbf{C}_f\mathbf{1})\mathbf{V}_g \tag{3.138}$$

Substituting (3.136) into (3.137), and the result into (3.138), we finally obtain

$$\begin{aligned}\mathbf{I}_2 = {} & (\mathbf{G}_m - j\Omega\mathbf{C}_f\mathbf{1})[j\Omega(\mathbf{C}_g + \mathbf{C}_f\mathbf{1})]^{-1} \\ & \cdot \{[j\Omega(\mathbf{C}_g + \mathbf{C}_f\mathbf{1})]^{-1} + \mathbf{R}_g\mathbf{1}\}^{-1}\mathbf{V}_1\end{aligned} \tag{3.139}$$

and $\mathbf{Y}_{2,1}$ is found by inspection. $\mathbf{Y}_{2,2}$ and $\mathbf{Y}_{1,2}$ are a little sticky algebraically but straightforward conceptually. From similar manipulations, we obtain

$$\mathbf{I}_2 = \left\{\mathbf{G}_d + \mathbf{Y}_f\left[\mathbf{1} + \mathbf{G}_m\left(\frac{1}{R_g}\mathbf{1} + j\Omega\mathbf{C}_g\right)^{-1}\right]\right\}\mathbf{V}_2 \tag{3.140}$$

and

$$\mathbf{I}_1 = -(\mathbf{1} + jR_g\Omega\mathbf{C}_g)^{-1}\mathbf{Y}_f\mathbf{V}_2 \tag{3.141}$$

from which $\mathbf{Y}_{2,2}$ and $\mathbf{Y}_{1,2}$ are easily identifiable. $\mathbf{Y}_f$ is defined as

$$\mathbf{I}_f = \mathbf{Y}_f\mathbf{V}_2 \tag{3.142}$$

where $\mathbf{I}_f$ is the current in C_f. $\mathbf{Y}_f$ is

$$\mathbf{Y}_f = \left[\left(\frac{1}{R_g}\mathbf{1} + j\Omega\mathbf{C}_g\right)^{-1} - \frac{j}{C_f}\Omega^{-1}\right]^{-1} \tag{3.143}$$

3.4.2.3 Example: Two-Port Formulation

We now calculate the input and output impedances, simultaneous conjugate-match impedances, transducer conversion gain, and maximum available conversion gain of the circuit of the previous example, at a specific pair of input and output frequencies. Figure 3.15 shows the circuit to be analyzed, where the two-port is described by the conversion matrix **Y**, derived in the previous example. The source and load impedances, generally functions of ω, are shown in series with the two-port; shorting either set of terminals loads the input or output port with the appropriate impedance.

We wish to calculate this circuit's gain and impedances at specific input and output frequencies. This means that, with the exception of the input port at the input frequency and the output port at the output frequency, we wish to terminate the ports in their source and load impedances at all mixing frequencies. The source and load impedances at the unwanted mixing frequencies are then absorbed into the network, and we are left with a conventional two-port, describable by a simple 2 × 2 Y matrix. The only feature that would distinguish this matrix from the Y matrix of a time-invariant network is that it represents input and output phasors at different frequencies, and if one of those frequencies is a lower sideband, its voltage and current are conjugate quantities.

We begin by putting the source and load impedances into a compatible two-port conversion matrix representation. This is

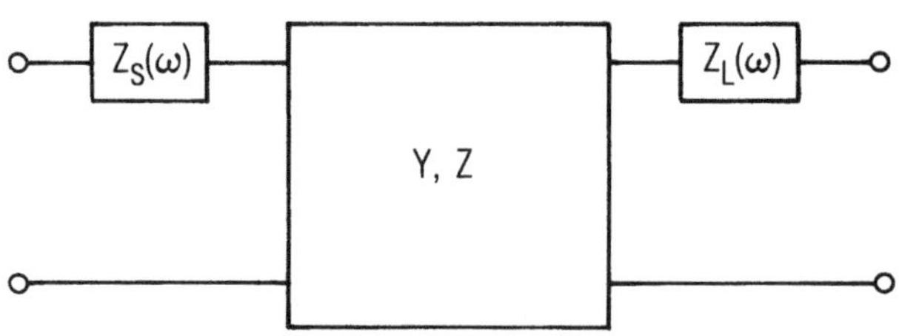

Figure 3.15 Circuit of the example. The block *Y, Z* is the circuit in Figure 3.14.

$$\mathbf{Z}_t = \begin{bmatrix} \mathbf{Z}_s & \mathbf{0} \\ \mathbf{0} & \mathbf{Z}_L \end{bmatrix} \tag{3.144}$$

where $\mathbf{Z}_s$ and $\mathbf{Z}_L$ are diagonal matrices of the form shown in (3.128). Following the notation for mixing products (3.110), we let ω_q be the input frequency and ω_r be the output frequency. The source and load impedances at these frequencies, $Z_s(\omega_q)$ and $Z_L(\omega_r)$, respectively, are set to zero in (3.144), because we want them to remain external to the circuit; the impedances at other frequencies are retained and are absorbed into the circuit. Following the rule for conventional two-ports, the impedance-form conversion matrix of the terminated network, $\mathbf{Z}_a$, is

$$\mathbf{Z}_a = \mathbf{Z}_t + \mathbf{Y}^{-1} \tag{3.145}$$

The admittance-form matrix for the combination of the FET and the source and load impedances is

$$\mathbf{Y}_a = \mathbf{Z}_a^{-1} \tag{3.146}$$

At this point, $\mathbf{Y}_a$ still relates two voltage vectors to two current vectors and has the form

$$\begin{bmatrix} \mathbf{I}_1 \\ \mathbf{I}_2 \end{bmatrix} = \begin{bmatrix} \mathbf{Y}_{a;1,1} & \mathbf{Y}_{a;1,2} \\ \mathbf{Y}_{a;2,1} & \mathbf{Y}_{a;2,2} \end{bmatrix} \begin{bmatrix} \mathbf{V}_1 \\ \mathbf{V}_2 \end{bmatrix} \tag{3.147}$$

We now reduce (3.147) to a simple 2×2 admittance matrix by terminating the ports at all unwanted mixing frequencies. To terminate the output port at all frequencies other than ω_r we set $\mathbf{V}_2$ to zero by shorting the output terminals at those frequencies; similarly, $\mathbf{V}_1$ is zeroed at all frequencies other than ω_q. Setting these voltage components to zero multiplies all the corresponding columns in $\mathbf{Y}_a$ by zero; therefore, those columns can be eliminated. Furthermore, because the input and output are shorted, the current components at those frequencies are not of interest, so the corresponding rows in $\mathbf{Y}_a$ and $\mathbf{I}$ can also be removed. The only terms left in $\mathbf{Y}_a$ can be put into the 2×2 matrix form,

$$\begin{bmatrix} I_1(\omega_q) \\ I_2(\omega_r) \end{bmatrix} = \begin{bmatrix} y_{1,1} & y_{1,2} \\ y_{2,1} & y_{2,2} \end{bmatrix} \begin{bmatrix} V_1(\omega_q) \\ V_2(\omega_r) \end{bmatrix} \tag{3.148}$$

where

$$\begin{aligned} y_{1,1} &= \mathbf{Y}_{a;1,1}(\omega_q, \omega_q) \\ y_{1,2} &= \mathbf{Y}_{a;1,2}(\omega_q, \omega_r) \\ y_{2,1} &= \mathbf{Y}_{a;2,1}(\omega_r, \omega_q) \\ y_{2,2} &= \mathbf{Y}_{a;2,2}(\omega_r, \omega_r) \end{aligned} \tag{3.149}$$

The rest is all downhill. Equation (3.148) can now be used with the usual assortment of Y-matrix relations. For example, if the load admittance is $Y_L(\omega_r)$, the input admittance has the familiar relation,

$$Y_{\text{in}}(\omega_q) = y_{1,1} - \frac{y_{1,2}\, y_{2,1}}{Y_L(\omega_r) + y_{2,2}} \tag{3.150}$$

and with a source admittance $Y_s(\omega_q)$, the output admittance is

$$Y_{\text{out}}(\omega_r) = y_{2,2} - \frac{y_{2,1}\, y_{1,2}}{Y_s(\omega_q) + y_{1,1}} \tag{3.151}$$

Note that if $r < 0$, the output admittance is conjugate; if $q < 0$, the input admittance is conjugate. In these cases, the conjugate of the load or source admittance must also be used in (3.150) and (3.151), respectively. The equation for transducer conversion gain, in terms of Y parameters, is

$$G_t = \frac{4\mathrm{Re}\{Y_s(\omega_q)\}\mathrm{Re}\{Y_L(\omega_r)\}|y_{2,1}|^2}{|[y_{1,1} + Y_s(\omega_q)][y_{2,2} + Y_L(\omega_r)] - y_{1,2}\, y_{2,1}|^2} \tag{3.152}$$

The Linvill stability factor, c, is

$$c = \frac{|y_{1,2}\, y_{2,1}|}{2\mathrm{Re}\{y_{1,1}\}\mathrm{Re}\{y_{2,2}\} - \mathrm{Re}\{y_{2,1}\, y_{1,2}\}} \tag{3.153}$$

If $c < 1$, the circuit is unconditionally stable, and no passive source impedance at ω_q or load at ω_r can cause oscillation. If $c < 1$, the maximum available conversion gain (MAG) and simultaneous conjugate match impedances $Y_{s,\,\mathrm{opt}}(\omega_q)$, $Y_{L,\,\mathrm{opt}}(\omega_r)$ are defined. They are

$$MAG = \frac{|y_{2,\,1}|^2}{2\mathrm{Re}\{y_{1,\,1}\}\mathrm{Re}\{y_{2,\,2}\} - \mathrm{Re}\{y_{2,\,1}y_{1,\,2}\} + T_y} \quad (3.154)$$

and

$$\mathrm{Im}\{Y_{s,\,\mathrm{opt}}(\omega_q)\} = -\mathrm{Im}\{y_{1,\,1}\} + \frac{\mathrm{Im}\{y_{2,\,1}y_{1,2}\}}{2\mathrm{Re}\{y_{2,\,2}\}} \quad (3.155)$$

$$\mathrm{Re}\{Y_{s,\,\mathrm{opt}}(\omega_q)\} = \frac{T_y}{2\mathrm{Re}\{y_{2,\,2}\}} \quad (3.156)$$

where

$$T_y = [(2\mathrm{Re}\{y_{1,\,1}\}\mathrm{Re}\{y_{2,\,2}\} - \mathrm{Re}\{y_{2,1}y_{1,\,2}\})^2 - |y_{1,\,2}y_{2,\,1}|^2]^{1/2} \quad (3.157)$$

The load impedance $Y_{L,\,\mathrm{opt}}(\omega_r)$ can be found from (3.155) and (3.156) by interchanging $y_{1,1}$ and $y_{2,2}$, and $y_{2,1}$ and $y_{1,2}$.

As is the case in a time-invariant circuit, unconditional stability at the excitation frequency and large-signal excitation level is not adequate to guarantee that the time-varying circuit is stable in a practical sense; for the circuit to be stable in practice, it must be unconditionally stable at all possible input frequencies and large-signal excitation levels. Varying the small-signal excitation frequency for which the Y parameters in (3.148) are determined also varies the higher-order mixing frequencies, and hence the embedding impedances at those frequencies. Stability, therefore, is a function of everything that affects the Y parameters, literally all the characteristics of the circuit and its large-signal excitation.

It is important to recognize that small-signal and large-signal stability are interrelated. To explain why this is so, we must note that a fundamental assumption in the conversion matrix theory is that small-signal voltages are small variations (in frequency as well as in magnitude and phase) in the large-signal voltage. The conversion matrix is in fact nothing more than the large-signal Jacobian, a matrix that relates the current and voltage deviations, evaluated at the mixing frequencies instead of the large-signal harmonics. Small-signal oscillation is a process where these variations build up spontaneously and without bound and eventually become indistinguishable from the large-signal voltage. If they occur at a different frequency from the large signal, they may appear as modulation, "snap" phenomena, parasitic oscillation, or other well-known manifestations of instability in nonlinear circuits.

The two-port conversion matrix of (3.148) is in admittance form only because an admittance-form conversion matrix is usually most convenient. It need not be expressed in this form, however; in fact, it can be converted to any two-port matrix form desired, such as an S matrix or even a T matrix (transfer-scattering matrix). The procedure for converting the Y matrix to one of these forms is precisely the same as for any other scalar matrix. For example, the S matrix is found from the Y matrix as

$$\mathbf{S} = (\mathbf{1} + \mathbf{Y}_{\text{norm}})^{-1}(\mathbf{1} + \mathbf{Y}_{\text{norm}}) \tag{3.158}$$

where $\mathbf{Y}_{\text{norm}}$ is the Y matrix (3.148) normalized to the S parameters' reference admittance. The interpretation of lower-sideband quantities ($q, r < 0$) in the S matrix may be a little confusing. For example, if $q = -1$ and $r = 0$, a common situation, the S matrix has the form

$$\begin{bmatrix} b_1^*(\omega_{-1}) \\ b_2(\omega_0) \end{bmatrix} = \begin{bmatrix} s_{1,1} & s_{1,2} \\ s_{2,1} & s_{2,2} \end{bmatrix} \begin{bmatrix} a_1^*(\omega_{-1}) \\ a_2(\omega_0) \end{bmatrix} \tag{3.159}$$

where $s_{1,1}$ is the conjugate of the input reflection coefficient:

$$\Gamma_{\text{in}}^* = s_{1,1} = \left.\frac{b_1^*(\omega_{-1})}{a_1^*(\omega_{-1})}\right|_{a_2(\omega_0) = 0} \tag{3.160}$$

and $|s_{2,1}|^2$ is, as usual, the transducer gain

$$G_t = |s_{2,1}|^2 = \left|\frac{b_2(\omega_0)}{a_1^*(\omega_{-1})}\right|^2_{a_2(\omega_0) = 0} \tag{3.161}$$

The fact that a_1 is conjugate in (3.161) does not change the magnitude of $s_{2,1}$. Fortunately, the fact that the definitions of $s_{2,1}$ and $s_{1,2}$ include one conjugate and one nonconjugate quantity rarely is a problem; the properties that are usually of most interest—gain, impedances, and stability—are scalar.

When the conversion-matrix formulation is used in this manner, it has significant advantages over multitone harmonic-balance analysis. Such characteristics as simultaneous conjugate match impedances and maximum available gain can be calculated easily; these would be much more difficult to determine with harmonic-balance analysis. Even calculating a set of two-port S parameters would require two harmonic-balance analyses. When conversion-matrix analysis is used, S parameters can be calculated with only one single-tone harmonic-balance analysis; the subsequent conversion-matrix manipulations are computationally inexpensive, especially compared to two-tone harmonic balance. A disadvantage is the lack of nonlinear calculations, but these can be included as well, as we shall see in Section 3.5.

3.4.3 Nodal Formulation

In order to use conversion matrices in a general-purpose circuit analysis program, we need a general-purpose method for formulating the equations. In static linear analysis, we often formulate the equations as an indefinite admittance matrix, which we then reduce to a conventional, nodal admittance matrix. We can do the same thing with conversion matrices. We end up with a set of equations that looks like (3.134), and we use manipulations identical to those of Section 3.4.2.3 to obtain S parameters, port reflection coefficients, gain, or other characteristics of interest.

Consider a time-varying admittance element, whose conversion matrix is $\mathbf{Y}_c$, connected between nodes i and j, as shown in Figure 3.16. Let $\mathbf{I}_i$ and $\mathbf{I}_j$ be the vectors of current in the element connected between nodes i and j, respectively, and $\mathbf{V}_i$ and $\mathbf{V}_j$ be the voltages. These voltages and currents have the form of the voltage and current vectors in (3.116) and (3.117). The current in the branch is

$$\mathbf{I}_i = \mathbf{Y}_c(\mathbf{V}_i - \mathbf{V}_j) \tag{3.162}$$

and

$$\mathbf{I}_j = \mathbf{Y}_c(\mathbf{V}_j - \mathbf{V}_i) \tag{3.163}$$

These show that $\mathbf{Y}_c$ can be added into the nodal matrix as

$$\begin{bmatrix} \cdots \\ \mathbf{I}_i \\ \mathbf{I}_j \\ \cdots \\ \cdots \end{bmatrix} = \begin{bmatrix} \cdots & \cdots & \cdots & \cdots & \cdots \\ \cdots & \cdots + \mathbf{Y}_c & \cdots - \mathbf{Y}_c & \cdots & \cdots \\ \cdots & \cdots - \mathbf{Y}_c & \cdots + \mathbf{Y}_c & \cdots & \cdots \\ \cdots & \cdots & \cdots & \cdots & \cdots \\ \cdots & \cdots & \cdots & \cdots & \cdots \end{bmatrix} \begin{bmatrix} \cdots \\ \mathbf{V}_i \\ \mathbf{V}_j \\ \cdots \\ \cdots \end{bmatrix} \tag{3.164}$$

which is entirely analogous to static nodal analysis. The process of generating this matrix is entirely mindless, so it is perfect for implementation on a computer.

3.4.3.1 Example: Nodal Formulation

We create the nodal matrix of the circuit in Figure 3.14(b). First, we number the nodes, as shown in Figure 3.17. As in (3.164), we add the admittances to the matrix in the appropriate locations. The result is

$$\mathbf{Y} = \begin{bmatrix} R_g^{-1}\mathbf{1} & 0 & -R_g^{-1}\mathbf{1} & 0 \\ 0 & j\Omega\mathbf{C}_g + \mathbf{G}_d + \mathbf{G}_m & -j\Omega\mathbf{C}_g - \mathbf{G}_m & -\mathbf{G}_d \\ -R_g^{-1}\mathbf{1} & -j\Omega\mathbf{C}_g & R_g^{-1}\mathbf{1} + j\Omega(\mathbf{C}_g + \mathbf{C}_f) & -j\Omega\mathbf{C}_f \\ 0 & -\mathbf{G}_m - \mathbf{G}_d & -j\Omega\mathbf{C}_f + \mathbf{G}_m & j\Omega\mathbf{C}_f + \mathbf{G}_d \end{bmatrix} \tag{3.165}$$

Note that the transconductance element has the same general form as a simple admittance, but the four $\mathbf{G}_m$ terms are off the main diagonal. In

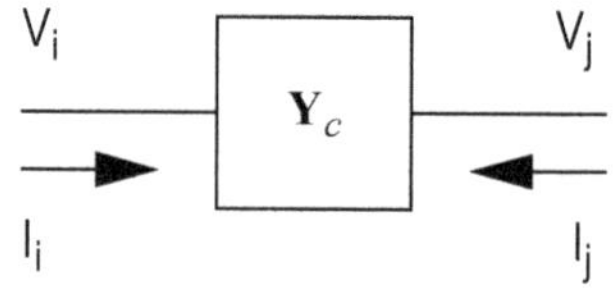

Figure 3.16 A time-varying admittance described by a conversion matrix, $\mathbf{Y}_c$.

general, however, the larger terms in the expressions, the dc Fourier components, are along the main diagonal in (3.165). This causes the conversion matrix to be diagonally dominant, so it is rarely ill conditioned. Ill conditioning can occur in devices that have large, nonlinear diffusion capacitances combined with a low minimum mixing frequency ω_0, and in negative-resistance components.

Reducing this matrix to the form of (3.148) requires no algebra, only numerical manipulations, so this method can be used to analyze very large circuits.

3.5 MULTITONE EXCITATION AND INTERMODULATION IN TIME-VARYING CIRCUITS

The small-signal analysis of the previous sections was based on the assumption that the excitation was vanishingly small. Accordingly, the nonlinear terms in the incremental Taylor series could be ignored, resulting in a linear, small-signal formulation. In this section, that assumption is discarded, and instead it is assumed only that the *incremental I/V* or *Q/V* characteristic is weakly nonlinear. This is not the same as assuming that the nonlinear device is weakly nonlinear; it means instead that the element is weakly nonlinear for small deviations from its instantaneous large-signal voltage. Virtually all nonlinear solid-state devices meet this condition, as long as they are not driven into saturation by the small-signal excitation.

The techniques in this section are most useful for determining intermodulation levels and spurious responses in heavily pumped circuits, such as mixers. The method is based on [3.15]; it also uses some concepts from the Volterra- and power-series theory in Chapter 4, and could be

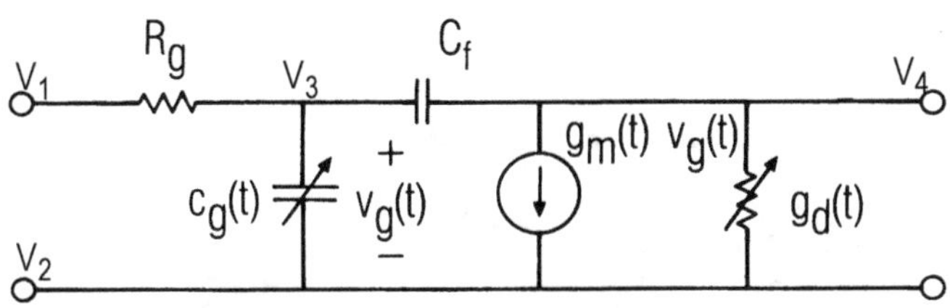

Figure 3.17 Simple FET equivalent circuit for the example. It is the same as Figure 3.14(b), except that the nodes are numbered for nodal analysis.

considered a time-varying application of the Volterra series. For these reasons the reader might do well to become conversant with Chapter 4 before continuing with this section.

In order to minimize unnecessary complications, the circuit model used in this section includes only a single set of terminals in the nonlinear subcircuit, with a resistive and capacitive nonlinearity. We do this for a couple of reasons: first, the analysis is complex, and simplifying the circuit provides lucidity. The results still include everything necessary to generalize the analysis to larger circuits. Second, the circuit itself is important: it describes the junction of a Schottky diode.

The linear part of our circuit can be described by a Norton equivalent, which consists of a single current source and embedding impedance. The model is shown in Figure 3.18; it is assumed in the figure that the large-signal nonlinear analysis has been performed, the large-signal voltages and currents have been recorded, and the currents and voltages shown are the small-signal, incremental ones. Except for the excitation source $i_s(t)$, these currents and voltages include intermodulation components as well as linear mixing products. The excitation $i_s(t)$ is a two-tone source having frequencies ω_1 and ω_2,

$$i_s(t) = I_{s1}\cos(m\omega_p + \omega_1) + I_{s2}\cos(m\omega_p + \omega_2) \tag{3.166}$$

where, as before, ω_p is the fundamental frequency of the large-signal excitation, and m is an integer; for the usual case of an upper-sideband input, $m = 1$. ω_1 and ω_2 can be upper- or lower-sideband components; for simplicity, we can assume them to be upper-sideband.

The spectrum of mixing frequencies is shown in Figure 3.19(a), and a detail of those closest to dc is shown in Figure 3.19(b). The spectrum shown in Figure 3.19(b) is mirrored on either side of each large-signal harmonic at positive and negative frequencies. ω_1 and ω_2 are the lowest-

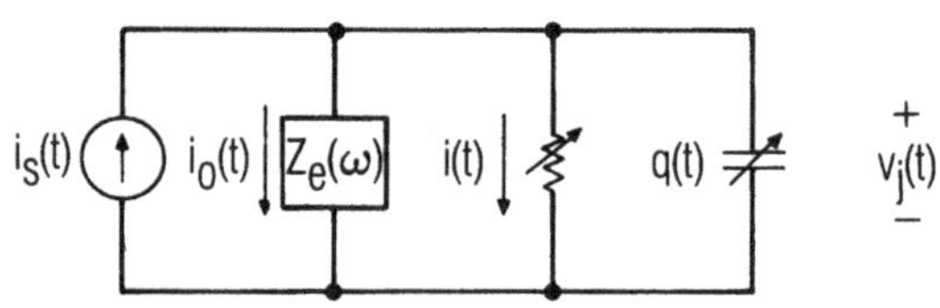

Figure 3.18 Small-signal, incremental, time-varying linear circuit derived from a pumped large-signal nonlinear circuit.

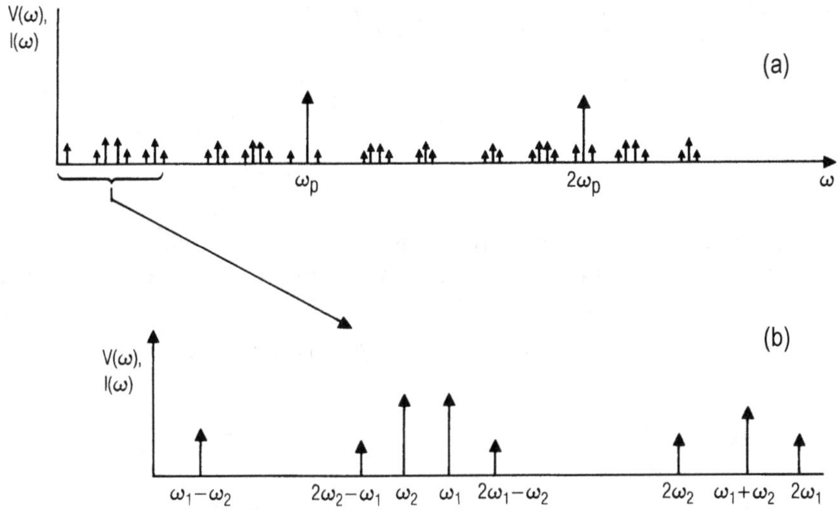

Figure 3.19 (a) Lowest-order mixing frequencies in the nonlinear time-varying circuit; (b) detail of the frequencies closest to dc.

frequency (usually IF) components of the excitation, and the rest are *intermodulation* (IM) products. The IM products shown in the figure are by far not the only ones possible; they are, instead, those of third or lower order that are closest to ω_1 and ω_2, and are consequently of greatest concern in practice.

Following the same process as in (3.94) through (3.98) and (3.105) through (3.109), but retaining the terms up to third degree, we have

$$i(t) = \left.\frac{d}{dV}f(V)\right|_{V = V_L(t)} v(t) + \left.\frac{1}{2}\frac{d^2}{dV^2}f(V)\right|_{V = V_L(t)} v^2(t) + \left.\frac{1}{6}\frac{d^3}{dV^3}f(V)\right|_{V = V_L(t)} v^3(t) + \ldots \tag{3.167}$$

and

$$q(t) = \left.\frac{d}{dV}f_Q(V)\right|_{V = V_L(t)} v(t) + \left.\frac{1}{2}\frac{d^2}{dV^2}f_Q(V)\right|_{V = V_L(t)} v^2(t) + \left.\frac{1}{6}\frac{d^3}{dV^3}f_Q(V)\right|_{V = V_L(t)} v^3(t) + \dots \tag{3.168}$$

where $f(V)$ and $f_Q(V)$, as before, are the large-signal I/V and Q/V characteristics, respectively. The above two equations can be expressed as

$$i(t) = g_1(t)v(t) + g_2(t)v^2(t) + g_3(t)v^3(t) + \dots \tag{3.169}$$

$$q(t) = c_1(t)v(t) + c_2(t)v^2(t) + c_3(t)v^3(t) + \dots \tag{3.170}$$

Limiting consideration to third-order components, we have

$$v(t) = v_1(t) + v_2(t) + v_3(t) \tag{3.171}$$

$$v^2(t) = v_1^2(t) + 2v_1(t)v_2(t) \tag{3.172}$$

$$v^3(t) = v_1^3(t) \tag{3.173}$$

where $v_n(t)$ is the nth-order voltage, the combination of all nth-order mixing products. Recall that an nth-order mixing product is any combination of n excitation frequencies, including both positive and negative frequencies. The square of the junction voltage obviously creates a second-order product from $v_1^2(t)$ and a third-order product by mixing the first-order $v_1(t)$ and second-order $v_2(t)$.

The differential equation describing Figure 3.18 is

$$\frac{dq}{dt} + i(t) + i_0(t) = i_s(t) \tag{3.174}$$

Substituting (3.169) through (3.173) into (3.174) and separating the equations into first, second, and third-order products, we have

$$\frac{d}{dt}[c_1(t)v_1(t)] + g_1(t)v_1(t) + i_{0,1}(t) = i_s(t) \tag{3.175}$$

$$\begin{aligned}\frac{d}{dt}[c_1(t)v_2(t) + c_2(t)v_1^2(t)] + g_1(t)v_2(t) \\ + g_2(t)v_1^2(t) + i_{0,2}(t) = 0\end{aligned} \tag{3.176}$$

and

$$\begin{aligned}\frac{d}{dt}[c_1(t)v_3(t) + 2c_2(t)v_1(t)v_2(t) + c_3v_1^3(t)] \\ + g_1(t)v_3(t) + 2g_2(t)v_1(t)v_2(t) \\ + g_3(t)v_1^3(t) + i_{0,3}(t) = 0\end{aligned} \tag{3.177}$$

where $i_{0,n}(t)$ is the nth-order current in $Z_e(\omega)$. These equations imply that a separate circuit can be generated for each mixing product; those circuits are shown in Figure 3.20. Figure 3.20(a), the linear, small-signal circuit, can be used to determine $v_1(t)$ by conversion-matrix techniques. The first-order voltage $v_1(t)$ is then used to find the excitation current in Figure 3.20(b), from which the second-order voltage $v_2(t)$ can be found. Note that the circuit in Figure 3.20(b) is *linear*; the only nonlinear process is in the formulation of the excitation current from $v_1(t)$. Therefore, once this current is determined, ordinary, linear conversion matrix analysis can be used to find the voltage across and current in $Z_e(\omega)$. Finally, $v_1(t)$ and $v_2(t)$ are used to find the third-order excitation current in Figure 3.20(c). In concept, these currents could be evaluated in the time domain or frequency domain; however, the rest of the circuit uses a frequency-domain characterization, so it is likely to be more convenient to express the source currents in the frequency domain as well. Furthermore, ω_1 and ω_2 are noncommensurate frequencies, so $v(t)$ is not periodic; this situation would introduce further difficulties into a time-domain analysis.

We now find frequency-domain expressions for both the junction voltages and the excitation currents. The voltage $v_1(t)$ is found from the small-signal linear analysis and has the form

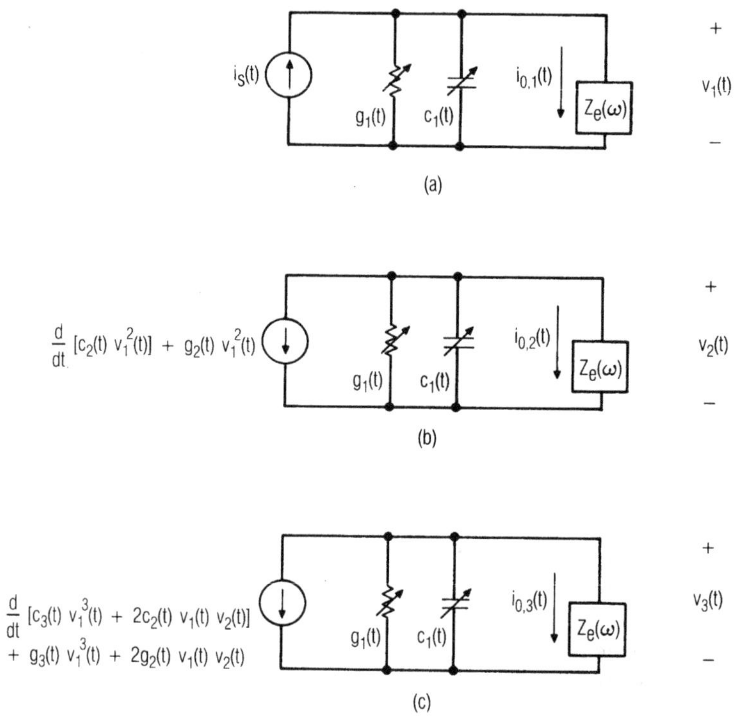

Figure 3.20 Linear circuits for determining the (a) first-order, (b) second-order, and (c) third-order IM components.

$$v_1(t) = \frac{1}{2} \sum_{m=-\infty}^{\infty} \sum_{\substack{q=-2 \\ q \neq 0}}^{2} V_{m,q} \exp[j(m\omega_p + \omega_q)t] \tag{3.178}$$

This expression is similar to the one used to express the mixing components in conversion-matrix analysis. Unlike that expression, however, it includes both upper- and lower-sideband frequency components. Because the conversion-matrix analysis was linear, we could ignore redundant frequency components; our present analysis is nonlinear, so we now must include components at all frequencies, both positive and negative.

$$v_1^2(t) = \frac{1}{4}\sum_{m=-\infty}^{\infty}\sum_{n=-\infty}^{\infty}\sum_{q=-2}^{2}\sum_{r=-2}^{2} V_{m,q}V_{n,r} \cdot \exp\{j[(m+n)\omega_p + \omega_q + \omega_r]t\} \quad (3.179)$$

In (3.179), $q, r \neq 0$; we shall assume this to be the case throughout the following analysis. The second-order terms of most interest are those at $k\omega_p + \omega_1 - \omega_2$ and $k\omega_p + 2\omega_1$. The components at $k\omega_p + \omega_1 + \omega_2$ and $k\omega_p + 2\omega_2$ can be found in a nearly identical manner, if they are of interest, so we will not consider them further. The terms of interest are designated by a and b subscripts, respectively:

$$v_{1a}^2(t) = \frac{1}{2}\sum_{m=-\infty}^{\infty}\sum_{n=-\infty}^{\infty} V_{m,1}V_{n,-2} \cdot \exp\{j[(m+n)\omega_p + \omega_1 - \omega_2]t\} \quad (3.180)$$

and

$$v_{1b}^2(t) = \frac{1}{4}\sum_{m=-\infty}^{\infty}\sum_{n=-\infty}^{\infty} V_{m,1}V_{n,1} \cdot \exp\{j[(m+n)\omega_p + 2\omega_1]t\} \quad (3.181)$$

The coefficient of (3.180) is 1/2 instead of 1/4 because there are two identical terms in the q, r summation in (3.179) at this frequency. Also, one should note that $v_{1a}^2(t)$ and $v_{1b}^2(t)$ are complex because they include only some of the terms in (3.179); thus, they do not represent real time functions.

The Taylor-series coefficients can be expressed by their Fourier series as

$$g_2(t) = \sum_{h=-\infty}^{\infty} G_{2,h}\exp(jh\omega_p t) \quad (3.182)$$

and

$$c_2(t) = \sum_{h=-\infty}^{\infty} C_{2,h} \exp(jh\omega_p t) \tag{3.183}$$

Substituting (3.180) through (3.183) into the parts of (3.176) that represent the source current gives the current-source components at these two frequencies, $i_{2a}(t)$ and $i_{2b}(t)$:

$$\begin{aligned} i_{2a}(t) = \frac{1}{2} \sum_{h=-\infty}^{\infty} \sum_{m=-\infty}^{\infty} \sum_{n=-\infty}^{\infty} & V_{m,1} V_{n,-2} \\ & \cdot \{G_{2,h} + C_{2,h} j[(h+m+n)\omega_p + \omega_1 - \omega_2]\} \\ & \cdot \exp\{j[(h+m+n)\omega_p + \omega_1 - \omega_2]t\} \end{aligned} \tag{3.184}$$

and

$$\begin{aligned} i_{2b}(t) = \frac{1}{4} \sum_{h=-\infty}^{\infty} \sum_{m=-\infty}^{\infty} \sum_{n=-\infty}^{\infty} & V_{m,1} V_{n,1} \\ & \cdot \{G_{2,h} + C_{2,h} j[(h+m+n)\omega_p + 2\omega_1]\} \\ & \cdot \exp\{j[(h+m+n)\omega_p + 2\omega_1]t\} \end{aligned} \tag{3.185}$$

$i_{2a}(t)$ and $i_{2b}(t)$ have the form

$$i_{2a}(t) = \frac{1}{2} \sum_{k=-\infty}^{\infty} I_{k,2a} \exp[j(k\omega_p + \omega_1 - \omega_2)t] \tag{3.186}$$

and

$$i_{2b}(t) = \frac{1}{2} \sum_{k=-\infty}^{\infty} I_{k,2b} \exp[j(k\omega_p + 2\omega_1)t] \tag{3.187}$$

Equating terms in (3.184) and (3.185) with those in (3.186) and (3.187), respectively, gives

$$I_{k,2a}(t) = \sum_{h=-\infty}^{\infty} \sum_{m=-\infty}^{\infty} \sum_{n=-\infty}^{\infty} V_{m,1} V_{n,-2} \cdot [G_{2,h} + C_{2,h} j(k\omega_p + \omega_1 - \omega_2)] \quad h + m + n = k \tag{3.188}$$

and

$$I_{k,2b}(t) = \frac{1}{2} \sum_{h=-\infty}^{\infty} \sum_{m=-\infty}^{\infty} \sum_{n=-\infty}^{\infty} V_{m,1} V_{n,1} \cdot [G_{2,h} + C_{2,h} j(k\omega_p + 2\omega_1)] \quad h + m + n = k \tag{3.189}$$

Limiting k to the range $(-K, ..., K)$ allows $I_{k,2a}$ and $I_{k,2b}$ to be expressed as column vectors:

$$\mathbf{I}_{2a} = \left[I^*_{-K,2a} \; I^*_{-K+1,2a} \; \cdots \; I^*_{-1,2a} \; I_{0,2a} \; I_{1,2a} \; \cdots \; I_{K,2a} \right]^T \tag{3.190}$$

and similarly for $\mathbf{I}_{2b}$. Finally, conversion-matrix analysis gives the vectors of second-order output currents:

$$\mathbf{I}_{0,2a} = -(1 + \mathbf{Y}_j \mathbf{Z}_{e,2a})^{-1} \mathbf{I}_{2a} \tag{3.191}$$

$$\mathbf{I}_{0,2b} = -(1 + \mathbf{Y}_j \mathbf{Z}_{e,2b})^{-1} \mathbf{I}_{2b} \tag{3.192}$$

where $\mathbf{Y}_j$ is the conversion matrix that represents the parallel combination of the time-varying conductance and capacitance, and $\mathbf{Z}_{e,2a}$ and $\mathbf{Z}_{e,2b}$ are the diagonal embedding impedance matrices (3.128) at their respective sets of mixing frequencies. The second-order voltages are

$$\mathbf{V}_{2a} = \mathbf{Z}_{e,2a} \mathbf{I}_{0,2a} \tag{3.193}$$

$$\mathbf{V}_{2b} = \mathbf{Z}_{e,\,2b}\mathbf{I}_{0,\,2b} \tag{3.194}$$

The third-order components are found analogously. The components of greatest interest are those at $2\omega_1 - \omega_2$ and $2\omega_2 - \omega_1$; both are derived identically, so only the former is considered here. The $v_1(t)v_2(t)$ terms in (3.177) have two components that generate $2\omega_1 - \omega_2$: $v_1(t)$ at ω_1 mixing with $v_{2a}(t)$ at $\omega_1 - \omega_2$, and $v_1(t)$ at $-\omega_2$ mixing with $v_{2b}(t)$ at $2\omega_1$. The components of $v_1^3(t)$ and $v_1(t)v_2(t)$ at this frequency are

$$v_1^3(t) = \frac{3}{8}\sum_{m=-\infty}^{\infty}\sum_{n=-\infty}^{\infty}\sum_{p=-\infty}^{\infty} V_{m,\,1}V_{n,\,1}V_{p,\,-2} \cdot \exp\{j[(m+n+p)\omega_p + 2\omega_1 - \omega_2]t\} \tag{3.195}$$

$$v_1(t)v_{2a}(t) = \frac{1}{4}\sum_{m=-\infty}^{\infty}\sum_{n=-\infty}^{\infty} V_{m,\,2a}V_{n,\,1} \cdot \exp\{j[(m+n)\omega_p + 2\omega_1 - \omega_2]t\} \tag{3.196}$$

and

$$v_1(t)v_{2b}(t) = \frac{1}{4}\sum_{m=-\infty}^{\infty}\sum_{n=-\infty}^{\infty} V_{m,\,2b}V_{n,\,-2} \cdot \exp\{j[(m+n)\omega_p + 2\omega_1 - \omega_2]t\} \tag{3.197}$$

The Fourier-series representations for the time-varying Taylor-series coefficients are

$$g_3(t) = \sum_{h=-\infty}^{\infty} G_{3,\,h}\exp(jh\omega_p t) \tag{3.198}$$

and

$$c_3(t) = \sum_{h=-\infty}^{\infty} C_{3,h}\exp(jh\omega_p t) \tag{3.199}$$

The resulting third-order components of the source current are

$$\begin{aligned} I_{k,3} = {} & \frac{3}{4}\sum_{h=-\infty}^{\infty}\sum_{m=-\infty}^{\infty}\sum_{n=-\infty}^{\infty}\sum_{\substack{p=-\infty \\ h+m+n+p=k}}^{\infty} V_{m,1}V_{n,1}V_{p,-2} \\ & \cdot[G_{3,h}+C_{3,h}j(k\omega_p+2\omega_1-\omega_2)] \\ & +\sum_{h=-\infty}^{\infty}\sum_{m=-\infty}^{\infty}\sum_{\substack{n=-\infty \\ h+m+n=k}}^{\infty}(V_{m,2a}V_{n,1}+V_{m,2b}V_{n,-2}) \\ & \cdot[G_{2,h}+C_{2,h}j(k\omega_p+2\omega_1-\omega_2)] \end{aligned} \tag{3.200}$$

The third-order current in $\mathbf{Z}_e(\omega)$ is

$$\mathbf{I}_{0,3} = -(1+\mathbf{Y}_j\mathbf{Z}_{e,3})^{-1}\mathbf{I}_3 \tag{3.201}$$

where $\mathbf{I}_{0,3}$ is the vector of output currents and $\mathbf{I}_3$ is the vector having the form of (3.190) whose components are $\mathbf{I}_{k,3}$ from (3.200). $\mathbf{Z}_{e,3}$ is the diagonal matrix of embedding impedances at the third-order mixing frequencies. Finally, the power of the third-order current component dissipated in the embedding network at the frequency $k\omega_p + 2\omega_1 - \omega_2$ is

$$P_{k,3} = 0.5|I_{0;k,3}|^2 Re\{Z_{e;k,3}\} \tag{3.202}$$

Equation (3.202) is the output power if the embedding network is lossless. If it is not (for example, if the diode series resistance has been included in it) it is necessary to subtract the real part of $\mathbf{Z}_{e;k,3}$ representing the loss from the impedance in (3.202).

3.6 MULTITONE HARMONIC-BALANCE ANALYSIS

We saw that harmonic-balance analysis was applicable to large-signal, single-tone problems, and that large-signal/small-signal analysis could be used to solve problems that involved multitone small-signal excitations and a single large-signal excitation. We shall see in the next chapter that power-series and Volterra-series techniques are very useful in analyzing weakly nonlinear circuits having multiple small-signal excitations at noncommensurate frequencies.

Although these cases cover a wide range of practical problems, there is still one remaining class of problems that has not been addressed: large-signal, excitation of strongly nonlinear circuits by several noncommensurate excitations. Examples of this type of problem are the calculation of intermodulation levels in power amplifiers and of large-signal intermodulation in mixers. This type of problem cannot be solved by large-signal/small-signal analysis or by Volterra-series techniques because both of these methods require that at least one signal, and sometimes all of them, be very weak. These problems can, however, be handled by a modified type of "harmonic" balance, which has been called, at various times, *generalized harmonic-balance analysis* or *spectral balance analysis*. Here we use the term, *multitone harmonic-balance analysis*. Finally, we examine *envelope analysis* (sometimes called *envelope transient analysis*), an alternate, approximate way to address certain kinds of multitone problems.

3.6.1 Generalizing the Harmonic-Balance Concept

The concept of harmonic-balance analysis is illustrated by Figure 3.3, which shows a nonlinear circuit partitioned into linear and nonlinear subcircuits. The voltages at the interconnections between the two subcircuits are variables which, when determined, define all the voltages and currents in the network. In the case of single-tone excitation, the voltages and currents are periodic, and thus have a fundamental-frequency component and a number of harmonics. There is nothing in that formulation, however, that requires the frequency components to be harmonically related. As we shall see, even the need for a Fourier transform does not limit the analysis to harmonic frequencies; we can easily generate a time-to-frequency transform (which, strictly, is not a Fourier transform, although, for lack of a better term, we will call it that) for noncommensurate frequencies.

We now consider the case where the excitation may have two or more noncommensurate frequencies, and the frequency components of the currents and voltages are no longer harmonically related. In general, the voltages and currents at each port in Figure 3.3 have a set of K frequency components:

$$\omega = \omega_k \qquad k = 0, 1, \ldots, K-1 \tag{3.203}$$

Usually $\omega_0 = 0$. These frequency components are mixing products, not harmonics; each mixing frequency ω_k arises as a linear combination of the excitation frequencies. In the case of a two-tone excitation,

$$\omega_k = m\omega_{p1} + n\omega_{p2} \tag{3.204}$$

where ω_{p1} and ω_{p2} are the frequencies of the two excitations, and each (m, n) pair maps into a unique k. All mixing frequencies up to some maximum value of m or n are included in the set of frequencies described by (3.204), although only positive ω_k need be included. (Negative-frequency components are included in the analysis implicitly when the Fourier transform converts the frequency-domain quantities to the time domain.) The number of frequency components retained in the set is subject to considerations similar to those that applied to single-tone harmonic balance. See Section 3.6.7 for more on this subject.

Of course, many problems require more than two noncommensurate excitation tones. We shall restrict the following discussion to the two-tone case; the extension to greater numbers of excitations is straightforward. It will also become apparent that the size of the harmonic-balance problem grows rapidly with the number of tones, and easily can become so large as to be impractical. This is a serious limitation of multitone harmonic-balance analysis.

The goal of the harmonic-balance analysis, as before, is to find a set of voltage components $V_{n,k}$ at the frequencies ω_k that satisfies (3.4). In this case, however, the components $I_{n,k}$ of the current vector and $Q_{n,k}$ of the charge vector represent the components at port n and at mixing frequency ω_k, where ω_k is not necessarily a harmonic of a single excitation frequency. The harmonic-balance equations are still valid in the multitone case; it is necessary only to replace the harmonics $k\omega_p$ with ω_k and to include all excitation tones in the excitation voltage vectors. Finally, the voltage, current, charge, and similar components are no longer harmonic components, but components at the frequency ω_k. They can no longer be determined by classical Fourier transform, but must be found by an alternative time-to-frequency transform. Finally, we must determine how to formulate the Jacobian for the nonharmonic case.

The development of multitone Fourier transformations has become a cottage industry for academics in the last decade. A large number of methods have been suggested. Instead of describing these in detail, we will focus on the nature of the problem and provide references for specific methods.

3.6.2 Reformulation and Fourier Transformation

One possibility is simply to find a common subharmonic for the two tones. In this case, the two-tone signal is periodic, and conventional Fourier transformation can be used. This is probably the most common and also the worst way to address the multitone problem; the reasons will be clear momentarily.

In order to use a Fourier transformation, the voltage and current waveforms must be periodic. This will be the case if and only if the excitation frequencies are commensurate; that is,

$$q\omega_{p1} = r\omega_{p2} \tag{3.205}$$

for some nonzero positive integers q and r. Then the waveforms have a period T, where

$$T = \frac{2\pi}{\omega_{p2} - \omega_{p1}} \tag{3.206}$$

In (3.206) we have assumed that $\omega_{p2} > \omega_{p1}$. In order to avoid aliasing errors, the waveforms must be sampled at a rate equal to twice the highest significant temporal (i.e., not radian) frequency; if that frequency is the Nth harmonic of the higher excitation frequency, $N\omega_{p2}/2\pi$, there must be $N\omega_{p2}/\pi$ samples per second. The number of samples S that must be made in each Fourier transformation is therefore the product of this quantity and T, or

$$S = \frac{2N\omega_{p2}}{\omega_{p2} - \omega_{p1}} \tag{3.207}$$

If ω_{p1} and ω_{p2} are closely spaced, S becomes a very large number. Furthermore, because the fast Fourier transform algorithm requires that S be a power of two, even the large number given by (3.207) must be increased to the next power of two. This large number of samples requires a comparably large—and often prohibitive—amount of computation time. It is especially frustrating to note that all but a few of the $S/2$ complex frequency components formed by the FFT are zero, and most of the computation time is expended in finding the magnitudes of these components, rather than the magnitudes of the K components of interest. Finally, a little consideration shows a fundamental flaw in this method: there is no reason why the frequency spacing, or the excitation frequency itself, should affect the size of the harmonic-balance problem.

The large amount of computation time is not the only problem that this method introduces. The large number of arithmetic operations necessary to form the Fourier transform reduces numerical precision, causing the result to be inaccurate. This is especially troublesome when the analysis includes both large and small frequency components, the usual situation in multitone analysis. Finally, because of the requirement that ω_{p1} and ω_{p2} be commensurate, it is not possible to use any frequencies of interest.

3.6.3 Discrete Fourier Transforms

One possible method for creating a multitone Fourier transform is to adapt a discrete Fourier transform (DFT). In fact, the fast Fourier transform (FFT) used commonly in harmonic-balance analysis is simply a method for performing a DFT while avoiding multiple, redundant arithmetic operations.

We wish to express the time waveform $x(t)$, which may represent either a voltage or a current, as

$$x(t) = \sum_{k=0}^{K-1} X_{c,k}\cos(\omega_k t) + X_{s,k}\sin(\omega_k t) \tag{3.208}$$

where ω_k are the set of mixing frequencies in the multitone problem. If the function $x(t)$ is sampled at the $S = 2K - 1$ time intervals $t_i = t_1, t_2, \ldots, t_{2K-1}$, the samples $x(t_i)$ can be expressed by a set of linear equations,

$$\begin{bmatrix} x(t_1) \\ x(t_2) \\ x(t_3) \\ \dots \\ x(t_S) \end{bmatrix} = \begin{bmatrix} 1 & \cos(\omega_1 t_1) & \sin(\omega_1 t_1) & \cos(\omega_2 t_1) & \sin(\omega_2 t_1) & \dots & \sin(\omega_{K-1} t_1) \\ 1 & \cos(\omega_1 t_2) & \sin(\omega_1 t_2) & \cos(\omega_2 t_2) & \sin(\omega_2 t_2) & \dots & \sin(\omega_{K-1} t_2) \\ 1 & \cos(\omega_1 t_3) & \sin(\omega_1 t_3) & \cos(\omega_2 t_3) & \sin(\omega_2 t_3) & \dots & \sin(\omega_{K-1} t_3) \\ \dots & \dots & \dots & \dots & \dots & \dots & \dots \\ 1 & \cos(\omega_1 t_S) & \sin(\omega_1 t_S) & \cos(\omega_2 t_S) & \sin(\omega_2 t_S) & \dots & \sin(\omega_{K-1} t_S) \end{bmatrix} \cdot \begin{bmatrix} X_0 \\ X_{c,1} \\ X_{s,1} \\ \dots \\ X_{s,K-1} \end{bmatrix} \tag{3.209}$$

or, in simpler notation,

$$\mathbf{x} = \Gamma^{-1}\mathbf{X} \tag{3.210}$$

Γ describes the transformation from the time to the frequency domain. In a classical DFT (3.209), the time samples are selected uniformly and the ω_k are harmonics. In the harmonic case, DFT or FFT generates little error in transforming between the time and frequency domains, because the rows of Γ^{-1} are orthogonal, and the matrix is well conditioned. If the frequencies are not harmonics, the rows are not orthogonal, and it is possible for some rows to be nearly linearly dependent; then the matrix is ill conditioned and large errors result. This is usually the case when two or more of the excitation frequencies, ω_{pn}, are closely spaced.

Because uniform time intervals often result in an ill-conditioned matrix when ω_k are not harmonics, nonuniformly spaced time samples provide better conditioning. In any case, the ω_k are fixed, so the choice of time samples is our only remaining degree of freedom in optimizing the DFT. But how do we select the sample points? Clearly, we select them to make the rows of Γ^{-1} orthogonal,[2] and developing methods for selecting a set of

2. Creating orthogonal rows guarantees orthogonal columns as well, so we do not need to consider orthogonality of the rows and columns separately.

sample points that results in orthogonal or nearly orthogonal rows is the key to developing our multitone transform.

Some methods for creating an optimum multitone DFT are the following:

- Almost-periodic Fourier transform (APFT) of Sorkin and Kundert [3.16];
- Two-dimensional FFT [3.17];
- Quasiorthogonal matrix method and filter-balance DFT [3.18];
- Time-mapped harmonic balance [3.19];
- APFT and mapping techniques of Rodrigues [3.20, 3.21];
- Artificial frequency mapping [3.22, 3.23];
- Determination of a low sampling frequency that prevents aliasing [3.24].

The simplest literature search undoubtedly will turn up many more such papers. We shall describe the APFT of Sorkin because it is an early, elegant method that is simple to understand and clearly illustrates the problems in selecting time points. (Unfortunately, its performance is not as good as later methods.) A second method we describe is the two-dimensional FFT, as it is an optimal method. Finally, we examine the use of artificial frequency mapping to provide our multitone transform.

3.6.4 Almost-Periodic Fourier Transform (APFT)

Although in the noncommensurate case the waveforms are not periodic, they are in some sense "almost" periodic, with a period given by (3.206). It is therefore possible to devise an "almost-periodic" transform that can be used to transform the waveforms between the time and frequency domains. the method we describe is one of the first for implementing such a transform [3.16].

One possibility for improving the conditioning of Γ^{-1} is to oversample; that is, to select more than $2K-1$ points. Selecting S according to (3.207) is an extreme example of this approach, but it happens that a set of $4K$ to $6K$ points, selected randomly over an interval T given by (3.206), usually provides a well-conditioned system. However, oversampling has the disadvantage that it increases the amount of computation time required to solve (3.209) and the equations describing the nonlinear subcircuit; it also introduces the minor problem of an overspecified system. We would therefore like to find a well-conditioned form of Γ^{-1} that does not require oversampling.

It is possible to create a well-conditioned system that is not oversampled by first choosing an excessive number of time points and reducing the number of points to the minimum. We begin by selecting approximately 1.5 times the minimum necessary sampling points, choosing the approximately $3K$ sample points randomly over an interval T given by (3.206). The resulting sine-cosine matrix is *tall*; it has more rows than columns. We then select $2K - 1$ rows of the matrix to form Γ^{-1} and note the corresponding time points; these are used as the sample points in Γ and $\mathbf{x}$. The rows we retain are rows of the matrix that, as closely as possible, form an orthogonal set of vectors.

The set of nearly orthogonal rows are chosen by a variation of the Gram-Schmidt orthogonalization procedure. Let γ_n represent the nth row of the matrix Γ^{-1}. We select one row arbitrarily, say γ_1, and remove the components in the direction of γ_1 of all the other vectors by forming

$$\gamma_n' = \gamma_n - \frac{\gamma_1^T\gamma_n}{\gamma_1^T\gamma_1}\gamma_1 \qquad n = 2, 3, \ldots, 2N \tag{3.211}$$

The set of vectors γ_n are all orthogonal to γ_1; because the vectors originally were the same length and had the same norm, the largest remaining vector (the one having the greatest norm) must have been the one most nearly orthogonal to γ_1. This row is retained and γ_1 is replaced by it in the next iteration. The process continues until the required number of vectors are selected.

3.6.5 Two-Dimensional FFT

Consider a two-tone excitation. The $x(t)$ vector can be expressed as

$$x(t) = \sum_m \sum_n X_{m,n} \exp[(m\omega_{p1} + n\omega_{p2})t] \tag{3.212}$$

where $X_{m,n}$ are the complex phasor magnitudes of the components at their respective frequencies. It is possible to treat the time as two independent time variables, so we can define ν_1 and ν_2, in which

$$v_1 = \omega_1 t = (r-1)\frac{2\pi}{N_m}$$
$$v_2 = \omega_2 t = (s-1)\frac{2\pi}{N_n} \quad (3.213)$$

where N_m and N_n are the number of sample points for the m and n series, respectively. These must be powers of two, and the number of samples must be twice the number of harmonics. This results in a two-dimensional grid of time samples, which can be processed with a two-dimensional FFT. The result is a two-dimensional set of frequency components, in which the component at $(m\omega_{p1}, n\omega_{p2})$ is the frequency component $m\omega_{p1} + n\omega_{p2}$.

The two-dimensional FFT is equivalent to a DFT in which we sample first at the rate determined by ω_{p1} and then sample at the ω_{p2} rate, beginning at each ω_{p1} sample. This is, of course, a large number of samples, and it would be prohibitive without the use of the FFT. An apparent disadvantage of the approach is the restriction of the sample set to powers of two, which invariably requires oversampling of the time waveform. Nevertheless, oversampling can be beneficial, as it reduces aliasing in the transform.

The method can be extended to any number of dimensions, but the time required to fill the multidimensional FFT matrix and to evaluate the transform increases exponentially with dimension. In practice, n-dimensional Fourier transformation is usually limited to $n \le 3$. The two-dimensional FFT is an optimal method; that is, it achieves the same conditioning as an orthogonal DFT. This is no surprise, as it consists of repeated FFT operations.

3.6.6 Artificial Frequency Mapping

Consider a simple resistive nonlinearity,

$$I = f(V) = aV^2 \quad (3.214)$$

Let $V(t)$ be the two-tone signal,

$$V(t) = V_1\cos(\omega_1 t) + V_2\cos(\omega_2 t) \quad (3.215)$$

Substituting (3.215) into (3.214) gives the unsurprising result,

$$I(t) = \frac{a}{2}\{V_1^2 + V_2^2 + V_1^2\cos(2\omega_1 t) + V_2^2\cos(2\omega_2 t) + 2V_1V_2[\cos((\omega_1+\omega_2)t) + \cos((\omega_1-\omega_2)t)]\} \tag{3.216}$$

Note that the coefficients in (3.216) are unrelated to the frequencies ω_1 and ω_2. Thus, the shape of the spectrum does not depend on the frequencies, as long as the nonlinearity is quasistatic. Therefore, for the kind of algebraic nonlinearities we normally encounter, we could solve the multitone problem by mapping the excitation frequencies in such a way that the resulting mixing products are equally spaced. We can then use a one-dimensional FFT to provide the Fourier transform.

For example, consider the simple frequency set,

$$\omega = m\omega_1 + n\omega_2 \tag{3.217}$$

where $0 \le m \le M$ and $|n| \le N$, with $m \ne 0$ when $n < 0$. (These complicated criteria merely prevent redundant frequency components.) We scale the frequencies with the coefficients,

$$\begin{aligned} s_1 &= 1 \\ s_2 &= \frac{\omega_1}{\omega_2(2N+1)} \end{aligned} \tag{3.218}$$

That is, $m\omega_1$ is multiplied by s_1 and $n\omega_2$ by s_2. This creates a uniform set of frequencies in ω, which can be Fourier transformed by a conventional FFT. Similar scaling functions can be used for other frequency sets and for more than two excitation tones.

An important advantage of artificial frequency mapping is its applicability to problems having a large number of noncommensurate excitations. An important disadvantage is that the time waveform returned by the FFT has no physical meaning. This is less of a disadvantage than it might seem; if time waveforms are desired as a final output from the simulation, the frequencies can be rescaled to their original values and the time waveforms calculated trigonometrically.

3.6.7 Frequency Sets

In single-tone harmonic balance, the selection of a frequency set is relatively simple: merely select the value of K, the highest harmonic in the series. We addressed this issue in Section 3.3.6. In multitone harmonic balance, the question becomes more complex.

We consider, for simplicity, the two-tone case. One option is to select some order, Q, and select frequencies such that

$$\omega = m\omega_{p1} + n\omega_{p2} \qquad |m| + |n| \le Q \tag{3.219}$$

If the values of m and n satisfying (3.219) are plotted on Cartesian axes, the pattern forms a diamond; therefore, it has been called a *diamond truncation* [3.22]. Similarly, selecting

$$\omega = m\omega_{p1} + n\omega_{p2} \qquad |m| \le M \qquad |n| \le N \tag{3.220}$$

results in a rectangular pattern, called the *rectangular* or *box truncation.* These criteria can be combined, resulting in something between those extremes. Neither, however, provides the precise set of frequencies shown in Figure 3.11 or 3.19; there is some justification, from Volterra theory, to believe that the latter sets are optimum for small-signal problems.

The optimization of frequency sets is an important unexplored problem in nonlinear circuit theory. Clearly, the selection of frequency sets depends on the nature of the nonlinearity and the excitation, but we can say little more about it. Useful research in this area would do much to enhance the performance of harmonic-balance simulation.

3.6.8 Determining the Jacobian

In single-tone harmonic-balance analysis, the Jacobian, given by (3.43), consists of the sum of the admittance matrix and terms representing the frequency domain I/V derivatives. In multitone analysis, (3.43) is still correct, but the admittance matrix is evaluated at the mixing frequencies used in the analysis, instead of the harmonic frequencies. The derivative matrices must be evaluated at those frequencies as well. We now derive the latter.

We begin by considering the nonlinear resistive elements; the reactive elements follow directly. The part of the Jacobian representing these elements is

$$\mathbf{J}_G = \frac{\partial \mathbf{I}_G}{\partial \mathbf{V}} \tag{3.221}$$

so

$$\partial \mathbf{I}_G = \mathbf{J}_G \partial \mathbf{V} \tag{3.222}$$

From (3.210),

$$\partial \Gamma \mathbf{i} = \mathbf{J}_G \partial \Gamma \mathbf{v} \tag{3.223}$$

and, since Γ is a constant matrix,

$$\Gamma \partial \mathbf{i} = \mathbf{J}_G \Gamma \partial \mathbf{v} \tag{3.224}$$

from which we obtain, by ordinary matrix manipulations,

$$\frac{\partial \mathbf{i}}{\partial \mathbf{v}} = \Gamma^{-1} \mathbf{J}_G \Gamma \tag{3.225}$$

or

$$\mathbf{J}_G = \Gamma \frac{\partial \mathbf{i}}{\partial \mathbf{v}} \Gamma^{-1} \tag{3.226}$$

The form of the matrix $\partial \mathbf{i} / \partial \mathbf{v}$ is analogous to that of the Jacobian. It is a set of diagonal submatrices:

$$\begin{bmatrix} \partial \mathbf{i}_1 \\ \\ \partial \mathbf{i}_2 \\ \\ \cdots \\ \\ \partial \mathbf{i}_N \end{bmatrix} = \begin{bmatrix} \frac{\partial \mathbf{i}_1}{\partial \mathbf{v}_1} & \frac{\partial \mathbf{i}_1}{\partial \mathbf{v}_2} & \cdots & \frac{\partial \mathbf{i}_1}{\partial \mathbf{v}_N} \\ \frac{\partial \mathbf{i}_2}{\partial \mathbf{v}_1} & \frac{\partial \mathbf{i}_2}{\partial \mathbf{v}_2} & \cdots & \frac{\partial \mathbf{i}_2}{\partial \mathbf{v}_N} \\ \cdots & \cdots & \cdots & \cdots \\ \frac{\partial \mathbf{i}_N}{\partial \mathbf{v}_1} & \frac{\partial \mathbf{i}_N}{\partial \mathbf{v}_2} & \cdots & \frac{\partial \mathbf{i}_N}{\partial \mathbf{v}_N} \end{bmatrix} \begin{bmatrix} \partial \mathbf{v}_1 \\ \\ \partial \mathbf{v}_2 \\ \\ \cdots \\ \\ \partial \mathbf{v}_N \end{bmatrix} \tag{3.227}$$

The individual $\partial \mathbf{i}_k / \partial \mathbf{v}_l$ submatrices have the form,

$$\begin{bmatrix} \partial i_k(t_1) \\ \\ \partial i_k(t_2) \\ \\ \cdots \\ \\ \partial i_k(t_S) \end{bmatrix} = \begin{bmatrix} \dfrac{\partial i_k(t_1)}{\partial v_l(t_1)} & 0 & \cdots & 0 \\ 0 & \dfrac{\partial i_k(t_2)}{\partial v_l(t_2)} & \cdots & 0 \\ \cdots & \cdots & \cdots & \cdots \\ 0 & 0 & \cdots & \dfrac{\partial i_k(t_S)}{\partial v_l(t_S)} \end{bmatrix} \begin{bmatrix} \partial v_l(t_1) \\ \\ \partial v_l(t_2) \\ \\ \cdots \\ \\ \partial v_l(t_S) \end{bmatrix} \tag{3.228}$$

The form of (3.228) is a diagonal because, in quasistatic nonlinear elements, $\partial i_k(t_n)/\partial v_l(t_m)$ is necessarily zero when $m \neq n$. It is possible, in some cases, to have nonzero off-diagonal elements; for example, if the control voltage is a delayed function of another control voltage. In such cases, it is necessary to modify the Jacobian in this manner, or convergence of the Newton process is poor.

Reactive elements follow the same pattern. For nonlinear capacitors, the Jacobian component, from (3.43), is $\mathbf{J}_Q = j\Omega\mathbf{C}$. The matrix $\mathbf{C} = \partial\mathbf{q}/\partial\mathbf{v}$ is given by (3.227) and (3.228), with q, of course, replacing i. The matrix Ω follows (3.21), with the mixing frequencies, instead of single-tone harmonics, along the main diagonal.

3.7 MODULATED WAVEFORMS AND ENVELOPE ANALYSIS

In linear systems, it is easy to determine, from single-tone analysis, how a component handles a modulated signal. One need only analyze the circuit at a number of frequencies, determine a transfer function, and multiply the excitation waveform by that function. In nonlinear circuits, it is not so simple; the effect of the circuit on the modulated waveform cannot be determined accurately from a single-tone analysis. Specialized methods, called *envelope transient analysis*, or simply *envelope analysis*, [3.25–3.27] have been developed to deal with modulated signals in an efficient manner. These methods are approximate. It is also possible to use multitone analysis in a more exact manner.

3.7.1 Modulated Signals

A narrowband modulated waveform, $s(t)$, can be represented as

$$s(t) = \mathrm{Re}\{s_e(t)\exp(j\omega_p t)\} \tag{3.229}$$

where $s_e(t)$ is the *complex envelope* of the signal, containing information about the magnitude and phase of the modulated waveform.[3] When such a waveform is distorted by a nonlinear circuit, harmonics of the carrier are generated and the envelope is distorted, causing its spectrum to spread. Each carrier harmonic is surrounded by modulation components. The distorted waveform can be represented as

$$s(t) = \sum_{m=-M}^{M} S_m(t)\exp(jm\omega_p t) \tag{3.230}$$

where the number of carrier harmonics, theoretically infinite, has been limited to $|m| \le M$. $S_m(t)$ represents the envelope function around the mth harmonic. We assume, for now, that the modulation is periodic and deterministic, so $S_m(t)$ can be expressed by

$$S_m(t) = \sum_{k=-K}^{K} S_{m,k}\exp(jk\omega_0 t) \tag{3.231}$$

and (3.230) becomes

$$s(t) = \sum_{m=-M}^{M}\sum_{k=-K}^{K} S_{m,k}\exp[j(k\omega_0 + m\omega_p)t] \tag{3.232}$$

Equation (3.232) shows that a modulated-waveform problem can be treated as a conventional, two-tone harmonic-balance analysis if the somewhat artificial requirement of a periodic modulating waveform is acceptable.

3. A *narrowband signal* has a fractional bandwidth that is small compared to the carrier frequency. Virtually all practical communication signals, even those considered "wideband" in some other sense (e.g., wideband CDMA systems) are narrowband in the sense we consider here.

3.7.2 Envelope Analysis

In a narrowband signal, the envelope function $S_m(t)$ varies slowly compared to the carrier; that is, we can always assume that $K\omega_0 \ll \omega_p$. Therefore, we can sample the waveform at a rate based on the envelope frequency, instead of the carrier frequency, performing harmonic-balance analyses at each sample point. We first perform a harmonic-balance analysis at $t = t_0$, giving $S_m(t_0)$ for all m; then, at $t = t_1$, giving $S_m(t_1)$, and so on. When the envelope functions at all M harmonics are known, they can be Fourier transformed to provide $S_{m,k}$. In effect, we are sampling $S_m(t)$ and determining the frequency components $S_{m,k}$ in (3.232). This process is inherently more efficient than multitone analysis, as it replaces a dimension of the multitone problem with a sequence of analyses. That sequence scales linearly with K; adding a dimension to the analysis would scale, at best, as $\sim K^{1.5}$ although $\sim K^2$ is more likely.

An even easier process would be to perform a number of analyses at a range of excitation amplitudes and phases to map the S_m space. Then, $S_m(t)$ could be found simply by applying the map to the real excitation. Indeed, this is the principle behind certain kinds of behavioral component models. Nothing we have said, so far, makes our envelope analysis superior to this latter approach. Properly implemented, however, an envelope analysis can account for phenomena that the latter approach cannot.

First, consider a nonlinear capacitor having the charge function $Q(V)$. From (3.231), its charge is

$$Q(t) = \sum_{m=-M}^{M} Q_m(t)\exp(jm\omega_p t) \tag{3.233}$$

Its current is found by differentiating:

$$\begin{aligned} I(t) &= \frac{d}{dt}Q(V) \\ &= \sum_{m=-M}^{M} \left(jm\omega_p Q_m(t) + \frac{d}{dt}Q_m(t)\right)\exp(jm\omega_p t) \end{aligned} \tag{3.234}$$

so the current components in the current-error vector, (3.25), are

$$\mathbf{I}_Q(t) = j\Omega\mathbf{Q}(t) + \frac{d\mathbf{Q}(t)}{dt} \tag{3.235}$$

The second term in (3.235), the derivative, is new. These derivatives are usually calculated by finite differences.

Second, we must account for the variation in the linear subcircuit's admittance over the frequency spectrum, however narrow, of the modulation. One way of many is simply to linearize the admittance in the vicinity of the carrier harmonics; then,

$$Y(m\omega_p + \delta\omega) = Y(m\omega_p) + \left.\frac{dY(\omega)}{d\omega}\right|_{\omega = m\omega_p} \delta\omega \tag{3.236}$$

Finally, (3.235) shows that we have created a nonquasistatic system. The solutions at each time point are linked, and this dependency must be included in the Jacobian. That is, we must have terms of the form

$$\mathbf{J}_{n,p} = \frac{\partial \mathbf{F}[\mathbf{V}(t_n)]}{\partial \mathbf{V}(t_p)} \tag{3.237}$$

If only the $(p-1)$ and $(p+1)$ time points are used to estimate the derivatives, three such submatrices are needed, and each harmonic-balance iteration must include them. Although this increases the dimension of the Jacobian by a factor of approximately three, the blocks lie along its main diagonal, so the damage is not too great. In some cases, the links can be removed entirely, at some cost of robustness, but improving solution speed [3.25].

References

[3.1] M. S. Nakhla and J. Vlach, "A Piecewise Harmonic Balance Technique for Determination of Periodic Response of Nonlinear Systems," *IEEE Trans. Circ. Syst.*, Vol. CAS-23, 1976, p. 85.

[3.2] S. W. Director and K. W. Current, "Optimization of Forced Nonlinear Periodic Currents," *IEEE Trans. Circ. Syst.*, Vol. CAS-23, 1976, p. 329.

[3.3] F. R. Colon and T. N. Trick, "Fast Periodic Steady-State Analysis for Large-Signal Electronic Circuits," *IEEE J. Solid-State Circ.*, Vol. SC-8, 1973, p. 260.

[3.4] J. M. Ortega and W. C. Rheinboldt, *Iterative Solution of Nonlinear Equations in Several Variables*, New York: Academic Press, 1981.

[3.5] R. G. Hicks and P. J. Khan, "Numerical Analysis of Nonlinear Solid-State Device Excitation in Microwave Circuits," *IEEE Trans. Microwave Theory Tech.*, Vol. MTT-30, 1982, p. 251.

[3.6] A. R. Kerr, "A Technique for Determining the Local Oscillator Waveforms in a Microwave Mixer," *IEEE Trans. Microwave Theory Tech.*, Vol. MTT-23, 1975, p. 828.

[3.7] V. Rizzoli et al., "Harmonic-Balance Simulation of Strongly Nonlinear Very Large-Size Microwave Circuits by Inexact Newton Methods," *IEEE MTT-S Int. Microwave Symp. Dig.*, 1996.

[3.8] M. Pernice and H. F. Walker, "NITSOL: A Newton Iterative Solver for Nonlinear Systems," *SIAM J. Sci. Comput.*, Vol. 19, 1998, p. 302.

[3.9] Y. Saad, *Iterative Methods for Sparse Linear Systems*, Boston: PWS Publishing, 1996.

[3.10] R. Barrett et al., *Templates for the Solution of Linear Systems: Building Blocks for Iterative Methods*, Philadelphia: SIAM, 1993.

[3.11] C. T. Kelly, *Iterative Methods for Linear and Nonlinear Equations*, Philadelphia: SIAM, 1995.

[3.12] H. Yeager and R. W. Dutton, "Improvement in Norm-Reducing Methods for Circuit Simulation," *IEEE Trans. Computer-Aided Design,* Vol. 8, 1989, p. 538.

[3.13] V. Rizzoli, "State-of-the-Art Harmonic-Balance Simulation of Forced Nonlinear Microwave Circuits by the Piecewise Technique," *IEEE Trans. Microwave Theory Tech.*, Vol. MTT-40, 1992, p. 12.

[3.14] K. S. Kundert and A. Sangiovanni-Vincentelli, "Simulation of Nonlinear Circuits in the Frequency Domain," *IEEE Trans. Computer-Aided Design*, Vol. CAD-5, 1986, p. 521.

[3.15] S. Maas, "Two-Tone Intermodulation in Diode Mixers," *IEEE Trans. Microwave Theory Tech.*, Vol. MTT-35, 1987, p. 307.

[3.16] G. B. Sorkin, K. S. Kundert, and A. Sangiovanni-Vincentelli, "An Almost-Periodic Fourier Transform for Use with Harmonic Balance," *IEEE MTT-S Int. Microwave Symp. Dig.*, 1987, p. 717.

[3.17] V. Rizzoli, C. Cecchetti, and A. Lipparini, "A General-Purpose Program for the Analysis of Nonlinear Microwave Circuits Under Multitone Excitation by Multidimensional Fourier Transform," *Proc. 17th European Microwave Conf.*, 1987.

[3.18] E. Ngoya et al., "Efficient Algorithms for Spectra Calculations in Nonlinear Microwave Circuits Simulators," *IEEE Trans. Circuits and Systems*, Vol. 37, 1990, p. 1339.

[3.19] O. Nastov and J. K. White, "Time-Mapped Harmonic Balance," *Proc. 36th Design Automation Conf.*, 1999.

[3.20] P. Rodrigues, "A General Mapping Technique for Fourier Transform Computation in Nonlinear Circuit Analysis," *IEEE Microwave and Guided Wave Letters*, Vol. 7, 1997, p. 374.

[3.21] P. Rodrigues, "An Orthogonal Almost-Periodic Fourier Transform for Use in Nonlinear Circuit Simulation," *IEEE Microwave and Guided Wave Letters*, Vol. 4, 1994, p. 74.

[3.22] K. S. Kundert, J. K. White, and A. Sangiovanni-Vincentelli, *Steady-State Methods for Simulating Analog and Microwave Circuits*, Boston: Kluwer, 1990.

[3.23] J. C. Pedro and N. Borges de Carvalho, "Artificial Frequency Mapping Techniques for Multitone Harmonic Balance," *IEEE MTT-S Int. Microwave Symp. Dig., Workshops*, 2000.

[3.24] V. Boric, J. East, and G. Haddad, "An Efficient Fourier Transform Algorithm for Multitone Harmonic Balance," *IEEE Trans. Microwave Theory Tech.*, Vol. MTT-47, 1999, p. 182.

[3.25] V. Rizzoli, A. Neri, and F. Mastri, "A Modulation-Oriented Piecewise Harmonic-Balance Technique Suitable for Transient Analysis and Digitally Modulated Signals," *Proc. 26th European Microwave Conf.*, 1996, p. 546.

[3.26] E. Ngoya, J. Sombrin, and J. Rousset, "Simulation de Circuits et Systemes: Methodes, Actuelles et Tendances," *Seminaire Antennes Actives-MMIC, Arles*, Arles, France, 1994.

[3.27] E. Ngoya and R. Larcheveque, "Envelop [sic] Transient Analysis: A New Method for the Transient and Steady-State Analysis of Microwave Communication Circuits and Systems," *IEEE MTT-S Int. Microwave Symp. Dig.*, 1996, p. 1365.

Chapter 4

Volterra-Series and Power-Series Analysis

The previous chapter was concerned with strongly nonlinear circuits having a single, large-signal excitation and sometimes one or more additional excitations that could be assumed to be vanishingly small. In this chapter we consider the opposite extreme, weakly nonlinear circuits having multiple, noncommensurate small-signal excitations. In these circuits, nonlinearities have a negligible effect upon their linear responses, but even low levels of nonlinear distortion can affect system performance. The problem of analyzing such circuits is sometimes called the *small-signal nonlinear problem*.

Two techniques will be examined. The first is *power-series analysis*, a technique that is relatively simple but requires a simplifying assumption that is often unrealistic: the circuit contains only ideal memoryless transfer nonlinearities. The power-series approach is useful in some instances, however, and gives the engineer a good intuitive sense of the behavior of many types of nonlinear circuits. The second technique is *Volterra-series analysis*, a very powerful method that does not require such restrictive assumptions. Volterra analysis is most useful for the analysis of intermodulation distortion and related phenomena from weakly nonlinear circuits and systems.

This chapter generally follows the approach of Weiner and Spina [4.1], modified as necessary for microwave applications. That book is based upon work reported in [4.2] and [4.3], performed in the early 1970s under U.S. Government sponsorship. References [4.4] and [4.5] are other excellent sources of information on the Volterra series.

4.1 POWER-SERIES ANALYSIS

4.1.1 Power-Series Model and Multitone Response

Many nonlinear systems and circuits are modeled as a filter, or other frequency-selective network, followed by a memoryless, broadband transfer nonlinearity. The model is shown in Figure 4.1, where the linear network has the transfer function $H(\omega)$ and the nonlinear one has the transfer function

$$w(t) = f(u(t)) = \sum_{n=1}^{N} a_n u^n(t) = a_1 u(t) + a_2 u^2(t) + \ldots \tag{4.1}$$

or

$$w(t) = \sum_{n=1}^{N} w_n(t) \tag{4.2}$$

where $w_n(t) = a_n u^n(t)$. For practical reasons, the series must be truncated at some value $n = N$.

In Figure 4.1 and (4.1), the transfer function variables $w(t)$ and $u(t)$ are small-signal, incremental quantities (i.e., small deviations around a dc bias point) and may be current or voltage. It is important that the transfer function $f(u)$ be single-valued, and that it be *weakly nonlinear*, expressing the nonlinearity adequately via a limited number of terms. In practice, N usually must be limited to 3 or at most 5, or the analysis becomes hopelessly laborious. The linear block $H(\omega)$ may represent a filter or, frequently, a matching network. To account for the effects of an output

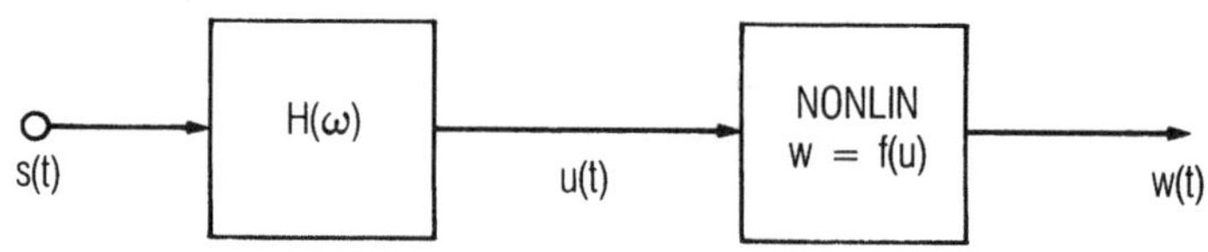

Figure 4.1 Power-series model of a nonlinear system: $H(\omega)$ is a linear circuit, and $f(u)$ is a memoryless nonlinear transfer function.

filter or a matching network, one may include another linear network at the output. The inclusion of such a network can be accomplished via ordinary linear circuit theory.

Figure 4.2 shows a simplified equivalent circuit of a FET, which can be described by this model. The elements of the circuit are readily identified with those of the block diagram in Figure 4.1: $v_s(t)$ corresponds to $s(t)$, $v(t)$ to $u(t)$, and $i(t)$ to $w(t)$. The input linear transfer function $H(\omega) = V(\omega) / V_s(\omega)$, where $V(\omega)$ and $V_s(\omega)$ are the frequency-domain equivalents of $v(t)$ and $v_s(t)$, respectively. Thus, in this example,

$$H(\omega) = \frac{V(\omega)}{V_s(\omega)} = \frac{1}{(R_s + R_i)C_i j\omega - L_s C_i \omega^2 + 1} \tag{4.3}$$

The only nonlinearity in the circuit is the transfer function $i = f(v)$ between the gate depletion voltage $v(t)$ and the drain current $i(t)$. The transfer function $f(v)$ is the power-series expansion of the current around the bias point; it is found by expanding the large-signal drain current/gate voltage characteristic $I_d = F(V)$ in a Taylor series:

$$\begin{aligned} f(v) &= F(V_{g0} + v) - F(V_{g0}) \\ &= \left.\frac{dF}{dV}\right|_{V=V_{g0}} v + \frac{1}{2}\left.\frac{d^2F}{dV^2}\right|_{V=V_{g0}} v^2 + \frac{1}{6}\left.\frac{d^3F}{dV^3}\right|_{V=V_{g0}} v^3 + \ldots \end{aligned} \tag{4.4}$$

where V_{g0} is the dc bias voltage across the capacitor. The coefficients a_n are found by comparing (4.1) to (4.4).

So far, nothing in this problem precludes the use of time-domain techniques for analyzing the nonlinear circuit of either Figure 4.1 or 4.2.

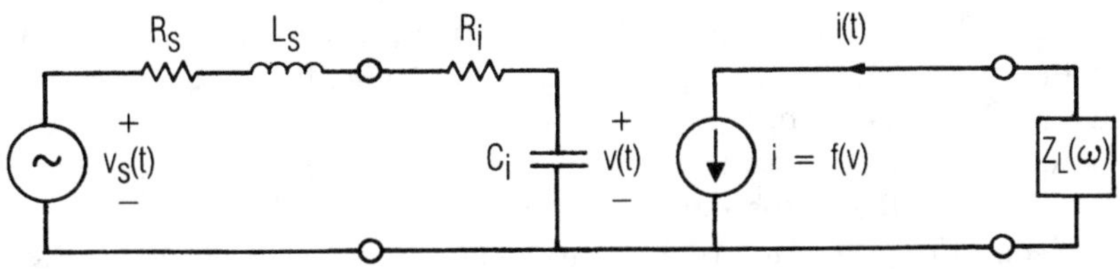

Figure 4.2 Simplified equivalent circuit of a FET for which the power-series model is applicable.

We could easily convert the linear transfer function $H(\omega)$ to an impulse response function $h(t)$, find $v(t)$ by convolution, and finally substitute $v(t)$ into (4.1) to obtain $i(t)$. Nevertheless, there are two problems inherent in using time-domain techniques to analyze this type of circuit. First, undertaking the solution of such problems is usually motivated by a need for frequency-domain, not time-domain, information, so a frequency-domain approach is the natural first choice. Second, if the excitations are at noncommensurate frequencies, the time-waveforms are not periodic, so obtaining valid frequency-domain data from a numerical time-domain result is often difficult. Finally, we use the frequency-domain approach because of the insight it gives us into a much more powerful technique, Volterra-series analysis.

The excitation $s(t)$ usually contains at least two noncommensurate frequencies. In Figure 4.2, $v_s(t)$ corresponds to $s(t)$ and can be expressed as

$$v_s(t) = \frac{1}{2}\sum_{q=1}^{Q} V_{s,q}\exp(j\omega_q t) + V_{s,q}^*\exp(-j\omega_q t) \tag{4.5}$$

where the asterisk indicates the complex conjugate. Equation (4.5) can be written in the following, more compact form:

$$v_s(t) = \frac{1}{2}\sum_{\substack{q=-Q \\ q\neq 0}}^{Q} V_{s,q}\exp(j\omega_q t) \tag{4.6}$$

where $V_{s,-q} = V_{s,q}^*$, $\omega_{-q} = -\omega_q$.

We assume in (4.6), and throughout the following analysis, that the excitation and the response have no dc component. In microwave circuits, the excitation invariably has no dc component, other than the bias, which we include implicitly by evaluating the a_n coefficients at the bias point. We noted in Chapter 1 that a nonlinear circuit may indeed generate dc components in its output in response to a sinusoidal excitation. When the excitation is small and the nonlinearities are weak, (which is, after all, our basic assumption), the dc components generated by the nonlinearities are very small, and invariably negligible. The practical effect of the generation of dc components is to offset the bias currents and voltages slightly from their quiescent values. In cases where significant bias offset occurs, for example, in a class-B amplifier, Volterra and power-series analyses usually are not applicable.

We shall assume throughout the remainder of this chapter that the excitations do not include dc components, and we will dispense with the notation to that effect ($q \neq 0$) in all the summations.

The output of the linear circuit is $v(t)$, which corresponds to $u(t)$ in Figure 4.1:

$$v(t) = \frac{1}{2} \sum_{q=-Q}^{Q} V_{s,q} H(\omega_q) \exp(j\omega_q t) \tag{4.7}$$

where $H(\omega_{-q}) = H^*(\omega_q)$. The output of the nonlinear stage is found by substituting $v(t)$, expressed by (4.7), into (4.1) for $u(t)$; the terms are all of the form $a_n v^n$, where

$$\begin{aligned} a_n v^n(t) &= a_n \left[\frac{1}{2} \sum_{q=-Q}^{Q} V_{s,q} H(\omega_q) \exp(j\omega_q t) \right]^n \\ &= \frac{a_n}{2^n} \sum_{q1=-Q}^{Q} \sum_{q2=-Q}^{Q} \cdots \sum_{qn=-Q}^{Q} V_{s,q1} \\ &\quad \cdot V_{s,q2} \cdots V_{s,qn} H(\omega_{q1}) H(\omega_{q2}) \ldots H(\omega_{qn}) \\ &\quad \cdot \exp[j(\omega_{q1} + \omega_{q2} + \ldots + \omega_{qn})t] \end{aligned} \tag{4.8}$$

The entire response is

$$i(t) = \sum_{n=1}^{N} a_n v^n(t) \tag{4.9}$$

where $i(t)$ is equivalent to $w(t)$ in (4.1). Equations (4.8) and (4.9) show that a large number of new frequencies can be generated by the nonlinearity; each frequency generated by the nth-degree term is a linear combination of n excitation frequencies, and the total response for each n is the sum of all possible linear combinations of n excitation frequencies. Figure 4.3 shows some of the lowest-order terms, those that are usually of most concern to system designers, when $Q = 2$ (two excitation tones) and $n \le 3$. Furthermore, the amplitude of each frequency component is proportional to the product of the amplitudes of all the contributing excitations.

It is important in the following analysis to distinguish between the concepts of *degree* and *order*. The *degree* of the nonlinearity refers simply to the power of $u(t)$ in the nonlinear transfer characteristic (4.1). An *n*th-*order* mixing frequency is defined as one that arises from the sum of n excitation frequencies. In general it is not possible to determine the order of a mixing product from its frequency; for example, the frequency $2\omega_1 - \omega_2$ appears at first to be of third order, that is, $2\omega_1 - \omega_2 = \omega_1 + \omega_1 - \omega_2$, but in reality it could be the fifth-order mixing product, $\omega_1 + \omega_1 + \omega_1 - \omega_1 - \omega_2$. In this example, our circuit contains only a single, ideal, transfer nonlinearity and no feedback, so an *n*th-degree nonlinearity generates mixing products up to only *n*th order. However, in more complicated circuits, a nonlinearity of degree n can generate mixing products of order equal to or greater than n. This situation exists because, in reality, the mixing products and the excitations generally are not limited to separate parts of the circuit, so a mixing product can mix with excitations or other mixing products. For example, if the circuit of Figure 4.2 included a feedback capacitance from the top of the current source to the top node of the input capacitor C_i, the control voltage $v(t)$ would

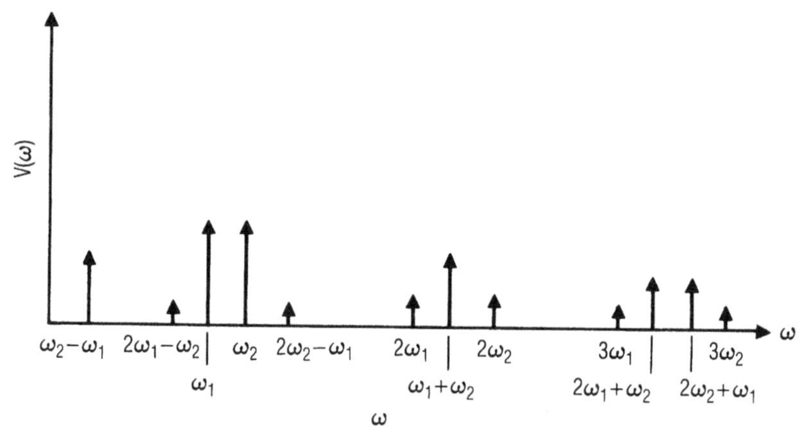

Figure 4.3 Spectrum of intermodulation frequencies resulting from two-tone excitation; the excitation frequencies are ω_1 and ω_2.

include mixing products as well as excitation-frequency components. Thus, $v(t)$ would consist of components at the excitation frequency and at mixing frequencies, and products of order greater than n would arise as nth-order combinations of any of those frequencies. In that situation, even if $n \le 3$, a $2\omega_1 - \omega_2$ product could mix with a component at $\omega_1 - \omega_2$ to form a fifth-order component at $3\omega_1 - 2\omega_2$.

To illustrate power-series analysis, we now consider a two-tone excitation ($Q = 2$) and find the part of the response associated with the second- and third-degree components of the output current, $n = 2$ and $n = 3$, respectively. The second-degree component is designated $i_2(t)$ because it corresponds to $w_2(t)$ in (4.1) and (4.2):

$$i_2(t) = \frac{a_2}{4} \sum_{q1=-2}^{2} \sum_{q2=-2}^{2} V_{s,q1} V_{s,q2} H(\omega_{q1}) H(\omega_{q2}) \cdot \exp[j(\omega_{q1} + \omega_{q2})t] \tag{4.10}$$

The summation in (4.8) generates $(2Q)^n$ terms; in this example, $Q = 2$, so there are 16 terms. The frequencies associated with the terms are the following:

$-\omega_2 - \omega_2$	$-\omega_2 - \omega_1$	$-\omega_2 + \omega_1$	$-\omega_2 + \omega_2$
$-\omega_1 - \omega_2$	$-\omega_1 - \omega_1$	$-\omega_1 + \omega_1$	$-\omega_1 + \omega_2$
$\omega_1 - \omega_2$	$\omega_1 - \omega_1$	$\omega_1 + \omega_1$	$\omega_1 + \omega_2$
$\omega_2 - \omega_2$	$\omega_2 - \omega_1$	$\omega_2 + \omega_1$	$\omega_2 + \omega_2$

We can readily see that these terms include harmonics of the input frequencies (e.g., $\omega_1 + \omega_1 = 2\omega_1$), repeated terms (e.g., $\omega_1 + \omega_2$ and $\omega_2 + \omega_1$), and dc terms (e.g., $\omega_1 - \omega_1$). Also, the frequencies occur in positive and negative pairs, so the terms in (4.10) occur in complex conjugate pairs. Thus, the frequency components represent real time waveforms, as they must in any real circuit. For example, the $\omega_1 - \omega_2$ component is

$$\begin{aligned} i_2'(t) &= a_2 v^2(t)\Big|_{\omega_1-\omega_2} \\ &= \frac{a_2}{4}\cdot 2\{V_{s,1}V_{s,2}^*(H(\omega_1)H^*(\omega_2)\exp[j(\omega_1-\omega_2)t]) \\ &\quad + V_{s,1}^*V_{s,2}(H^*(\omega_1)H(\omega_2)\exp[-j(\omega_1-\omega_2)t])\} \end{aligned} \tag{4.11}$$

The coefficient of 2 ahead of the brackets indicates that there are two terms at $\omega_1 - \omega_2$ and two terms at $-(\omega_1 - \omega_2)$ in the double summation: $q1 = 1$, $q2 = -2$ and $q1 = -2$, $q2 = 1$ give identical terms at $\omega_1 - \omega_2$; $q1 = -1$, $q2 = 2$ and $q1 = 2$, $q2 = -1$ give identical terms at $-(\omega_1 - \omega_2)$. Finally, (4.11) can be expressed in cosine form:

$$i_2'(t) = a_2\left|V_{s,1}V_{s,2}H(\omega_1)H(\omega_2)\right|\cos[(\omega_1-\omega_2)t+\phi_2] \tag{4.12}$$

where ϕ_2 is the phase angle associated with the complex coefficients in (4.11). The purpose of intermodulation (IM) analysis is usually to determine output power at the mixing frequency, so ϕ_2 is rarely of interest. Phase angles are important in Volterra analysis and in power-series analysis only when two components that have different orders but the same frequency are combined. For example, saturation effects can be analyzed by combining the linear output at ω_1 and the third-order component at $\omega_1 + \omega_1 - \omega_1 = \omega_1$; the phase difference between these components affects the circuit's amplitude saturation (AM-to-AM) and phase (AM-to-PM) characteristics.

The current component generated by the third-degree term, $i_3(t)$, can be found in a similar manner. From (4.8), with $n = 3$, we have

$$\begin{aligned} i_3(t) &= a_3 v^3(t) \\ &= \frac{a_3}{8}\sum_{q1=-2}^{2}\sum_{q2=-2}^{2}\sum_{q3=-2}^{2} V_{s,q1}V_{s,q2}V_{s,q3} \\ &\quad \cdot H(\omega_{q1})H(\omega_{q2})H(\omega_{q3})\exp[j(\omega_{q1}+\omega_{q2}+\omega_{q3})t] \end{aligned} \tag{4.13}$$

The $i_3(t)$ summation has $(2Q)^n = 4^3 = 64$ terms, although not all represent different mixing frequencies. Half of the terms in (4.13) are simply conjugates of the others and, as in the second-degree case, many terms are identical. Some (but definitely not all!) of the frequencies in (4.13) are

$$
\begin{aligned}
\omega_2 + \omega_2 - \omega_1 &= 2\omega_2 - \omega_1 \\
\omega_1 + \omega_1 - \omega_2 &= 2\omega_1 - \omega_2 \\
\omega_1 + \omega_1 - \omega_1 &= \omega_1 \\
\omega_2 + \omega_2 - \omega_2 &= \omega_2 \\
\omega_1 + \omega_1 + \omega_1 &= 3\omega_1 \\
\omega_2 + \omega_2 + \omega_2 &= 3\omega_2
\end{aligned}
\tag{4.14}
$$

The first two of these mixing frequencies are important because they often occur at frequencies close to ω_1 and ω_2. There are three identical terms at $2\omega_2 - \omega_1$ and three at $2\omega_1 - \omega_2$; the terms at $2\omega_2 - \omega_1$ occur when

$$
\begin{aligned}
q1 &= 2, \quad & q2 &= 2, \quad & q3 &= -1 \\
q1 &= 2, \quad & q2 &= -1, \quad & q3 &= 2
\end{aligned}
\tag{4.15}
$$

and

$$
q1 = -1, \qquad q2 = 2, \qquad q3 = 2
$$

Because there are three terms at this frequency in (4.13), the coefficient 3 is used in the expression for the current component at $2\omega_2 - \omega_1$:

$$
\begin{aligned}
i_3'(t) &= a_3 v^3(t)\big|_{2\omega_2 - \omega_1} \\
&= \frac{a_3}{8} \cdot 3\{V_{s,1}^* V_{s,2}^2 H^*(\omega_1) H^2(\omega_2) \exp[j(2\omega_2 - \omega_1)t] \\
&\quad + V_{s,1} V_{s,2}^{*2} H(\omega_1) H^{*2}(\omega_2) \exp[-j(2\omega_2 - \omega_1)t]\}
\end{aligned}
\tag{4.16}
$$

The cosine form of (4.16) is

$$
i_3'(t) = \frac{3a_3}{4} \left| V_{s,1} V_{s,2}^2 H(\omega_1) H^2(\omega_2) \right| \cos[(2\omega_2 - \omega_1)t + \phi_3] \tag{4.17}
$$

Again, ϕ_3 represents the combined phases of the complex coefficients in (4.16) and may be ignored in determining the power levels of third-order mixing components.

4.1.2 Frequency Generation

The new frequencies generated by a transfer nonlinearity, expressed by (4.8), are easy to predict. A large number of frequencies are possible; we shall assume that K nth-order frequencies are of interest, and any one of them, $\omega_{n,k}$, $k = 1...K$, can be expressed as follows:

$$\begin{aligned} \omega_{n,k} = {} & m_{-Q}\omega_{-Q} + \ldots + m_{-2}\omega_{-2} + m_{-1}\omega_{-1} \\ & + m_1\omega_1 + m_2\omega_2 + \ldots + m_Q\omega_Q \end{aligned} \tag{4.18}$$

where m_i is the number of times the frequency ω_i occurs in generating the mixing frequency $\omega_{n,k}$. Because exactly n terms are generated by an nth-degree nonlinearity, the set of values of m_i that defines any single mixing frequency is subject to the constraint

$$\sum_{i=-Q}^{Q} m_i = n \tag{4.19}$$

Some of the nth-order terms in (4.18) may not appear to be a combination of n frequency components; for example, if $n = 3$ and $Q = 2$, some of those terms are the following:

$$\begin{aligned} \omega_1 + \omega_2 + \omega_{-2} &= \omega_1 \\ \omega_1 + \omega_1 + \omega_{-1} &= \omega_1 \\ \omega_2 + \omega_2 + \omega_{-2} &= \omega_2 \end{aligned} \tag{4.20}$$

and these seem to involve only one frequency. This is an illustration of the fact, stated in the previous section, that it is not generally possible to determine the order of a mixing product from its frequency. The evidence that these products are indeed third-order is their cubic dependence on $V_{s,q}$ and $H(\omega_q)$ in (4.8).

To obtain the correct magnitude of each IM component, determining the number of terms at each frequency is clearly necessary. For a mixing frequency given by (4.18), the number of terms is given by the multinomial coefficient

$$t_{n,k} = \frac{n!}{m_{-Q}! \dots m_{-2}! m_{-1}! m_1! m_2! \dots m_Q!} \tag{4.21}$$

For example, in the second-order case of $\omega_1 - \omega_2 = \omega_1 + \omega_{-2}$ examined earlier, $n = 2$, $m_1 = 1$, $m_{-2} = 1$, so

$$t_{n,k} = \frac{2!}{1!1!} = 2 \tag{4.22}$$

as we had determined by brute force. Similarly, for the $n = 3$ case, $\omega_{n,k} = 2\omega_2 - \omega_1$ so $m_2 = 2$, $m_{-1} = 1$, and

$$t_{n,k} = \frac{3!}{2!1!} = 3 \tag{4.23}$$

which agrees with the coefficient in (4.16) and (4.17).

4.1.3 Intercept Point and Power Relations

A *two-tone test* is a common method for determining the intermodulation properties of a nonlinear or quasilinear circuit. In such a test, two excitations of equal amplitude and separated only slightly in frequency are applied to the circuit, and the powers of the resulting IM components are measured. Figure 4.4 shows the test setup for such measurements. The two excitations are combined at the input of the nonlinear component, a variable attenuator is used to adjust the input level, and the output-frequency components are observed on the display screen of a spectrum analyzer. A spectrum similar to that shown in Figure 4.3 is normally observed. In a two-tone test, $V_{s,1} = V_{s,2} = V_s$, $\omega_1 \approx \omega_2 = \omega$, and $H(\omega_1) \approx H(\omega_2) = H(\omega)$; V_s can be assumed real without loss of generality.[1] These approximations are almost always valid for third-order IM at $2\omega_1 - \omega_2$ and $2\omega_2 - \omega_1$, but they may be somewhat strained when applied

1. While true of a two-tone excitation, this reasoning is not valid for IM problems involving a large number of excitation tones. In that case, the envelope of the composite waveform depends on the initial phases of the excitation tones and the magnitude of the mixing products may, as well. This is especially true of large-signal IM problems, which are outside the scope of this chapter.

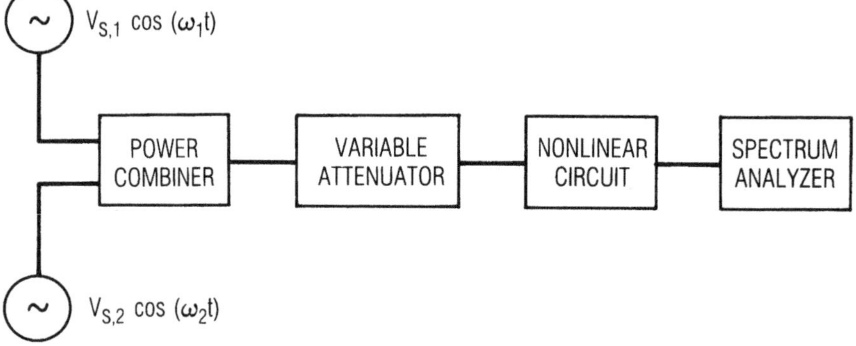

Figure 4.4 Two-tone test circuit. In general $V_{s,1}$ and $V_{s,2}$ are equal, and ω_1 and ω_2 are nearly equal.

to second-order products, like $\omega_2 - \omega_1$ and $2\omega_2$, which can easily be out of band. We justify them by recognizing that second-order products are of primary concern when they are in-band; thus, $H(\omega)$ is approximately constant, and in any case the qualitative results of the following analysis are more important than the quantitative ones.

When these approximations are made and the phase angle ϕ_2 is ignored, (4.12) becomes

$$i_2'(t) = a_2 V_s^2 |H(\omega)|^2 \cos[(\omega_1 - \omega_2)t] \tag{4.24}$$

The second-order IM output power, the power dissipated in the real part of $Z_L(\omega_1 - \omega_2)$, is

$$P_{\text{IM2}} = \frac{1}{2} |i_2'(t)|^2 \text{Re}\{Z_L(\omega_1 - \omega_2)\} \tag{4.25}$$

We assume for simplicity that $\text{Re}\{Z_L(\omega_1 - \omega_2)\} = R_L$, a constant. Then

$$P_{\text{IM2}} = \frac{1}{2} a_2^2 V_s^4 |H(\omega)|^4 R_L \tag{4.26}$$

The available power of each input tone is

$$P_{\text{av}} = \frac{V_s^2}{8R_s} \tag{4.27}$$

and the output IM power can be written in terms of the available input power:

$$P_{\text{IM2}} = 32a_2^2|H(\omega)|^4 R_s^2 R_L P_{\text{av}}^2 \tag{4.28}$$

The same can be done with the third-order IM component at $2\omega_2 - \omega_1$. The output current at that frequency is

$$i_3'(t) = \frac{3}{4} a_3 V_s^3 |H(\omega)|^3 \cos[(2\omega_2 - \omega_1)t] \tag{4.29}$$

and the IM output power is

$$P_{\text{IM3}} = \frac{9}{32} a_3^2 V_s^6 |H(\omega)|^6 R_L \tag{4.30}$$

As with the second-order component, the third-order IM output power can be expressed as a function of available input power P_{av}:

$$P_{\text{IM3}} = 144\, a_3^2 |H(\omega)|^6 R_s^3 R_L P_{\text{av}}^3 \tag{4.31}$$

It is normally most convenient to express the IM powers P_{IM2} and P_{IM3} in dBm, not linear quantities. Thus, with P_{IM2}, P_{IM3}, and P_{av} expressed in terms of dBm, (4.28) and (4.31) become

$$P_{\text{IM2}} = 10 \log(32a_2^2|H(\omega)|^4 R_s^2 R_L) + 2P_{\text{av}} - 30 \tag{4.32}$$

$$P_{\text{IM3}} = 10 \log(144a_3^2|H(\omega)|^6 R_s^3 R_L) + 3P_{\text{av}} - 60 \tag{4.33}$$

The IM output power given by (4.32) and (4.33), along with the linear output power, are graphed in Figure 4.5. At low levels, the second- and third-order IM powers vary by 2 dB/dB and 3 dB/dB, respectively, with input power level, while the linear output varies at the expected 1 dB/dB

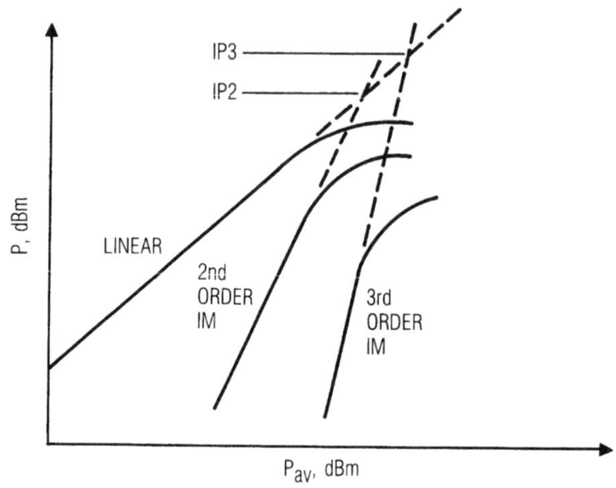

Figure 4.5 Input/output power curves for linear and intermodulation components. By tradition, the power shown in these curves is the power in each tone of the linear or IM product.

rate. In fact, further analysis shows that *n*th-degree IM products always vary by *n* dB/dB with input power level. At some point, the linear output power saturates, because the output power available from any real component is always finite; the IM characteristics also saturate at approximately the same input level. Below this saturation level, however, the IM power curves, in terms of dBm, are straight lines.

The straight-line behavior of the IM characteristic can be used to predict IM levels at any input power level. It is necessary only to know the output level at one point in order to define the entire curve. A convenient point is the extrapolated point at which the IM and linear output powers are equal; this point is different, in general, for each order of IM product (and, in fact, for each different mixing frequency of a given order) and is called the *intermodulation intercept point*. This point is useful because it defines not only the IM curve, but its relation to the linear curve as well. Therefore, it can be used to find not only the IM output power, but the ratio of linear to IM power level, often a more important quantity.

The IM characteristic follows the equation for any straight-line function:

$$P_{\mathrm{IM}n} = nP_{\mathrm{av}} + P_0 \tag{4.34}$$

where P_0 is a constant that we will evaluate. In terms of linear output power,

$$P_{\mathrm{IM}n} = n(P_{\mathrm{lin}} - G) + P_0 = nP_{\mathrm{lin}} + P_0' \tag{4.35}$$

At the nth-order intercept point IP_n,

$$P_{\mathrm{IM}n} = P_{\mathrm{lin}} = IP_n \tag{4.36}$$

Substituting (4.36) into (4.35) gives

$$P_0' = (1 - n)IP_n \tag{4.37}$$

Substituting (4.37) into (4.35) finally gives the result,

$$P_{\mathrm{IM}n} = nP_{\mathrm{lin}} - (n - 1)IP_n \tag{4.38}$$

Equation (4.38) gives the relationship between the linear power, P_{lin}, the nth-order IM output power, $P_{\mathrm{IM}n}$, and the nth-order intercept point, IP_n, at input levels below saturation. The only quantity that must be determined in order to define (4.38) is the intercept point IP_n. To determine IP_n, an expression for IM level must be found by an analysis similar to the one above; IP_n can then be found by comparison. A straightforward analysis of the circuit in Figure 4.2 gives the transducer gain G_t in decibels:

$$G_t = 10\log(4a_1^2|H(\omega)|^2 R_s R_L) \tag{4.39}$$

Substituting this expression and $P_{\mathrm{av}} = P_{\mathrm{lin}} - G_t$ into (4.32) and (4.33), and doing some straightforward algebra, gives the expressions

$$P_{\mathrm{IM2}} = 10\log\left(\frac{2a_2^2}{a_1^4 R_L}\right) + 2P_{\mathrm{lin}} - 30 \tag{4.40}$$

and

$$P_{\mathrm{IM3}} = 10\log\left(\frac{9a_3^2}{4a_1^6R_L^2}\right) + 3P_{\mathrm{lin}} - 60 \tag{4.41}$$

By comparing (4.40) and (4.41) to (4.38), we find IP_2 and IP_3 in dBm:

$$IP_2 = 10\log\left(\frac{a_1^4}{a_2^2}\frac{R_L}{2}\right) + 30 \tag{4.42}$$

$$IP_3 = 10\log\left(\frac{2}{3}\frac{a_1^3}{a_3}R_L\right) + 30 \tag{4.43}$$

We again note that (4.42) and (4.43) apply only to the specific second- and third-order intermodulation products $\omega_1 - \omega_2$ and $2\omega_2 - \omega_1$, respectively. Although the intercept points are also valid for the $\omega_2 - \omega_1$ and $2\omega_1 - \omega_2$ products, they are generally not valid for other products of the same orders, for example, $2\omega_1$ and $3\omega_2$. These latter components have different intercept points. They have the same dependence on a_1, a_2, a_3, and R_L, but the fractional coefficient within the parentheses in (4.42) or (4.43) is different; also, some of the assumptions used in generating (4.42) and (4.43) may not be valid for other IM products. Equation (4.38) is valid for all IM products, as long as the correct intercept point is used for IP_n.

Although the power-series concept is simple to implement and gives a good intuitive sense of the IM performance of a quasilinear circuit, it is severely limited. The most obvious limitation of power-series analysis is in the difficulty or, frequently, the impossibility of applying it to circuits that are not unilateral; the circuit must be described by a cascade of linear blocks and a transfer nonlinearity. Most real circuits are not adequately described by such simple models. A second limitation is that a power-series analysis requires memoryless nonlinearities; in particular, it cannot include nonlinear capacitances, which are often significant contributors to IM distortion in solid-state devices. One of the effects caused in part by nonlinear reactances is that the "straight-line" portion of the IM characteristic, shown in Figure 4.5, is not precisely straight; it often includes curvature and small ripples, and, in some cases, sharp nulls are observed at power levels close to saturation. Even so, the IM characteristic of most quasilinear circuits often includes a dominant straight-line component, and an intercept point can be defined in such a way that it describes the component's intermodulation characteristics with reasonable accuracy.

We must also remember that the intercept point concept, as described here, is directly applicable only to two-tone excitations, and that the power relations are based upon the assumption that the levels of both excitations vary in tandem. In practice, one signal may vary and the other may not; in this case the variation in the power of any IM output tone with variation in the power of one excitation tone will differ from the previous case. The variation in output level with variations in a single excitation can be found via the following rule: if the level of a single excitation tone at frequency ω_i is varied while all the other tones remain constant, the IM output power at $\omega_{n,k}$ varies by m_i dB/dB, where $\omega_{n,k}$, ω_i, and m_i are as used in (4.18). We consider the third-order IM product at $2\omega_2 - \omega_1$ as an example. The IM frequency component at $2\omega_2 - \omega_1$ varies by 2 dB/dB with variations in the level of the ω_2 excitation and by 1 dB/dB with variations in the level of the ω_1 excitation. This rule can be used to find the level of an IM product when the excitation levels are dissimilar, as long as the two-tone IM intercept point for excitations of identical levels is known.

4.1.4 Intermodulation Measurement

The system shown in Figure 4.4 does not fully illustrate the difficulty in making good IM measurements. Therefore, before leaving this subject, we offer a few suggestions for making such measurements accurately:

- The outputs of the signal generators must be well filtered. A harmonic output from one signal generator can mix in the test device with the fundamental of the other. The result is a second-order mixing product occurring at the same frequency as a third-order product.
- It is possible for the output of one generator to leak into the input of the other, generating IM products. Those IM products are then applied to the input of the nonlinear circuit and amplified by it. For this reason, it is wise to use isolators or attenuators at the output ports of the signal generators.
- The signal generators must be very stable; usually, they must be frequency synthesizers. Frequency instability limits the resolution bandwidth that can be used in the spectrum analyzer, and thus the sensitivity of the measurement.
- Spurious signals from the frequency synthesizers can interfere with the IM measurement. Usually, these are related to the instruments' 5-MHz time base. For this reason, it is best to use a frequency spacing that is not a har-

monic of the time-base frequency. A frequency spacing of 11 MHz, or some other odd number, is often ideal.

- The input intercept point of the spectrum analyzer must be much greater than the output intercept point of the test device. (If it is not, the measurement will be that of the analyzer, not the test device.) The simplest way to improve the spectrum analyzer's intercept point is to increase its input attenuation, but this decreases its sensitivity as well. In tests of linear power amplifiers, it may be impossible to keep the IM products above the spectrum analyzer's noise level when enough input attenuation is used. In that case, it is necessary to use a narrowband filter to reject the fundamental components at the output of the test device.

4.1.5 Interconnections of Weakly Nonlinear Components

Equation (4.38) is useful for finding the two-tone intermodulation levels in a single, quasilinear circuit. Microwave systems, however, comprise a number of such circuits interconnected in a variety of ways, and it is invariably necessary to have an IM characterization of the entire system. In this section we derive the intercept point of a cascade interconnection of stages. Having that intercept point, we can use (4.38) to find the IM levels at the output of the cascade. The effect on IP_n of the parallel or hybrid interconnection of identical components, a much simpler subject, is considered in Chapter 5; in most cases, we find that all the intercept points are increased by $10 \log_{10}(M)$ dB, where M is the number of identical components in the combination.

Figure 4.6 shows the cascade interconnection of several two-ports. These two-ports may be amplifiers, mixers, control components, or any other type of weakly nonlinear component. We can accommodate linear components by assigning them intercept points that are much greater than those of the other elements. The stages are designated A_m, $m = 1...M$, and their transducer gains and nth-order intercept points are G_m and $IP_{n,m}$, respectively. We assume that the gain and input/output impedances of each stage are constant over a frequency range that includes all IM products of interest, and that the stages' nonlinearities do not interact.

Under these conditions, the system operates as follows: a two-tone signal is applied to the input of A_1, and A_1 passes the linear signal to the output and generates distortion products. These are applied to the input of A_2; A_2 again passes the linear and IM outputs from A_1 to its output, and generates some IM distortion of its own. These distortion products occur at the same frequencies as those generated by A_1, and thus are combined with those from A_1. This process is repeated throughout the rest of the cascade.

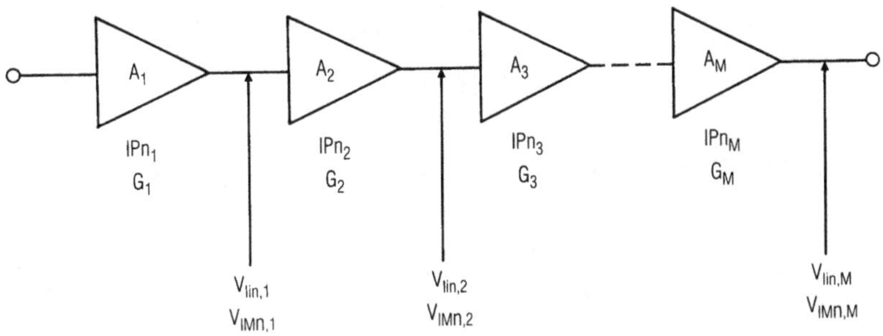

Figure 4.6 Cascade connection of weakly nonlinear components. Although the stages shown here are amplifiers, they can be any type of unilateral two-port.

In general, the phases of the IM distortion products generated in any stage and those passed from its input are unknown. Thus, one generally does not know whether those IM components will be combined in phase (increasing the IM level) or out of phase (decreasing it) or with some other phase. To circumvent this problem, we make a worst-case assumption: that all distortion products combine in phase. This assumption results in an upper bound to the intercept point. That bound inevitably is close to what we measure in practice, because it is likely that, somewhere in the system's passband, all the IM components will have nearly the same phase.

The bound is valid for another reason. A Volterra analysis of a cascade of components shows that, in many situations, the IM products always combine in phase, so the worst-case result is not a bound, but a precise calculation [4.6]. Those cases include such ordinary ones as a cascade of identical components.

The voltage of the linear products at the output of A_1 is designated $V_{\text{lin},1}$ and the IM voltage generated in A_1 is $V_{\text{IM}n,1}$; at the output of A_2, the linear component is $G_2{}^{1/2}V_{\text{lin},1}$ and the IM voltage is $G_2{}^{1/2}V_{\text{IM}n,1} + V_{\text{IM}n,2}$. Thus, at the output of the last stage, A_M,

$$V_{\text{lin},M} = V_{\text{lin},1}(G_2G_3 \ldots G_M)^{1/2} \tag{4.44}$$

and

$$V_{\mathrm{IM}n} = V_{\mathrm{IM}n,M} + V_{\mathrm{IM}n,M-1}G_M^{1/2} + V_{\mathrm{IM}n,M-2}(G_M G_{M-1})^{1/2} + \ldots + V_{\mathrm{IM}n,1}(G_2 G_3 \ldots G_M)^{1/2} \tag{4.45}$$

Converting (4.38) from dBm to units of power gives

$$IP_n^{(n-1)} = P_{\mathrm{lin}}^n P_{\mathrm{IM}n}^{-1} \tag{4.46}$$

We now square $V_{\mathrm{lin},M}$ and $V_{\mathrm{IM}n}$ to obtain $P_{\mathrm{lin},M}$ and $P_{\mathrm{IM}n}$, and substitute the squared forms of (4.44) and (4.45) into (4.46). We also substitute $P_{IMn,m}^{1/2}$ for $V_{\mathrm{IM}n,m}$ in (4.45). This substitution and a little algebra give the result

$$IP_n^{(1-n)/2} = (G_2 \ldots G_M)^{-n/2} P_{\mathrm{lin},1}^{-n/2} [P_{\mathrm{IM}n,M}^{1/2} + (P_{\mathrm{IM}n,M-1} G_M)^{1/2} + \ldots + (P_{\mathrm{IM}n,1} G_2 \ldots G_M)^{1/2}] \tag{4.47}$$

Equation (4.46) shows that, at the output of any stage,

$$P_{\mathrm{IM}n,m} = IP_{n,m}^{(1-n)} P_{\mathrm{lin},1}^n (G_2 \ldots G_m)^{1/2} \tag{4.48}$$

Finally, we substitute (4.48) into (4.47) and, again, grind through the algebra. The result is the cascade relation for intercept point,

$$IP_n^{(1-n)/2} = IP_{n,M}^{(1-n)/2} + (G_M IP_{n,M-1})^{(1-n)/2} + (G_M G_{M-1} IP_{n,M-2})^{(1-n)/2} + \ldots + (G_2 \ldots G_M IP_{n,1})^{(1-n)/2} \tag{4.49}$$

Equation (4.49) shows that the amount each stage contributes to the output intercept point of the cascade is a function of the intercept point of that stage multiplied by the gain of all the stages following it. It is important to note that the gain G_m and nth-order intercept point $IP_{n,m}$ in (4.49) are in units of watts or mW, not dBm.

4.2 VOLTERRA-SERIES ANALYSIS

4.2.1 Introduction to the Volterra Series

The power-series analysis of Section 4.1 was based on the system model shown in Figure 4.1, in which the frequency-sensitive linear part of the circuit and the memoryless nonlinear elements were clearly separate from each other. The model used for the Volterra-series analysis, shown in Figure 4.7, is similar, but the separation between the memoryless and the reactive parts of the circuit has been eliminated. Thus, the nonlinear block may contain a mix of linear and nonlinear elements, with or without memory and feedback. The nonlinear elements may be either resistive or reactive, and they are characterized by power series having the same form as (4.1). As in the previous section, the excitation $s(t)$ contains, in general, a number of individual sinusoidal excitation components having noncommensurate frequencies.

In the previous section we showed that the response $w(t)$ to the excitation $s(t)$, found via the power-series model, could be expressed by (4.8) and (4.9). Those expressions can be recast as follows:

$$
\begin{aligned}
w(t) &= \sum_{n=1}^{N} a_n \left[\frac{1}{2} \sum_{q=-Q}^{Q} V_{s,q} H(\omega_q) \exp(j\omega_q t) \right]^n \\
&= \sum_{n=1}^{N} \frac{a_n}{2^n} \sum_{q1=-Q}^{Q} \sum_{q2=-Q}^{Q} \cdots \sum_{qn=-Q}^{Q} V_{s,q1} \\
&\quad \cdot V_{s,q2} \cdots V_{s,qn} H(\omega_{q1}) H(\omega_{q2}) \cdots H(\omega_{qn}) \\
&\quad \cdot \exp[j(\omega_{q1} + \omega_{q2} + \ldots + \omega_{qn})t]
\end{aligned}
\tag{4.50}
$$

where $H(\omega)$ was the transfer function of the linear part of the circuit, and a_n, $n = 1 \ldots N$, were the coefficients of the terms in the power series that characterized the memoryless nonlinear block. The excitation $s(t)$ was a small-signal, incremental voltage,

Figure 4.7 Weakly nonlinear circuit model for Volterra-series analysis. The circuit may have both reactive and resistive nonlinearities.

$$s(t) = \frac{1}{2}\sum_{q=-Q}^{Q} V_{s,q} \exp(j\omega_q t) \tag{4.51}$$

(By the use of the term $V_{s,q}$, we have implied that the signal is a voltage, as is usually the case in microwave circuits, but of course it could be a current as well.) As before, $q \neq 0$. In Volterra-series analysis, we assume that the excitation has the same form as (4.51), and we find that the response can be expressed as the following function of the excitation:

$$\begin{aligned} w(t) = \sum_{n=1}^{N} \frac{1}{2^n} \sum_{q1=-Q}^{Q} \sum_{q2=-Q}^{Q} \cdots \sum_{qn=-Q}^{Q} & V_{s,q1} \\ & \cdot V_{s,q2} \cdots V_{s,qn} H_n(\omega_{q1}, \omega_{q2}, \ldots, \omega_{qn}) \\ & \cdot \exp[j(\omega_{q1} + \omega_{q2} + \ldots + \omega_{qn})t] \end{aligned} \tag{4.52}$$

The only formal difference between (4.52) and (4.50) is that (4.52) contains a single function, $H_n(\omega_{q1}, \omega_{q2}, \ldots, \omega_{qn})$, instead of the product of linear transfer functions, $a_n H(\omega_{q1}) H(\omega_{q2}) \ldots H(\omega_{qn})$. The former function is called the *nth-order nonlinear transfer function*. Knowing it, we can find individual mixing and distortion products in a manner identical to that used in the power-series analysis.

Volterra-series analysis, like power-series analysis, is based on the assumptions that the circuit is weakly nonlinear and that the multiple excitations are small and noncommensurate. In some cases, the two approaches are equivalent; comparing (4.50) and (4.52), we can see that power-series analysis is a special case of Volterra-series analysis, one in which the nonlinear transfer function can be expressed as

$$H_n(\omega_{q1}, \omega_{q2}, \ldots, \omega_{qn}) = a_n H(\omega_{q1}) H(\omega_{q2}) \ldots H(\omega_{qn}) \tag{4.53}$$

All the previous work in Section 4.1 regarding frequency mixing in circuits characterized by a power series is valid for the Volterra series; the primary difference is in the form of the nonlinear transfer function.

4.2.2 Volterra Functionals and Nonlinear Transfer Functions

A fundamental tenet of linear system and circuit theory is that the output $w(t)$ of a linear system or circuit having excitation $s(t)$ can be expressed by the convolution integral

$$w(t) = \int_{-\infty}^{\infty} h(\tau) s(t - \tau)\, d\tau \tag{4.54}$$

where $h(t)$ is the *impulse response*, the response to a pulse having infinitesimal width and infinite amplitude but unit energy. This pulse is called the *Dirac delta function*, $\delta(t)$. It has the property that

$$f(t_0) = \int_{-\infty}^{\infty} f(t) \delta(t - t_0)\, dt \tag{4.55}$$

Equation (4.54) is valid only for linear circuits and systems. An extension of (4.54) to the case of nonlinear systems was proposed by Norbert Wiener [4.7, 4.8], who applied the work of Volterra [4.9] to the problem of analyzing nonlinear systems. Wiener showed that the response of a weakly nonlinear circuit having small excitations can be described by the functional series

$$\begin{aligned} w(t) = & \int_{-\infty}^{\infty} h_1(\tau) s(t-\tau_1) d\tau_1 \\ & + \int_{-\infty}^{\infty}\int_{-\infty}^{\infty} h_2(\tau_1, \tau_2) s(t-\tau_1) s(t-\tau_2) d\tau_1 d\tau_2 \\ & + \int_{-\infty}^{\infty}\int_{-\infty}^{\infty}\int_{-\infty}^{\infty} h_3(\tau_1, \tau_2, \tau_3) s(t-\tau_1) \\ & \cdot s(t-\tau_2) s(t-\tau_3) d\tau_1 d\tau_2 d\tau_3 + \ldots \end{aligned} \tag{4.56}$$

In (4.56), the multidimensional function $h_n(\tau_1, \tau_2, \ldots, \tau_n)$ is called the *nth-order kernel* or the *nth-order nonlinear impulse response*. Just as the linear frequency-domain transfer function $H(\omega)$ is the Fourier transform of $h(t)$, the nonlinear transfer function $H_n(\omega_{q1}, \omega_{q2}, \ldots, \omega_{qn})$ is the n-dimensional Fourier transform of $h_n(\tau_1, \tau_2, \ldots, \tau_n)$. The excitation function $s(t)$ may be any finite small-signal voltage or current waveform, although in microwave circuits we will be interested exclusively in the case where $s(t)$ is the sum of several sinusoidal components, given by (4.51).

Equation (4.56) can be expressed in more compact form as

$$w(t) = \sum_{n=1}^{N} w_n(t) \tag{4.57}$$

where

$$\begin{aligned} w_n(t) = & \int_{-\infty}^{\infty}\int_{-\infty}^{\infty} \cdots \int_{-\infty}^{\infty} h_n(\tau_1, \tau_2, \ldots, \tau_n)\, s(t-\tau_1) \\ & \cdot s(t-\tau_2) \ldots s(t-\tau_n) d\tau_1 d\tau_2 \ldots d\tau_n \end{aligned} \tag{4.58}$$

and we have limited the series to N terms (Nth order). The frequency-domain form of the response can be found by substituting (4.51) into (4.58). The result is

$$
\begin{aligned}
w_n(t) = \frac{1}{2^n} \int_{-\infty}^{\infty} \int_{-\infty}^{\infty} \cdots \int_{-\infty}^{\infty} h_n(\tau_1, \tau_2, \ldots, \tau_n) \sum_{q1=-Q}^{Q} V_{s,\,q1} \\
\cdot \exp[j\omega_{q1}(t-\tau_1)] \sum_{q2=-Q}^{Q} V_{s,\,q2} \\
\cdot \exp[j\omega_{q2}(t-\tau_2)] \ldots \sum_{qn=-Q}^{Q} V_{s,qn} \\
\cdot \exp[j\omega_{qn}(t-\tau_n)]\ d\tau_1 d\tau_2 \ldots d\tau_n
\end{aligned}
\tag{4.59}
$$

As before, we assume in (4.59) and in the following expressions that all summations over $qi = -Q \ldots Q$ do not include $qi = 0$, and for clarity we will not make this point explicitly. Rearranging the terms in (4.59) and interchanging the order of integration and summation gives

$$
\begin{aligned}
w_n(t) = \frac{1}{2^n} \sum_{q1=-Q}^{Q} \sum_{q2=-Q}^{Q} \cdots \sum_{qn=-Q}^{Q} V_{s,\,q1} V_{s,\,q2} \cdots V_{s,\,qn} \\
\cdot \exp[j(\omega_{q1} + \omega_{q2} + \ldots + \omega_{qn})t] \\
\cdot \int_{-\infty}^{\infty} \int_{-\infty}^{\infty} \cdots \int_{-\infty}^{\infty} h_n(\tau_1, \tau_2, \ldots, \tau_n) \\
\cdot \exp[-j(\omega_{q1}\tau_1 + \omega_{q2}\tau_2 + \ldots + \omega_{qn}\tau_n)] \\
\cdot d\tau_1 d\tau_2 \ldots d\tau_n
\end{aligned}
\tag{4.60}
$$

The terms from the integral sign to the end of (4.60) can be recognized as a multidimensional Fourier transform:

$$H_n(\omega_{q1}, \omega_{q2}, \ldots, \omega_{qn}) = \int_{-\infty}^{\infty}\int_{-\infty}^{\infty}\cdots\int_{-\infty}^{\infty} h_n(\tau_1, \tau_2, \ldots, \tau_n) \cdot \exp[-j(\omega_{q1}\tau_1 + \omega_{q2}\tau_2 + \ldots + \omega_{qn}\tau_n)] \cdot d\tau_1\, d\tau_2 \ldots d\tau_n \tag{4.61}$$

Of course, we could find the nonlinear impulse response from the frequency-domain nonlinear transfer function by inverse-Fourier transforming:

$$h_n(\tau_1, \tau_2, \ldots, \tau_n) = \frac{1}{(2\pi)^n}\int_{-\infty}^{\infty}\int_{-\infty}^{\infty}\cdots\int_{-\infty}^{\infty} H_n(\omega_{q1}, \omega_{q2}, \ldots, \omega_{qn}) \cdot \exp[j(\omega_{q1}\tau_1 + \omega_{q2}\tau_2 + \ldots + \omega_{qn}\tau_n)] \cdot d\omega_{q1}\, d\omega_{q2} \ldots d\omega_{qn} \tag{4.62}$$

Calculating multidimensional Fourier transforms is a nasty business; for this and other reasons, we work entirely in the frequency domain. Replacing the integrals in (4.62) with the nonlinear transfer function $H_n(\omega_{q1}, \omega_{q2}, \ldots, \omega_{qn})$ gives the following expression for $w_n(t)$:

$$w_n(t) = \frac{1}{2^n}\sum_{q1=-Q}^{Q}\sum_{q2=-Q}^{Q}\cdots\sum_{qn=-Q}^{Q} V_{s,q1}V_{s,q2}\cdots V_{s,qn} \cdot H_n(\omega_{q1}, \omega_{q2}, \ldots, \omega_{qn}) \cdot \exp[j(\omega_{q1} + \omega_{q2} + \ldots + \omega_{qn})t] \tag{4.63}$$

and summing this expression for $w_n(t)$ over n, $n = 1 \ldots N$, to obtain $w(t)$, gives the expected result, (4.52).

It is worthwhile at this point to examine (4.63) and (4.52) and to note some of their important implications. First, as in the power series analysis, the total response in the Volterra case is simply the sum of the individual nth-order responses. In the power-series case, this result was guaranteed by the separation of the linear from the nonlinear parts of the circuit, and by the limitation of the analysis to a single transfer nonlinearity. In the Volterra case, this result is an obvious consequence of the form of (4.56), but it is not obvious from the nature of the circuit model, where the nonlinear and linear parts of the circuit are freely intermingled. Our ability to separate even orders of mixing products, as well as different mixing products of the same order, is the key to the practicality of the Volterra series. Without that ability, the analysis of weakly nonlinear circuits would be hopelessly laborious.

Although it is beyond the scope of this book to do so, it is possible to show in several different ways that the series (4.56) is convergent, and that the magnitude of each successive term is smaller than the previous one. Because each of the integral terms in (4.56) represents a single order of mixing products $w_n(t)$ in the circuit's total response $w(t)$, the power in the higher-order response components must be less than that in the lower-order response components. This result is consistent with the experimental observation that higher-order nonlinear distortion products are invariably weaker than lower-order ones.

A second property of the nonlinear transfer function is that it must be symmetrical in ω. The reason for this symmetry is obvious from a practical standpoint: there is no order associated with the different tones in the multitone excitation of (4.51), so one must be able to permute the frequencies in, for example, (4.63) without changing the response. Equation (4.62) implies that $h_n(\tau_1, \tau_2, \ldots, \tau_n)$ must also be symmetrical in τ if $H_n(\omega_{q1}, \omega_{q2}, \ldots, \omega_{qn})$ is symmetrical in ω.

4.2.3 Determining Nonlinear Transfer Functions by the Harmonic Input Method

The nonlinear transfer function can be found via a technique called the *harmonic input*, or *probing* method. This method is not very different in concept from the process of finding the frequency-domain transfer function $H(\omega)$ of a linear circuit: we assume that the circuit has the simplest possible excitation, find the response, substitute both into the input/output equation, in this case (4.63), and finally solve algebraically for $H_n(\omega_{q1}, \omega_{q2}, \ldots, \omega_{qn})$.

In linear analysis, we find the linear transfer function $H(\omega)$ by assuming that the input voltage is $1 \cdot \exp(j\omega t)$ and manipulating the output into the form $H(\omega)\exp(j\omega t)$. The ratio of these quantities is the linear transfer function $H(\omega)$. In the case of a nonlinear circuit the situation is, as usual, a little more complicated, but the concepts are much the same. In order to find the nth-order part of the response, we assume the excitation to be

$$s(t) = \exp(j\omega_1 t) + \exp(j\omega_2 t) + \ldots + \exp(j\omega_n t) \tag{4.64}$$

The excitation used to find the nth-order transfer function is the sum of n positive-frequency phasors of unit magnitude; the negative-frequency components are not included. The excitation $s(t)$ need not be a real function of time, because that excitation is used only to determine the transfer functions. From (4.63) the nth-order response component at $\omega_1 + \omega_2 + ... + \omega_n$ has the form

$$w_n(t)\Big|_{\omega = \omega_1 + \omega_2 + \dots + \omega_n} = n! H_n(\omega_1, \omega_2, \dots, \omega_n) \cdot \exp[j(\omega_{q1} + \omega_{q2} + \dots + \omega_{qn})t] \tag{4.65}$$

This expression for $w_n(t)$ is substituted into the circuit equations; only the terms of nth order are retained; terms of other orders do not contribute to the nth-order response, so they can be ignored. The nonlinear transfer function $H_n(\omega_{q1}, \omega_{q2}, \dots, \omega_{qn})$ is found algebraically.

In all cases, the nth-order nonlinear transfer function is found to be a function of the transfer functions of order less than n. Thus, we first use $s(t) = \exp(j\omega_1 t)$ to find $H_1(\omega_1)$, the linear transfer function, then use $s(t) = \exp(j\omega_1 t) + \exp(j\omega_2 t)$ to find the second-order transfer function $H_2(\omega_1, \omega_2)$ as a function of $H_1(\omega_1)$ and $H_1(\omega_2)$. We continue this process until transfer functions of all desired orders have been determined. When all n transfer functions have been found, (4.63) and (4.57) are used to find the levels of the frequency components of interest in the total response. It is not necessary in evaluating (4.63) to find the levels of all possible frequency components; it is necessary only to determine those of interest at each order.

4.2.3.1 Example: Harmonic-Input Method

Figure 4.8 shows a simple, weakly nonlinear circuit consisting of a nonlinear capacitor, a linear resistor, and a voltage source. We shall find the nonlinear transfer function between the excitation, the voltage $v_s(t)$, and the response, the current $i(t)$.

We assume that the capacitor can be characterized adequately by a second-degree polynomial, so the small-signal, incremental voltage v across the capacitor can be expressed as follows:

$$v = S_1 q + S_2 q^2 \tag{4.66}$$

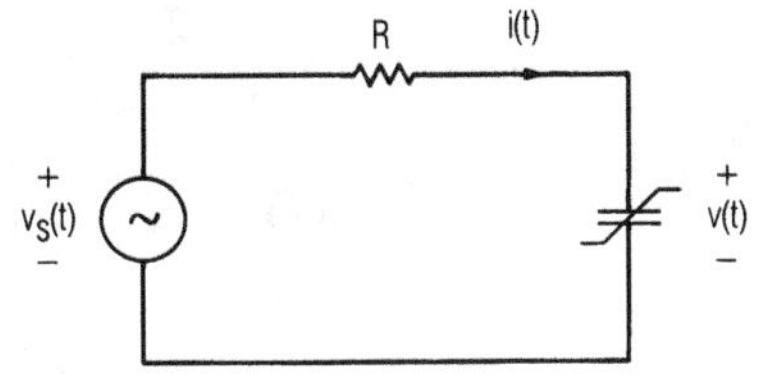

Figure 4.8 Circuit of the example, consisting of a resistor and a nonlinear capacitor.

where S_1 is the capacitor's linear, small-signal, incremental elastance and S_2 is the second-degree Taylor-series coefficient for the capacitor's $V(Q)$ expansion. It is usually most convenient to represent the capacitor's charge as a function of voltage; however, if the voltage can be expressed as a single-valued function of charge over the range of voltages the capacitor will encounter, the charge/voltage function can be inverted to obtain (4.66). (If the voltage is not a single-valued function of charge, the circuit is not amenable to Volterra-series analysis, and also may be unstable.)

The loop voltage equation of the circuit is

$$v_s(t) = Ri(t) + S_1 q(t) + S_2 q^2(t) \tag{4.67}$$

where the charge waveform $q(t)$ is the time-integral of the current:

$$q(t) = \int_{-\infty}^{t} i(\tau) d\tau \tag{4.68}$$

and τ is a variable of integration. We assume in all cases that $q(t)$, $t \to -\infty$, is zero. The nth-order current component, $i_n(t)$, is given by (4.63), where $i_n(t) = w_n(t)$:

$$\begin{aligned} i_n(t) = \frac{1}{2^n} \sum_{q1=-Q}^{Q} \sum_{q2=-Q}^{Q} \cdots \sum_{qn=-Q}^{Q} & V_{s,\,q1} V_{s,\,q2} \cdots V_{s,\,qn} \\ & \cdot H_n(\omega_{q1}, \omega_{q2}, \ldots, \omega_{qn}) \\ & \cdot \exp[j(\omega_{q1} + \omega_{q2} + \ldots + \omega_{qn})t] \end{aligned} \tag{4.69}$$

and the current $i(t)$ is

$$i(t) = \sum_{n=1}^{N} i_n(t) \tag{4.70}$$

We begin by finding the first-order transfer function. Following the form of (4.51), we set

$$v_s(t) = \exp(j\omega_1 t) = \sum_{q=1}^{1} V_{s,q} \exp(j\omega_q t) \tag{4.71}$$

which implies that

$$V_{s,1} = 2.0 \tag{4.72}$$

Substituting (4.71) and (4.72) into (4.69), and (4.69) into (4.70), we obtain

$$i(t) = \sum_{n=1}^{N} \frac{1}{2^n} H_n(\omega_1, \omega_1, \ldots, \omega_1) \exp(jn\omega_1 t) \tag{4.73}$$

The only first-order component in $i(t)$ is the term $H_1(\omega_1)\exp(j\omega_1 t)$. Furthermore, the integration of this term in (4.68) to form $q(t)$ is a linear process, so if $i(t)$ is limited to first order, $S_1 q(t)$ must be of first order as well. The term $q_2(t)$ generates components of second order and above, but no first-order terms. Accordingly,

$$i(t) = H_1(\omega_1)\exp(j\omega_1 t) \tag{4.74}$$

We now substitute (4.71) and (4.73) into (4.67) and equate terms of first order; terms other than first-order do not affect the first-order transfer function, so they can be ignored. Because $q_2(t)$ contains no such terms, it is eliminated, and the remaining terms in (4.67) contain only $i_1(t)$. Equation (4.67) becomes

$$\exp(j\omega_1 t) = RH_1(\omega_1)\exp(j\omega_1 t) + S_1 \int_{-\infty}^{t} H_1(\omega_1)\exp(j\omega_1\tau)d\tau \tag{4.75}$$

Performing the integration and solving for $H_1(\omega_1)$ gives the first-order transfer function,

$$H(\omega_1) = \frac{j\omega_1}{Rj\omega_1 + S_1} \tag{4.76}$$

Equation (4.76) is, of course, nothing more than the linear admittance of the series-connected resistor and capacitor; the first-order transfer function is equivalent to the linear transfer function.

The second-order transfer function is found by setting

$$v_s(t) = \exp(j\omega_1 t) + \exp(j\omega_2 t) \tag{4.77}$$

and by finding the component of $i_2(t)$ at $\omega_1 + \omega_2$. Comparing (4.77) to (4.51), we see that

$$V_{s,\,q} = 2.0 \qquad q = 1, 2 \tag{4.78}$$

and we note that $i(t) = i_1(t) + i_2(t)$. Because we need only second-order terms to form the second-order transfer function, no current components of order greater than two need be included. The excitation has two frequency components, so the first-order current $i_1(t)$ also has two frequency components:

$$i_1(t) = \frac{1}{2}\sum_{q=1}^{2} V_{s,q}\, H_1(\omega_q)\exp(j\omega_q t) \tag{4.79}$$

Substituting for $V_{s,\,q}$ gives

$$i_1(t) = H_1(\omega_1)\exp(j\omega_1 t) + H_1(\omega_2)\exp(j\omega_2 t) \tag{4.80}$$

$H_2(\omega_1, \omega_2)$ relates the excitation voltages to the second-order current $i_2(t)$:

$$i_2(t) = \frac{1}{4}\sum_{q1=1}^{2}\sum_{q2=1}^{2} V_{s,q1}V_{s,q2}H_2(\omega_{q1},\omega_{q2})\exp[j(\omega_{q1}+\omega_{q2})t] \quad (4.81)$$

There are two identical terms in the summation at $\omega_1 + \omega_2$: $q_1 = 1$, $q_2 = 2$; and $q_1 = 2$, $q_2 = 1$. Substituting for $V_{s,q}$ via (4.78) and performing the summation gives

$$i_2'(t) = 2H_2(\omega_1,\omega_2)\exp[j(\omega_1+\omega_2)t] \quad (4.82)$$

where the prime indicates that only the components at a single frequency of (4.81), $\omega_1 + \omega_2$, are included.

We now substitute $i(t)$ into (4.67) to find the second-order transfer function. As before, all terms in the equation of order other than two and at frequencies other than $\omega_1 + \omega_2$ do not contribute to $H_2(\omega_1, \omega_2)$, so they can be ignored. Equation (4.77) shows that $v_s(t)$ contains only first-order components, so it is ignored; only the second-order current components $i_2(t)$ contribute to second-order terms in the linear terms $Ri(t)$ and $S_1q(t)$, so in these terms $i_1(t)$ is ignored. However, only $i_1(t)$ contributes to second-order terms in $S_2q^2(t)$, so

$$\begin{aligned} S_1q(t) &= S_1\int_{-\infty}^{t} i_2'(\tau)\,d\tau \\ &= S_1\int_{-\infty}^{t} 2H_2(\omega_1,\omega_2)\exp[j(\omega_1+\omega_2)\tau]d\tau \end{aligned} \quad (4.83)$$

and carrying out the integration in (4.83) gives

$$S_1q(t) = \frac{2S_1}{j(\omega_1+\omega_2)}H_2(\omega_1,\omega_2)\exp[j(\omega_1+\omega_2)t] \quad (4.84)$$

The squared term, $S_2q^2(t)$, is

$$S_2q^2(t) = S_2\left[\int_{-\infty}^{t} i_1(\tau)d\tau\right]^2 \tag{4.85}$$

Substituting (4.80) into (4.85) gives

$$S_2q^2(t) = S_2\left[\int_{-\infty}^{t}\sum_{q=1}^{2} H_1(\omega_q)\exp(j\omega_q\tau)\,d\tau\right]^2 \tag{4.86}$$

Changing the order of integration and summation in (4.86), performing the integration, and then squaring (4.86) gives the result

$$\begin{aligned} S_2q^2(t) = {} & S_2\sum_{q1=1}^{2}\sum_{q2=1}^{2}\frac{1}{j\omega_{q1}}\frac{1}{j\omega_{q2}}H_1(\omega_{q1})H_1(\omega_{q2}) \\ & \cdot\exp[j(\omega_{q1}+\omega_{q2})t] \end{aligned} \tag{4.87}$$

The summation in (4.87) has two identical terms at $\omega_1 + \omega_2$; therefore,

$$S_2q^2(t)\Big|_{\omega_1+\omega_2} = S_2\left(\frac{-2}{\omega_1\omega_2}\right)H_1(\omega_1)H_1(\omega_2)\exp[j(\omega_1+\omega_2)t] \tag{4.88}$$

Substituting (4.82), (4.84), and (4.88) into (4.67) gives the circuit equation for the second-order components,

$$\begin{aligned} 0 = {} & 2RH_2(\omega_1,\omega_2)\exp[j(\omega_1+\omega_2)t] \\ & + \frac{2S_1}{j(\omega_1+\omega_2)}H_2(\omega_1,\omega_2)\exp[j(\omega_1+\omega_2)t] \\ & - \frac{S_2}{\omega_1\omega_2}H_1(\omega_1)H_1(\omega_2)\exp[j(\omega_1+\omega_2)t] \end{aligned} \tag{4.89}$$

Solving for $H_2(\omega_1, \omega_2)$ gives the expression for the second-order transfer function:

$$H_2(\omega_1,\omega_2) = \frac{j(\omega_1+\omega_2)}{\omega_1\omega_2}\frac{S_2H_1(\omega_1)H_1(\omega_2)}{j(\omega_1+\omega_2)R+S_1} \tag{4.90}$$

The third- and higher-order transfer functions are found in an analogous manner. To find the third-order transfer function, we set

$$v_s(t) = \exp(j\omega_1 t) + \exp(j\omega_2 t) + \exp(j\omega_3 t) \tag{4.91}$$

that is,

$$V_{s,q} = 2.0 \qquad q = 1, 2, 3 \tag{4.92}$$

and find the component of $i_3(t)$ at $\omega_1 + \omega_2 + \omega_3$. It has the form

$$i_3(t) = \frac{1}{8}\sum_{q1=1}^{3}\sum_{q2=1}^{3}\sum_{q3=1}^{3} V_{s,q1}V_{s,q2}V_{s,q3}H_3(\omega_{q1},\omega_{q2},\omega_{q3}) \cdot \exp[j(\omega_{q1}+\omega_{q2}+\omega_{q3})t] \tag{4.93}$$

Again, because they are linear terms, $Ri(t)$ and $S_1q(t)$ generate third-order mixing products from $i_3(t)$ only. From (4.21), there are six components at $\omega_1+\omega_2+\omega_3$ in (4.93), so

$$i_3'(t) = 6H_3(\omega_1,\omega_2,\omega_3)\exp[j(\omega_1+\omega_2+\omega_3)t] \tag{4.94}$$

and

$$\begin{aligned} S_1q(t) &= S_1\int_{-\infty}^{t} i_3(\tau)\,d\tau \\ &= \frac{6}{j(\omega_1+\omega_2+\omega_3)}H_3(\omega_1,\omega_2,\omega_3) \cdot \exp[j(\omega_1+\omega_2+\omega_3)t] \end{aligned} \tag{4.95}$$

The excitation $v_s(t)$ has only first-order terms, so it is again eliminated from consideration. The mixing products generated by $S_2 q_2(t)$ are not obvious, however, as they were in the second-order case. Evaluating $q_2(t)$ is conceptually straightforward but algebraically messy:

$$q^2(t) = \left[\int_{-\infty}^{t} \sum_{n=1}^{N} \sum_{q1=1}^{3} \cdots \sum_{qn=1}^{3} H_n(\omega_{q1}, \omega_{q2}, \ldots, \omega_{qn}) \cdot \exp[j(\omega_{q1} + \ldots + \omega_{qn})\tau] d\tau\right]^2 \tag{4.96}$$

Interchanging the order of summation and integration and performing the integration gives

$$q^2(t) = \left[\sum_{n=1}^{N} \sum_{q1=1}^{3} \cdots \sum_{qn=1}^{3} \frac{H_n(\omega_{q1}, \omega_{q2}, \ldots, \omega_{qn})}{j(\omega_{q1} + \ldots + \omega_{qn})} \cdot \exp[j(\omega_{q1} + \ldots + \omega_{qn})t]\right]^2 \tag{4.97}$$

Squaring (4.97) produces the nasty expression

$$\begin{aligned} q^2(t) = \sum_{n=1}^{N} \sum_{m=1}^{N} \sum_{q1=1}^{3} \cdots \sum_{qn=1}^{3} \sum_{p1=1}^{3} \cdots & \\ \cdot \sum_{pm=1}^{3} \frac{1}{j(\omega_{q1} + \ldots + \omega_{qn})} \frac{1}{j(\omega_{p1} + \ldots + \omega_{pm})} & \\ \cdot H_n(\omega_{q1}, \omega_{q2}, \ldots, \omega_{qn}) H_m(\omega_{p1}, \omega_{p2}, \ldots, \omega_{pm}) & \\ \cdot \exp[j(\omega_{q1} + \ldots + \omega_{qn} + \omega_{p1} + \ldots + \omega_{pm})t] & \end{aligned} \tag{4.98}$$

A careful inspection of (4.98) shows that third-order terms exist only when $m = 1$, $n = 2$ and $m = 2$, $n = 1$, and because of the symmetry in (4.98), both of these cases give identical results. Thus, we need consider only one of them, say, $n = 1$, $m = 2$, and double the result. Evaluating (4.98) and retaining the terms at $\omega_1 + \omega_2 + \omega_3$, we have

$$\begin{aligned} q^2(t) = 2\Big[& \frac{1}{j\omega_1} \frac{2}{j(\omega_2+\omega_3)} H_1(\omega_1) H_2(\omega_2,\omega_3) \\ & + \frac{1}{j\omega_2} \frac{2}{j(\omega_1+\omega_3)} H_1(\omega_2) H_2(\omega_1,\omega_3) \\ & + \frac{1}{j\omega_3} \frac{2}{j(\omega_1+\omega_2)} H_1(\omega_3) H_2(\omega_1,\omega_2) \Big] \\ & \cdot \exp[j(\omega_1+\omega_2+\omega_3)t] \end{aligned} \tag{4.99}$$

Substituting (4.94), (4.95), and (4.99) into (4.67), we obtain

$$\begin{aligned} 0 = \; & 6RH_3(\omega_1,\omega_2,\omega_3) + \frac{6S_1 H_3(\omega_1,\omega_2,\omega_3)}{j(\omega_1+\omega_2+\omega_3)} \\ & - 4S_2\Big[\frac{H_1(\omega_1)H_2(\omega_2,\omega_3)}{\omega_1(\omega_2+\omega_3)} + \frac{H_1(\omega_2)H_2(\omega_1,\omega_3)}{\omega_2(\omega_1+\omega_3)} \\ & + \frac{H_1(\omega_3)H_2(\omega_1,\omega_2)}{\omega_3(\omega_1+\omega_2)} \Big] \end{aligned} \tag{4.100}$$

Solving (4.100) for the third-order transfer function, we have

$$H_3(\omega_1,\omega_2,\omega_3) = \frac{2}{3} \frac{j(\omega_1+\omega_2+\omega_3)S_2\{H_1 H_2\}}{j(\omega_1+\omega_2+\omega_3)R + S_1} \tag{4.101}$$

where $\{H_1 H_2\}$ represents the term in parentheses multiplied by $4S_2$ in (4.100).

It is important to note that we obtained a third-order transfer function, indicating the presence of third-order mixing products, even though the nonlinearity was only of second degree. This result was predicted in Section 4.1 and was illustrated in Chapter 1.

4.2.4 Applying Nonlinear Transfer Functions

When the nonlinear transfer functions $H_n(\omega_1, \omega_2, ..., \omega_n)$, $n = 1 \ldots N$, have been determined, the frequency components of interest can be found in a straightforward manner from (4.52). It is important to recognize that there are many identical terms in (4.52), as explained in Section 4.1.2, and to find the number of identical terms from (4.21). Furthermore, each mixing frequency may have significant current or voltage components at more than one order. We illustrate these points, and the application of the nonlinear transfer functions, by the following examples.

4.2.4.1 Example: Application of Nonlinear Transfer Functions

We wish to find the current component of the third-order mixing frequency $\omega_{3,k} = 2\omega_1 - \omega_2$ in the previous example. The excitation is

$$v_s(t) = V_{s,1}\cos(\omega_1 t) + V_{s,2}\cos(\omega_2 t) \tag{4.102}$$

or

$$v_s(t) = \frac{1}{2}\sum_{q=-2}^{2} V_{s,q}\exp(j\omega_q t) \tag{4.103}$$

We begin by recognizing that the product of interest occurs as a mixing frequency of all odd orders of three or greater; that is,

$$\begin{aligned} 2\omega_1 - \omega_2 &= \omega_1 + \omega_1 - \omega_2 && (n = 3) \\ &= \omega_1 + \omega_1 + \omega_1 - \omega_1 - \omega_2 && (n = 5) \\ &= \omega_1 + \omega_1 + \omega_2 - \omega_2 - \omega_2 && (n = 5) \end{aligned}$$

and so on. However, we assumed earlier that orders above three contribute negligibly to this component, a safe assumption when the nonlinearity is less than third degree and the excitation is very weak.

From (4.69) we obtain

$$i_3(t) = \frac{1}{8}\sum_{q1=-2}^{2}\sum_{q2=-2}^{2}\sum_{q3=-2}^{2} V_{s,q1}V_{s,q2}V_{s,q3} \cdot H_3(\omega_{q1},\omega_{q2},\omega_{q3})\exp[j(\omega_{q1}+\omega_{q2}+\omega_{q3})t] \tag{4.104}$$

Where $H_3(\omega_{q1},\omega_{q2},\omega_{q3})$ is given by (4.101). From (4.21) we find the number of terms at $\omega_{3,k} = 2\omega_1 - \omega_2$, where $n = 3$, $m_1 = 2$, and $m_{-2} = 1$:

$$t_{n,k} = \frac{n!}{m_1!m_{-2}!} = \frac{3!}{2!1!} = 3 \tag{4.105}$$

Evaluating (4.104) at the frequency $\omega_{3,k}$, and including the coefficient $t_{n,k} = 3$, gives

$$i_3'(t) = \frac{3}{8}\{V_{s,1}^2 V_{s,-2}H_3(\omega_1,\omega_1,-\omega_2)\exp[j(2\omega_1-\omega_2)t] + V_{s,-1}^2 V_{s,2}H_3(-\omega_1,-\omega_1,\omega_2)\exp[-j(2\omega_1-\omega_2)t]\} \tag{4.106}$$

The prime indicates that (4.106) represents only a single frequency component, not all the frequency components of $i_3(t)$. We note that

$$H_3(-\omega_1,-\omega_1,\omega_2) = H_3^*(\omega_1,\omega_1,-\omega_2) \tag{4.107}$$

In this case $V_{s,1}$ and $V_{s,2}$ have arbitrary phases, so without losing generality we can set the phase of both equal to zero; then

$$V_{s,-1} = V_{s,1}^* = V_{s,1} \tag{4.108}$$

and

$$V_{s,-2} = V_{s,2}^* = V_{s,2} \tag{4.109}$$

With this assumption, (4.106) becomes

$$i_3'(t) = \frac{3}{4} V_{s,1}^2 V_{s,2}|H_3(\omega_1,\omega_1,-\omega_2)|\cos[(2\omega_1-\omega_2)t+\phi_3] \tag{4.110}$$

where ϕ_3 is the phase of $H_3(\omega_1, \omega_1, -\omega_2)$.

From (4.100) and (4.101) we see that

$$H_3(\omega_1, \omega_1, -\omega_2) = \frac{2}{3}\frac{j(2\omega_1 - \omega_2)S_2\{H_1H_2\}}{j(2\omega_1 - \omega_2)R + S_1} \tag{4.111}$$

and at this mixing frequency,

$$\{H_1H_2\} = \frac{2H_1(\omega_1)H_2(\omega_1, -\omega_2)}{\omega_1(\omega_1 - \omega_2)} - \frac{H_1^*(\omega_2)H_2(\omega_1, \omega_1)}{2\omega_1\omega_2} \tag{4.112}$$

4.2.4.2 Example: Application to Gain Compression or Enhancement

We wish to find the current in the circuit at ω_1 when the excitation is

$$v_s(t) = V_{s,1}\cos(\omega_1 t) = \frac{1}{2}[V_{s,1}\exp(j\omega_1 t) + V_{s,-1}\exp(-j\omega_1 t)] \tag{4.113}$$

where, again, we have assumed $V_{s,1}$ to be real. This is, of course, a trivial problem unless we include the effects of the capacitor's nonlinearity. The effect of this nonlinearity is to add a third-order component at ω_1, the term $\omega_1 = \omega_1 + \omega_1 - \omega_1$.

Equations (4.69) and (4.70) are evaluated, as in the earlier example. there are three identical terms at ω_1 in the three-fold summation; we could use (4.21) to determine that fact, or simply recognize that they are obviously $\omega_1 + \omega_1 - \omega_1, \omega_1 - \omega_1 + \omega_1$, and $-\omega_1 + \omega_1 + \omega_1$. In any case, we obtain

$$\begin{aligned} i(t) = {} & \frac{1}{2}[V_{s,1}H_1(\omega_1)\exp(j\omega_1 t) + V_{s,-1}H_1(-\omega_1)\exp(-j\omega_1 t)] \\ & + \frac{3}{8}[V_{s,1}^3 H_3(\omega_1, \omega_1, -\omega_1)\exp(j\omega_1 t) \\ & + V_{s,-1}^3 H_3(-\omega_1, -\omega_1, \omega_1)\exp(-j\omega_1 t)] \end{aligned} \tag{4.114}$$

Converting (4.114) into cosine form gives

$$
\begin{aligned}
i(t) = {} & V_{s,1}|H_1(\omega_1)|\cos(\omega_1 t + \phi_1) \\
& + \frac{3}{4}V_{s,1}^3|H_3(\omega_1, \omega_1, -\omega_1)|\cos(\omega_1 t + \phi_3)
\end{aligned}
\tag{4.115}
$$

The phase angles ϕ_1 and ϕ_3 in (4.115) are the phase angles of $H_1(\omega_1)$ and $H_3(\omega_1, \omega_1, -\omega_1)$, respectively. Because both terms in (4.115) have the same frequency, these phases cannot be ignored; they are important in establishing the behavior of $i(t)$ with changes in $V_{s,1}$. The results show that, when $V_{s,1}$ is very small, the current depends upon the linear transfer function only, but as $V_{s,1}$ increases, the third-order transfer function rapidly becomes more significant. As a result, $i(t)$ does not increase linearly with $V_{s,1}$. If $\phi_3 \cong \phi_1 + \pi$, a common situation, the current increases progressively less rapidly with $V_{s,1}$ and at some point may even decrease. If $\phi_3 \cong \phi_1$, $i(t)$ increases more rapidly than $V_{s,1}$, and gain enhancement is observed. (Gain enhancement is infrequently encountered, and then only over a limited range of input levels.) We see that saturation effects can be attributed to the progressively greater significance of high-order mixing components, at the fundamental excitation frequency, as excitation level is increased.

4.2.5 The Method of Nonlinear Currents

Another approach to Volterra-series analysis is called the *method of nonlinear currents*. In this technique, current components are calculated from voltage components of lower order, much as are transfer functions in the harmonic-input method. Voltage components of the same order are then determined from those currents, and the next higher-order currents are found. It is necessary only to calculate the frequency components that are of interest, or that contribute to a higher-order component of interest; it is rarely necessary to calculate the entire nonlinear transfer function.

In many cases, the method of nonlinear currents is easier to use than the transfer function approach: it is a little easier to apply to circuits having multiple nodes and is more amenable to computer-aided design techniques. However, the nonlinear-current method is not based on transfer functions, (although it can be used to generate the nonlinear transfer functions of circuits numerically), so it is not directly useful for system analysis. Nevertheless, it is possible to show that the recursive process at the heart of this method is equivalent to the explicit generation of nonlinear transfer functions.

Figure 4.9 shows a simple nonlinear circuit consisting of a voltage source, a linear resistor, and a nonlinear conductance. The conductance has voltage $v(t)$ across it and current $i(t)$; its I/V relation is

$$i = g_1 v + g_2 v^2 + g_3 v^3 + \ldots \tag{4.116}$$

where, as before, the g_n are Taylor-series coefficients. As in the previous cases, $i(t)$ and $v(t)$ in (4.116) represent the small-signal incremental current and voltage in the nonlinear conductance—that is, the ac deviation around a bias point. The voltage $v(t)$ consists of all orders of mixing products:

$$v(t) = v_1(t) + v_2(t) + v_3(t) + \ldots \tag{4.117}$$

where $v_n(t)$ represents the sum of all nth-order mixing products.

Using the substitution theorem (Section 2.2.1), we can redraw the circuit of Figure 4.9 as shown in Figure 4.10, in which the nonlinear conductance has been replaced by a linear conductance and several current sources. The linear conductance represents the linear part of (4.116), and the current sources each represent a nonlinear term in (4.116). If we limit (4.116) to the third degree, and limit consideration to third-order mixing products, then

$$v(t) = v_1(t) + v_2(t) + v_3(t) \tag{4.118}$$

$$v^2(t) = v_1^2(t) + 2v_1(t)v_2(t) \tag{4.119}$$

$$v^3(t) = v_1^3(t) \tag{4.120}$$

The $v_1^2(t)$ term on the right side of (4.119) generates only second-order mixing products, and the second term, $2v_1(t)v_2(t)$, represents third-order products. The circuit of Figure 4.10 can be rearranged as shown in Figure

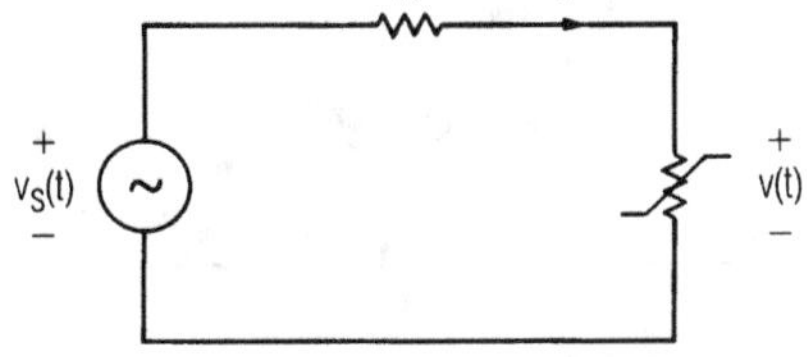

Figure 4.9 A simple series circuit that includes a weakly nonlinear resistor.

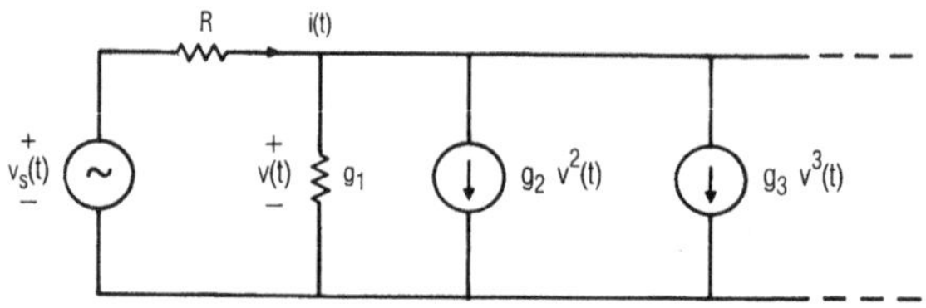

Figure 4.10 The circuit of Figure 4.9, in which the nonlinear resistor has been converted, via the substitution theorem, to a linear resistor and a set of current sources.

4.11, so that each current source represents the same order of mixing frequency; then

$$i(t) = i_{\text{lin}}(t) + i_2(t) + i_3(t) \tag{4.121}$$

where

$$i_{\text{lin}}(t) = g_1 v(t) = g_1[v_1(t) + v_2(t) + v_3(t)] \tag{4.122}$$

$$i_2(t) = g_2 v_1^2(t) \tag{4.123}$$

$$i_3(t) = 2g_2 v_1(t)v_2(t) + g_3 v_1^3(t) \tag{4.124}$$

The current sources $i_2(t)$ and $i_3(t)$ in Figure 4.11 represent all the second- and third-order current components in the nonlinear element that arise from the terms in (4.116) of degree greater than one. The linear part of (4.116),

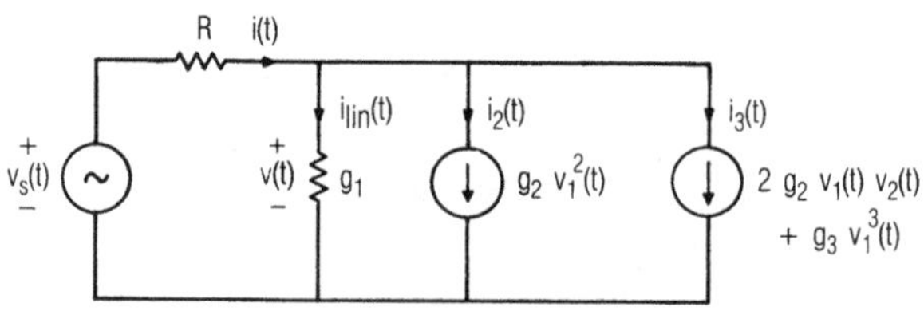

Figure 4.11 A circuit equivalent to that of Figure 4.10, except the current sources have been rearranged so that each represents a single order of mixing products.

expressed by (4.122), accounts for all the other first- and higher-order current components. Rearranging the circuit this way has two remarkable results: first, the circuit in Figure 4.11 is linear, although the current sources are nonlinear functions of the various-order voltage components. Second, the first-order voltage components $v_1(t)$ are generated by the first-order source $v_s(t)$, the second-order current $i_2(t)$ is a function of the first-order voltages, and the third-order current $i_3(t)$ is a function of the first- and second-order voltages. We find that the currents of each order greater than one are always functions of lower-order voltages. These facts suggest a method of solution:

1. Find the first-order components by setting the current sources equal to zero and finding $v_1(t)$ under $v_s(t)$ excitation; this is an ordinary linear analysis.
2. Find the second-order current, $i_2(t) = g_2 v_1^2(t)$, from the voltages $v_1(t)$ found in the previous step. Then set $v_s(t)$ equal to zero, with $i_2(t)$ the only excitation, and find the second-order voltages, $v_2(t)$, from a linear analysis of the circuit.
3. Find the third-order current $i_3(t)$ from $v_1(t)$, $v_2(t)$, $g_2(t)$, and $g_3(t)$. Then, with $v_s(t)$ and $i_2(t)$ set to zero, find the third-order voltage components.

Because the circuit in Figure 4.11 is linear, we can find the total voltage, $v(t)$, as the superposition of the responses to each individual excitation source.

4.2.5.1 Example: Nonlinear Current Method

We use the nonlinear current method to find the response of the circuit of Figures 4.9 to 4.11 to the multitone excitation,

$$v_s(t) = V_{s,1}\cos(\omega_1 t) + V_{s,2}\cos(\omega_2 t) + \ldots + V_{s,Q}\cos(\omega_Q t) \tag{4.125}$$

or

$$v_s(t) = \frac{1}{2}\sum_{q=-Q}^{Q} V_{s,q}\exp(j\omega_q t) \tag{4.126}$$

where, as usual, $q \neq 0$. We set the current sources equal to zero and find

$$v_1(t) = \frac{1}{Rg_1 + 1} v_s(t) \tag{4.127}$$

From (4.123), (4.126), (4.127), and Figure 4.11, the current source $i_2(t)$ is

$$\begin{aligned} i_2(t) &= g_2 v_1^2(t) \\ &= g_2 \frac{1}{(Rg_1 + 1)^2} \frac{1}{4} \sum_{q_1 = -Q}^{Q} \sum_{q2 = -Q}^{Q} V_{s,\,q1} V_{s,\,q2} \\ &\quad \cdot \exp[j(\omega_{q1} + \omega_{q2})t] \end{aligned} \tag{4.128}$$

Setting all sources except $i_2(t)$ to zero, we find easily that

$$\begin{aligned} v_2(t) &= \frac{-R}{Rg_1 + 1} i_2(t) \\ &= \frac{-g_2 R}{(Rg_1 + 1)^3} \frac{1}{4} \sum_{q_1 = -Q}^{Q} \sum_{q2 = -Q}^{Q} V_{s,\,q1} V_{s,\,q2} \\ &\quad \cdot \exp[j(\omega_{q1} + \omega_{q2})t] \end{aligned} \tag{4.129}$$

The third-order current $i_3(t)$ consists of two separate terms:

$$\begin{aligned} i_3(t) &= 2g_2 v_1(t) v_2(t) + g_3 v_1^3(t) \\ &= i_{3a}(t) + i_{3b}(t) \end{aligned} \tag{4.130}$$

$$i_{3a}(t) = \frac{2g_2}{Rg_1 + 1} \frac{1}{2} \sum_{q1=-Q}^{Q} V_{s,q1} \exp(j\omega_{q1} t)$$
$$\cdot \frac{-g_2 R}{(Rg_1 + 1)^3} \frac{1}{4} \sum_{q2=-Q}^{Q} \sum_{q3=-Q}^{Q} V_{s,q2} V_{s,q3} \tag{4.131}$$
$$\cdot \exp[j(\omega_{q2} + \omega_{q3})t]$$

which can be simplified to

$$i_{3a}(t) = \frac{-g_2^2 R}{4(Rg_1 + 1)^4} \sum_{q_1=-Q}^{Q} \sum_{q2=-Q}^{Q} \sum_{q3=-Q}^{Q} V_{s,q1} V_{s,q2} V_{s,q3} \cdot \exp[j(\omega_{q1} + \omega_{q2} + \omega_{q3})t] \tag{4.132}$$

Substituting (4.127) into the second term of (4.130), $i_{3b}(t)$, gives

$$i_{3b}(t) = \frac{g_3}{8(Rg_1 + 1)^3} \sum_{q_1=-Q}^{Q} \sum_{q2=-Q}^{Q} \sum_{q3=-Q}^{Q} V_{s,q1} V_{s,q2} V_{s,q3} \cdot \exp[j(\omega_{q1} + \omega_{q2} + \omega_{q3})t] \tag{4.133}$$

The third-order voltage is

$$v_3(t) = -i_3(t) \frac{R}{Rg_1 + 1} \tag{4.134}$$

and the explicit form of $v_3(t)$ can be found by substituting (4.132) and (4.133) into (4.134).

The expressions (4.127), (4.129), and (4.134) can be evaluated to determine any mixing product of interest. In some cases, however, it is easier to perform a type of ad hoc analysis to determine these products; then, only the minimum number of frequency components necessary to obtain a particular mixing product are evaluated, instead of general expressions for all mixing products. This approach is possible because only a limited number of lower-order mixing products contribute to any specific higher-order product. This analysis is illustrated by the following example.

4.2.5.2 Example: Two-Tone Intermodulation

We wish to find the $2\omega_1 - \omega_2$ third-order current in the previous example, where the excitation source has two tones. The excitation is given by (4.126) with $Q = 2$. From (4.126) and (4.127),

$$v_1(t) = \frac{1}{2}\sum_{q=-2}^{2} V_q \exp(j\omega_q t) \tag{4.135}$$

where

$$V_q = \frac{V_{s,q}}{Rg_1 + 1} \tag{4.136}$$

Clearly, we need only find the positive-frequency component at $2\omega_1 - \omega_2$; the negative-frequency component is just its complex conjugate. Therefore, we need only find the lower-order mixing products that contribute to this positive-frequency component. The third-order component we wish to find is generated by both terms in (4.130), $i_{3a}(t)$ and $i_{3b}(t)$. The term $i_{3b}(t)$ is an obvious contributor, generating the mixing product $\omega_1 + \omega_1 - \omega_2$, but $i_{3a}(t)$ also contributes to $2\omega_1 - \omega_2$ via two mixing products: the second-order frequency $2\omega_1$ in $v_2(t)$ mixing with the first-order frequency $\omega_{-2} = -\omega_2$ in $v_1(t)$, and the second-order frequency $\omega_1 - \omega_2$ mixing with the first-order frequency ω_1. All other mixing products of order three or lower that contribute to $2\omega_1 - \omega_2$ are just the negative-frequency components of these or are repeated, identical terms. Thus, in order to find $2\omega_1 - \omega_2$, we need only find the first-order components at ω_1 and $-\omega_2$, the second-order components at $2\omega_1$ and $\omega_1 - \omega_2$, and the third-order component from $i_{3b}(t)$ at $2\omega_1 - \omega_2$.

The second-order current is $i_2(t) = g_2 v_1^2(t)$; we designate the two second-order current components of interest at $2\omega_1$ and $\omega_1 - \omega_2$, $i_{2a}(t)$ and $i_{2b}(t)$, respectively. From (4.128) and (4.136),

$$i_{2a}(t) = \frac{g_2}{4} V_1^2 \exp(j2\omega_1 t) \tag{4.137}$$

and

$$i_{2b}(t) = \frac{g_2}{2} V_1 V_2^* \exp(j(\omega_1 - \omega_2)t) \tag{4.138}$$

We note in (4.138) that $V_{-2} = V_2^*$ (= V_2, in this example, because, in this purely resistive circuit, all voltages are real), and that there are two terms in (4.128) at $\omega_1 - \omega_2$ but only one at $2\omega_1$; thus the difference in the coefficients. The voltage components at these frequencies, which we designate $v_{2a}(t)$ and $v_{2b}(t)$, are

$$v_{2a}(t) = \frac{-R}{Rg_1 + 1} i_{2a}(t) \tag{4.139}$$

$$v_{2b}(t) = \frac{-R}{Rg_1 + 1} i_{2b}(t) \tag{4.140}$$

The third-order current $i_3'(t)$ at $2\omega_1 - \omega_2$ is

$$\begin{aligned} i_3'(t) &= i_{3a}(t) + i_{3b}(t) \\ &= 2g_2 v_1(t) v_2(t) + g_3 v_1^3(t) \end{aligned} \tag{4.141}$$

where $v_1(t)$ is given by (4.135), $v_2(t) = v_{2a}(t) + v_{2b}(t)$, and only the terms at $2\omega_1 - \omega_2$ are retained in $i_3'(t)$. Then,

$$\begin{aligned} i_{3a}(t) = 2g_2 \bigg\{ &\frac{V_1}{2} \exp(j\omega_1 t) \frac{-Rg_2}{2(Rg_1 + 1)} V_1 V_2^* \exp[j(\omega_1 - \omega_2)t] \\ &+ \frac{V_2^*}{2} \exp(-j\omega_2 t) \frac{-Rg_2}{4(Rg_1 + 1)} V_1^2 \exp(j2\omega_1 t) \bigg\} \end{aligned} \tag{4.142}$$

or, by simplifying (4.142),

$$i_{3a}(t) = \frac{3}{4} \frac{-Rg_2^2}{(Rg_1 + 1)} V_1^2 V_2^* \exp[j(2\omega_1 - \omega_2)t] \tag{4.143}$$

The contribution from the cubic term is

$$i_{3b}(t) = \frac{3}{8} g_3 V_1^2 V_2^* \exp[j(2\omega_1 - \omega_2)t] \tag{4.144}$$

Finally,

$$i_3'(t) = \frac{3}{4}\left[\frac{-Rg_2^2}{(Rg_1 + 1)} + \frac{g_3}{2}\right] V_1^2 V_2^* \exp[j(2\omega_1 - \omega_2)t] \tag{4.145}$$

The cosine form of (4.145) is found by including the negative-frequency part of $i_3'(t)$; it is simply the conjugate of (4.145). Finally,

$$i_3'(t) = \frac{3}{2}\left[\frac{-Rg_2^2}{(Rg_1 + 1)} + \frac{g_3}{2}\right] \left|V_1^2 V_2\right| \cos[(2\omega_1 - \omega_2)t] \tag{4.146}$$

We note that (4.143) and (4.144) are equivalent to (4.132) and (4.133) when (4.136) is used to substitute $V_{s,1}$ and $V_{s,2}$ for V_1, V_2, and when (4.21) is used to calculate the number of identical terms in the triple summation at $2\omega_1 - \omega_2$. As before,

$$v_3'(t) = i_3'(t) \frac{-R}{Rg_1 + 1} \tag{4.147}$$

4.2.5.3 Example: Application to Nonlinear Capacitance

To show how the nonlinear-current method can be applied to nonlinear capacitances, we shall find an expression for the current in the circuit of Figure 4.8 through the second order. In the example of Section 4.2.3.1, the capacitor was characterized in (4.66) as

$$v = S_1 q + S_2 q^2 \tag{4.148}$$

We now need an expression of the form $q = f(v)$; this can be found, as shown in [1.1], by series reversion:

$$q = C_1 v + C_2 v^2 + C_3 v^3 + \ldots \tag{4.149}$$

where

$$C_1 = \frac{1}{S_1} \tag{4.150}$$

$$C_2 = -\frac{S_2}{S_1^3} \tag{4.151}$$

and

$$C_3 = \frac{2S_2^2}{S_1^5} \tag{4.152}$$

The current $i(t)$ is

$$i(t) = \frac{d}{dt}q(t) = C_1\frac{d}{dt}v(t) + C_2\frac{d}{dt}v^2(t) + C_3\frac{d}{dt}v^3(t) \tag{4.153}$$

After separating the current components of different orders, as in (4.118) through (4.124), we write the first- through third-order current components:

$$i_1(t) = C_1\frac{d}{dt}v(t) \tag{4.154}$$

$$i_2(t) = C_2\frac{d}{dt}v_1^2(t) \tag{4.155}$$

$$i_3(t) = C_2\frac{d}{dt}[2v_1(t)v_2(t)] + C_3\frac{d}{dt}v_1^3(t) \tag{4.156}$$

Again we express $v_s(t)$ by (4.126). Then

$$i_1(t) = \frac{1}{2} \sum_{q=-Q}^{Q} \frac{C_1 j\omega_q}{RC_1 j\omega_q + 1} V_{s,q} \exp(j\omega_q t) \tag{4.157}$$

and

$$v_1(t) = \frac{1}{2} \sum_{q=-Q}^{Q} \frac{1}{RC_1 j\omega_q + 1} V_{s,q} \exp(j\omega_q t) \tag{4.158}$$

Substituting (4.158) into (4.155) gives the second-order current $i_2(t)$:

$$i_2(t) = \frac{C_2}{4} \sum_{q1=-Q}^{Q} \sum_{q2=-Q}^{Q} \frac{j(\omega_{q1} + \omega_{q2})}{(RC_1 j\omega_{q1} + 1)(RC_1 j\omega_{q2} + 1)} \cdot V_{s,q1} V_{s,q2} \exp[j(\omega_{q1} + \omega_{q2})t] \tag{4.159}$$

The second-order voltage is found from the linear circuit, in which $i_2(t)$ is the only excitation:

$$v_2(t) = \frac{-C_2}{4} \sum_{q1=-Q}^{Q} \sum_{q2=-Q}^{Q} \cdot \frac{Rj(\omega_{q1} + \omega_{q2}) V_{s,q1} V_{s,q2} \exp[j(\omega_{q1} + \omega_{q2})t]}{[RC_1 j(\omega_{q1} + \omega_{q2}) + 1](RC_1 j\omega_{q1} + 1)(RC_1 j\omega_{q2} + 1)} \tag{4.160}$$

The astute reader may recognize parts of (4.157) and (4.159) as first- and second-order nonlinear transfer functions; the fractional quantity in (4.157) is clearly equivalent to (4.76) if (4.150) replaces C_1 with S_1. Similarly, the terms in (4.159) are equivalent to those in (4.90), the second-order transfer function, after $H_1(\omega)$ from (4.76) is substituted. If we desired third-order current or voltage components, we could use (4.156) and follow the same process used to obtain the second-order components. The procedure is almost identical to the one employed in the conductance examples; the only difference is that it is necessary to differentiate the multiple summation. Because we are limited to sinusoidal, steady-state excitations, this differentiation simply involves multiplication by $j(\omega_{q1}+\omega_{q2}+\ldots+\omega_{qn})$.

4.2.6 Application to Large Circuits

By now the reader is probably astounded by the enormous amount of effort we have devoted to the analysis of thoroughly trivial circuits. At this point, one might begin to suspect that Volterra-series analysis of large circuits would be prohibitively laborious. Fortunately, the complexity of the analysis increases approximately in proportion to the number of nonlinear elements, not in proportion to the overall complexity of the circuit, so with a little careful bookkeeping, one can apply the Volterra series successfully to remarkably large circuits.

We now consider the circuit of Figure 4.12(a), which consists of a linear network, an excitation source $v_s(t)$, a load admittance $Y_L(\omega)$, and $P-1$ nonlinear elements (the source impedance is treated as a part of the linear network). The nonlinear elements are separated from the linear network, and terminals are added so that each element is in parallel with a port. As with the single-element circuit of the example in Section 4.2.5.1, we use the substitution theorem to replace each nonlinear element by a linear element and a set of current sources representing the current components of order greater than one. The linear elements, including the load admittance $Y_L(\omega)$, are then included in the linear part of the network. As before, we are left with an equivalent circuit of the nonlinear network, one that consists of a linear network and current sources for each order of the mixing products greater than one; these currents are nonlinear functions of the lower-order mixing voltages. The linear network has $P+1$ ports, where $v_L(t)$ and $v_s(t)$ are observed at the Pth and $(P+1)$th ports respectively, and is described by its admittance matrix.

The port voltages and currents can be expressed by the admittance equations, in matrix form:

$$-\begin{bmatrix} I_{1,n} \\ I_{2,n} \\ \cdots \\ I_{P,n} \end{bmatrix} = \begin{bmatrix} Y_{1,1} & Y_{1,2} & \cdots & Y_{1,P} \\ Y_{2,1} & Y_{2,2} & \cdots & Y_{2,P} \\ \cdots & \cdots & \cdots & \cdots \\ Y_{P,1} & Y_{P,2} & \cdots & Y_{P,P} \end{bmatrix} \begin{bmatrix} V_{1,n} \\ V_{2,n} \\ \cdots \\ V_{P,n} \end{bmatrix} + \begin{bmatrix} Y_{1,P+1} \\ Y_{2,P+1} \\ \cdots \\ Y_{P,P+1} \end{bmatrix} \quad (4.161)$$

where $I_{p,n}$ is the phasor representation of the current at some specific nth-order mixing frequency in $i_{p,n}(t)$, the nth-order current source at port p in

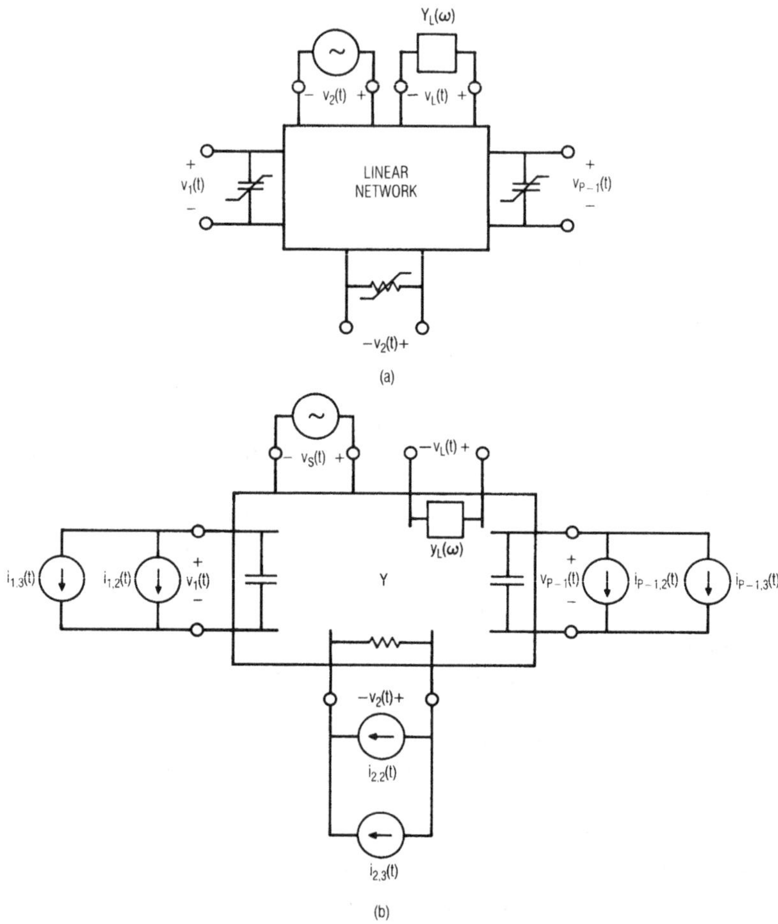

Figure 4.12 (a) A nonlinear circuit divided into a multiport linear network and a set of nonlinear elements. Each element is in parallel with a separate port. (b) The circuit at (a) converted into a linear circuit and a set of current sources.

Figure 4.12(b), and $V_{p,\,n}$ is the corresponding voltage. (There are, of course, many such mixing frequencies at each order; we have left off the frequency subscript in (4.161) for simplicity.) The Y matrix is evaluated at the mixing frequency of interest. We note that $V_{s,\,n} = 0$ when $n > 1$ and $I_{p,\,n} = 0$ when $n = 1$; that is, $V_{s,\,n}$ represents $v_s(t)$, the first-order excitation, and the current sources all represent mixing products. Also, in general $I_{p,\,n} = 0$ for all n unless the output port includes a nonlinear element.

The first-order voltages at all the ports can be found readily by setting **I**, the current vector in (4.161), to zero. We also set $n = 1$ and form

$$\begin{bmatrix} V_{1,\,1} \\ V_{2,\,1} \\ \cdots \\ V_{P,\,1} \end{bmatrix} = -\begin{bmatrix} Y_{1,\,1} & Y_{1,\,2} & \cdots & Y_{1,\,P} \\ Y_{2,\,1} & Y_{2,\,2} & \cdots & Y_{2,\,P} \\ \cdots & \cdots & \cdots & \cdots \\ Y_{P,\,1} & Y_{P,\,2} & \cdots & Y_{P,\,P} \end{bmatrix}^{-1} \begin{bmatrix} Y_{1,\,P+1} \\ Y_{2,\,P+1} \\ \cdots \\ Y_{P,\,P+1} \end{bmatrix} [V_{s,\,1}] \tag{4.162}$$

In general $v_s(t)$ is a multitone excitation, so the admittance matrix in (4.162) must be formulated at each frequency. For orders $n \geq 2$ we set $V_{s,n} = 0$ and $\mathbf{I} \neq \mathbf{0}$; then

$$\begin{bmatrix} V_{1,\,n} \\ V_{2,\,n} \\ \cdots \\ V_{P,\,n} \end{bmatrix} = -\begin{bmatrix} Y_{1,\,1} & Y_{1,\,2} & \cdots & Y_{1,\,P} \\ Y_{2,\,1} & Y_{2,\,2} & \cdots & Y_{2,\,P} \\ \cdots & \cdots & \cdots & \cdots \\ Y_{P,\,1} & Y_{P,\,2} & \cdots & Y_{P,\,P} \end{bmatrix}^{-1} \begin{bmatrix} I_{1,\,n} \\ I_{2,\,n} \\ \cdots \\ I_{P,\,n} \end{bmatrix} \tag{4.163}$$

In order to find the output power at any mixing frequency, we need only evaluate the currents $I_{p,\,n}$; we then use (4.163) to obtain $V_{p,\,n}$, the voltage across the load admittance at each mixing frequency of interest. For simplicity, we use the ad hoc evaluation of mixing products described in the example of Section 4.2.5.2 and apply it to a specific port in Figure 4.12(b). To minimize confusion, we streamline the notation somewhat in the following derivation; we eliminate the port subscript, p, but retain the order subscript, n. The reader should recognize that the voltage and current variables in the following derivation refer to any one port.

The excitation $v_s(t)$ is given by (4.126), and the first-order voltages at each port are found by applying (4.162) at each of the Q excitation frequencies. The first-order voltage at any one port is

$$v_1(t) = \frac{1}{2} \sum_{q=-Q}^{Q} V_{1,\,q} \exp(j\omega_q t) \tag{4.164}$$

We have included the additional subscript 1 in $V_{1,q}$, indicating first order, to distinguish it from higher-order voltages (this subscript was not necessary earlier). If the nonlinear element at that port is a conductance, its incremental I/V characteristic is

$$i = g_1 v + g_2 v^2 + g_3 v^3 + \dots \tag{4.165}$$

and if it is a capacitor, its Q/V characteristic is

$$q = C_1 v + C_2 v^2 + C_3 v^3 + \dots \tag{4.166}$$

The second-order current, in the case of a conductance, is

$$\begin{aligned} i_2(t) &= g_2 v_1^2(t) \\ &= \frac{g_2}{4} \sum_{q_1=-Q}^{Q} \sum_{q2=-Q}^{Q} V_{1,q1} V_{1,q2} \exp[j(\omega_{q1} + \omega_{q2})t] \end{aligned} \tag{4.167}$$

and, in the case of a capacitor,

$$\begin{aligned} i_2(t) &= C_2 \frac{d}{dt} v_1^2(t) \\ &= \frac{C_2}{4} \sum_{q_1=-Q}^{Q} \sum_{q2=-Q}^{Q} j(\omega_{q1} + \omega_{q2}) V_{1,q1} V_{1,q2} \\ &\quad \cdot \exp[j(\omega_{q1} + \omega_{q2})t] \end{aligned} \tag{4.168}$$

We now limit the summation in (4.167) and (4.168) to the current components in $i_2(t)$ at frequencies of interest. The components of interest are not only those whose levels we wish to know, but those that contribute to third- and higher-order mixing products of interest. Current components in $i_2(t)$ at other frequencies may be ignored. Thus, there may be several components in (4.167) and (4.168) that must be evaluated.

The components of $i_2(t)$ that are retained from (4.167) and (4.168) each can be put in the form

$$i_{2,k}(t) = \frac{1}{2}[I_{2,k}\exp(j\omega_{2,k}t) + I^*_{2,k}\exp(-j\omega_{2,k}t)] \tag{4.169}$$

and the total current that these terms represent is

$$i_2'(t) = \sum_{k=1}^{K} i_{2,k}(t) \tag{4.170}$$

As before, the prime indicates that not all the terms in (4.167) or (4.168) are represented in (4.170), although in this case $i_2'(t)$ is real. There are K second-order mixing frequencies of interest in (4.167), (4.168) where

$$\omega_{2,k} = \omega_{q1} + \omega_{q2} \qquad k = 1 \ldots K \tag{4.171}$$

Each $\omega_{2,k}$ is the sum of some two excitation frequencies ω_{q1} and ω_{q1}. We let $t_{2,k}$, given by (4.21), represent the number of terms in (4.168) at frequency $\omega_{2,k}$; then, in the case of a conductance,

$$\begin{aligned} i_{2,k}(t) = {} & \frac{g_2 t_{2,k}}{4}\{V_{1,q1}V_{1,q2}\exp[j(\omega_{q1}+\omega_{q2})t] \\ & + V^*_{1,q1}V^*_{1,q2}\exp[-j(\omega_{q1}+\omega_{q2})t]\} \end{aligned} \tag{4.172}$$

and in a capacitor,

$$\begin{aligned} i_{2,k}(t) = {} & \frac{C_2 t_{2,k}}{4}\{j(\omega_{q1}+\omega_{q2})V_{1,q1}V_{1,q2} \\ & \cdot \exp[j(\omega_{q1}+\omega_{q2})t] \\ & - j(\omega_{q1}+\omega_{q2})V^*_{1,q1}V^*_{1,q2} \\ & \cdot \exp[-j(\omega_{q1}+\omega_{q2})t]\} \end{aligned} \tag{4.173}$$

Equating (4.172) and (4.173) with (4.169) at each frequency $\omega_{2,k}$ gives an expression for $I_{2,k}$:

$$I_{2,k} = \frac{g_2 t_{2,k}}{2} V_{1,q1} V_{1,q2} \tag{4.174}$$

for conductances, and

$$I_{2,k} = \frac{C_2 t_{2,k}}{2} j\omega_{2,k} V_{1,q1} V_{1,q2} \tag{4.175}$$

for capacitors.

This process must be repeated at all the ports having nonlinear elements (i.e., all except the Pth port). The second-order currents at all the ports, and at the first mixing frequency ($\omega_{2,1}$), are then substituted into (4.163) and the port voltages at that frequency are found. The admittance matrix is then reformulated at the next mixing frequency ($\omega_{2,2}$) and (4.163) determines all the port voltages at that mixing frequency. The process is repeated K times, until all the port voltages at all the K second-order mixing frequencies, $\omega_{2,1}$ to $\omega_{2,K}$, are determined.

We now have the second-order voltage components of interest. At each port,

$$v_{2,k}(t) = \frac{1}{2}[V_{2,k}\exp(j\omega_{2,k}t) + V_{2,k}^{*}\exp(-j\omega_{2,k}t)] \tag{4.176}$$

and the second-order voltage at these frequencies is

$$v_2'(t) = \sum_{k=1}^{K} v_{2,k}(t) \tag{4.177}$$

Next, we find the third-order current components in terms of the first- and second-order voltages $V_{1,q}$ and $V_{2,k}$, respectively. These third-order mixing frequencies are designated $\omega_{3,m}$, $m = 1 \ldots M$. We continue to assume that the degree of the nonlinearity in (4.165) and (4.166) is limited to three; thus, in the case of a conductance, the current at some specific port and frequency $\omega_{3,m}$ is

$$i_{3,m} = 2g_2 v_1'(t) v_2'(t) + g_3 v_1'^3(t) \tag{4.178}$$

where $v_1'(t)$ and $v_2'(t)$ include all the first- and second-order frequency components that contribute to $\omega_{3,m}$. Then,

$$\begin{aligned} i_{3,m} = & \sum_{\omega_q + \omega_{2,k} = \omega_{3,m}} 2g_2 \frac{1}{2}[V_{1,q} \exp(j\omega_q t) + V_{1,q}^* \exp(-j\omega_q t)] \\ & \cdot \frac{1}{2}[V_{2,k} \exp(j\omega_{2,k} t) + V_{2,k}^* \exp(-j\omega_{2,k} t)] \\ & + \frac{g_3}{8} \sum_{\substack{q1=-Q \\ (\omega_{q1}+\omega_{q2}+\omega_{q3}=\omega_{3,m})}}^{Q} \sum_{q2=-Q}^{Q} \sum_{q3=-Q}^{Q} V_{s,q1} V_{s,q2} V_{s,q3} \\ & \cdot \exp[j(\omega_{q1} + \omega_{q2} + \omega_{q3})t] \end{aligned} \tag{4.179}$$

The first term in (4.179) is summed over all k and q that give the desired mixing frequency $\omega_{3,m}$ where

$$\omega_{3,m} = \omega_q + \omega_{2,k} = \omega_{q1} + \omega_{q2} + \omega_{q3} \tag{4.180}$$

and the triple summation is evaluated only at the same frequencies. Again we designate $t_{3,m}$, given by (4.21), as the number of terms in the triple summation at frequency $\omega_{3,m}$, and

$$
\begin{aligned}
i_{3,m}(t) = & \left[\sum_{\omega_q + \omega_{2,k} = \omega_{3,m}} \frac{g_2}{2} V_{1,q} V_{2,k} \right. \\
& \left. + \frac{g_3 t_{3,m}}{8} V_{1,q1} V_{1,q2} V_{1,q3} \right] \exp(j\omega_{3,m} t) \\
& \quad (\omega_{q1} + \omega_{q2} + \omega_{q3} = \omega_{3,m}) \\
& + \left[\sum_{\omega_q + \omega_{2,k} = \omega_{3,m}} \frac{g_2}{2} V_{1,q}^{*} V_{2,k}^{*} \right. \\
& \left. + \frac{g_3 t_{3,m}}{8} V_{1,q1}^{*} V_{1,q2}^{*} V_{1,q3}^{*} \right] \exp(-j\omega_{3,m} t) \\
& \quad (\omega_{q1} + \omega_{q2} + \omega_{q3} = \omega_{3,m})
\end{aligned}
\tag{4.181}
$$

In the case of a nonlinear capacitor,

$$
i_{3,m}(t) = 2C_2 \frac{d}{dt}[v_1'(t) v_2'(t)] + C_3 \frac{d}{dt} v'^3_1(t) \tag{4.182}
$$

By the same approach, we obtain

$$
\begin{aligned}
i_{3,m}(t) = {} & j\omega_{3,m}\Bigg[\sum_{\omega_q+\omega_{2,k}=\omega_{3,m}} \frac{C_2}{2} V_{1,q}V_{2,k} \\
& + \frac{C_3 t_{3,m}}{8} V_{1,q1}V_{1,q2}V_{1,q3}\Bigg]\exp(j\omega_{3,m}t) \\
& \quad (\omega_{q1}+\omega_{q2}+\omega_{q3}=\omega_{3,m}) \\
& - j\omega_{3,m}\Bigg[\sum_{\omega_q+\omega_{2,k}=\omega_{3,m}} \frac{C_2}{2} V^*_{1,q}V^*_{2,k} \\
& + \frac{C_3 t_{3,m}}{8} V^*_{1,q1}V^*_{1,q2}V^*_{1,q3}\Bigg]\exp(-j\omega_{3,m}t) \\
& \quad (\omega_{q1}+\omega_{q2}+\omega_{q3}=\omega_{3,m})
\end{aligned}
\tag{4.183}
$$

The third-order current at $\omega_{3,m}$ can be expressed in the form

$$
i_{3,m}(t) = \frac{1}{2}[I_{3,m}\exp(j\omega_{3,m}t) + I^*_{3,m}\exp(-j\omega_{3,m}t)] \tag{4.184}
$$

Comparing (4.181) and (4.183) to (4.184), we obtain

$$
i_{3,m}(t) = \sum_{\omega_q+\omega_{2,k}=\omega_{3,m}} g_2 V_{1,q}V_{2,k} + \frac{g_3 t_{3,m}}{4} V_{1,q1}V_{1,q2}V_{1,q3} \quad (\omega_{q1}+\omega_{q2}+\omega_{q3}=\omega_{3,m}) \tag{4.185}
$$

for the conductance, and

$$
\begin{aligned}
i_{3,m}(t) = {} & j\omega_{3,m}\Bigg[\sum_{\omega_q+\omega_{2,k}=\omega_{3,m}} C_2 V_{1,q}V_{2,k} \\
& + \frac{C_3 t_{3,m}}{4} V_{1,q1}V_{1,q2}V_{1,q3}\Bigg] \\
& \quad (\omega_{q1}+\omega_{q2}+\omega_{q3}=\omega_{3,m})
\end{aligned}
\tag{4.186}
$$

for the capacitor.

Equations (4.185) and (4.186) represent a single mixing frequency $\omega_{3,m}$ at a single port. Either (4.185) or (4.186) must be evaluated for each nonlinear element, and then (4.163) must be used to find the voltage components $V_{3,m}, m = 1 \ldots M$. The Y matrix is then reformulated at the next third-order mixing frequency of interest, and the currents and voltages are again determined.

4.2.7 Controlled Sources

In all the previous sections we ignored the possibility that the circuit might include controlled sources, because Volterra-series modeling of controlled sources is not significantly different from the cases examined above. When a controlled source is included in Figure 4.12(b), the current is simply a function of a voltage at another port instead of the voltage at the current source's terminals. Thus, equations such as (4.185) and (4.186) remain valid as long as the voltages are those of the appropriate port, the one that defines the source's control voltage.

4.2.8 Spectral Regrowth and Adjacent-Channel Power

Many types of modern communication systems organize users into a number of contiguous channels. In such systems, modulating waveforms are carefully filtered to prevent energy from one user spreading into an adjacent channel, causing interference. In spite of such filtering, however, third- and higher-order distortion can cause broadening of the modulated spectrum, called *spectral regrowth.* In weakly nonlinear circuits, this phenomenon can be modeled by Volterra methods.

We assume that our input waveform, $x(t)$, which can represent either voltage or current, is modulated by a periodic signal. In this case, the waveform has a spectrum of discrete frequency components. Then,

$$x(t) = \frac{1}{2}\sum_{q=-K}^{K} X(\omega_q)\exp(j\omega_q t) \tag{4.187}$$

As before, the summation does not include $q = 0$. The system has the linear transfer function $H_1(\omega)$ and the third-order nonlinear transfer function, $H_3(\omega_1, \omega_2, \omega_3)$. Each component of the first-order, linear output $y_1(t)$ is

$$y_1(t) = \frac{1}{2}\sum_{q=-K}^{K} X(\omega_q)H_1(\omega_q)\exp(j\omega_q t) \tag{4.188}$$

and the third-order output is

$$\begin{aligned} y_3(t) = \frac{1}{8}\sum_{q1=-K}^{K}\sum_{q2=-K}^{K}\sum_{q3=-K}^{K} & X(\omega_{q1})X(\omega_{q2})X(\omega_{q3}) \\ & \cdot H_3(\omega_{q1},\omega_{q2},\omega_{q3})\exp[j(\omega_{q1}+\omega_{q2}+\omega_{q3})t] \end{aligned} \tag{4.189}$$

The components that contribute to spectral regrowth are nth order, where n is odd and $(n-1)/2$ components are negative. We assume that third-order terms dominate and one frequency is negative. Thus, we are interested in terms of the form,

$$\begin{aligned} Y_3(\omega_{q1}+\omega_{q2}-\omega_{q3}) = & \frac{3t_k}{4}X(\omega_{q1})X(\omega_{q2})X^*(\omega_{q3}) \\ & \cdot H_3(\omega_{q1},\omega_{q2},-\omega_{q3}) \end{aligned} \tag{4.190}$$

where t_k, as before, represents the number of identical terms at a particular mixing product, k. For narrowband systems, we can assume that both the linear and nonlinear transfer functions are approximately constant, so

$$\begin{aligned} H_3(\omega_{q1},\omega_{q2},-\omega_{q3}) &\approx H_3 \\ H_1(\omega_q) &\approx H_1 \end{aligned} \tag{4.191}$$

Under this assumption, we need only perform a single Volterra analysis to determine H_1 and H_3. Then, we sum (4.189) over all the terms satisfying (4.190). Some of these terms occur at the excitation frequencies, causing compression, while others fall immediately outside the band, causing spectral regrowth.

Figure 4.13 shows a calculation involving a modulated waveform having 10 frequency components. The growth of the interference sidebands and the compression of the modulated waveform are clearly evident. The analysis required less than 1 second on a 200-MHz Pentium computer.

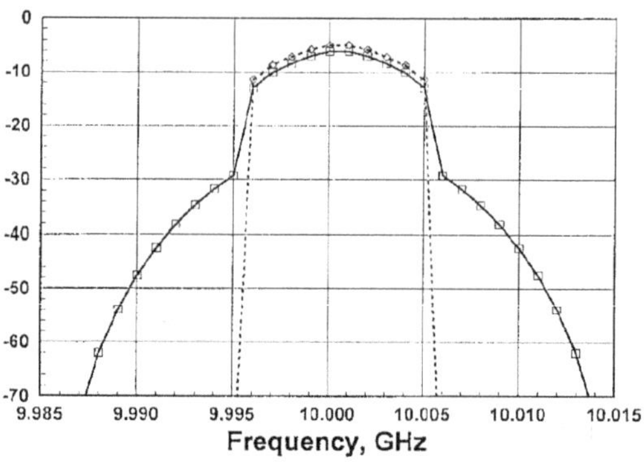

Figure 4.13 An example of spectral regrowth analysis by Volterra series. The dashed line is the linear output, and the solid line includes third-order distortion.

References

[4.1] D. D. Weiner and J. F. Spina, *Sinusoidal Analysis and Modeling of Weakly Nonlinear Circuits*, New York: Van Nostrand, 1980.

[4.2] J. W. Graham and L. Ehrman, "Nonlinear System Modeling and Analysis with Applications to Communications Receivers," *Rome Air Development Center Technical Report No. RADC-TR-73-178*, 1973.

[4.3] J. J. Bussgang, L. Ehrman, and J. W. Graham, "Analysis of Nonlinear Systems with Multiple Inputs," *Proc. IEEE*, Vol. 62, 1974, p. 1088.

[4.4] M. Schetzen, *The Volterra and Wiener Theories of Nonlinear Systems*, New York: Wiley, 1980.

[4.5] M. Schetzen, "Multilinear Theory of Nonlinear Networks," *Journal of the Franklin Inst.*, Vol. 320, 1985, p. 221.

[4.6] S. A. Maas, "Third-Order Intermodulation Distortion in Cascaded Stages," *IEEE Microwave and Guided-Wave Letters*, Vol. 5, 1995, p. 189.

[4.7] N. Wiener, *Nonlinear Problems in Random Theory*, New York: Technology Press, 1958.

[4.8] N. Wiener, "Response of a Nonlinear Device to Noise," *MIT Radiation Lab. Rpt. V-16S*, April 6, 1942.

[4.9] V. Volterra, *Theory of Functionals and of Integral and Integro-Differential Equations*, New York: Dover, 1959.

Chapter 5

Balanced and Multiple-Device Circuits

Single solid-state devices have limitations that may be troublesome in certain applications. One of these is output power; a single device is not always adequate to supply sufficient power or dynamic range. In other cases, a circuit generates potentially troublesome harmonics or intermodulation products, or has spurious responses, and these often cannot be eliminated by filtering. Some of these problems can be solved by a balanced circuit. Balanced circuits connecting two or more solid-state devices or two-port components have many attractive properties beyond improved power and dynamic range; in some cases they can improve bandwidth and input or output match.

Solid-state devices can be combined in many ways. The simplest technique is to connect one or more transistors in parallel. Direct interconnection is often impractical, however, because it changes impedance levels, requires nearly identical devices, or can lead to strange types of spurious oscillation; for example, odd-mode oscillations in power amplifiers. Sometimes it is preferable to employ power-combining components such as hybrids and power dividers, which isolate the individual devices from each other and preserve input and output impedance levels. In some cases, devices can be combined by hybrids at the input or output and directly connected at the opposite port. This chapter examines the types and properties of interconnected devices and discusses the trade-offs involved in the design of balanced and multiple-device circuits.

5.1 Balanced Circuits Using Microwave Hybrids

5.1.1 Properties of Ideal Hybrids

A microwave hybrid coupler is a lossless, four-port, passive component. Each port is matched, and the power applied to any input port is split equally between a pair of output ports. The remaining port is isolated; that is, none of the input power is transferred to it. It is possible to show, by using the properties of the S matrix, that only two types of ideal hybrids are possible: the 180-degree hybrid, in which one path between ports has a phase reversal, and the 90-degree or quadrature hybrid, which has two 90-degree phase shifts.

Figure 5.1 shows, schematically, both types of hybrids. The lines between ports show the possible power transfers and phase shifts; for example, power applied to port 1 of the 180-degree hybrid emerges 3 dB lower and with identical phase at ports 3 and 4, and port 2 is isolated. If port 4 is excited, the outputs are ports 1 and 2, and the voltages at those ports differ in phase by 180 degrees. Similarly, power applied to port 1 of the quadrature hybrid emerges at ports 3 and 4, with 90-degree phase difference, and port 2 is isolated.

The S matrices of ideal 180-degree and quadrature hybrids, $\mathbf{S}_{180}$ and $\mathbf{S}_{90}$, are

$$\mathbf{S}_{180} = \frac{1}{\sqrt{2}}\begin{bmatrix} 0 & 0 & 1 & 1 \\ 0 & 0 & 1 & -1 \\ 1 & 1 & 0 & 0 \\ 1 & -1 & 0 & 0 \end{bmatrix} \tag{5.1}$$

and

$$\mathbf{S}_{90} = \frac{1}{\sqrt{2}}\begin{bmatrix} 0 & 0 & -j & 1 \\ 0 & 0 & 1 & -j \\ -j & 1 & 0 & 0 \\ 1 & -j & 0 & 0 \end{bmatrix} \tag{5.2}$$

Equations (5.1) and (5.2) imply an absolute phase shift of 0, 90, or 180 degrees between the input port and the output ports. However, in real hybrids, the phase difference between a pair of output ports, not between

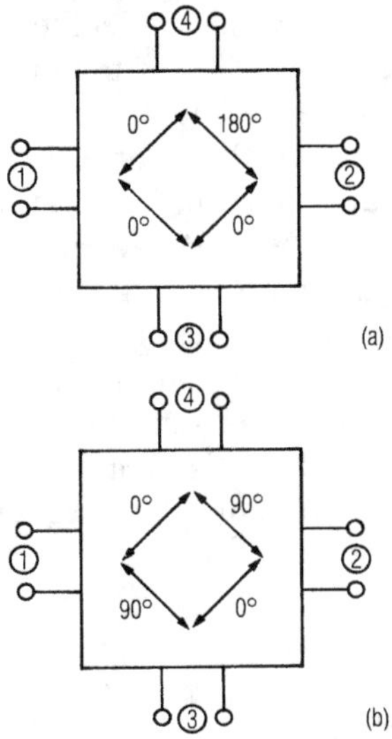

Figure 5.1 Ideal (a) 180-degree and (b) 90-degree, or quadrature hybrids.

the input and output, is of most importance. For example, in Figure 5.1(a) the difference in phase between ports 3 and 4, when driven at port 2, must be 180 degrees; the phase difference between ports 2 and 4, and between 2 and 3, is rarely of concern. The S matrices verify that, in both hybrids, transmission between ports 1 and 2, or between ports 3 and 4, is impossible; these ports are called *mutually isolated pairs*.

Hybrids can be used as power combiners as well as power dividers if inputs are applied to mutually isolated ports. If, for example, waveforms are applied to ports 1 and 2 of the 180-degree hybrid in Figure 5.1(a), the output at port 3 is proportional to the sum of the inputs, and the output at port 4 is proportional to their difference. When the 180-degree hybrid is used this way, port 3 is called the sum, or *sigma*, port, and port 4 is called the difference, or *delta* port. If the other ports, 3 and 4, are used as inputs, then port 1 is the sigma port and port 2 is the delta.

Practical hybrid couplers do not exhibit these ideal characteristics and have only a limited bandwidth over which they approximate the ideal response. The nonidealities of greatest concern are isolation, phase balance, amplitude balance, loss, and port VSWR. Phase balance is the deviation from the ideal phase difference at a pair of output ports; amplitude balance, usually expressed in decibels, is the ratio of the amplitude levels at the output ports. Isolation, also expressed in decibels, is the loss between a pair of mutually isolated ports, and the loss is the ratio of available input power to the sum of the powers at the two output ports. The loss accounts for the dissipation and reflection loss in the hybrid, including power delivered to the termination of the isolated port; it does not include the unavoidable 3-dB power-split loss. The port VSWRs are invariably imperfect, not only because of manufacturing limitations, but also because VSWR, isolation, and loss are not independent quantities; for example, in all hybrids, if even one port VSWR is imperfect, isolation cannot, in theory, be perfect. The VSWRs of the individual ports (as functions of frequency) generally are not the same unless the hybrid is symmetrical. A hybrid's balance, isolation, and VSWR, as a function of frequency, usually establish its bandwidth.

5.1.2 Practical Hybrids

5.1.2.1 Transformer Hybrid

The transformer hybrid is a practical structure for use at frequencies between a few megahertz and approximately 500 MHz, although careful design occasionally allows operation above 2 GHz. This hybrid uses the symmetry properties of a transformer to achieve 180-degree hybrid operation. Its simplest form is shown in Figure 5.2, in which the ports are numbered in a manner that corresponds to Figure 5.1(a). In this configuration the impedance at ports 3 and 4 is not the same as that of ports 2 and 3; however, it is possible to devise more complex transformer hybrid circuits that have equal port impedances.

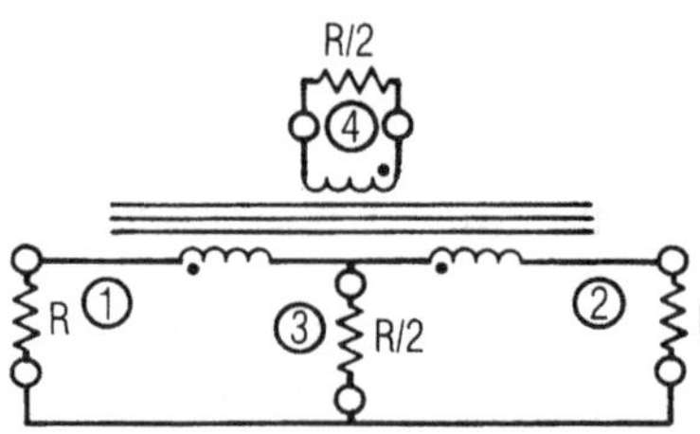

Figure 5.2 The transformer hybrid. All three windings are have the same number of turns, and the port numbering follows Figure 5.1(a).

Figure 5.3 illustrates the operation of the transformer hybrid. In Figure 5.3(a), power is applied to port 4 and is split between the load resistors at ports 1 and 2. Because of the symmetry of the structure, no voltage appears across the resistor at port 3, so it is isolated from port 4. In Figure 5.3(b), port 3 is excited. The currents in the secondary windings (those connected to ports 1 and 2) must be equal and opposite because of the symmetry of the structure, so no voltage is generated across any of the windings, and the loads at ports 1 and 2 are effectively in parallel with port 3. Figure 5.3(c) shows the operation of the hybrid with port 2 excited. Because the windings all have an equal number of turns, the current generated in the primary (port 4) winding is equal to that generated in the winding connected to port 2, causing a power division between those ports. These currents also induce equal but opposite currents in the remaining winding, isolating port 1. Note that the voltage polarities in Figures 5.3(a) and 5.3(c) imply that the 180-degree phase division is between ports 2 and 4.

Transformer hybrids are often realized as shown in Figure 5.4, by a set of trifilar windings on a toroidal core. The core is usually made of a ferrite material. This structure is favored because it confines the magnetic field within the windings and thus provides very good coupling over a wide bandwidth. Transformers realized as so-called *transmission-line transformers* [5.1, 5.2] are used primarily as baluns, not hybrids.

An important property of the transformer hybrid is that its phase and amplitude balance are determined by the structure of the circuit, and not by frequency-sensitive elements such as half-wavelength transmission lines. Accordingly, the hybrid's balance is usually very good over a broad frequency range, and its bandwidth is generally limited by loss and degradation of isolation. This degradation occurs at high frequencies because of stray inductance and capacitive coupling between the transformer windings. Operation is limited at low frequencies by a standard requirement for transformers that the self inductances of the windings have reactances much greater than the load and source impedances.

5.1.2.2 Ring (Rat-Race) Hybrid

Figure 5.5 shows the ring or rat-race hybrid. Unlike the transformer hybrid, the ring hybrid requires frequency-sensitive elements, namely transmission lines of a precise length, that make it a narrow-band component. Figure 5.5 shows a ring hybrid in a form that can be realized in microstrip or stripline; ring hybrids have also been realized in a wide variety of other transmission media, including waveguide.

Power applied to any port of the ring hybrid is divided equally between the two adjacent ports. The remaining port is isolated because there are

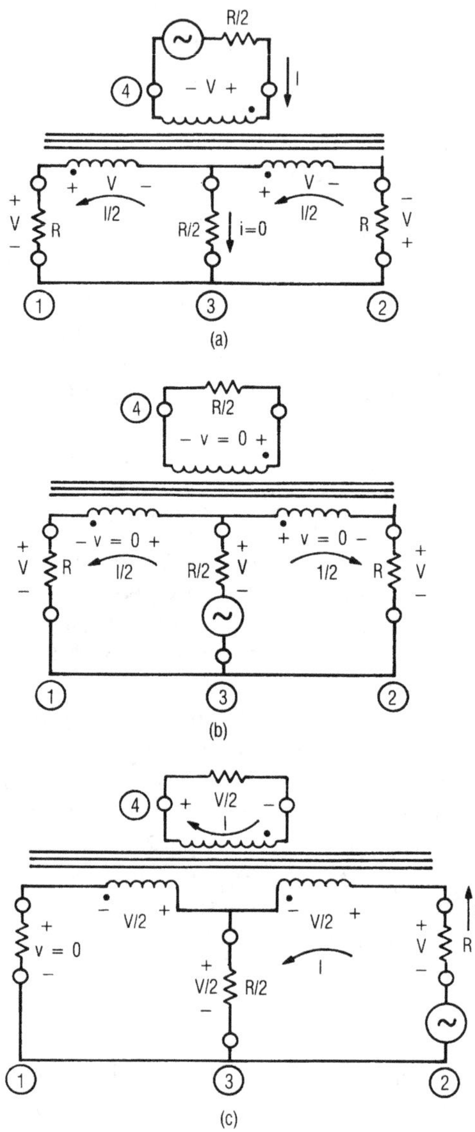

Figure 5.3 Currents and voltages in the transformer hybrid when different ports are excited: (a) port 4 excited; (b) port 3 excited; and (c) port 2 excited.

always two paths between the input port and the isolated port: going around the ring in one direction leads from the input to the isolated port over a 0.5-wavelength path; in the other direction the path is 1.0 wavelength, or 0.5-wavelength longer. The longer path introduces a phase reversal that cancels the voltage at the isolated port and creates a virtual ground at its point of connection to the ring. Because of the extra half wavelength of transmission line, the path from port 4 to port 2 has the 180-degree phase shift.

Because of its relatively low loss and the simplicity of its design and fabrication, the ring hybrid is a very popular design. All the ports have the same impedance, and the ring's characteristic impedance is $\sqrt{2}$ times the port impedance. If transmission line dispersion and junction effects are negligible, the VSWR of each port is less than 2.0 over a nearly 100% bandwidth; however, the transmission bandwidth is much narrower than the VSWR bandwidth, 10% to 20% at most.

5.1.2.3 Wilkinson Hybrid

The Wilkinson hybrid of Figure 5.6 is another simple but effective design. It is usually used as a combiner or power splitter, but it is actually a type of 180-degree hybrid that has a built-in resistor termination on port 2. The resistor's value is $2R$, where R is the impedance of the other three ports. The Wilkinson hybrid uses quarter-wavelength transmission lines, but its phase and amplitude balance depend primarily upon circuit symmetry and thus are broadband. Ports 1 and 2 are mutually isolated, as are ports 3 and 4.

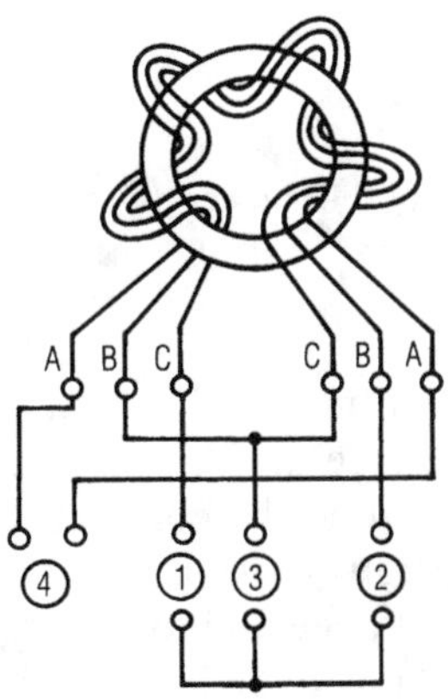

Figure 5.4 The transformer hybrid realized by a trifilar winding on a toroidal core.

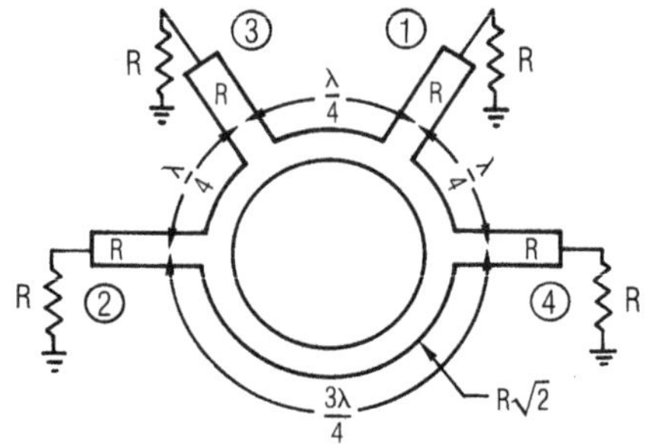

Figure 5.5 The ring hybrid in microstrip or stripline form.

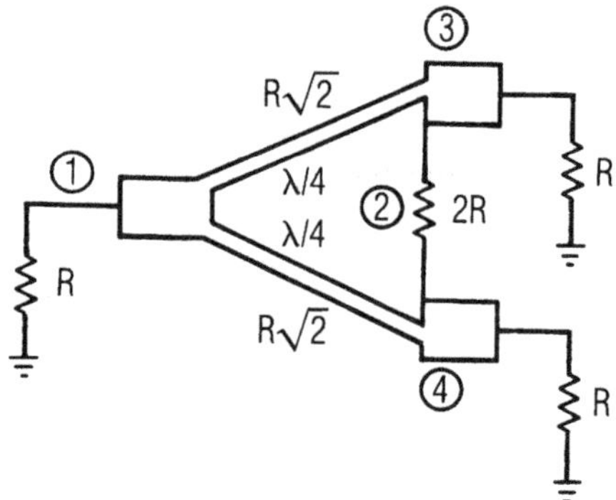

Figure 5.6 The Wilkinson hybrid or power divider.

The Wilkinson hybrid is often used as a power combiner for individual amplifier stages. An advantage in this application is that its port termination, the $2R$ resistor, need not be connected to ground, and because of its excellent balance, large trees of Wilkinson combiners and dividers can be made with good overall balance and low loss. Furthermore, it is possible to make broadband, well-balanced, Wilkinson-like dividers having multiple outputs.

5.1.2.4 Branch-Line Quadrature Hybrid

A branch-line quadrature hybrid is shown in Figure 5.7. It consists of two quarter-wave transmission lines connected by quarter-wave branches; the series lines have characteristic impedances $R/\sqrt{2}$, and the branch impedances are simply R. This hybrid is simple to design and fabricate, and it has very low loss. Because it does not require bond wires or narrow microstrip lines, it can be fabricated successfully on soft substrates to realize low-cost circuits.

The branch-line hybrid has a relatively narrow bandwidth, approximately 10% at the 20-dB isolation points. The port return loss is nearly identical to the isolation. The transmission bandwidths of different pairs of ports are not the same; the excess loss in transmission from port 1 to port 3 is only 0.4 dB over a 40% bandwidth, while the excess loss from port 1 to port 4 is 2.2 dB. The low impedance lines, which provide low loss, also create large discontinuities at the junctions, so performance degrades at high frequencies. When the branch lengths are on the order of the line widths, the hybrid becomes completely impractical. Multisection designs have much greater bandwidth; see [5.3, 5.4].

5.1.2.5 Coupled-Line Quadrature (Lange) Hybrid

One of the most popular types of quadrature directional couplers consists of a pair of coupled microstrip lines, one-quarter wavelength long, as shown in Figure 5.8. This coupler is designed by selecting the even- and odd-mode characteristic impedances of the coupled lines so that

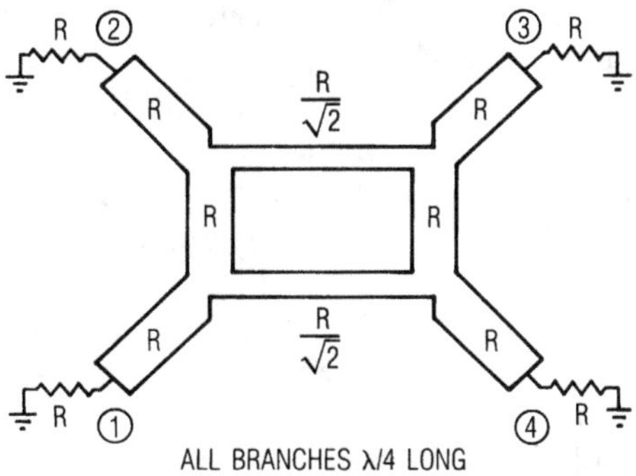

Figure 5.7 The branch-line quadrature hybrid in microstrip or stripline form. All branches are one-quarter wavelength long.

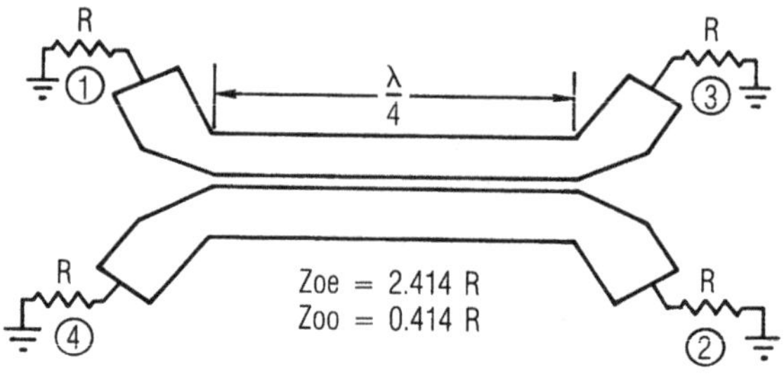

Figure 5.8 The simplest form of the coupled-line hybrid. This structure is used to realize directional couplers having low coupling coefficients; it cannot provide sufficient coupling to be used as a 3-dB hybrid.

$$Z_{0e} = R\left(\frac{1+c}{1-c}\right)^{1/2} \tag{5.3}$$

and

$$Z_{0o} = R\left(\frac{1-c}{1+c}\right)^{1/2} \tag{5.4}$$

where R is the port impedance and c is the voltage coupling ratio, the square root of the power coupling ratio. If such a coupler is designed to achieve a 3-dB power division, it is a quadrature hybrid. In order to achieve 3-dB coupling ($c = 0.707$), (5.3) and (5.4) imply that $Z_{0e} = 2.414R$ and $Z_{0o} = 0.414R$, or 120.7Ω and 20.7Ω, respectively, in a 50Ω system.

Unfortunately, it is virtually impossible in practice to obtain 3-dB coupling from a single pair of edge-coupled lines, even on substrates having high dielectric constants, because the required spacing between the microstrips is impractically small. Furthermore, the coupler has the practical disadvantage that the output ports are always on opposite sides of the structure, so a symmetrical circuit is impossible.

Both of these problems can be solved in a remarkably simple and elegant manner. A solution to the coupling problem is to split the two coupled lines into four, as shown in Figure 5.9. The four strips now have three pairs of adjacent edges, instead of only one in the two-strip case, so the capacitance between the strips is approximately tripled. This modification allows the even- and odd-mode characteristic impedances required to realize a 3-dB coupler to be achieved successfully. The output ports can

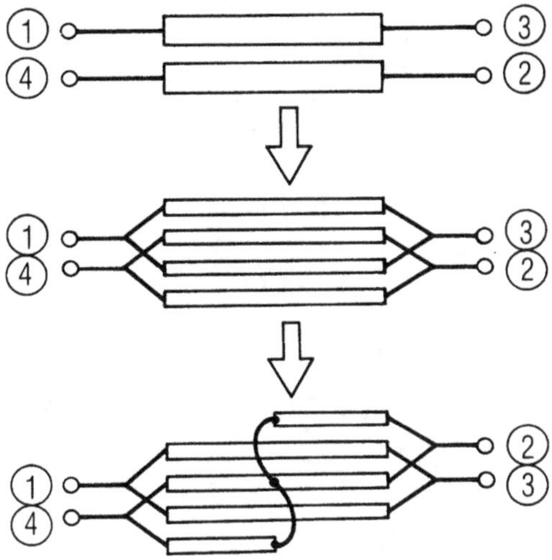

Figure 5.9 Evolution of the Lange hybrid. The simple coupler of Figure 5.8 is split into four strips to increase its coupling, and the strips are rearranged to place both outputs on the same side of the structure.

be interchanged by dividing one of the outer strips and moving half of it to the other side of the coupler; this modification moves the port connected to that strip to the desired location. It is necessary to interconnect the separated strips via wires. This hybrid, named after its inventor, J. Lange [5.5], is one of the most popular and broadband quadrature couplers in use. For more information on their design and operation, see [5.4] and [5.6].

The ideal coupled-line hybrid has a 0.5-dB coupling bandwidth of approximately 50%. Furthermore, the phase balance, isolation, and port VSWRs of a coupled-line hybrid are theoretically perfect and independent of frequency. Perfect isolation implies that any two output ports are complementary; that is, the sum of the powers at the output ports equals the input power. However, the balance is not frequency-independent. If port 1 is excited with voltage V_1, the voltages at the terminated output ports, V_3 and V_4, are, respectively,

$$\frac{V_4}{V_1} = \frac{jc\sin(\theta)}{(1-c^2)^{1/2}\cos(\theta) + j\sin(\theta)} \tag{5.5}$$

and

$$\frac{V_3}{V_1} = \frac{(1 - c^2)^{1/2}}{(1 - c^2)^{1/2}\cos(\theta) + j\sin(\theta)} \tag{5.6}$$

The electrical length θ of the coupler is

$$\theta = \frac{\pi}{2}\frac{\omega}{\omega_0} \tag{5.7}$$

where ω_0 is the hybrid's center frequency. If $c = 0.707$ (a 3-dB hybrid), dividing (5.5) by (5.6) gives

$$\frac{V_4}{V_3} = j\sin(\theta) \tag{5.8}$$

showing that V_4 leads V_3 by 90 degrees at all frequencies. The balance is

$$\left|\frac{V_4}{V_3}\right|^2 = \sin^2(\theta) \tag{5.9}$$

Practical Lange couplers have significant nonidealities. The parasitic inductances of the wires needed for the crossover connections and the unequal phase velocities of even and odd modes on microstrip coupled lines are the dominant effects that limit the coupler's performance. These effects are especially severe at high frequencies. Even so, it is relatively easy to minimize the effects of these factors, and to realize Lange hybrids having remarkably good performance over broad bandwidths. In applications that are not sensitive to amplitude balance, Lange hybrids can be used over very wide bandwidths, often greater than one octave. Conversely, by overcoupling the lines, the imbalance at the band edges can be reduced, and the bandwidth increased, at the cost of imbalance at the bandcenter.

5.1.3 Properties of Hybrid-Coupled Components

Figure 5.10 shows a pair of two-ports connected in parallel by microwave hybrids. The two-ports can be of any type, and the hybrids can be 180-

degree or quadrature designs. The requirements for the interconnection are (1) that the ports of each hybrid that are connected to the two-ports be mutually isolated pairs; (2) that the phase shifts from the input to the output through each branch (i.e., through the input hybrid, one of the two-ports, and the output hybrid) be equal; and (3) that the two-ports be identical. If these conditions are met, the coupled pair of two-ports has the same gain as either two-port, but twice the output-power capability. The coupled pair may have other useful properties, depending upon the type of hybrid used for the interconnection.

5.1.3.1 180-Degree Hybrid-Coupled Components

Figure 5.11 shows a pair of identical two-port components connected in parallel by 180-degree hybrids. The components can be connected to in-phase or out-of-phase pairs of the hybrid's ports, but the characteristics are different for the two interconnections.

This configuration is frequently used to power-combine amplifier stages. Because of their structural simplicity and excellent amplitude and phase balance, Wilkinson hybrids are a natural choice for power-combining; they are configured with port 1 as the input or output and the amplifiers connected to ports 3 and 4 (see Figure 5.6). In order to achieve high output power, Wilkinson-like power dividers having multiple outputs are often used to combine a large number of amplifiers.

Figure 5.12 illustrates the operation of the input and output hybrids. In Figure 5.12(a), the input port has voltage V_s and current I_s; the available power at the input is divided in half by the hybrid, so the voltage V_i and current I_i at the two outputs of the hybrid, the inputs of N_1 and N_2, is

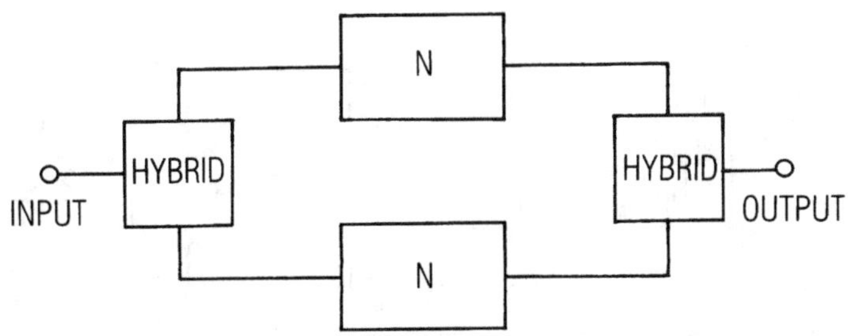

Figure 5.10 General configuration of hybrid-coupled two-ports.

$$|V_i| = \frac{|V_s|}{\sqrt{2}} \tag{5.10}$$

and

$$|I_i| = \frac{|I_s|}{\sqrt{2}} \tag{5.11}$$

We have assumed that the hybrid is ideal, the two-ports are identical, and both two-ports have the same input impedance Z_i. One can show from (5.1) that the input impedance of the hybrid under these conditions must also be Z_i; the input reflection coefficient of the hybrid-coupled components is that of the individual components. The voltages and currents at the outputs of N_1 and N_2 (i.e., those at the inputs of the output hybrid), V_o and I_o, are identical. The 3-dB power division in the output hybrid reduces V_o by $\sqrt{2}$, but two input voltages are combined in phase, so the output voltage and current V_L and I_L are

$$|V_L| = \frac{2|V_o|}{\sqrt{2}} \tag{5.12}$$

and

$$|I_L| = \frac{2|I_o|}{\sqrt{2}} \tag{5.13}$$

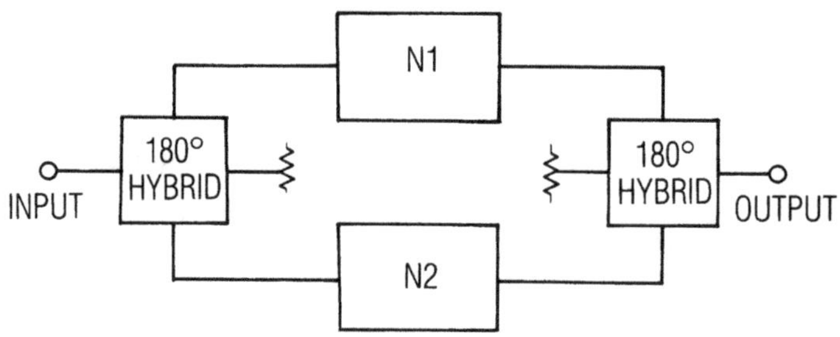

Figure 5.11 Hybrid-coupled two-ports using 180-degree hybrids or power dividers.

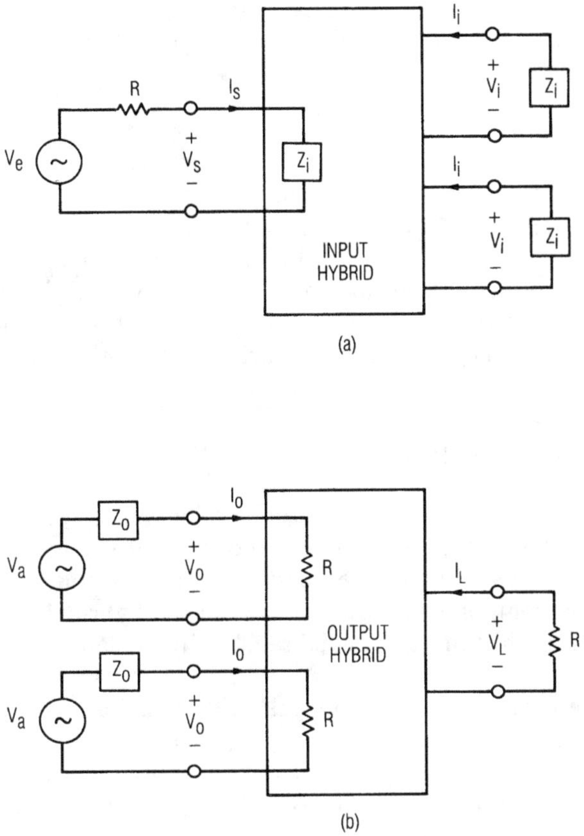

Figure 5.12 180-degree hybrid-coupled components. This configuration has limited second-order harmonic and spurious rejection as long as the output hybrid's amplitude and phase balance are maintained at the spurious output frequency.

Unsurprisingly, the output power is 3 dB greater than the powers at either input. Thus, the hybrid introduces a 3-dB loss in available power at its inputs, but reclaims that loss at its output by combining the voltages in phase. The transducer gain (Section 1.5) of the hybrid-coupled pair is, therefore, the same as that of the individual stages, but the available output power is 3 dB greater.

A consequence of coupling the components via a 180-degree hybrid is a modest improvement in the worst-case VSWR. It can be shown that the reflection coefficient at the input of the ideal hybrid is

$$\Gamma_{in} = 0.5(\Gamma_3 + \Gamma_4) \tag{5.14}$$

where Γ_{in} is the input impedance looking into port 1, and Γ_3 and Γ_4 are the reflection coefficients of the terminations on the input hybrid's output ports, ports 3 and 4, respectively, in Figure 5.1. Thus, if the two-ports are not precisely identical, and at some frequency the input reflection coefficient of one two-port is much poorer than that of the other, the averaging effect of the 180-degree hybrid reduces the worst-case input reflection coefficient. The same property is evident at the output.

5.1.3.2 Quadrature-Coupled Components

Figure 5.13 shows a pair of two-port components coupled via quadrature hybrids. The crossover paths between ports in the ideal hybrids have identical phase delays of 90 degrees, and the straight-through paths have no phase delay; the components are connected to the hybrids' ports in such a way that the phase shift through each branch of the balanced structure is the same.

The magnitudes of the port voltages in the quadrature hybrids are the same as those of the 180-degree hybrid; (5.10) through (5.13) apply to Figure 5.13 as well as to Figure 5.12. The only consequence of the different phase shifts is that the quadrature hybrid's ports must be configured as

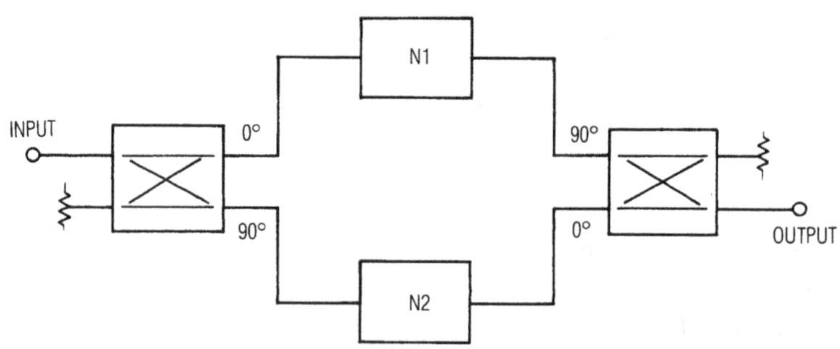

Figure 5.13 Hybrid-coupled two-ports using quadrature hybrids. The paths in the hybrid that have the 90-degree phase shift are the crossover paths.

shown in Figure 5.13, so that the output voltages of the individual components N_1 and N_2 combine in phase. Thus, the gain of the quadrature-coupled components is equal to that of the individual two-ports, and the output power capability is 3 dB greater.

The most attractive property of the quadrature-coupled configuration is that, in the ideal case, the input reflection coefficient is zero, regardless of the input reflection coefficients of the individual two-ports. One can derive this property by using (5.2) and by assuming that port 1 is the input and that ports 3 and 4 have terminations with reflection coefficient Γ. The terminations constrain the *a* and *b* waves at ports 3 and 4 as follows:

$$a_3 = \Gamma b_3 \tag{5.15}$$

and

$$a_4 = \Gamma b_4 \tag{5.16}$$

Substituting (5.15) and (5.16) into (5.2) gives

$$\begin{bmatrix} b_1 \\ b_2 \end{bmatrix} = \Gamma \begin{bmatrix} 0 & -j \\ -j & 0 \end{bmatrix} \begin{bmatrix} a_1 \\ a_2 \end{bmatrix} \tag{5.17}$$

Equation (5.17) implies that all the input power reflected from the individual components is dissipated in the load at port 2, and none emerges from port 1; in simple terms, the input port is matched. The same considerations apply to the output port, which is also matched. An intuitive explanation of this phenomenon is that a wave reflected from port 4 returns to port 1 without phase shift, but a wave reflected from port 3 passes through the hybrid's 90-degree path twice, returning to port 1 with 180-degree phase shift. Thus, reflected waves cancel at port 1. However, the reflected waves undergo identical phase shifts between the input and port 4, and therefore combine in phase. Similarly, one can show that, when the terminations on ports 3 and 4 are unequal and the termination on port 2 is ideal,

$$\Gamma_{in} = 0.5(\Gamma_3 - \Gamma_4) \tag{5.18}$$

where $\Gamma_{in} = b_1/a_1$, and Γ_3, Γ_4 are the reflection coefficients of the terminations at ports 3 and 4, respectively. Thus, even when the port terminations are not precisely equal, the input VSWR may still be very low.

This property of quadrature-coupled components is indeed delightful; equally delightful is the fact that, if the quadrature hybrid is an ideal coupled-line hybrid, good gain is achieved over a very broad bandwidth. The reason for the broadband operation is that the coupled-line hybrid's phase balance, port VSWR, and isolation are theoretically perfect and frequency independent. Its amplitude balance is imperfect, varying as $\sin^2(\omega / \omega_0)$, but this imperfection is not as important as it may seem at first. Because the hybrid's isolation and input VSWR are perfect, all the available input power must appear at the output ports; that is, if the loss from the input to one port of the hybrid is L ($L < 1$), the loss from the input to the other port must be $1 - L$. Figure 5.13 shows that the coupled stages are connected to the hybrids in such a way that a signal must experience loss L through one hybrid and loss $1 - L$ in the other. Therefore, the gain through the input hybrid, either component, and the output hybrid is $L\,(1 - L)G_t$, where G_t is the transducer gain of the identical two-port components. The gain of the coupled pair of components is $4L\,(1 - L)G_t$; even if the imbalance is fairly large, this gain is very close to the ideal gain G_t. For example, at 0.5 and 1.5 times the center frequency, the coupling of an ideal hybrid drops to 0.33 (i.e., $L = 0.33$) and its amplitude balance ($L / (1 - L)$) is a seemingly horrific 3 dB; however, the gain reduction of the coupled pair of components over this 3:1 frequency range is only 0.8 dB. Even greater bandwidth can be achieved by designing the hybrid to have a bandcenter power division other than 3 dB: if the coupling between port 4 and port 1 is made greater than 3 dB at center frequency, the band-edge balance of the hybrid is improved, as is the worst-case imbalance over the entire band. Similarly, one can show that the effect of imperfect amplitude balance is to raise the magnitude of the input reflection coefficient to $|(2L - 1)\,\Gamma|$, where Γ is the component's input reflection coefficient in the absence of the hybrids.

5.1.3.3 Effect of Imperfect Balance

In the previous sections we frequently assumed that the hybrids were ideal and the two-ports were identical. Such perfection never occurs in practice, of course, so it is important to be able to estimate the effects of imperfection. Estimating the effects of phase, amplitude, and gain imbalance is not difficult as long as the hybrids are not too far from ideal.

In any balanced circuit, two voltage components combine in phase after traveling through different paths (each consisting of the input hybrid, one of the two parallel two-port components, and the output hybrid) between the circuit's input and output. The effect of dissimilar gains in the two-ports and amplitude imbalance in the hybrids is that the amplitudes of those voltage components are not identical. Similarly, phase imbalance arising in either the components or the hybrids causes the two output voltage components to have different phases.

The two voltage components at the output of the output hybrid can be described as phasors, as shown in Figure 5.14. The phase difference between the two components is θ, and the amplitudes of the voltage components, $V_{o,1}$ and $V_{o,2}$, generally are also different. If the voltage components had equal amplitude and $\theta = 0$, the output power would be

$$P_{o,e} = \frac{1}{2}\frac{|V_{o,1} + V_{o,2}|^2}{R} \tag{5.19}$$

or

$$P_{o,e} = 2\frac{|V_{o,1}|^2}{R} \tag{5.20}$$

where R is the hybrid's output termination resistance and $V_{o,1} = V_{o,2}$ is the amplitude of either component. When they are unequal and $\theta \neq 0$,

$$P_{o,u} = \frac{1}{2}\frac{|V_{o,1} + V_{o,2}\exp(j\theta)|}{R} \tag{5.21}$$

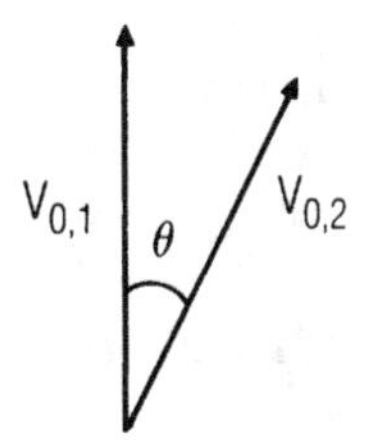

Figure 5.14 Voltage phasors at the output of the hybrid-coupled circuit. $V_{o,1}$ is the voltage of the signal that passed through the upper path in Figure 5.11 or 5.13, through the input hybrid, N_1, and the output hybrid; $V_{o,2}$ is the signal that followed the lower path.

or, letting $\delta = V_{o,2}/V_{o,1}$ with $V_{o,2} < V_{o,1}$,

$$P_{o,u} = \frac{1}{2}\frac{|V_{o,1}|^2[1 + 2\cos(\theta) + \delta^2]}{R} \tag{5.22}$$

If we assume that $V_{o,1}$ is a reference voltage, so that it has the same value for the cases of perfect and imperfect balance,

$$\frac{P_{o,u}}{P_{o,e}} = \frac{1}{4}[1 + 2\cos(\theta) + \delta^2] \tag{5.23}$$

Equation (5.23) indicates that a 20-degree phase imbalance and 1-dB gain imbalance between the two branches reduces the overall gain of the circuit by 0.6 dB. This degree of balance is not particularly difficult to maintain in most cases, even over broad bandwidths, so one may conclude that the penalty, in terms of gain, for phase and amplitude imbalance is not particularly severe.

5.1.3.4 Harmonics and Spurious Signals

Hybrid-coupled circuits may provide limited rejection of spurious signals and harmonics generated in the parallel two-ports. Such rejection is by no means guaranteed, because it depends upon the type of hybrid used in the balanced circuit and that hybrid's properties at the harmonic or spurious frequency. The spurious signals of greatest concern are usually close to the frequency of the desired signal; harmonics invariably are not close to the desired signal, but may still be of concern in broadband systems. If the phase and amplitude balance of the hybrid are uniform over a wide frequency range (an acceptable assumption for certain hybrid types, e.g., the transformer hybrid), then it is possible for the balanced structure to have significant spurious rejection.

The only balanced structure having significant spurious- and harmonic-rejection properties is shown in Figure 5.15. In this figure the two-ports are combined by 180-degree hybrids and are connected to mutually isolated, out-of-phase ports. If a two-tone signal is applied to the circuit, the input voltage $v_{i,1}(t)$ at N_1 is

$$v_{i,1}(t) = V_1\cos(\omega_1 t) + V_2\cos(\omega_2 t) \tag{5.24}$$

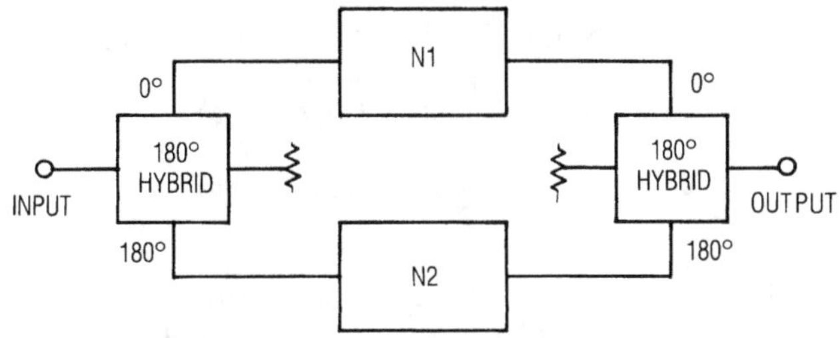

Figure 5.15 180-degree hybrid-coupled components. This configuration has limited second-order harmonic and spurious rejection as long as the output hybrid's amplitude and phase balance are maintained at the spurious output frequency.

and $v_{i,2}(t)$, the input voltage at N_2, is

$$v_{i,2}(t) = V_1\cos(\omega_1 t + \pi) + V_2\cos(\omega_2 t + \pi) \tag{5.25}$$

For simplicity we can model the networks by the power-series approach of Section 4.1, wherein each network consists of a linear two-port having the transfer function $H(\omega)$, followed by a nonlinear, frequency-independent element having the transfer function

$$f(V) = a_1 V + a_2 V^2 + a_3 V^3 + \ldots \tag{5.26}$$

The second-order voltage components at the output of N_1, $v_{o,1}(t)$, are

$$\begin{aligned} v_{o,1}(t) = {} & a_2|H(\omega_1)H(\omega_2)|\,V_1V_2\cos((\omega_1+\omega_2)t) \\ & + a_2|H(\omega_1)H^*(\omega_2)|\,V_1V_2\cos((\omega_1-\omega_2)t) \\ & + 0.5a_2|H(\omega_1)|\,V_1^2\cos(2\omega_1 t) \\ & + 0.5a_2|H(\omega_2)|\,V_2^2\cos(2\omega_2 t) \end{aligned} \tag{5.27}$$

The second-order outputs at N_2 are

$$\begin{aligned} v_{o,2}(t) &= a_2\left|H(\omega_1)H(\omega_2)\right| V_1 V_2 \cos((\omega_1+\omega_2)t+2\pi) \\ &+ a_2\left|H(\omega_1)H^*(\omega_2)\right| V_1 V_2 \cos((\omega_1-\omega_2)t) \\ &+ 0.5a_2\left|H(\omega_1)\right| V_1^2 \cos(2\omega_1 t+2\pi) \\ &+ 0.5a_2\left|H(\omega_2)\right| V_2^2 \cos(2\omega_2 t+2\pi) \end{aligned} \tag{5.28}$$

which is clearly the same as $v_{o,1}(t)$. The signal $v_{o,2}(t)$ undergoes an additional 180-degree phase shift in the output hybrid, but $v_{o,1}(t)$ does not; thus, all the second-order voltages cancel in the output as long as the bandwidth of the hybrid is broad enough to include them. Similar analysis shows that all even-order mixing products and harmonics are rejected as long as the hybrid's balance and phase properties are the same at the mixing or harmonic frequency as they were at the excitation frequency. Conversely, all odd-order harmonics and mixing products differ in phase by 180 degrees at the outputs of N_1 and N_2 and combine in phase after the 180-degree phase shift in the output hybrid. Thus, odd-order products are not rejected.

The spurious-rejection properties of the quadrature-coupled circuit are significantly different from those of the 180-degree hybrid-coupled circuit. Applying the same analysis to the quadrature-coupled circuit shows that second-order mixing products are 180 degrees out of phase at the outputs of N_1 and N_2. These voltage components are applied to the quadrature output hybrid, so the second-order mixing products would be expected to have a 90-degree phase difference when they are combined, providing only 3-dB rejection. However, most quadrature hybrids do not have the same amplitude or phase characteristics at the second harmonic as at the intended operating frequency, so it is usually not possible to make general statements about their second-order rejection properties. Third harmonics at the output of N_2 are delayed by 270 degrees, and the third-harmonic properties of most quadrature hybrids are approximately the same as the fundamental-frequency properties. Thus, the voltage components have a 180-degree phase difference when they are combined in the output hybrid, so third harmonics are rejected. Some, but not all, third-order mixing products are rejected; the third-order intermodulation products at $2\omega_1 - \omega_2$ and $2\omega_2 - \omega_1$ are not rejected, but those at $2\omega_1 + \omega_2$ and $2\omega_1 + \omega_2$ are rejected. These rejections, of course, are largely theoretical, as they depend on the characteristics of the coupler, near the third harmonic, being identical to those at the fundamental. This is, at best, only approximately the case.

5.1.3.5 Intermodulation Intercept Point

Section 4.1.3 introduced the intermodulation (IM) intercept point, the point at which the extrapolated two-tone IM levels and linear output levels are identical. Because of the balanced circuit's power-combining effect, interconnecting two components in a balanced structure gives the combination a greater intercept point than that of the individual components. We saw, in Section 4.1.3, that the level of nth-order IM products, $P_{\mathrm{IM}n}$, can be found from the linear output level and the intercept point as follows:

$$P_{\mathrm{IM}n} = nP_{\mathrm{lin}} - (n-1)IP_n \tag{5.29}$$

where P_{lin} is the level of the linear output power and IP_n is the nth-order intercept point. All power levels are in dBm. Equation (5.29) can be rearranged to express IP_n:

$$IP_n = \frac{nP_{\mathrm{lin}} - P_{\mathrm{IM}n}}{n-1} \tag{5.30}$$

If the two-ports are operated at identical output levels, they have identical IM output levels. At the output of the balanced circuit, both the IM and linear output levels are 3 dB higher than those of the individual two-ports. The IP_n of the balanced circuit must differ from that of the individual components, so for the balanced circuit (5.30) becomes

$$P_{\mathrm{IM}n,\,C} = \frac{n(P_{\mathrm{lin}} + 3) - (P_{\mathrm{IM}n} + 3)}{n-1} \tag{5.31}$$

Equation (5.31) can be rearranged to give

$$P_{\mathrm{IM}n,\,C} = P_{\mathrm{IM}n} + 3 \tag{5.32}$$

The intercept point of the balanced structure is 3 dB greater than that of the individual two-ports, regardless of order. Furthermore, one can show via the same approach that combining m two-ports via any power-combining technique increases the intercept point by $10\log_{10}(m)$ dB.

5.1.3.6 Noise Figure

The noise figure of a pair of identical components connected by ideal hybrids, either 90- or 180-degree, is equal to that of the individual components. This result may be paradoxical, as the input hybrids introduce 3-dB loss, and anyone familiar with Friis' formula [5.7] might expect them to increase the noise figure by minimum of 3 dB at room temperature.

The paradox can be resolved by viewing the problem in terms of signal-to-noise ratio (SNR) instead of noise figure. At the input, the hybrid splits the power of the signal, so the SNR, in the components, is indeed 3 dB lower that it would be without the hybrid. However, at the output, the signals combine voltage-wise in the output hybrid, while the noise from the two amplifiers, being uncorrelated, combines power-wise. The result is a 3-dB increase in SNR in the output combiner, which restores the SNR lost at the input.

Two phenomena can increase the noise figure of hybrid-coupled components. First, the input hybrid's excess loss affects the noise figure in the same way as input loss in a single component. Second, the input hybrid's termination applies its noise directly to the inputs of the components. Ideally, this noise is cancelled in the output hybrid. However, at frequencies near the edge of the hybrid's passband, where the isolation is imperfect, not all of this noise is rejected. The phenomenon is relatively minor, however, usually increasing the noise temperature by only a few degrees.

5.2 Direct Interconnection of Microwave Components

It is not always necessary or desirable to use some type of hybrid to interconnect solid-state devices or components. Although the previous section was concerned primarily with ideal hybrids, real, nonideal hybrids must be used in practical circuits, and real hybrids are not perfect. Hybrids introduce additional loss, which may not be tolerable in power circuits, and their imperfect balance and VSWR may also degrade a circuit's performance. They also increase the circuit's size and weight and make it more expensive to design and fabricate. The direct interconnection of components circumvents some of these problems, but at the expense of losing the inherent isolation between stages that hybrids provide.

The primary purpose of connecting solid-state devices directly is to increase output power without adding complexity; for example, high-power microwave bipolar and field-effect transistors are realized by directly connecting many smaller devices, called *cells*, in parallel. A second purpose is to eliminate undesired harmonics and intermodulation products; even- or odd-order products sometimes can be eliminated by appropriately connecting devices together. These rejection properties are often exploited in the design of frequency converters such as frequency multipliers and mixers.

5.2.1 Harmonic Properties of Two-Terminal Device Interconnections

Nonlinear two-terminal circuit elements such as diodes are often connected in parallel or series in order to eliminate certain harmonics or mixing products. Because spurious signals often are in-band and cannot be removed by filtering, the ability to eliminate such products without resorting to filters is often valuable. In circuits designed to have very wide bandwidths, harmonics may be within the output passband, but even in narrowband circuits, certain spurious mixing products may be in-band. Two of the most important interconnections of nonlinear devices having spurious-rejection properties are called the *antiparallel* and the *antiseries*, or, more commonly but less elegantly, the *push-push* interconnection. A third type of interconnection is called the *series interconnection*; it is a variation of the antiparallel interconnection but has different properties.

5.2.1.1 Antiparallel Interconnection

Figure 5.16(a) shows a two-terminal nonlinear conductance having the I/V characteristic

$$I = f(V) = aV + bV^2 + cV^3 + dV^4 + eV^5 + \dots \tag{5.33}$$

The element's I/V characteristic is generally not symmetrical, so it is marked with a + sign at one terminal in order to indicate its polarity. In Figure 5.16(b), the applied voltage is reversed. In this case,

$$I = f(-V) = -aV + bV^2 - cV^3 + dV^4 - eV^5 + \dots \tag{5.34}$$

and the odd-degree components of the power series are negative. Finally, if the element is reversed, but the voltage and current conventions remain as in Figure 5.16(a),

$$I = -f(-V) = aV - bV^2 + cV^3 - dV^4 + eV^5 + \dots \tag{5.35}$$

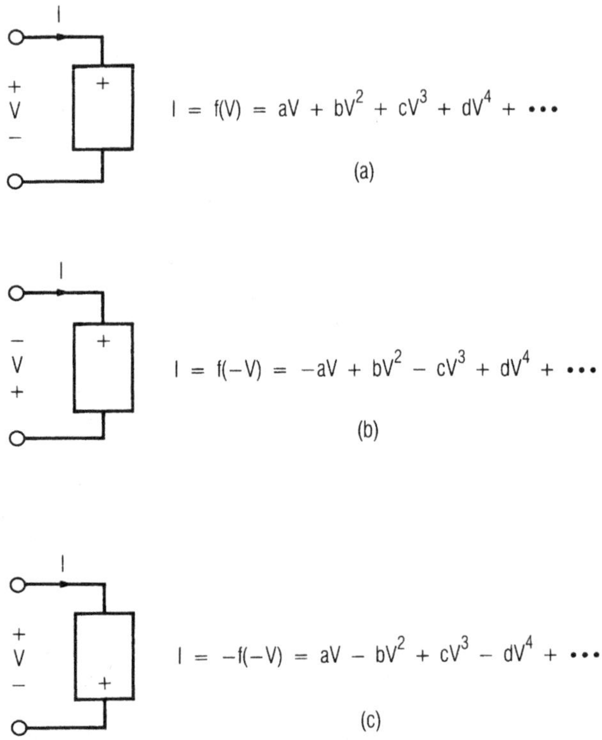

Figure 5.16 Voltage/current relations in a conductive nonlinearity with three different voltage and current polarities.

This case, illustrated in Figure 5.16(c), is the converse of the previous one: the even-degree current components are negative. We conclude from (5.35) that reversing the terminals of a nonlinear circuit element changes the sign of the even-degree terms in its power series.

Figure 5.17 shows the antiparallel interconnection of two ideal conductive nonlinear elements described by (5.33) through (5.35). The current in element A, I_A, is found from (5.33):

$$I_A = f(V) = aV + bV^2 + cV^3 + dV^4 + eV^5 + \dots \tag{5.36}$$

I_B, the current in element B, is found from (5.35):

$$I_B = -f(-V) = aV - bV^2 + cV^3 - dV^4 + eV^5 + \dots \tag{5.37}$$

and finally the total external current, I, is

$$I = I_A + I_B = f(V) - f(-V) = 2aV + 2cV^3 + \dots \tag{5.38}$$

From (5.38) we can see that the external current does not include any even-degree components. Therefore, the antiparallel pair of nonlinear elements operates as if it were a single element having only odd-degree nonlinearities; indeed, plotting the I/V characteristic would show that it is a purely odd function of voltage. It was shown in Chapter 1 that even-degree nonlinearities generate even-order mixing products, and odd-degree nonlinearities generate odd-order mixing products. Thus, antiparallel-connected nonlinear elements generate no even-order mixing products from the frequencies in their terminal voltages.

This result may at first seem impossible because, from (5.36) and (5.37), the even-degree current components still exist in I_A and I_B. In order to examine this mystery further, we consider the loop current, I_{loop}, in Figure 5.17. The loop current must consist only of the components for which

$$I_{\text{loop}} = I_A = -I_B = bV^2 + dV^4 + \dots \tag{5.39}$$

The odd-degree components in I_{loop} must be zero because it is impossible to have $-aV = aV$, $-CV^3 = CV^3$, ..., for all V. Equation (5.39) shows that I_{loop} contains the missing even-degree current components and does not include any of the odd-degree terms. We now can see what has happened: the even- and odd-order mixing components have been separated; the even-order current components circulate in the loop, while the odd-order current components circulate in the external circuit.

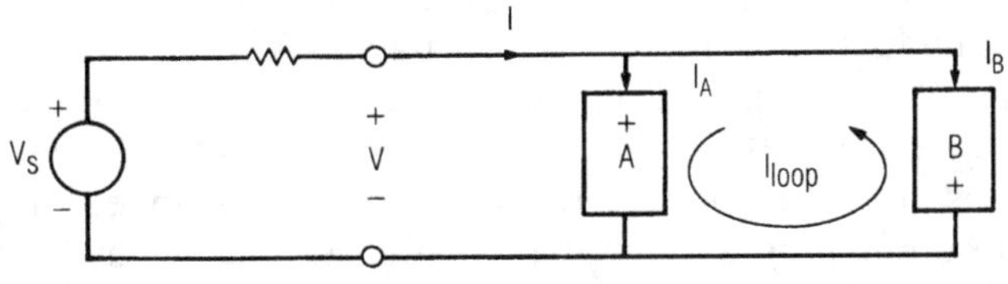

Figure 5.17 Antiparallel connection of two identical nonlinear elements.

The lack of even-order components circulating in the external circuit can be used to an advantage. For example, antiparallel diodes can realize a frequency tripler having inherently low second- and fourth-harmonic output. Such a tripler has no inherent rejection of fundamental-frequency output. The antiparallel pair also can be employed as a mixer that has no mixing response between the RF frequency and the fundamental component of the local oscillator. A mixer using antiparallel diodes achieves efficient mixing between the RF and the second LO harmonic, a third-order mixing product. Such *subharmonically pumped mixers* are in wide use; they are particularly valuable at millimeter wavelengths where fundamental-frequency LO power may be difficult or expensive to obtain.

The separation of the even- and odd-order frequency components of the current has another implication, one that is subtle but very important. Because no even-order currents circulate in the external circuit, no even-order voltages are generated between the elements' terminals. Thus, the even-order components of the terminal voltage, as well as the external even-order currents, are zero. Furthermore, each nonlinear element generates even-order currents that are equal to those of the other and are opposite in direction. The existence in each component of a circulating current and zero terminal voltage implies a short circuit at the terminals; each element in effect short circuits the other at all even-order mixing frequencies.

The fact that each element short circuits the other at even-order harmonics and mixing frequencies allows considerable simplification of the analysis of such components. It is not necessary to include both nonlinear components in the analysis, or the embedding impedances at even-order harmonics and mixing frequencies. Instead, we need only include one element in the circuit, express the I/V characteristic of the single element as $I = 2f(V)$, and set all the even-order embedding impedances to zero. In this way, we obtain a single-device equivalent circuit that describes the two-device circuit completely. The results will be the same as those of an analysis that includes both nonlinear elements.

Scaling the nonlinear element as $I = 2f(V)$ is not always wise or even possible because, in some cases, the nonlinear element may not be a simple two-terminal conductive nonlinearity; it may have a relatively complex equivalent circuit, one described by a model that is difficult to modify. Furthermore, modifying the model may require modifying the computer program that is used to analyze the circuit; such changes may not be possible if the source program is not available and, in any case, may not be advisable because of the possibility of introducing errors. Instead, it may be preferable to generate a single-device equivalent circuit by modifying the external circuit.

To scale the external circuit, all odd-order embedding impedances are artificially set to twice their true values, even-order embedding impedances are set to zero, and the antiparallel pair is replaced by a single device. This process is illustrated in Figure 5.18, in which $Z'(\omega) = 2Z(\omega)$ at odd-order mixing frequencies and $Z'(\omega) = 0$ at even-order frequencies. On completing the analysis of the single-device circuit, all absolute power levels (e.g., LO power of a mixer, input and output powers of a multiplier) must be doubled, but all relative levels (e.g., conversion loss or gain) remain unchanged.

A limitation of the single-device equivalent circuit is that it is valid only in the case of perfect balance and identical nonlinear elements. This limitation is not as severe as it seems, however, because, as with hybrid-coupled circuits, performance in many respects is not highly sensitive to balance. More importantly, a single-device equivalent circuit provides great intuitive insight into the operation of a balanced circuit.

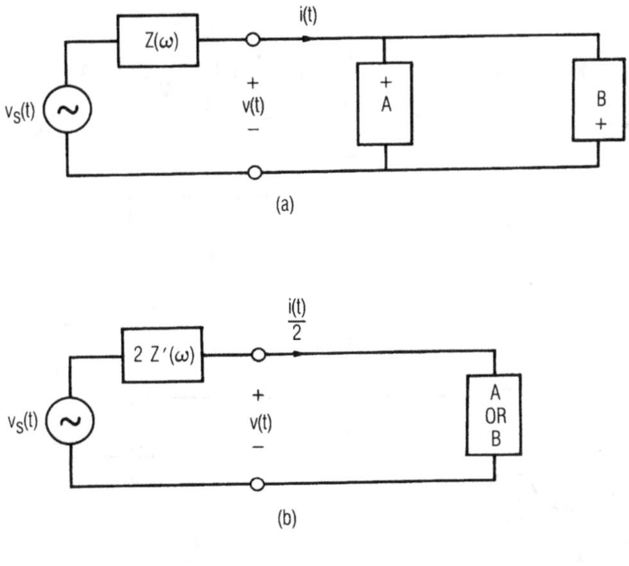

Figure 5.18 Generation of the single-device equivalent circuit of the antiparallel-connected elements: (a) the complete circuit; (b) the single-device equivalent. $Z'(\omega)$ equals $Z(\omega)$ for odd-order frequencies; it is zero for even-order frequencies.

5.2.1.2 Antiseries Interconnection

A dual case of the antiparallel connection is the antiseries connection shown in Figure 5.19. Because of the circuit's symmetry, $V_A = V_B = V$ and

$$I_L = I_A + I_B \tag{5.40}$$

I_A is found from (5.33) and I_B from (5.34). Adding these gives the output current,

$$I_L = 2bV^2 + 2dV^4 + \dots \tag{5.41}$$

The output current in the load, R_L, is an even-degree function of the voltage across the nonlinear elements. Thus, under sinusoidal excitation, the load current and voltage contain only even-order mixing frequencies and harmonics. We find the loop current in a manner similar to that of the antiparallel case, by recognizing that $I_A = -I_B$ for current components in I_{loop}:

$$I_{\text{loop}} = I_A = -I_B = aV + cV^3 + \dots \tag{5.42}$$

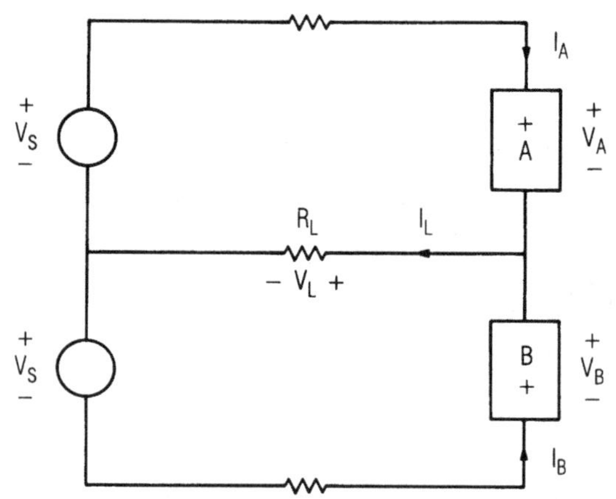

Figure 5.19 Antiseries connection of two identical nonlinear elements. The dual sources must be realized via a transformer or other 180-degree hybrid.

I_{loop} is an odd-degree function of V, so the loop current must contain the odd-order mixing components and harmonics of the frequencies in its terminal voltage waveform. Like the antiparallel interconnection, the antiseries interconnection separates even- and odd-order frequency components, but in the latter case, the even-order components circulate in the external circuit.

Figure 5.20 shows how the even- and odd-order voltage components, $v_e(t)$ and $v_o(t)$, respectively, and the even- and odd-order current components, $i_e(t)$ and $i_o(t)$, are distributed in the circuit [$v_o(t)$ does not include the fundamental-frequency component]. The load current $I_L(t)$ has only even-order components, so only even-order voltages $v_e(t)$ exist at its terminals. Although element B does not directly short-circuit element A at the odd-order frequencies, as it did in the antiparallel case, the entire lower half of the loop short-circuits the entire upper half of the loop, a fact evidenced by the lack of odd-order voltage components across R_L. This observation can be used to decompose the antiseries-connected pair of devices into a single-device equivalent circuit; the process is illustrated in Figure 5.21.

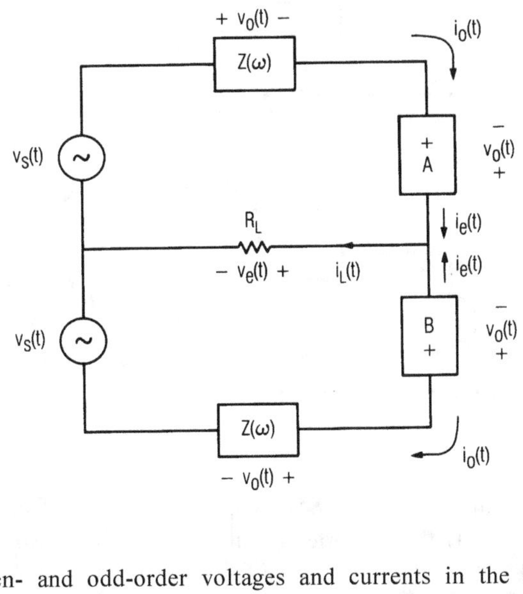

Figure 5.20 Even- and odd-order voltages and currents in the antiseries circuit. Although the odd-order voltage components exist only across A, B, and both $Z(\omega)$, the voltages across these elements do not consist solely of odd-order components.

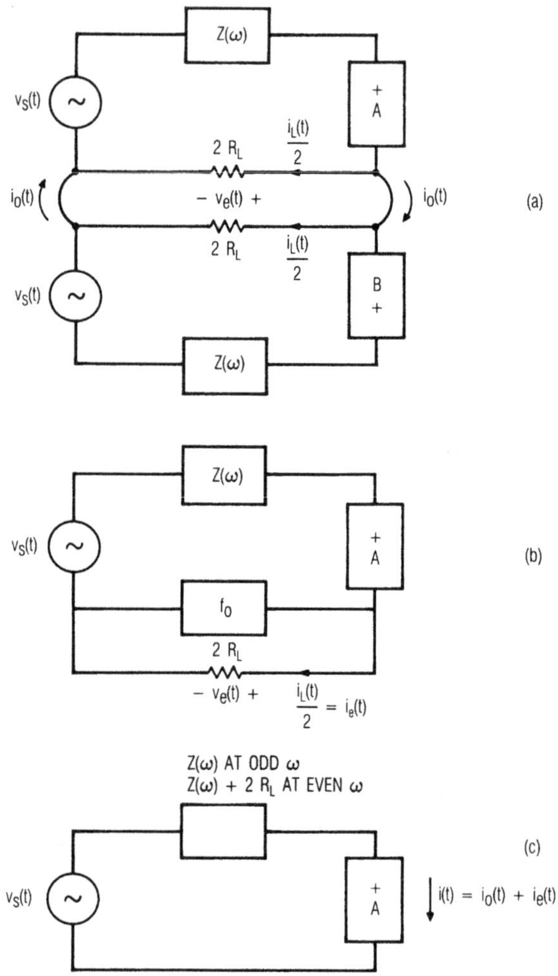

Figure 5.21 Generation of the single-device equivalent circuit of the antiseries circuit: (a) the antiseries circuit is split into two half-circuits; (b) the lower half-circuit is replaced by a short circuit f_o at the odd-order frequencies; (c) the single-device equivalent is formed by including the load resistor $2R_L$ in $Z(\omega)$.

In Figure 5.21(a), the load resistor R_L has been split into two resistors, each having resistance $2R_L$; each resistor also has half the load current, $I_L(t)/2$. In this circuit, $I_L(t)/2 = i_e(t)$, where $i_e(t)$ is the even-order current in either half of the divided circuit. The odd-order current components do not circulate in R_L, but pass through the links between the half-circuits; because the odd-order current in the top half of the circuit passes through the bottom half without generating any voltage across it, the bottom half-circuit effectively shorts the upper half-circuit at the odd-order frequencies. For this reason, the lower half-circuit can be replaced by an ideal filter, f_o shown in Figure 5.21(b), that short circuits the load resistor at odd-order frequencies and is an open circuit at even-order frequencies. This circuit is a valid single-device equivalent of the circuit in Figure 5.20.

The circuit in Figure 5.21(b) is not in the form we prefer, however; it is not in the canonical form that we have assumed to exist in the previous chapters. We prefer an equivalent circuit that consists of the device, embedding impedance, and voltage source in series. That circuit is shown in Figure 5.21(c), in which the load resistance is absorbed into the embedding network. The embedding impedance of this circuit therefore is $Z(\omega) + 2R_L$ at even-order frequencies and $Z(\omega)$ at odd-order frequencies. The loop current, $i(t)$ in Figure 5.21(c), consists of both the even- and odd-order components; that is, $i(t) = i_e(t) + i_o(t)$. These components can be separated easily in the frequency domain, and the output power is found from the desired frequency component of $i_e(t)$ and $2R_L$. As before, the output power of the single-device equivalent circuit is half that of the complete circuit.

The dual excitation sources shown in Figures 5.19 through 5.21 could be realized by some type of balun or 180-degree hybrid. The transformer hybrid is often employed for this task; port 4 in Figure 5.2 is the input, and ports 2 and 3 are the outputs. Because port 3 is in series with R_L, port 3 is usually shorted and connected to ground instead of terminated. Other types of 180-degree hybrids can also be used to provide the dual sources, as long as an out-of-phase pair of ports is used as output.

Although the antiseries connection of two-terminal devices has important uses in microwave electronics, one of the most familiar applications is in a low-frequency circuit, the full-wave rectifier shown in Figure 5.22(a). Fourier analysis of the output current and voltage waveforms, Figure 5.22(b), shows that the waveforms contain no excitation-frequency component; the output frequencies are only dc and even harmonics of the excitation frequency. This same circuit can be scaled to microwave frequencies and used as a second-harmonic frequency multiplier having minimal fundamental and third-harmonic output. One such multiplier is described in [5.8].

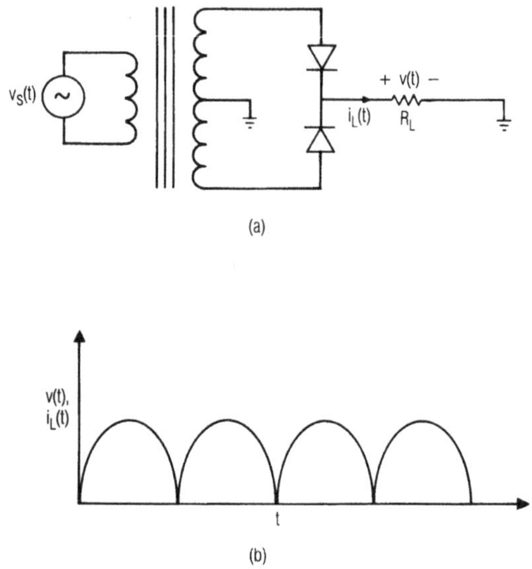

Figure 5.22 A common antiseries circuit: the full-wave rectifier. (a) Circuit; (b) voltage and current waveforms.

5.2.1.3 Series Interconnection

Figure 5.23 shows an interconnection of two nonlinear elements with an output transformer that couples them to the load, R_L; for lack of a better term, we call this a *series interconnection.* From the descriptions of the antiparallel and antiseries circuit, it should be clear that the even- and odd-order currents in the nonlinear elements are as shown in the figure.

The primary circuit (the tapped side) of the output transformer is excited in phase by the odd-order currents in the nonlinear elements; these currents induce equal but opposing currents in the secondary side, and consequently there is no odd-order current in the transformer's secondary winding. Because there is no secondary current, the odd-order voltage across both the secondary and primary must be zero; thus, the nonlinear elements are connected to ground through the transformer at odd-order mixing frequencies. Furthermore, the even-order currents are equal and opposite at the node connecting A, B, and $Z(\omega)$; this node is a virtual ground for even-order products.

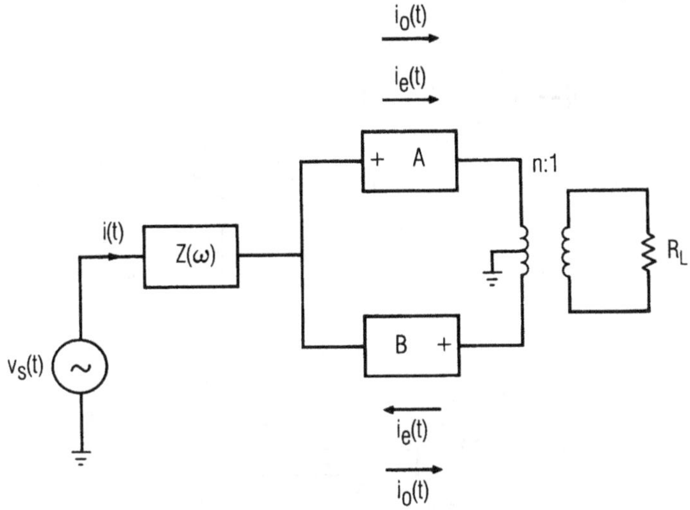

Figure 5.23 Series connection of two nonlinear elements. In a microwave circuit, the transformer is realized by a 180-degree hybrid.

The single-device equivalent circuits are found in the usual manner, by splitting $Z(\omega)$ and the transformer into two parallel elements. The circuit can then be separated into two separate circuits, each of which has the form shown in Figure 5.24(a). In this figure, the load impedance has been transferred to the primary side of the transformer, and the elements f_o and f_e, ideal filters that are short circuits at the odd- and even-order mixing frequencies, respectively, provide the short circuits at the virtual ground points. The elements can be consolidated further as shown in Figure 5.24(b), in which $Z(\omega)$, f_o, f_e, and the load have been expressed as a single impedance: $2Z'(\omega) = n^2/\, 2R_L$ at even-order frequencies, and $2Z'(\omega) = 2Z(\omega)$ at odd-order frequencies. Except for the change in impedances and the fact that the even-order products are the output quantities, this circuit is identical to the single-device equivalent circuit of the antiparallel interconnection in Figure 5.18. The series circuit operates in a manner similar to that of the antiparallel circuit; in the former, however, the output is coupled to the devices via the circulating odd-order current instead of the even-order terminal current. Consequently, in the series circuit, the even-order products are the output, not the odd-order.

In a microwave realization of the series circuit, some type of 180-degree hybrid would be used in place of the transformer. Depending upon

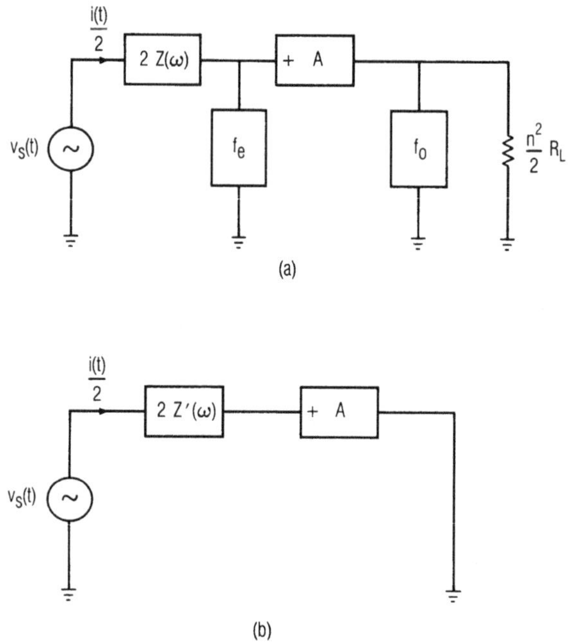

Figure 5.24 Single-device equivalents of the circuit in Figure 5.23: (a) half-circuit representation where f_o and f_e are ideal series resonators tuned to the odd- and even-order frequencies, respectively; (b) representation in which $Z(\omega)$ has been modified to account for f_o and f_e.

the characteristics of the hybrid, it may be necessary to change the description of the series impedance $2Z'(\omega)$ in Figure 5.24(b), because many microstrip hybrids do not present short circuits to all odd-order currents, as does the transformer. Many hybrids present a short circuit to the odd-order currents at the excitation frequency, but they present open circuits to other odd-order products. In this case, it is necessary to modify $Z'(\omega)$ in Figure 5.24(b) appropriately.

5.2.1.4 Properties of Direct Parallel Interconnection of Two-Ports

Two-port components can be connected in parallel at their input and output ports, as shown in Figure 5.25. Direct parallel interconnection is very common in microwave circuits; power FET and bipolar devices, and even

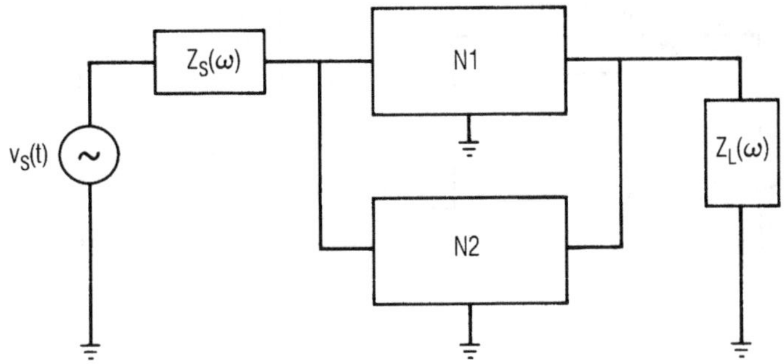

Figure 5.25 Direct parallel connection of a pair of two-ports.

some small-signal devices, are realized as parallel combinations of many low-power cells. Characterizing parallel-connected linear two-ports is straightforward: the Y matrix of parallel two-ports equals the sum of their individual Y matrices. Nonlinear two-ports cannot be described by Y parameters, however, nor by any similar linear two-port equations.

The simplest approach to the analysis of parallel-connected two-ports is to generate a single-component equivalent circuit. Generating the equivalent circuit requires changing only the source and load impedances at all harmonics or mixing frequencies.

Figure 5.26 illustrates the process of generating the single-device equivalent circuit. Because of the symmetry of the structure, it is possible to split $Z_s(\omega)$ and $Z_L(\omega)$ into two parallel impedances of $2Z_s(\omega)$ and $2Z_L(\omega)$. This operation preserves the voltage levels in the circuit but reduces the input and output currents by a factor of two compared to the currents of the combined pair. The input and output power levels in the single-component equivalent circuit are, therefore, half those of the parallel-connected pair of two-ports. However, because both available input power and output power are reduced by the same factor, the gain is the same.

The single-component equivalent circuit in Figure 5.26 shows that each component is effectively terminated by an impedance twice that of the actual load; thus, the impedance levels necessary to match a parallel-coupled pair of two-ports is half that required by a single two-port. This is the major disadvantage of such interconnections and the primary limitation in achieving high power by paralleling individual solid-state devices; the impedance necessary to match the parallel combination drops by a factor

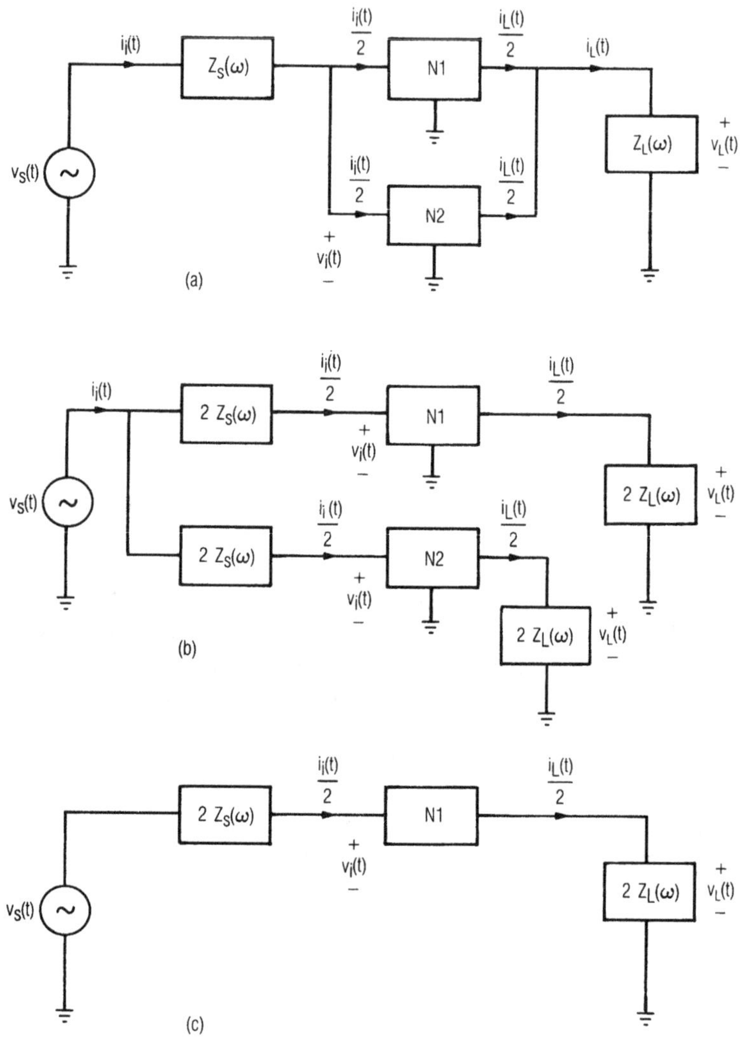

Figure 5.26 Evolution of the single-element equivalent circuit of the directly connected two-ports: (a) directly connected circuit; (b) splitting the source and load impedances does not change the voltages or total currents; (c) single-element equivalent.

equal to the number of devices, while the total current increases by the same factor. As we parallel more devices, we eventually reach a point where it is no longer possible to match the combination by circuits having practical element values. Furthermore, because of the low impedances, parallel-connected devices have high combined input and output currents, so I^2R losses in the matching circuits may be substantial. Other practical problems, such as maintaining phase and amplitude balance in a large number of separate devices, and especially thermal problems, also limit the number that can be connected in parallel. We shall explore some of these practical matters further in Chapter 9.

Another implication of Figure 5.26, one particularly relevant to the design of mixers and frequency multipliers, is that the embedding impedance presented to each device is twice the actual terminating impedance. Because of this property, it is sometimes possible to achieve optimum terminating impedances more easily in a balanced structure than in a single-device structure. For example, the optimum IF load impedance for each diode in a balanced diode mixer is usually approximately 100Ω. This impedance can be achieved without transformation by connecting the IF outputs of the two individual mixers in parallel in a balanced pair. The shared 50Ω IF load is equivalent to each mixer having an individual load of 100Ω at its IF port.

Because the phase shifts are identical in both two-ports in the parallel-coupled circuit, the circuit does not reject any harmonics or mixing frequencies of any order, even or odd. It does, however, achieve the same 3-dB improvement in intermodulation intercept point as do the hybrid-coupled two-ports. Accordingly, in applications where harmonic or spurious rejection is important, it is best to use circuits that are hybrid-coupled.

References

[5.1] J. Sevick, *Transmission-Line Transformers*, Atlanta: Noble Publishing, 1996.

[5.2] P. L. D. Abrie, *The Design of Impedance-Matching Networks for Radio-Frequency and Microwave Amplifiers*, Norwood, MA: Artech House, 1985.

[5.3] G. Matthaei, L. Young, and E. M. T. Jones, *Microwave Filters, Impedance-Matching Networks, and Coupling Structures*, Norwood, MA: Artech House, 1980.

[5.4] K. Chang (ed.), *Handbook of Microwave and Optical Components*, Vol. 1, New York: Wiley, 1989.

[5.5] J. Lange, "Interdigitated Stripline Quadrature Hybrid," *IEEE Trans. Microwave Theory Tech.*, Vol. MTT-17, 1969, p. 1150.

[5.6] R. Mongia, I. Bahl, and P. Bhartia, *RF and Microwave Coupled-Line Circuits*, Norwood, MA: Artech House, 1999.

[5.7] R. E. Collin, *Foundations for Microwave Engineering*, Second Edition, New York: McGraw-Hill, 1992.

[5.8] S. A. Maas and Y. Ryu, "A Broadband, Planar, Monolithic Resistive Frequency Doubler," *IEEE Microwave and Millimeter-Wave Monolithic Circuits Symposium Digest*, 1994, p. 443.

Chapter 6

Diode Mixers

The most common type of microwave-frequency mixer uses a Schottky-barrier diode. Diode mixers are useful over a remarkably broad range of frequencies: inexpensive doubly balanced mixers can be obtained for use at the lower microwave frequencies (below 20 GHz), and mature single-diode mixer designs are available for millimeter-wave applications as well. This chapter is concerned with the practical aspects of designing mixers in both frequency ranges. For further information on the subject of microwave and millimeter-wave mixers, the author shamelessly suggests his own books on the subject [2.5, 6.1].

6.1 MIXER DIODES

Virtually all modern diode mixers employ Schottky-barrier diodes as the mixing elements. Inexpensive silicon diodes are used in most prosaic mixer applications; these diodes, available in a wide variety of chip and packaged forms, are adaptable to virtually any transmission medium. In particular, they can be obtained in so-called *quads*, consisting of four diodes in a ring, or, occasionally, cross configuration. These are useful in balanced mixers and frequency multipliers. Silicon diodes are available in a wide variety of packages and barrier heights (a term we shall define in Section 6.1.1.7), making them extremely versatile devices.

GaAs diodes are more expensive than silicon, but they can provide better conversion loss and noise performance, especially at high frequencies. GaAs diodes have higher breakdown voltages than silicon. At low frequencies their performance advantage is minimal, so they generally are not obtainable in the low-cost packages sometimes used for silicon diodes. GaAs diodes are available as chips, as beam-lead devices, and in miniature ceramic and quartz packages.

Schottky-barrier diodes are discussed generally in Section 2.4. Modeling of Schottky devices is discussed in Section 2.4.2, and mixer diodes in particular are examined in Section 2.4.3. Those sections describe the theory and modeling of the Schottky junction; this section is concerned with the use of such diodes in practical mixer circuits.

6.1.1 Mixer-Diode Types

6.1.1.1 Chip Diodes

Perhaps the simplest diode used commonly in mixers is an unpackaged chip having the cross section shown in Figure 2.10. Such chips can be mounted in ceramic or plastic packages or may be used unpackaged in hybrid circuits. Chip diodes often have multiple anodes; several anodes of different diameters may be defined on a single chip in order to compensate for manufacturing tolerances or to allow one type of diode to be used at widely differing frequencies. However, in order to allow bonding of a wire or ribbon to the anode, the anode must be at least 10 to 15 microns in diameter, or it must have a metal overlay to increase its area. Such large anodes have relatively high junction capacitance and thus are not optimum for millimeter-wave operation; reducing the capacitance, without reducing junction area, requires low epilayer doping levels that increase series resistance.

If a different method is used to connect the anode, the diodes' anodes can be smaller. One such method, which facilitates connections to very small anodes, is to use a pointed spring wire, or *whisker*. Because of the difficulty of creating reliable, whisker-contacted diodes, considerable effort has been devoted to the development of high-frequency beam-lead diodes that have integral leads and low parasitics. Today, whisker-contacted diodes are virtually obsolete.

6.1.1.2 Beam-Lead Devices

An important and very versatile type of diode is the Schottky-barrier beam-lead device shown in cross section in Figure 6.1. A beam-lead device has integral ribbon leads connected to its anode and cathode, and the cathode lead is on the same side of the chip as the anode. The beam-lead diode has many of the desirable characteristics of both packaged and chip devices: it can have a small anode and low series inductance, and no wire bonds or other special methods are needed to connect it to a circuit.

The beam-lead diode requires two major modifications of the Schottky-barrier diode structure shown in Figure 2.10: first, the top surface must somehow be connected to the substrate; and second, the anode ribbon must be so designed that it does not cover a large area above the epilayer, or a large parasitic "overlay" capacitance results. The most common method of minimizing overlay capacitance is to locate the anode close to the edge of the chip. The resulting structure is not entirely satisfactory for many purposes, however, because the very small anode connection close to the edge of the chip is delicate, and the overlay capacitance may still be substantial. The cathode connection is made by etching away the oxide layer and epilayer in a region near the anode. Because of the long, thin current path, the series resistance may also be high (see Section 6.1.1.6).

Many ingenious structures have been proposed to circumvent the problems inherent in beam-lead diodes; the research literature of the 1980s is full of examples [6.2–6.4]. Today, the usual approach is to use an "air bridge" connection; that is, a conductor that connects the anode to its lead with an air gap underneath. The air insulating region can be fabricated as a trench [6.4] or, more commonly, the metal can arch over it. Less effective, but perhaps cheaper and more rugged, is the use of oxide isolation to minimize overlay capacitance.

Beam-lead diodes are available as pairs and quads for balanced mixers and frequency multipliers, as well as single devices.

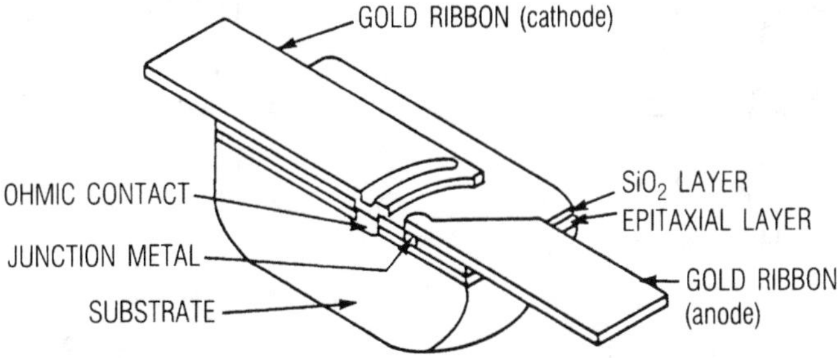

Figure 6.1 Cross section of a beam-lead diode. The diode's most serious limitation for high-frequency operation is the overlay capacitance between the anode ribbon and the epilayer.

6.1.1.3 Packaged Diodes

Inexpensive silicon diodes—single devices, pairs, and quads—are available in a wide variety of packages. Figure 6.2(a) shows one of the most common, consisting of a circular ceramic substrate, on which the diode and leads are mounted, coated with a dome of epoxy. The substrate can be as small as 1.25 mm in diameter. Although the package parasitics are substantial and can vary considerably between devices, such diodes are used regularly at frequencies up to 18 GHz, and occasionally even to 26 GHz. They are the most commonly used type of diode for commercial mixers.

Silicon diodes are also available in a wide variety of standard surface-mount packages. Surface-mount packages, although small, have large parasitics, making them useful only to approximately 5 to 6 GHz.

Both silicon and GaAs diodes are available in more expensive ceramic packages, which are usually hermetically sealed and thus acceptable for use in high-reliability applications or in severe environments [Figure 6.2(b)]. Ceramic packages are generally larger than the smallest epoxy packages, so their parasitics are the same or even greater. Chip diodes are often installed in a so-called *pill package*, a cylindrical ceramic package having metal end caps, that may be either flat or have pins (Figure 6.3). A pill package introduces additional parasitics, largely the capacitance of the end caps and inductance of the bond ribbon.

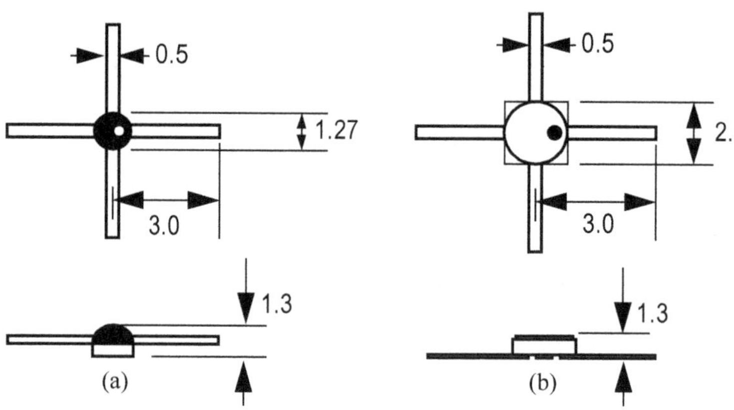

Figure 6.2 Common diode packages: (a) ceramic/epoxy package; (b) hermetic ceramic package. Dimensions are in millimeters.

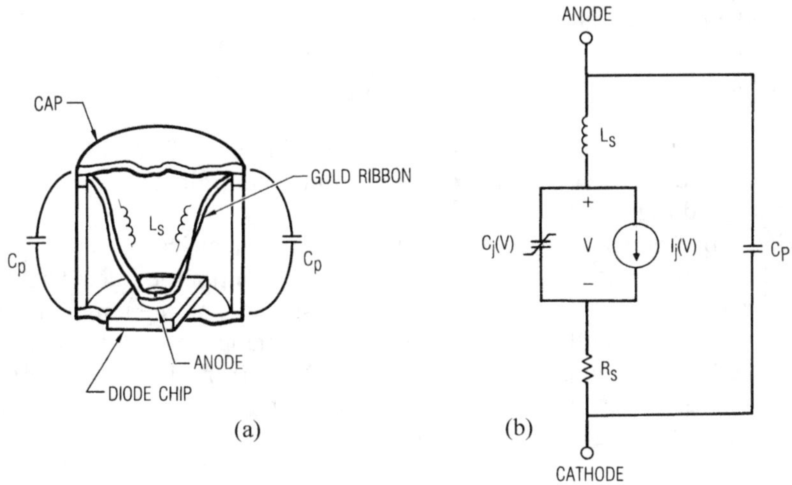

Figure 6.3 (a) Cutaway view and (b) equivalent circuit of a chip diode in a "pill" package. L_s is the inductance of the bonding wire or ribbon; C_p is the capacitance of the package resulting from the metallic top and bottom and ceramic sidewalls.

6.1.1.4 Flip-Chip and Leadless Devices

The fragility of the leads on beam-lead devices has prompted the development of various kinds of leadless devices, often described by the illogical term *leadless beam-lead diodes*. Such chips have a ball or dome of gold or some other solderable metal in the corners of the chip, where beam leads would otherwise be attached. The chip is mounted upside down and bonded to the circuit by the ball or dome. Parasitics of such devices are approximately the same as those of beam-lead diodes, but handling is much easier. Inverted devices cannot be inspected after installation, so they are often unacceptable for high-reliability applications.

6.1.1.5 Monolithic Devices

Diodes for use in RF or microwave monolithic circuits can be fabricated in a number of ways. Monolithic devices must be compatible with MESFET, HEMT, or HBT technology and should not require extra processing steps or mask layers. As a result, the diode design is invariably compromised to some degree. Rarely, a monolithic process includes a diode that is independent of the transistor design.

Gate Diode

In MESFET or HEMT processes, a diode can be made from a FET's gate-to-channel junction. The gate is the anode, and the drain and source, which are connected together, form the cathode.

Gate diodes usually have relatively poor electrical characteristics. First, the anode shape is not appropriate for a mixer diode. The long, narrow gate metallization has significant resistance, and the long periphery, for a given area, results in high fringing capacitance. Second, as with all planar diodes, the component of series resistance from the ohmic region is relatively high. It is unusual for gate diodes to have cutoff frequencies (Section 2.4.3) above a few hundred gigahertz.

HBT Diode

Schottky diodes can be fabricated in heterojunction bipolar transistor (HBT) technology by depositing a metal anode on a collector mesa, and the anode is usually connected to the rest of the circuit by an air bridge. The result is a kind of mesa diode, whose electrical characteristics can be quite good, although rarely as good as achieved with discrete diodes or diodes fabricated in an independent process.

The HBT collector mesa usually is lightly doped, well below the optimum for a mixer diode. Since a diode's series resistance is largely that of the undepleted active layer, HBT diodes can have high series resistance. To reduce the series resistance, the collector mesa may be etched thinner before the anode is deposited. This requires an extra process step and mask layer, but it significantly improves the performance of the diode. It also causes the active layer to be nearly depleted at zero bias, so the junction capacitance has little variation with voltage. In the past, so-called *Mott diodes* were purposely given thin, lightly doped epilayers to minimize capacitance variation. Mott diodes were used primarily for low-temperature millimeter-wave operation [6.3], but they have no particular advantages for more ordinary applications.

Independent Diode Process

Occasionally a designer is fortunate enough to work with a process in which the diodes are fabricated independently; that is, they are not constrained by the requisites of the transistor processes. The most common type of diode used in such a process is a mesa diode. Mesa diodes are similar to the HBT diode described above, but the dimensions and doping are optimized for Schottky-barrier mixer devices.

Mesa diodes have a small component of capacitance between the ohmic region and the substrate metallization that connects to the outer end of the air bridge. This component can be reduced by the use of an empty trench [6.4, 6.5].

6.1.1.6 Limitations of Planar Diodes

Series Resistance

Planar diodes usually have high series resistance per junction area, which makes their cutoff frequencies relatively low, compared to chip devices. In a chip diode, the current direction is largely vertical in the substrate and spreads rapidly, so the substrate resistance is small, and the undepleted epilayer dominates the series resistance. In planar devices, the current is largely horizontal, confined to a thin region near the substrate/epilayer interface, which is relatively long, making its resistance significant.

In planar diodes, the series resistance can be minimized by forming the cathode ohmic contact to surround the anode over most of its circumference. Although necessary for low series resistance, this structure creates a fringing component of capacitance between the anode and cathode metallizations. The resulting intermetallic capacitance is not in parallel with the junction capacitance, so its effect is not as severe as an increase in junction capacitance; it is like the package capacitance C_p in Figure 6.3.

In lightly doped diodes (e.g., HBT diodes), electrons may approach saturation velocity in the active layer. This phenomenon limits the junction current, so the diode behaves very much as if the series resistance were increasing, in a nonlinear fashion, with junction current. When this phenomenon occurs, it may be necessary to treat R_s as a nonlinear element.

Junction Capacitance

We have already mentioned several phenomena in planar diodes that introduce additional capacitive parasitics and reduce the capacitance variation with junction voltage. As a result, (2.59) may not describe the capacitance accurately. Often, (2.59) can be corrected simply by adding a constant component of capacitance or by limiting the junction voltage in the expression in some numerically acceptable manner (Section 2.3.6). Sometimes it is necessary to use an entirely different expression.

6.1.1.7 Barrier Height

Silicon diodes are available in high, medium, or low barrier heights. Low-barrier diodes turn on around 0.3V, medium-barrier around 0.5V, and high-barrier around 0.6 to 0.7V.[1] Barrier height is varied through the use of different anode materials and epilayer doping, changing the quantity ϕ_b in (2.63) and thus I_{sat}. GaAs devices are available in only a single barrier height.

Low-barrier diodes usually have higher series resistance than medium- or high-barrier devices. When used in mixers, low-barrier diodes require less local oscillator (LO) power than high-barrier devices, but have greater intermodulation distortion. In resistive frequency multipliers, low-barrier diodes operate at lower input and output levels; for reasons discussed in Chapter 7, they may also have lower conversion efficiency. Very low-barrier diodes, which turn on around 0.1V, are necessary for unbiased detectors. Detector diodes often use *p* epilayers to achieve low barrier height and to minimize low-frequency noise.

6.2 NONLINEAR ANALYSIS OF MIXERS

The analysis of diode mixers is a straightforward application of harmonic-balance analysis. Either multitone harmonic-balance or large-signal/small-signal analysis can be used, depending on the needs of the design. In most cases, large-signal/small-signal analysis is adequate and considerably more efficient than multitone analysis. Simple and straightforward, it provides conversion efficiency, frequency response, input and output impedances, and noise figure. Multitone harmonic-balance analysis can provide more information about the mixer's performance, including saturation, spurious responses, and intermodulation levels, but it must be used with caution.

6.2.1 Multitone Harmonic-Balance Analysis of Mixers

Multitone harmonic-balance analysis of mixers is not nearly as straightforward as large-signal/small-signal analysis. It is easy to obtain results that are either inaccurate or even completely erroneous, especially for multitone excitations. In this section, we examine several important considerations.

---

1. By *turn on*, we mean a junction current on the order of 1 mA.

6.2.1.1 Frequency Set

A mixer excited by a large-signal LO at frequency ω_p and a small-signal RF at ω_R generates the frequencies

$$\omega = k\omega_p \pm \omega_0 \tag{6.1}$$

where k is the harmonic number and ω_0 is the lowest-frequency mixing product, usually $\omega_0 = |\omega_R - \omega_p|$. This spectrum consists of a number of harmonics surrounded by a pair of sidebands. The conversion-matrix formulation in Section 3.4 shows that we need at least $2K$ LO harmonics, plus dc, to establish sidebands around the first K LO harmonics. The resulting spectrum, when $K = 4$, is shown in Figure 6.4. In that case, 17 frequency components are needed.

In most harmonic-balance simulators, the frequency set for multitone harmonic-balance analysis is chosen according to

$$\omega = n\omega_R + m\omega_p \qquad |n| + |m| \le H \tag{6.2}$$

where H is a user-selectable constant. To obtain all the frequency components in (6.1), for the same case of $K = 4$, we require $H = 8$. The number of frequency components is $0.5\,H(H+1) = 36$, a much larger—and unnecessarily large—set of frequencies. Some simulators, fortunately, allow limits on the harmonics of the individual excitations. Setting $n \le 1$ and $m \le 8$ gives 23 components, a significant improvement but still too many.

The problem becomes much worse with two or more RF excitation tones. Unless the simulator can set the number of RF and LO harmonics

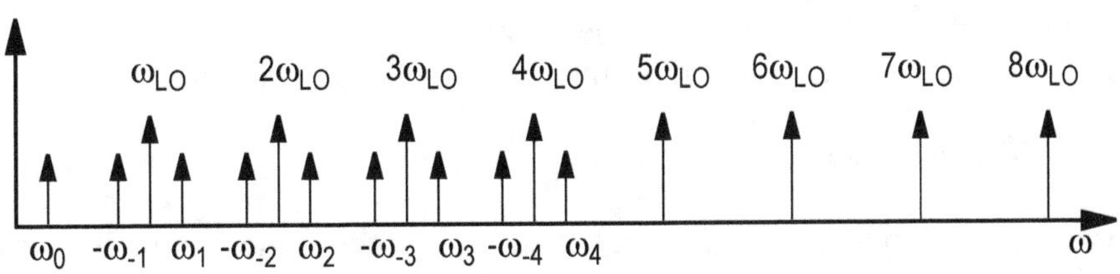

Figure 6.4 Mixer frequency spectrum when $K = 4$. This set is adequate for analysis of conversion efficiency and port impedances, but not for intermodulation, saturation, or other phenomena involving harmonics of the RF excitation.

independently, the problem becomes impractically huge. The solution is simple: the simulator should offer a special frequency set specifically designed for mixer analysis.

The above considerations apply to the analysis of conversion loss, port impedances, and similar pseudolinear effects. For analysis of intermodulation distortion, spurious responses, saturation, and similar effects, the optimum frequency set is less clear. For two-tone intermodulation analysis, the following set works well:

$$\omega = k\omega_p \pm (n\omega_{01} + m\omega_{02}) \qquad |n| + |m| \le O, \quad k \le K \tag{6.3}$$

In (6.3), n, m, and k are integers, O is the order of the intermodulation, K is the maximum LO harmonic, and ω_{01} and ω_{02} are the IF frequencies of the two excitation tones. This creates a set of sidebands around each LO harmonic, $k\omega_p$, that look like the spectrum in Figure 4.3. In this case, K must be relatively large, much larger than the value used in conversion-loss analysis. A similar frequency set, with $O = 3$, should be adequate for compression analysis.

6.2.1.2 Model Limitations

Diode Model

In Section 2.3.2, we noted that a device model used for nth-order intermodulation analysis must have both an accurate I/V (or Q/V) characteristic and accurate derivatives through the nth order. Fortunately, the exponential I/V function used to characterize Schottky diodes meets this criterion, at least up to the third derivative, and possibly higher. The junction capacitance, however, may not be so accurate, especially in planar diodes having thin epilayers. In such devices, the modeling problem becomes especially acute, as the static Q/V characteristic, and its derivatives, must be accurate over the entire LO voltage range.

Unrealistically Low Conversion Loss

Analyses of diode mixers frequently produce unrealistically low predictions of convergence loss. Several phenomena can produce this result:

1. The series resistance, R_s, may not have been modeled properly. We noted in Section 2.8.1 that thermal effects can cause the measured R_s

to be low. We also saw, in Section 6.1.1.6, that velocity saturation in the undepleted epilayer can increase R_S. Both phenomena cause R_S to be underestimated, leading to optimistic calculations of conversion loss.

2. In all types of mixers, but especially in upconverters, the pumped junction capacitance can improve the conversion efficiency. In the past, this phenomenon has been exploited, in parametric upconverters using varactors, to provide gain. When the conventional Schottky junction expression (2.59) models the capacitance, but the capacitance variation is not as great as the expression implies, the calculated conversion loss may be unrealistically low. To minimize series resistance, most modern diodes have thin epilayers, and (2.59) does not hold at reverse voltages beyond those that fully deplete the epilayer. Worse, (2.59) implies infinite capacitance as $V \rightarrow \phi$. The parameter FC [see (2.60) and (2.61)] in the SPICE diode model is designed to prevent this error; it must be chosen with great care. When FC is too large, the capacitance variation used in the simulation is also much too great, and fictitious parametric conversion-loss enhancement may occur.
3. Mixers are sometimes surprisingly sensitive to losses in the embedding network at high-order mixing frequencies. When circuit losses are ignored in the simulation, losses at these frequencies, not only the RF and IF, are removed. The result is an unrealistically low estimate of the mixer's conversion loss.

These errors are easy to make, so they are quite common. For this reason, predicted conversion loss below 5 dB, in any type of mixer, should be viewed with skepticism.

Passive Element Models

The effect of passive circuit-element models (e.g., transmission-line discontinuities) is always a concern in mixer design. In principle, it would appear that passive-element models must be accurate to the highest frequency used in the mixer analysis, so the required range, in even ordinary mixer designs, could be over 100 GHz. In fact, several realities ameliorate the situation. First, the magnitudes of the current and voltage components in the diode decrease with order, so errors in embedding impedances have progressively smaller effect on the solution, especially on the lower-order products. Second, while high harmonics and high-order mixing products may be necessary for an accurate analysis, the accuracy of those frequency components is not as important because the diode's capacitance tends to short circuit the junction at high frequencies. Therefore, at high frequencies, little of the junction current actually circulates in the external circuit.

These considerations apply to all kinds of nonlinear circuits, not only to diode mixers.

6.2.1.3 Convergence Criteria

In large-signal/small-signal analysis, the harmonic-balance analysis involves only the LO, and its convergence is not particularly critical. In multitone harmonic balance, however, the analysis includes both the large LO harmonics and many much smaller components, including RF and intermodulation products. Intermodulation components of interest may be 100 dB or more below the largest LO component. It is essential that the harmonic-balance simulator examine each frequency component for adequate convergence; it is not adequate, for example, to look only at the magnitude of the current-error vector, as it is inevitably dominated by the largest components.

6.3 SINGLE-DIODE MIXER DESIGN

Diode mixer design is primarily a process of matching the pumped diode to the RF input and IF output, terminating the diode properly at LO harmonics and unwanted mixing frequencies (i.e. those other than the RF and IF), and adequately isolating the input, LO, and output ports. That isolation, and in some cases the termination, can be provided by filters, a balanced structure, or both. The choice depends on the frequencies and the intended application.

Single-diode mixers are worth examining for two reasons. First, single-diode mixers are used, albeit infrequently, in a number of applications. Second, a single-diode mixer is a prototype for a balanced mixer, which may consist of nothing more than a hybrid-coupled pair of single-diode mixers. If a designer understands single-diode mixers, he will recognize that balanced mixers are simply another variation on a familiar theme. The design process for single-diode mixers is fundamental to all diode mixers.

6.3.1 Design Approach

In this section we introduce a partially empirical process for the design of a single-diode mixer. The process is applicable to many kinds of nonlinear circuits, not only mixers. It is as follows:

1. Using linear circuit techniques, design all the parts (primarily baluns and hybrids) that do not require nonlinear analysis.
2. Estimate the source and load impedances that must be presented to the device.
3. Design matching circuits that present those impedances to the diode and provide appropriate filtering functions.
4. "Build" the circuit on the computer and analyze and optimize it.

The first three steps are precisely those used to design nonlinear circuits before the advent of general-purpose nonlinear-circuit simulators. In those days, step 4 was to build the circuit and tune it in the lab.[2]

Our goal is to develop an approximate but accurate design before resorting to nonlinear simulation. If steps 1 to 3 are performed adequately, the computer simulation should not require huge modifications of the circuit. Nonlinear analysis, even under the best circumstances, is time-consuming, and thus an expensive part of the design task. We want to avoid as much of it as possible. Especially, we wish completely to avoid numerical optimization of the circuit, as it is more difficult to implement, more time-consuming, and less likely to be successful than in linear circuit design.

6.3.2 Design Philosophy

Figure 6.5 shows the circuit of a generic single-diode mixer. It consists of a diode and three filter/matching circuits that match the RF, IF, and LO terminations to the diode. It is clearly necessary in any mixer that these circuits do not interact; that is, no circuit affects the tuning of the others. They must also provide the appropriate termination to the diode at LO harmonics and at unwanted mixing frequencies. These requirements—termination and noninteraction—imply a filtering function as well as a matching one. Filtering alone, however, is not the only requirement; the

---

2. And step 5 often was to discard the circuit and start over.

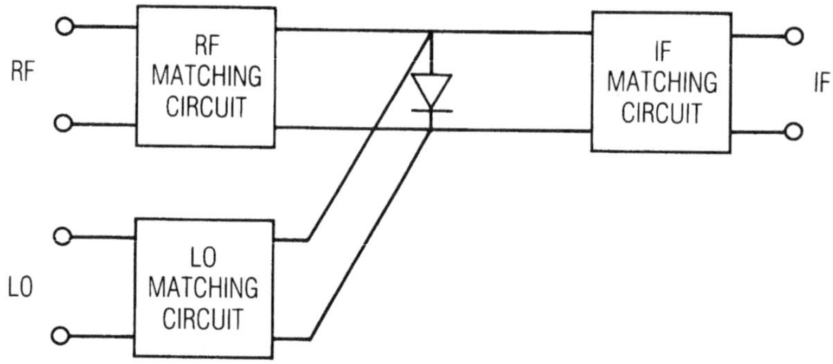

Figure 6.5 Single-diode mixer circuit.

impedance presented to the diode, over a wide range of frequencies, must be controlled by the design.

This is a large number of requirements, and it is difficult to meet all of them simultaneously. Some of these requirements are met automatically by the structure of the mixer. For example, in a mixer having a waveguide RF input and a coaxial IF output, interaction between the RF and IF matching circuits at the IF frequency is usually obviated because the IF frequency, which is ordinarily below cutoff, cannot propagate in the waveguide. The IF circuit, however, must still be designed to reject the RF and LO frequencies. Almost all practical single-diode mixers exploit some characteristic of their structures to satisfy the above requirements; it is virtually impossible to have a practical, single-diode mixer that depends solely on the electrical characteristics of the matching circuits for all the above requirements.

Before we can design a mixer, we need to know the following:

1. The input impedance of the pumped diode at the RF frequency;
2. The output impedance of the pumped diode at the IF frequency;
3. The LO input impedance;
4. The diode's optimum termination at undesired mixing frequencies and LO harmonics;
5. The required port-to-port isolation.

If these five parameters are known, designing a mixer requires only designing the matching/filtering networks. However, the matching networks inevitably affect the parameters in the list. How do we handle this dilemma? The best way is to estimate what we need to know, design the matching circuits, and calculate the performance. The estimate, made correctly, is usually close enough to allow good performance with minor modifications of the circuit. If not, we can obtain a better estimate, redesign the matching circuits, and repeat the process.

The process is easier than it sounds. For example, suppose we estimate an LO input impedance, for the diode, of 50Ω. After designing the matching circuits, we enter the circuit into a harmonic-balance circuit-simulation program, and find that the input impedance, with our matching circuit, is actually 70Ω. In this case, we can modify the matching circuit slightly. Alternatively, we can increase the LO level slightly, reducing the input impedance. Finally, we might choose to accept the 1.4 VSWR, which is, after all, thoroughly acceptable for most applications.

Frequently, to minimize cost and facilitate manufacture, the mixer may have no RF, LO, or IF matching at all. In this case, which is actually the norm for commercial mixers, the diode itself is used as a tuning element. First, we decide what the source impedance shall be (invariably 50Ω), and then select a diode that (1) has negligible junction reactance at the RF and LO frequency, and (2) exhibits the required input impedance at the desired LO level. We accomplish the former by the diode's anode size, and the latter by the barrier height. If the frequency is high, it may impossible to select a diode having negligible junction reactance; in that case, simple tuning (e.g., a series or shunt inductance) may be needed.

A long history of doing large-signal/small-signal mixer calculations allows one to make some generalizations about the embedding impedances at unwanted mixing frequencies and input/output impedances of the pumped diode. Exceptions to the following observations can, of course, be found; nevertheless, they are generally valid for practical mixers:

1. The pumped diode can be modeled at the RF frequency as a resistor and capacitor in parallel. The resistor is usually in the range of 50Ω to 150Ω and the capacitor is between C_{j0} and 1.5 C_{j0}. The IF output impedance is usually between 75Ω and 150Ω. At low IF frequencies the reactive part of the output impedance is almost always negligible.

2. Unusual embedding impedances at other mixing frequencies (especially at the sum frequency, $\omega_{RF} + \omega_{LO}$) can cause the RF input impedance to be surprisingly high. If the RF input impedance is not in the range stated above, this condition should be suspected.

3. The large-signal LO input impedance is usually close to the small-signal RF input impedance, especially if the IF frequency is low and the diode is short-circuited at its LO harmonics and unwanted mixing frequencies. As LO level increases, both RF and the LO input impedances decrease. RF impedance usually levels off at some particular LO level, while LO impedance continues to decrease.
4. The RF and IF input impedances are close to the high end of the impedance range given in (1) if the embedding impedances are open circuits at the unwanted mixing frequencies and LO harmonics. They are close to the low end of the range if the impedances are short circuits at those frequencies.
5. Open-circuit embedding impedances give the lowest conversion loss, highest noise temperature, and worst intermodulation performance; short-circuit embedding impedances result in slightly greater conversion loss but lower noise and distortion.
6. Short-circuit embedding impedances, at LO harmonics and unwanted mixing frequencies, generally result in the best overall performance. It may be possible to improve one aspect of the mixer's performance by using other embedding impedances, but other characteristics suffer.

The source and load impedances of interest are those presented to the terminals of the intrinsic diode; that is, they do not include package or other parasitics. Such parasitics must be treated as part of the embedding circuits. These observations indicate that the best overall mixer performance results from a short-circuit diode termination at LO harmonics and unwanted mixing frequencies. In that case, the IF output impedance is usually close to 100Ω and the RF and LO input impedances are approximately those of a 100Ω resistor in parallel with C_{j0}.

The diode's termination at the image frequency is the most critical of all the terminations at unwanted mixing frequencies. In many mixers the image frequency is close to the RF frequency, so the image termination is the same as the RF source impedance. However, if the IF frequency is relatively high, it is possible to use a filter (or the filtering properties of the RF matching circuit) to terminate the diode in a reactance at the image frequency. This practice, called *image enhancement*, can improve the conversion efficiency of the mixer. Image enhancement must be used with care, however, because it is possible for an image-enhanced mixer to achieve only a modest improvement in conversion loss at the expense of poor intermodulation and noise performance.

It may be surprising that these generalizations apply to any diode. They are possible because the I/V characteristics of all diodes are fundamentally the same: they are exponentials. The only apparently significant difference in the *I*/*V* characteristics of different diodes is in their values of I_{sat}; however, even differences in I_{sat} are less important than they might appear, because of the current's exponential dependence on voltage. One property of an exponential *I*/*V* characteristic is that the same conductance waveform can be achieved with any value of I_{sat}, simply by scaling the bias voltage and LO power appropriately. The junction capacitance, of course, has a significant effect on the LO waveform, but most diodes have the same *C*/*V* dependence, and C_{j0} in most well-designed mixers is selected to have approximately the same reactance at the RF frequency. The similarity in *I*/*V* and *C*/*V* characteristics between diodes causes all well-designed mixers to have conversion losses within a few decibels of each other, regardless of frequency, structure, or intended application.

6.3.3 Diode Selection

The selection of an appropriate diode for a specific design is important in achieving good performance and minimizing cost. Most mixers intended for use at the lower microwave frequencies are not designed to achieve the lowest possible conversion loss; instead, they are designed to exhibit good overall performance, including low port VSWRs, flat frequency response, stability, low distortion, and, especially, low-cost manufacture. In these mixers, the important trade-offs are usually economic, not technical. The electrical parameters of the diode are more critical in millimeter-wave mixers, where maintaining good conversion and noise performance is much more difficult.

An initial consideration is the selection of a silicon or a GaAs device. GaAs devices can achieve better conversion performance than silicon, but their advantage at low frequencies is minimal, and their cost is greater. Therefore, GaAs devices probably should be reserved for use at higher frequencies, primarily millimeter-wave applications. Compared to silicon devices, GaAs devices generally have higher cutoff frequencies (Section 2.4.3), higher breakdown voltages, and better resistance to ionizing radiation. In single-diode and singly balanced mixers, GaAs diodes' higher breakdown voltages provide a wider range of optimum conversion loss and noise figure with LO power variation. The greater I_{sat} of GaAs devices implies that these diodes require higher LO power.

The diode's cutoff frequency is an important consideration in diode selection. Although a first-order analysis indicates that the cutoff frequency is independent of junction area, second-order effects cause a diode's cutoff frequency to increase as its area decreases. Small, well-made GaAs Schottky-barrier diodes often have cutoff frequencies above 2,500 GHz.

Inexpensive silicon diodes usually have cutoff frequencies of a few hundred gigahertz.

It is important to recognize that capacitive parasitics affect f_c only if they are directly in parallel with the junction. Parasitics, (e.g., intermetallic or package capacitance) connected to the diode's external terminals may affect matching, but they do not affect f_c.

Minimizing both R_s and C_{j0} is necessary to achieve low conversion loss and distortion, but they are inverse trade-offs. Thus, a large part of selecting a diode is making an appropriate trade-off between these quantities. One consideration is the *conversion-loss degradation factor*, δ, which accounts for the loss in the series resistance at the RF frequency. It is

$$\delta = 1 + \frac{R_s}{Z_s} + \frac{Z_s f_{\mathrm{RF}}^2}{R_s f_c^2} \tag{6.4}$$

where f_{RF} is the RF frequency and f_c is the cutoff frequency. (In this case, R_s is evaluated at f_{RF}, not dc.) Z_s is the source impedance at the RF frequency and at the terminals of the resistive junction (i.e., the terminals of the current source $I(V)$ in Figure 2.8). Z_s is assumed to be real. The cutoff frequency usually remains approximately constant with small changes in R_s, so f_c can be treated as constant and (6.4) minimized. The value of R_s that minimizes δ is

$$R_s = Z_s \frac{f_{\mathrm{RF}}}{f_c} \tag{6.5}$$

For example, a mixer operating at 20 GHz, using a diode having a cutoff frequency of 1,000 GHz and $Z_s = 100\Omega$, has an optimum R_s of 2Ω. This is a very low series resistance, and a diode having such a low R_s would have a large anode area. The resulting value of C_{j0}, 0.08 pF, would be uncomfortably large. We noted earlier that second-order effects (associated with the nonuniform junction electric field near the edge of the anode) generally cause large-area diodes to have lower cutoff frequencies than small diodes. Consequently, such a large diode would not have optimum f_c, and the high junction capacitance might introduce matching difficulties. Thus, a 20-GHz mixer could probably have a higher R_s, perhaps 4 to 6 ohms, and lower C_{j0}, and still achieve close to the optimum δ. In general, considerations of matching and cutoff frequency dictate a lower C_{j0} than the optimum given by (6.5). For this reason, the best R_s/C_{j0} trade-off is usually to use a relatively large C_{j0}, consistent with matching limitations.

The parameters in the *I*/*V* characteristic are either of secondary importance or are not under the designer's control. Obviously, it is desirable to minimize the ideality factor, η; η depends primarily upon the quality of the diode manufacturing process. The parameter I_{sat} in (2.62) is proportional to the junction area; in small diodes, I_{sat} is small, implying that the diode has high current density for moderate total current. The conductance waveform, however, is proportional to total junction current, not to current density; thus, in small diodes, current-density limitations in the junction may prevent the peak current from being great enough to achieve a high peak conductance. This situation complicates the design of millimeter-wave mixers by raising impedance levels and increasing conversion loss.

The package or diode structure is dictated by the type of circuit in which the diode is used. Surface-mount or epoxy-packaged diodes are most often used on soft composite substrates or printed circuit boards. Beam-lead and chip diodes can be mounted on soft substrates, but because of their fragility, care in attaching them is necessary. Microstrip circuits on alumina or other hard substrates provide better support. Pill-packaged diodes are best reserved for waveguide or stripline applications.

6.3.4 dc Bias

It is sometimes advantageous to apply dc bias to the diode in a single-diode mixer. The dc voltage forward-biases the junction, but, in the absence of LO power, is not enough to cause appreciable junction current. dc bias has two advantages: (1) it can reduce the required LO power, and (2) it provides a degree of freedom for adjusting the diode's conductance waveform, (and therefore the mixer's input and output impedances) and for optimizing conversion efficiency. dc bias is common in high-performance millimeter-wave mixers; it is rarely used in more ordinary applications.

6.3.5 Design Example

The circuit shown in Figure 6.6 is used frequently in block downconverters, receiver front ends for commercial satellite television receivers. It consists of a ring resonator for LO injection, an RF filter (primarily to reject the image frequency), a diode, and an IF filter. The ring resonator, a narrowband structure, is a convenient means for injecting a fixed-frequency LO. Because its bandwidth is narrow, it is unsuitable for tunable LOs.

For this design, the RF frequency range is 12.0 to 12.5 GHz, IF is 1.0 to 1.5 GHz, and the LO is fixed at 11.0 GHz. The LO power level should be

10 dBm or less. In a commercial circuit, conversion efficiency is not the most important characteristic of the mixer. More important are low cost and consistent performance over a large number of manufactured units. Because manual tuning is expensive, the mixer must require little or no tuning.

We begin by assuming that the RF and LO input impedances, and IF load impedance, all must be 50Ω. In this case, we design primarily for a match to the diode over the required frequency ranges, not to optimize conversion efficiency; we are designing for flatness and bandwidth, not conversion loss. This approach is acceptable because (1) the most important characteristics of this mixer are to achieve adequate bandwidth and flat response, (2) the conversion efficiency is likely to be adequate, and (3) the 50Ω source obviously does not require tuning. Of course, if the initial design surprises us with high conversion loss, we can modify it.

The ring resonator and RF filter are designed in a straightforward manner. Therefore, we focus on the diode—in fact, the mixer—and defer the design of the former circuits until later. We begin by selecting a diode. To minimize LO power requirements, we select a low-barrier silicon Schottky device. To minimize the effects of the junction capacitance, it must have a junction reactance of 100Ω or more at 12.5 GHz. A diode having $C_{j0} = 0.1$ pF and $R_s = 12\Omega$ seems appropriate. (Even less junction capacitance would be desirable, but few diodes having lower C_{j0} are readily available, and they have even higher R_s.) This diode has $I_{sat} = 10^{-8}$ A and $\eta = 1.25$. From (2.66), the cutoff frequency is 132 GHz and from (6.4) $\delta = 1.27$, or 1.05 dB. This value of δ is only 0.3 dB greater than the optimum given by (6.5). These are reasonable results for this kind of mixer.

Figure 6.7 shows the circuit used to simulate the mixer alone. The IF filter consists of a parallel L-C resonator, which limits the bandwidth and

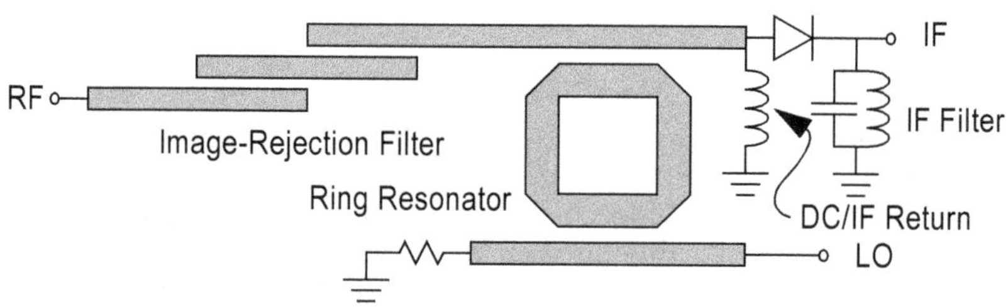

Figure 6.6 A practical single-diode mixer. This type of mixer is used commonly in commercial and consumer microwave equipment.

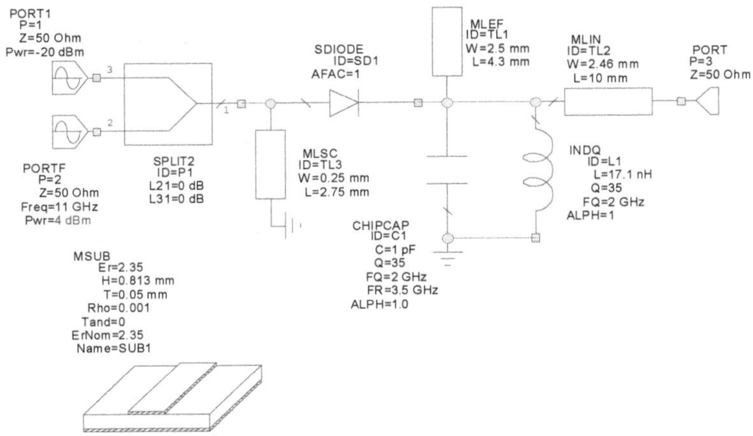

Figure 6.7 The mixer part of the circuit in Figure 6.6 is designed separately from the rest of the circuit. This allows us to optimize the mixer without the potentially confusing effects of the filters.

provides a dc return for the rectified diode current. An important function of the IF filter is to ground the cathode of the diode at the RF and LO frequencies. Because of its parasitics, the lumped-element circuit probably will not be a short circuit at high frequencies, so we add an open-circuit stub. To minimize LO leakage into the IF, the stub is tuned to 11.0 GHz. A high-impedance stub from the anode to ground provides a return path for dc and IF currents. That stub is less than one-quarter wavelength long, making it inductive and providing a small degree of tuning. It is expected that this tuning will not have to be adjusted in manufacturing.

Figure 6.8 shows the calculated conversion loss and port reflection coefficients of the mixer element in Figure 6.7. The RF and LO port impedances are somewhat higher than desired, but the VSWRs are still low enough to provide good performance. The predicted conversion loss is approximately 6 dB over the 12- to 12.5-GHz band.

The rest of the design is straightforward. The RF filter is a simple, parallel-coupled structure, designed according to conventional methods (see, for example, [5.3, 6.6]). The RF filter is designed primarily to reduce noise from the front-end amplifier in the IF band. For this purpose, it requires approximately 13-dB rejection. A filter having two sections is needed. The ring resonator is one wavelength in circumference at the LO frequency and is coupled to the RF and LO source by quarter-wavelength sections of the ring. The coupling is adjusted empirically, on the computer.

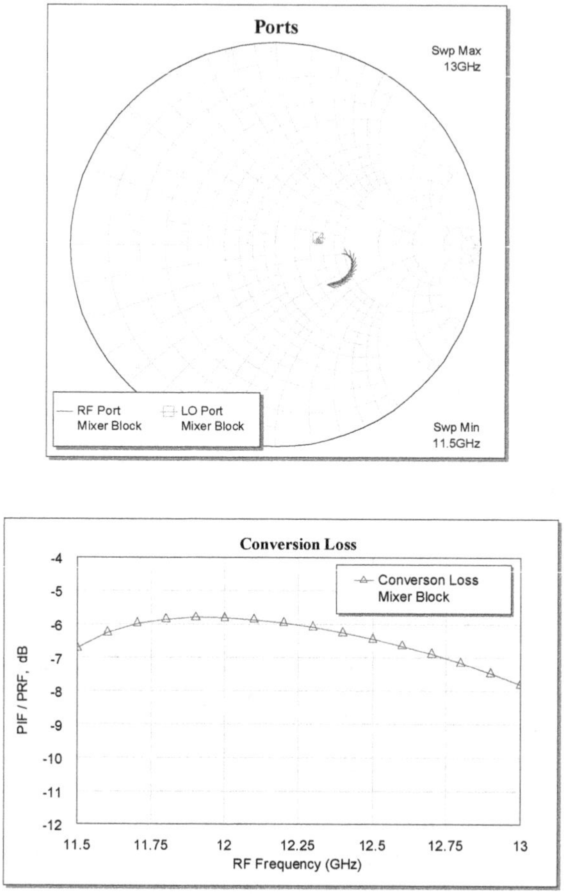

Figure 6.8 Port reflection coefficients and conversion loss of the mixer element in Figure 6.7. $P_{LO} = 4$ dBm.

If the resonator were lossless, it would have a very narrow bandwidth, only a few tens of megahertz. Line losses, however, increase its bandwidth and introduce considerable transmission loss. A loss of 3 to 5 dB in the LO path is not unusual for such a structure, and that loss must be overcome by LO power. The ring does not couple to the RF path, so RF losses are low.

Figure 6.9 shows the calculated conversion loss of the entire mixer. The conversion loss is 7.4 to 7.8 dB across the band. The increased loss, in comparison to that of the mixer element alone, is largely from the RF filter

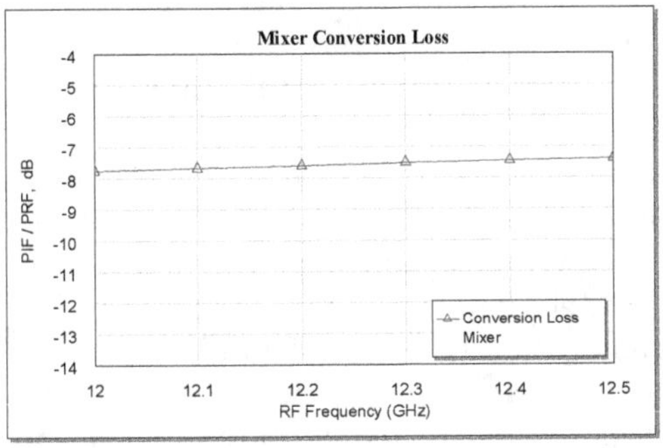

Figure 6.9 Conversion loss of the complete mixer.

and changes in the embedding impedances when the power divider in Figure 6.7 is replaced by the RF and LO filters.

6.4 BALANCED MIXERS

Most diode mixers used at microwave and even millimeter-wave frequencies are balanced. The advantages of balanced mixers over single-diode mixers are (1) rejection of spurious responses and intermodulation products, (2) inherent LO-to-RF isolation, (3) in some cases, inherent LO-to-IF or RF-to-IF isolation, and (4) rejection of AM noise in the LO. The most important disadvantage of balanced mixers is their greater LO power requirements.

Commercially available balanced mixers are frequently small, lightweight, inexpensive, broadband components. In many applications, their good spurious-response properties are essential. Furthermore, in systems where the LO and RF bands overlap, balanced mixers must be used, because it is impossible to separate the LO from the RF by filtering.

6.4.1 Singly Balanced Mixers

A singly balanced mixer consists of two single-diode mixing elements, which may be nothing more than two individual diodes, combined by either a 180-degree or a 90-degree hybrid. The LO and RF are applied to one pair of mutually isolated ports, and the mixing elements, which we shall simply call *mixers*, are connected to the other pair of ports. The diodes in the two mixers must be connected to the ports in such a way that their polarities are opposite. The IF outputs of the individual mixers can be combined by another hybrid, or, more commonly, connected in parallel. The properties of hybrid-connected nonlinear elements are described in detail in Chapter 5.

Figure 6.10 shows a singly balanced mixer that uses a 180-degree hybrid. The RF and LO are connected to one pair of mutually isolated ports; the single-diode mixers, represented by diode symbols in the figure, are connected to the other pair.

In a singly balanced mixer, it is essential that the dc path through the diodes be continuous. If the diodes are open-circuited at dc, the mixer simply will not work. Often, the hybrid provides that path. In Figure 6.10 the inductors L1 and L2 realize the so-called IF return; they ground their respective ends of the diodes at the IF frequency. The inductors also provide a dc return in cases where the hybrid does not. dc bias, if desired, can be provided to both diodes by a voltage source in series with either of these inductors. If bias is used, dc blocks between the hybrid and diodes also may be necessary.

Because the IF ports are connected in parallel, the impedance presented to each single-diode mixer at the IF frequency is *twice* that of the actual IF load. If the IF load impedance is 50Ω, each diode has, in effect, a 100Ω load. Section 5.2 explains this point in detail; however, one can see that this is the case by thinking of the IF load as two loads in parallel, each having twice the impedance of the actual load. Because of the symmetry of the circuit, the two loads can be separated so that each is connected to only one

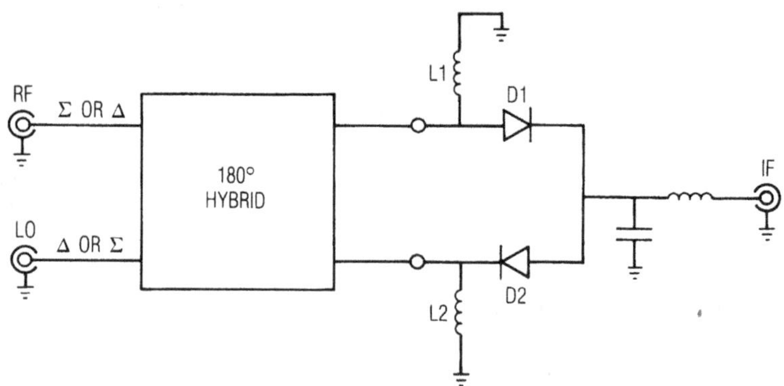

Figure 6.10 180-degree, singly balanced mixer. The diodes D1 and D2 can be unmatched diodes or complete, individual, single-diode mixers. The mixer can be configured with either the sigma or delta port as the RF; the other is the LO.

mixer. Since the optimum IF load impedance is usually close to 100Ω, this is a beneficial property.

We can estimate the conversion loss of the balanced mixer by analyzing the individual single-diode mixers, each having the doubled IF load impedance, and from Chapter 5 the conversion loss of the balanced mixer must be the same as that of the individual single-diode mixers (plus, of course, hybrid losses, and we must remember that the balanced mixer requires twice the LO power of the individual mixers). When the individual diode mixers are designed and optimized, they are connected to a hybrid, and the entire structure is analyzed. If the single-diode mixers and the hybrid have been designed well, little or no further optimization should be needed. Thus, the design process for a single diode mixer, described in Section 6.3, is directly applicable to the design of a balanced mixer.

A singly balanced mixer can also be realized by replacing the 180-degree hybrid in Figure 6.10 by a quadrature hybrid. The individual mixers are the same as those used with the 180-degree hybrid, and, as before, they are connected to mutually isolated ports.

The main differences in operating characteristics between the 180-degree and quadrature hybrid mixers are in the port's VSWRs, isolation, and spurious response properties. In the 180-degree mixer, the input VSWR at the LO and RF ports is dominated by the VSWRs of the individual mixers, and the RF/LO isolation is dominated by the isolation of the hybrid. The quadrature hybrid, however, operates in a very different manner (see Section 5.1). LO power reflected from the individual mixers does not return to the LO port, but instead exits the RF port; similarly, reflected RF power exits the LO port. The LO/RF and RF/LO isolation is therefore equal to the input return loss of the individual mixers at the LO and RF frequencies, respectively; the port isolation of the quadrature-hybrid mixer depends primarily on the input VSWRs of the two individual mixers, not on the isolation of the hybrid itself. As long as the LO and RF source VSWRs are good, the mixer's LO and RF input VSWRs are also good. However, if the RF port termination has a poor VSWR at the LO frequency, the circuit's balance can be upset and the LO pumping of the individual mixers becomes unequal; similarly, a poor LO port termination at the RF frequency can upset RF balance.

The spurious-response properties of the 180-degree and quadrature-hybrid mixers also differ. If the sigma port of the hybrid is used as the LO port, the 180-degree hybrid mixer rejects spurious responses involving even harmonics of the LO; if the sigma port is the RF port, even harmonics of the RF that mix with any harmonics of the LO are rejected. The quadrature-hybrid mixer, however, does not reject the even harmonics of one signal, either the RF or LO, mixing with the odd harmonics of the other. Both types of mixers, however, reject the even LO harmonics that mix with the even RF harmonics.

It is worth noting that while many seemingly different types of singly balanced mixers have been developed, all are fundamentally realizations of either the 180-degree or quadrature structures. Nothing else exists. An example is the crossbar mixer shown in Figure 6.11(a), which is in fact a type 180-degree hybrid mixer. In the crossbar mixer two diodes are connected in series across the RF waveguide, and the LO is coupled to the diodes via a metallic strip (the crossbar) that acts as a coupling probe in the LO waveguide. The probe is also used for the IF output. The orientation of the probe and the RF and LO waveguides is such that the probe does not couple the LO and RF waveguides.

The fact that the crossbar mixer is a type of 180-degree hybrid balanced mixer is evident from the polarities of the LO and RF voltages at the diode, as shown in Figure 6.11(b). The RF voltage applied to the diodes has the same phase it would have if the RF signal had been applied to the delta port of a 180-degree hybrid, and the LO voltage pumps the diodes out of phase, as if it had been applied to the sigma port.

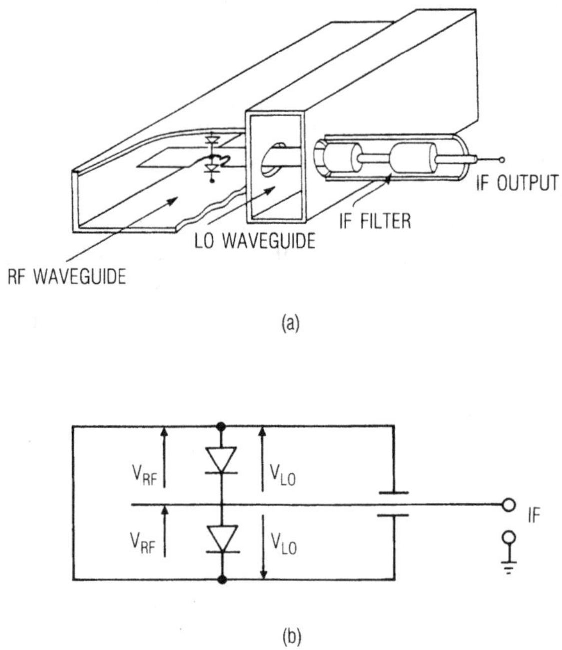

Figure 6.11 (a) A crossbar mixer; (b) polarities of the LO and RF voltages at the diodes.

Singly balanced mixers have many of the desirable properties of balanced mixers, yet can be treated in many ways like single-diode mixers. It is practical for singly balanced mixers to have matching circuits and dc bias, giving them good conversion efficiency, flat bandwidth, and low VSWR. The structures used for doubly balanced mixers, described in the next section, do not allow for practical matching circuits and dc bias. Accordingly, it can be more difficult to optimize a doubly balanced mixer. Doubly balanced mixers are used primarily in applications where their superior spurious-response properties are essential, and those applications comprise most types of modern microwave systems.

6.4.2 Singly Balanced Mixer Example

As an example, we create a singly balanced mixer from the single-diode mixer described in Section 6.3.5. This involves connecting two of the mixing elements shown in Figure 6.7 to a rat-race hybrid (Section 5.1.2.2).

The single-diode mixer in Figure 6.7 must be modified slightly, by changing the load impedance to 100Ω. This change accounts for the parallel connection of the individual mixers' IF ports. Rerunning the analysis of the single mixing element, we find that the conversion loss is nearly identical to the 50Ω case; the mixer is not very sensitive to the IF load impedance.

Next, we design the hybrid. The combined RF and LO bandwidth of the mixer is approximately 13%; this is close to the limit for a rat-race hybrid, but (as we shall see) is achievable with a little empirical modification of the basic design. The hybrid is modeled as shown in Figure 6.12, with several microstrip transmission line sections and junction discontinuities. The lengths of the transmission lines are not exactly the ideal values of one-quarter and three-quarters of a wavelength; they are modified slightly to account for the junction parasitics and to optimize the bandwidth. We keep the one-quarter wavelength sections equal in length, but allow the three-quarter wavelength section to be varied independently. A little tuning gives the result shown in Figure 6.13.

We now connect two of the mixer elements to ports 2 and 3 of the hybrid, remembering, of course, to reverse the diode in one of them. The RF excitation is connected to port 1 and the LO to port 4. The LO power must be increased 3 dB (compared to the single-diode mixer), plus the hybrid's minimal excess loss, shown in Figure 6.13. The resulting conversion loss is shown in Figure 6.14. Again, the increase in conversion loss of approximately 1 dB, compared to the loss of the single-diode mixing element, is caused largely by differences in the embedding impedances at high-order mixing products.

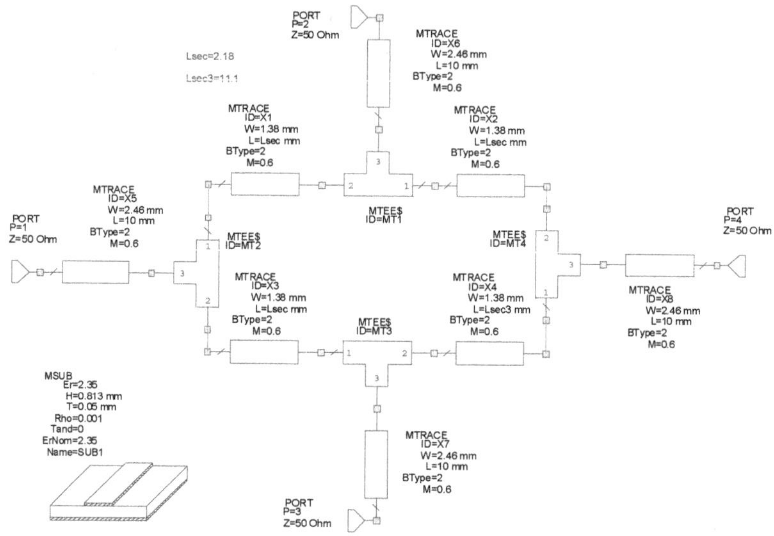

Figure 6.12 Circuit model of a 180-degree rat-race hybrid used for the singly balanced mixer.

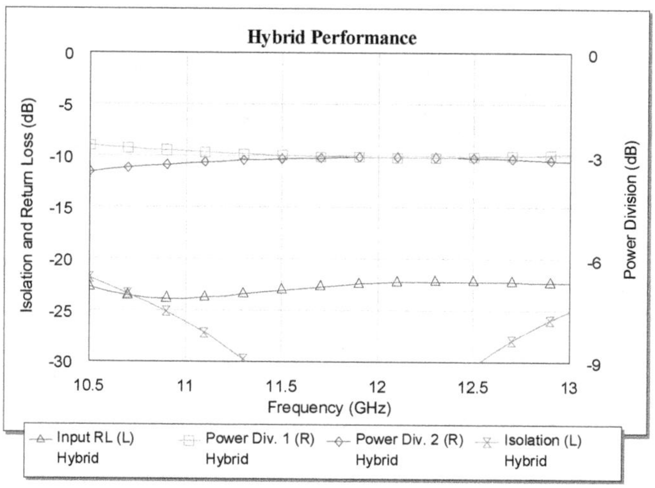

Figure 6.13 Performance of the hybrid in Figure 6.12.

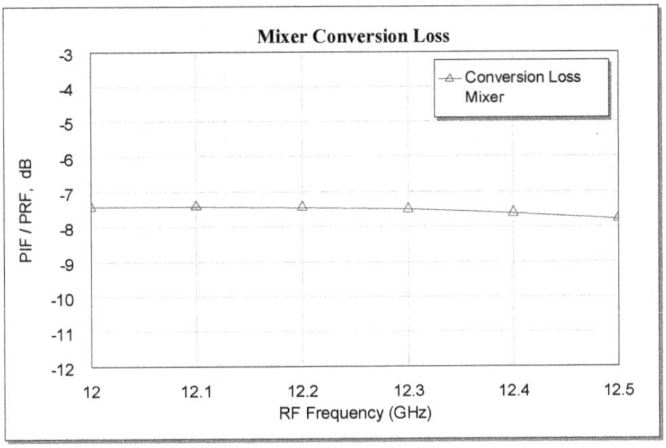

Figure 6.14 Conversion loss of the complete mixer.

6.4.3 Doubly Balanced Mixers

The two most common types of doubly balanced mixers are the ring mixer and the star mixer. The ring mixer is more amenable to low-frequency applications, in which transformers can be used, but it is also practical at high frequencies. The star mixer is used primarily in microwave applications, as it is better suited to operation with microwave baluns. There is no significant fundamental difference in the electrical properties or performance of both mixer types; as we shall see, both are polarity-switching, or *commutating*, mixers.

6.4.3.1 Ring Mixer

The classical ring mixer circuit, sometimes called a *ring modulator*, is shown in Figure 6.15. The circuit consists of a ring of four diodes, designated D1 through D4, and two transformers, T1 and T2. The transformers are identical to the transformer hybrid described in Section 5.1.2.1 and are often realized as separate trifilar windings on toroidal cores. (One winding is used as the primary, and the other two are connected in series to form the secondary.) The secondaries of these transformers are connected to the nodes of the diode ring, labeled A through D.

The operation of the mixer can be described very simply if the diodes are viewed as ideal, LO-driven switches. When the polarity of the LO

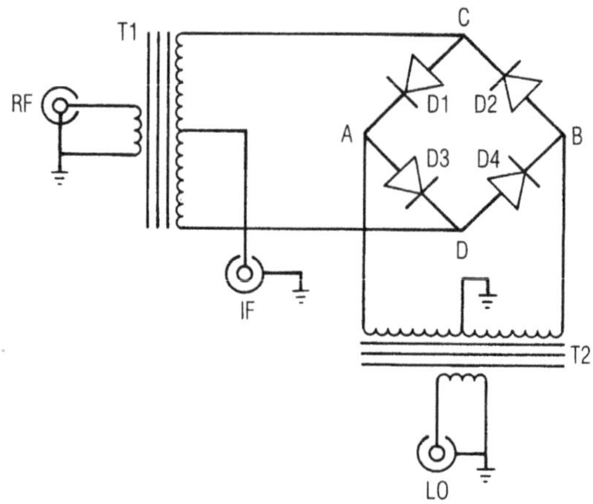

Figure 6.15 The ring mixer. The IF port can be the center tap of either transformer; however, LO-to-IF isolation is usually better if the RF transformer's center tap is used.

voltage is such that the right side of the secondary of T2 is positive, diodes D1 and D2 are turned on and D3 and D4 are turned off. During this half of the LO cycle, D1 and D2 short circuit T2 so that node C is connected to ground through the center tap of T2. The upper half of T1 is thus connected through these diodes to the IF port, and the RF port is momentarily connected to the IF port. When the LO voltage reverses, D3 and D4 are turned on and D1 and D2 are turned off. Then the lower half of T1 is connected to the IF, so the RF is again connected to the IF, but with its polarity now reversed. The mixer therefore acts as a polarity-reversing switch, connecting the RF port to the IF but reversing its polarity every half LO cycle.

The IF voltage is

$$v_{\mathrm{IF}}(t) = s(t) v_{\mathrm{RF}}(t) \tag{6.6}$$

or

$$v_{\mathrm{IF}}(t) = \sum_{n=1}^{\infty} b_n \sin(n\omega t)\, v_{\mathrm{RF}}(t) \tag{6.7}$$

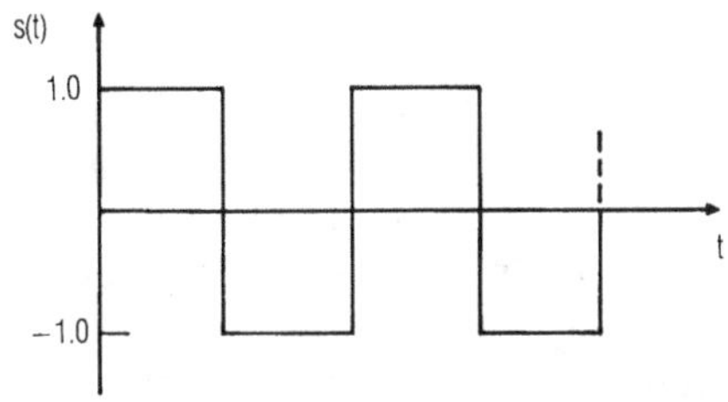

Figure 6.16 Switching waveform of the ideal ring mixer. This waveform also is valid for the star mixer.

where n is odd and $s(t)$, the switching waveform in Figure 6.16, has been represented as a Fourier series. Downconversion occurs via the product of the fundamental-frequency component of $s(t)$ and the sinusoidal $v_{RF}(t)$. Because $s(t)$ is a symmetrical square wave, it has no dc component in its Fourier-series representation; (6.7) shows that $v_{IF}(t)$ can have no RF-frequency component. Consequently, the RF and IF are isolated, even though no filters are used. The waveform $s(t)$ also has no even-harmonic components ($b_n = 0$, n even), so there can be no spurious responses associated with even LO harmonics. Because of the symmetry of the circuit, spurious responses associated with the even harmonics of the RF also are rejected. Furthermore, at all times either D1 and D2 are shorted or D3 and D4 are shorted; this short-circuit prevents the coupling of RF voltage to the LO port or LO voltage to the RF port. Even if the diodes are not ideal switches, the RF-to-LO isolation is theoretically perfect because the instantaneous RF voltage at A and B must always be the same, as long as the voltage drops in the pairs (D1, D2) and (D3, D4) are identical. Since the voltage drops across these diodes are identical, the symmetry of the circuit causes nodes C and D, the terminals of the T1 secondary, to be virtual ground points for the LO. The RF transformer secondary is connected to these LO ground points, so the LO-to-RF isolation is likewise theoretically perfect.

Ring mixers using transformers are best for broadband applications at frequencies up to a few hundred megahertz, although careful transformer design (involving so-called *transmission-line transformers* [5.1, 5.2]) can sometimes raise the frequency limit above 2 GHz. A ring mixer's bandwidth is limited primarily by the bandwidth of its RF and LO transformers; the diodes rarely limit performance in this frequency range. Ring mixers can also be used as modulators, phase detectors, and even voltage-controlled attenuators; they are very versatile components.

6.4.3.2 Microwave Ring Mixer

Replacing the transformers in Figure 6.15 with baluns realizes a microwave ring mixer. Microwave baluns, however, do not have anything comparable to the transformer's center tap, so some other provision must be made for the IF connection. A common microwave implementation of the ring mixer is shown in Figure 6.17.

The mixer in Figure 6.17 uses parallel-strip baluns. These can be viewed either as a pair of coupled transmission lines or as a single, balanced transmission line; in fact, the two are equivalent when

$$Z_{0o} = \frac{Z_{0b}}{2} \qquad Z_{0e} \rightarrow \infty \tag{6.8}$$

Z_{0o} and Z_{0e} are the odd- and even-mode impedances of the coupled line, and Z_{0b} is the characteristic impedance of the balanced line. In the ideal case, as $Z_{0e} \rightarrow \infty$, the structure cannot support an even mode, so all the energy applied to its unbalanced terminal must propagate on the line in the odd mode.

In practice, it is difficult to achieve a large value of Z_{0e}. Usually the balun is realized as a suspended substrate with an air gap between the lines

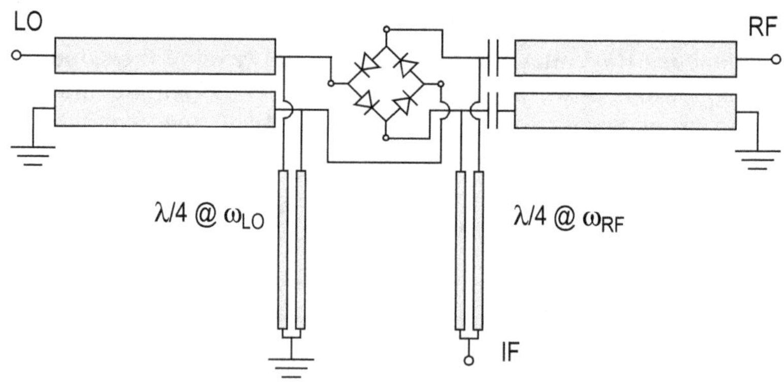

Figure 6.17 A microwave realization of the ring mixer. The baluns consist of parallel coupled strips, and the IF is extracted by means of quarter-wavelength stubs. Blocking capacitors are used to prevent IF leakage from the RF port.

and ground surfaces, and the substrate is made very thin (often as little as 125 μm) to minimize the line widths for a desired value of Z_{0o}.

Because the balun is equivalent to a balanced transmission line, its impedance Z_{0o} can be selected so that the line operates as a quarter-wavelength transformer. Then

$$Z_{0b} = \sqrt{Z_0 Z_d} \tag{6.9}$$

where Z_0 is the port impedance (usually 50Ω) and Z_d is a single diode's junction impedance, which we have previously estimated as 50Ω to 100Ω. Using the line as a transformer limits its bandwidth considerably, however, so we might instead choose to set Z_{0b} = 50Ω and tolerate an input VSWR that may be as high as 2.0.

As with most balanced mixers, the bandwidth is limited primarily by the balun. The high-frequency limit occurs when the balun's length approaches one-quarter wavelength in terms of the even-mode phase velocity. At this point, the finite Z_{0e} creates a resonance and a characteristic "glitch" in the passband. The low-frequency limit is established by Z_{0e}; the even-mode input impedance, at the balanced end, must be kept high. This requires that

$$Z_{0e} \tan(\theta_e) \gg Z_0 \tag{6.10}$$

where θ_e is the balun's even-mode length in degrees of phase.

These considerations show that the balun's bandwidth—and therefore the mixer's bandwidth—are established by the even-mode impedance of the balun; the greater the balun's Z_{0e}, the wider its bandwidth. The bandwidth of such mixers can exceed a decade; a 2- to 26-GHz mixer is not unusual.

The IF is extracted by a pair of stubs. These are one-quarter wavelength long at the centers of the RF and LO frequency ranges, and realize high-impedance stubs in parallel with the balanced ends of the baluns. At the IF, the conductors represent a substantial inductance, so a wide IF bandwidth is not possible. The RF balun must have IF blocks, often realized by capacitors. The L-C combination can create resonances in the IF band if the designer does not select their values carefully.

6.4.3.3 Ring Mixer Design Example

To illustrate the design of a ring mixer, we design a 2- to 26-GHz mixer using parallel-strip baluns. Because of the high maximum frequency, we use beam-lead diodes having minimal parasitics; however, cost and LO power requirements do not allow the use of a GaAs device.

Because of the high frequency, we need a diode having the lowest C_{j0} available. Minimizing LO power requirements dictates a low-barrier device, but low-barrier devices often have high series resistance. We select a medium-barrier device as a compromise. The best device available has $I_{sat} = 1.0{\cdot}10^{-12}$ A, $C_{j0} = 0.07$ pF, and $R_s = 12\Omega$. Its external parasitics consist of an overlay capacitance of 0.02 pF and a series inductance of 0.3 nH.

We begin with the design of the balun. For best bandwidth, we want it to have a characteristic impedance of $Z_{0b} = 50\Omega$, so from (6.8) $Z_{0o} = 25\Omega$. To minimize the line width, we must minimize the thickness of the suspended substrate used for the balun. The minimum thickness available is 125 μm, but to improve the support for the fragile beam-lead device, we select 250 μm. Parallel-strip baluns work best when the even- and odd-mode phase velocities are as close to equal as possible; this requirement dictates that the substrate have a low dielectric constant. A composite substrate having $\varepsilon_r = 2.35$ is chosen. Finally, standard mixer packages allow a clearance of 2.3 mm between the substrate and the top- and bottom-plates. Using transmission-line design software, we find that strips having a width of 0.81 mm provide the desired odd-mode impedance, and the resulting even-mode impedance is approximately 260Ω and the effective dielectric constant is 1.125.

In evaluating the bandwidth, we encounter a difficulty. Making the balun one-quarter wavelength long at 26 GHz means that it will be 0.012 wavelengths, or 6.9 degrees, long at 2 GHz. We find that $Z_{0e}\tan(6.9) = 32\Omega$, which does not satisfy (6.10). However, by tapering the ground-plane side of the balun, we might be able to achieve the desired bandwidth. The taper causes the excitation, which normally consists of equal even- and odd-mode components, to excite the odd mode more than the even. Although this improves the bandwidth, electromagnetic simulation is required to determine the balun's characteristics.

Figure 6.18 shows the balun. The taper is designed empirically; its width at the input must be several times the top-line width, and the taper must become more gradual as its width approaches that of the line. With a little tuning on the computer, we obtain a balun design having the input return loss and balance shown in Figure 6.19. In this analysis, the balun is

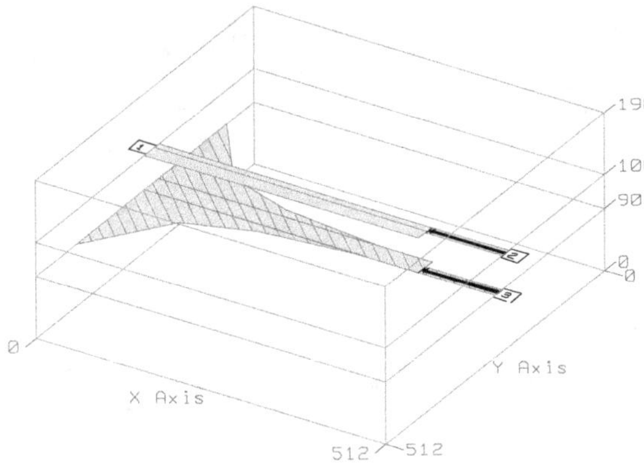

Figure 6.18 Tapered balun used in the ring-mixer design example. The thickness of the dielectric layer is artificially increased for ease of viewing.

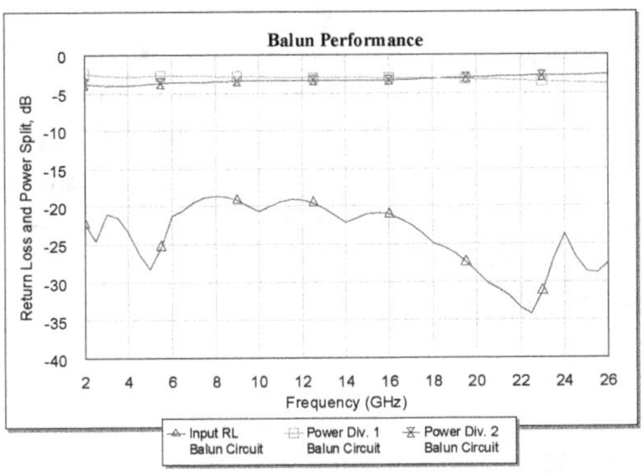

Figure 6.19 Calculated performance of the tapered balun.

treated as a power divider; the two output terminals are treated as a pair of ports, and we plot the power division at those two output ports. The imbalance is a measure of the balun's even-mode output. The calculation shows an imbalance of 1.2 dB at the band edges; although not exciting, this result is not unusual for this type of balun.

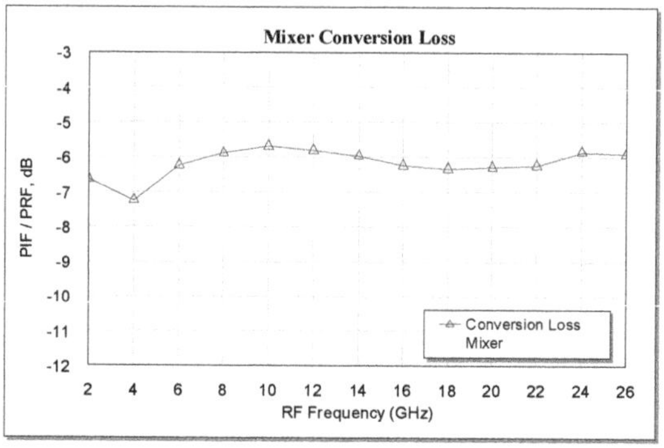

Figure 6.20 Calculated conversion loss of the ring mixer.

Finally, we connect the diodes to the balun, add transmission lines for IF extraction, and simulate the circuit in its entirety. To optimize the circuit, we have relatively few degrees of freedom; the most important is LO power. We can, of course, modify the balun slightly or select a different diode.

The mixer's conversion performance is shown in Figure 6.20. Conversion loss is quite acceptable, 6 to 7 dB over most of the band. The input return loss is modest, 5 to 10 dB across the band, and the use of four medium-barrier diodes causes the LO power to be rather high, 16 dBm. This level of performance is typical of this kind of mixer. Other types of mixers, such as the "horseshoe" balun mixer [6.1], may offer improved performance, including much wider IF bandwidth.

6.4.3.4 Star Mixer

In a star mixer, one terminal of each of four diodes is connected to a common node, which is used as the IF terminal. The mixer's operation can be described by the transformer realization shown in Figure 6.21, which, unlike the ring mixer, is rarely used in practical circuits. The mixer uses two transformers, each of which has two windings; T2a and T2b are the secondaries of one transformer, while T1a and T1b are those of the other.

This mixer, like the ring mixer, operates as a polarity-reversing switch. When the dotted sides of the LO transformer secondaries T2a and T2b are positive, D1 and D2 are turned on, D3 and D4 are turned off, and the dotted sides of T1a and T1b are connected to the IF port. The RF port is thus connected to the IF through the transformer T1 and the diodes. When the LO polarity reverses, D1 and D2 are off and D2 and D3 are on; then the undotted side of both T1 windings is connected to the IF port, and the RF is again connected to the IF but its polarity is reversed. The RF polarity is therefore applied to the IF port, but its polarity is reversed at the LO frequency. Because the star mixer operates by the same principles as the ring mixer, it should be no surprise that the spurious-response properties of the star mixer are the same as those of the ring mixer.

Figure 6.22 shows a version of a microwave star mixer. The balun, a variation of a structure known as the *Marchand balun*, is described in detail in [6.1, 6.7, 6.8]. The Marchand balun is remarkably broadband; it is theoretically capable of decade bandwidths, but the form used in mixers is limited to bandwidths of a little over an octave.

The star mixer has low parasitic IF inductance, giving it a broad IF bandwidth. Unfortunately, the Marchand balun is an open circuit for even-mode excitation, and the IF excites the baluns in an even mode. For this reason, the IF cannot overlap the RF or LO bands, but it can sometimes approach 70% to 80% of the lower end of the RF/LO frequency range. Another limitation of this circuit is that the RF and LO baluns must cover the same frequency range; thus, each balun must cover the entirety of both the RF and LO bands.

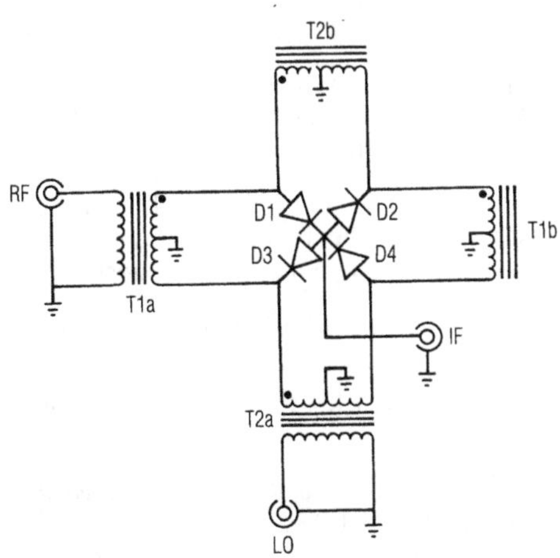

Figure 6.21 Transformer equivalent circuit of the star mixer.

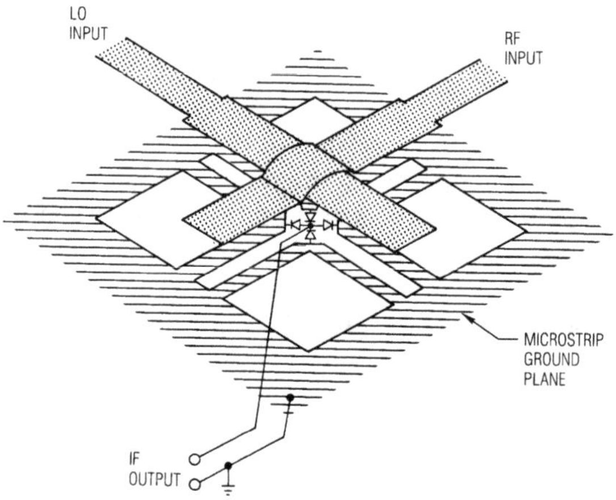

Figure 6.22 Doubly balanced star mixer for use at microwave frequencies.

References

[6.1] S. A. Maas, *The RF and Microwave Circuit-Design Cookbook*, Norwood, MA: Artech House, 1999.

[6.2] A. G. Cardiasmenos, "New Diodes Cut the Cost of MMW Mixers," *Microwaves*, Sept. 1978, p. 78.

[6.3] B. J. Clifton, "Schottky Diode Receivers for Operation in the 100–1000 GHz Region," *Radio and Electronic Engineer*, Vol. 49, 1979, p. 333.

[6.4] W. L. Bishop et al., "A Novel Whiskerless Schottky Diode for Millimeter and Submillimeter Wave Applications," *IEEE MTT-S International Microwave Symposium Digest*, 1987, p. 607.

[6.5] W. L. Bishop et al., "A Micron-thickness, Planar Schottky Diode Chip for Terahertz Applications with Theoretical Minimum Parasitic Capacitance" *IEEE MTT-S International Microwave Symposium Digest*, 1990, p. 1305.

[6.6] R. Rhea, *HF Filter Design and Computer Simulation*, Tucker, GA: Noble Publishing, 1994.

[6.7] N. Marchand, "Transmission Line Conversion Transformers," *Electronics*, Vol. 17, No. 12, 1979, p. 52.

[6.8] R. Mongia, I. Bahl, and P. Bhartia, *RF and Microwave Coupled-Line Circuits*, Norwood, MA: Artech House, 1999.

Chapter 7

Diode Frequency Multipliers

A large part of the electronics in any microwave communications system is devoted to generating signals at specific frequencies. Frequently, signals of high stability and low noise are needed; these are sometimes obtained by generating harmonics from a very stable low-frequency source, such as a crystal oscillator. For better or worse, harmonic generation is one of the things that nonlinear circuits do best, so it should be no surprise that varactor, step-recovery, and Schottky-barrier diodes are employed widely in frequency-generating systems.

Diode circuits that use varactors or step-recovery diodes (SRDs) are frequently employed as harmonic generators at microwave frequencies. These are reactive multipliers: they make use of the diode's nonlinear capacitance characteristic. Varactors are used primarily to multiply microwave signals to low harmonics, rarely over four times the source frequency; in contrast, SRDs are used to multiply signals in the UHF or low microwave range to very high harmonics. Both components are inherently narrowband and, when properly designed, have good efficiency and low noise.

Resistive diodes—Schottky-barrier diodes—are sometimes used in low-order frequency multipliers. Resistive multipliers are less efficient than reactive multipliers, but they can be made very broadband. Furthermore, it is usually easier to develop a resistive multiplier than a reactive one; reactive multipliers are sensitive to even slight mistuning, and therefore have a well-deserved reputation of being difficult to optimize. In contrast, resistive multipliers are relatively easy to adjust and are not nearly as sensitive.

7.1 VARACTOR FREQUENCY MULTIPLIERS

7.1.1 Noise Considerations

In the past it was common to use varactor frequency multipliers to generate moderate to high levels of RF power. Solid-state sources now available have greater efficiency, fewer components, and greater bandwidth than varactor multipliers. Furthermore, GaAs MESFET frequency multipliers, described in Chapter 10, are capable of greater efficiency and bandwidth than are diode multipliers. One might wonder why varactor frequency multipliers are still used at all.

The major advantage of a varactor frequency multiplier is that, because it is a reactive device, it generates very little noise. This property is particularly valuable where low phase noise is desired: local oscillator (LO) sources for radar applications and many types of phase- or phase/amplitude-modulated communication systems. The dominant noise source in a varactor multiplier is the thermal noise of its series resistance and of its circuit losses; both are very small in a well-designed device and circuit. Frequency multipliers using Schottky-barrier varactors can achieve high efficiency and low noise at output frequencies of several hundred gigahertz; such multipliers, driven by Gunn or FET sources, can generate adequate LO power for single-diode mixers, with very low AM noise levels.

Even if a multiplier introduces no phase noise of its own, the process of frequency multiplication—even by an ideal, noiseless multiplier—inevitably increases phase noise. The reason for this unfortunate characteristic is that a frequency multiplier is in fact a phase multiplier, so it multiplies the phase deviations as well as the frequency of the input signal. The minimum carrier-to-noise degradation, ΔCNR, in decibels, caused by an ideal frequency multiplier is

$$\Delta CNR = 20 \log_{10}(n) \tag{7.1}$$

where n is the multiplication factor. Thus, a frequency doubler ($n = 2$) degrades the CNR of the input signal by at least 6 dB; a quadrupler degrades the CNR by at least 12 dB. If the multiplier is noisy, it can add even more phase noise to the input signal, and ΔCNR can be even greater.

In addition to phase noise, AM noise is a concern in many types of systems. AM noise in the LO can be an especially serious problem in low-noise receivers: if the LO signal has AM noise sidebands at the RF frequency, that noise can be downconverted to the IF, significantly increasing the noise temperature of the receiver.

Solid-state devices used in frequency sources or in multiplier chains are usually driven into saturation, and their limiting effects usually remove much of the signal's AM noise. However, these effects do not completely eliminate AM noise, so sources that are inherently noisy (e.g., IMPATT devices and klystron tubes) often generate signals that have high AM noise levels, in spite of any limiting that may occur. Using an amplifier to increase the power level of a signal, even if the amplifier is driven into saturation, also can introduce a large amount of AM noise. The use of balanced mixers or narrowband filters at the mixer's LO port can do much to reduce the effects of such noise; however, in some cases, balanced mixers or filtering may not be possible. An LO system consisting of a Gunn or FET source followed by a varactor multiplier usually has minimal AM noise and is therefore a preferred configuration for millimeter-wave systems.

7.1.2 Power Relations and Efficiency Limitations

Manley and Rowe [7.1] developed a set of general relations between the real powers at all mixing frequencies in a nonlinear capacitor. The *Manley-Rowe relations* are valid for any nonlinear capacitor driven by one or two signals having noncommensurate frequencies. The relations are remarkable in that they do not depend directly upon the capacitor's *Q*/*V* characteristic or the levels of the applied excitations. (They do require, however, that voltages and currents exist at certain frequencies, and this requirement contains implicit assumptions about the *Q*/*V* characteristic and embedding circuit.) The Manley-Rowe relations have been applied to parametric amplifiers and upconverters as well as to varactor frequency multipliers, and they establish limits to the gain or loss of such components.

The two Manley-Rowe relations are

$$\sum_{m=0}^{\infty} \sum_{n=-\infty}^{\infty} \frac{mP_{m,n}}{mf_1 + nf_2} \tag{7.2}$$

$$\sum_{n=0}^{\infty} \sum_{m=-\infty}^{\infty} \frac{nP_{m,n}}{mf_1 + nf_2} \tag{7.3}$$

where f_1 and f_2 are the frequencies of the two excitation signals, and $P_{m,n}$ is the average real power into the capacitor at the frequency $|m f_1 + n f_2|$. (Note that, in this case, the input powers at the excitation frequencies, $P_{1,0}$ and $P_{0,1}$, are the powers absorbed by the network, not available powers from the sources.) These relations can be derived from the sole considerations that the capacitor is lossless, and that the capacitor's Q/V characteristic is single-valued.

A frequency multiplier has only a single excitation, f_1, so $f_2 = 0$, the summation over n can be eliminated, and all the terms of (7.3) become zero. Equation (7.2) becomes

$$\sum_{m=0}^{\infty} P_m = 0 \tag{7.4}$$

where P_m is the power in the diode at the frequency $m f_1$. Equation (7.4) indicates that all the input power must be converted into output power at the harmonics of f_1; none can be dissipated in the reactive junction (note that (7.4) does not say where the output power must be dissipated; in practice much of it may be dissipated in circuit losses or in the series resistance). In an Mth-harmonic multiplier, the highest possible value of P_m occurs when only P_1 and P_m are nonzero; then $P_m = -P_1$. In that case, the output power at P_m is equal to the input power at P_1, and if the input power equals the available power of the source, the multiplier has 100% efficiency.

For this optimum efficiency to be achieved, there must be no real power in the circuit at any of the unwanted harmonic frequencies. This condition is guaranteed when the diode's junction is terminated in a pure reactance at all harmonics other than the desired one. In practice, however, the diode's series resistance makes a pure reactance impossible; this resistance is always in series with the terminating impedance, and thus dissipates power at all harmonics. To eliminate power dissipation in the series resistance, one might be tempted to open-circuit the diode at all unwanted harmonics; then the unwanted harmonic currents in the series resistance would be zero and no power would be dissipated. The next best approach would be to short-circuit the diode at all unwanted harmonics; a short circuit would not eliminate the dissipation in the series resistance, but would prevent harmonic power dissipation in the output network.

It happens, however, that in diodes having C/V characteristics close to that of the ideal Schottky or *pn* junction, and in frequency multipliers that generate harmonics greater than the second, short-circuit terminations at unwanted harmonics are preferred. The diode's voltage, as a function of charge, is the cause of this counterintuitive situation. The *V*/*Q* function has a square-law characteristic and therefore cannot generate voltage components beyond the second harmonic, unless harmonic current components also exist. Let us suppose that the diode is driven at the excitation frequency by an ideal current source and thus has only open-circuit harmonic terminations. The *Q*/*V* characteristic of an ideal, uniformly doped junction, given by (2.58), with $\gamma = 0.5$, is

$$Q(V) = -2C_{j0}\phi\left(1 - \frac{V}{\phi}\right)^{0.5} \tag{7.5}$$

which can be rearranged to express *V* as a function of *Q*:

$$V = \phi\left(\frac{Q_\phi^2 - Q^2}{Q_\phi^2}\right) \tag{7.6}$$

where $Q_\phi = 2C_{j0}\phi$, a constant. If the diode is open-circuited at all harmonics, the current can have no harmonic components and thus must be sinusoidal at the fundamental frequency. Because the current is sinusoidal, the charge also varies sinusoidally at the same frequency; if the voltage has a square-law dependence on *Q*, it must also have a square-law dependence upon the current. Squaring this sinusoid produces only second harmonics; therefore, if the varactor is open-circuited at all harmonics, there can be no voltage components across the junction at any harmonics beyond the second, and the multiplier is limited to second-harmonic operation. In order to have a third-harmonic output, it is necessary to have a large second-harmonic component of junction current; then the third "harmonic" arises as a second-order mixing product between the fundamental excitation and the second-harmonic current. In order to have this large second-harmonic current, there must be a short circuit across the junction—called a *short-circuit idler*—at the second harmonic. Similarly, for higher-harmonic outputs, idlers must be provided at the intermediate harmonics. For example, a quadrupler could have a second-harmonic idler, or both a second-harmonic and a third-harmonic idler; a quintupler would likely have at least second- and third-harmonic idlers.

The idea that varactor multipliers can generate only second harmonics directly is strictly valid only for varactors that have the ideal *Q*/*V* characteristic of (7.5). The *Q*/*V* characteristics of real varactors normally deviate somewhat from (7.5), and some second-harmonic current is generated by second-harmonic voltage dropped across the finite embedding impedance. Furthermore, because the charge-storage properties of a p^+n varactor increase its V/Q nonlinearity beyond second degree, overdriving the diode generates current and voltage components at harmonics greater than the second. An extreme case is that of the step-recovery diode, which has a very strong C/V nonlinearity; idlers are not normally needed in SRD multipliers. However, both theory and experimental evidence indicate that the use of idlers can improve the efficiency of all reactive frequency multipliers, even those using SRDs.

Idlers are usually realized as short-circuit resonators that are separate from the input and output matching circuits. In practice, idlers are usually realized by a series resonance that is chosen more for its convenience than for high performance; frequently, the series resonance of the varactor's package is used as an idler at high frequencies, and tuning elements are often included to tune the resonance precisely to the desired harmonic. This technique results in a very compact multiplier that can be realized easily in strip transmission media, but probably has a lower Q than would a waveguide-mounted diode having a separate idler cavity. It is important to maximize idler Q in multipliers designed to have high efficiency, because the large idler currents must circulate in the idler resonator's loss resistance; this resistance, like the diode's series resistance, can generate significant power losses. Low series resistance and high unloaded Q are therefore clear requirements of high-efficiency frequency multiplication.

In theory, high-order multipliers are most efficient when they have idlers at all intermediate harmonics. Unfortunately, it is rarely practical at microwave frequencies to provide more than one idler, and harmonics up to the fourth can be generated efficiently in such multipliers. The difficulty in realizing several idlers is one of the factors that limits the order of multiplication of a varactor multiplier.

Finally, we examine two important details. The first is that the efficiency limitation established by the Manley-Rowe relations is only part of the story. These relations show that all the input power must be converted to output power at the fundamental and harmonic frequencies; this result is obvious because a reactive element—linear or nonlinear—cannot dissipate power. Thus, the power gain of a frequency multiplier using an ideal diode can be unity. However, we are really interested in the *transducer* gain of the multiplier, not the *power* gain, and the Manley-Rowe relations do not prove anything regarding the transducer gain. To show that the transducer gain can be as great as the power gain, we must show that it is possible to achieve a conjugate match at the multiplier's input; then, the available power equals the input power, and the Manley-Rowe limit is valid for transducer gain as well as power gain.

Proving that the input can be matched is not a simple task; indeed, we can show that in many nonlinear circuits, the input can not be matched. For example, a circuit consisting of an ideal diode (one having zero resistance in the "on" state) in series with a load resistor cannot be matched, and has the property that the total power at all frequencies delivered to the load can be no more than half the available power. Although no power can be dissipated in the diode, and all the input power is delivered to the load, the maximum *transducer* gain is –3 dB! The loss is a reflection loss at the input. Fortunately, we can see from harmonic-balance calculations and other evidence that the input of a reactive frequency multiplier can indeed be matched; we will not attempt to prove this point in any general way.

The second detail is that the Manley-Rowe relations do not simply establish a limit to the efficiency of a varactor frequency converter; they describe a fundamental characteristic of any pumped nonlinear reactance. In the case of a frequency multiplier, that characteristic, expressed by (7.4), is consistent with intuition: the sum of all the harmonic power components must equal the input power. This relationship is precisely valid for the reactive junction of the diode and does not depend in any way upon the excitation level or the external circuit of the multiplier. If the multiplier is badly designed, the input power may be low, and the output power is dissipated in the series resistance and wasted in unwanted harmonics; if the circuit is well designed, input power is coupled efficiently to the diode junction, loss in the series resistance is minimized, and significant real power exists in the diode only at the input and output frequencies. In both cases, however, the Manley-Rowe relations are satisfied. Thus, although these relations can establish limits to a multiplier's efficiency, they do not guarantee that efficiency; achieving optimum efficiency in a frequency multiplier requires using a varactor that has low series resistance, selecting the varactor that is appropriate for the frequency and power level at which it is to be operated, using idlers, and matching the input and output impedances.

7.1.3 Design of Varactor Frequency Multipliers

Figure 7.1 shows the structure of a varactor frequency multiplier. It consists of input and output matching circuits, a varactor, and M idler resonators, $f_{i,1}, \ldots, f_{i,M}$. Designing the multiplier requires estimating the parameters of a diode that is appropriate for the multiplier's frequency and power level and determining the source and load impedances. When these

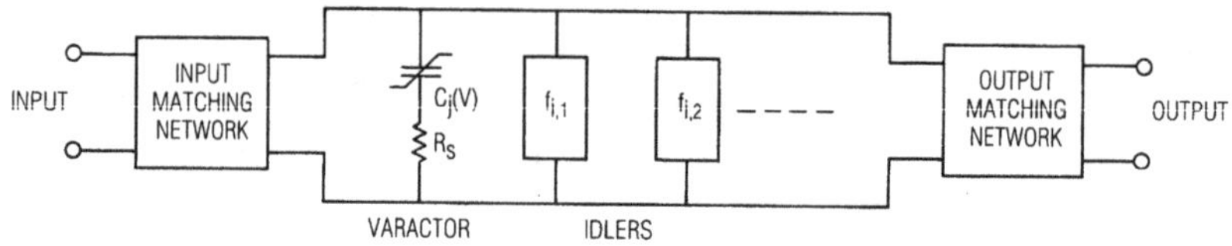

Figure 7.1 Varactor frequency multiplier. The blocks marked $f_{i,\,n}$ are the idlers at the nth harmonic. The input and output matching circuits must not interact.

quantities are known, the matching circuits and idler resonators can be realized in the conventional manner.

A classic paper by Burkhardt [7.2] has been the basis for the design of many frequency multipliers. Burkhardt's analysis is not unlike the harmonic-balance analysis described in Chapter 3, but his results are presented in a normalized and tabulated form, so they can be used to design a wide variety of multipliers. The assumptions and limitations in Burkhardt's work are that (1) the idlers are lossless series resonators (short-circuit idlers); (2) only input, output, and idler currents in the diode are considered; (3) idlers and input/output circuits resonate with the average diode elastance; (4) the diode's junction voltage varies between the reverse-breakdown voltage and ϕ, although the varactor may be overdriven; and (5) the varactor's dynamic Q (2.69), evaluated at the output frequency, is greater than 50.

The diode's normalized drive level D is defined as

$$D = \frac{q_{\max} - Q_B}{q_\phi - Q_B} \tag{7.7}$$

where Q_B is the depletion charge at breakdown and q_ϕ is the charge when the junction voltage just reaches ϕ. The charge $q_{\max}$ is the maximum stored charge; if the junction voltage just barely reaches ϕ, $q_{\max} = q_\phi$ and $D = 1.0$. This is the maximum drive level possible in a Schottky-barrier varactor, although, in practice, the junction voltage is usually limited by resistive conduction to a value less than ϕ, so $D < 1.0$. In a p^+n varactor, $q_{\max}$ may be greater than q_ϕ, so D can be greater than unity; however, in this case the positive excursion of the junction voltage is clamped at ϕ. Burkhardt gives data for drive levels from $D = 1.0$ to $D = 1.6$.

Tables 7.1 and 7.2 give the important parameters necessary for designing varactor doublers and triplers. The optimum conversion efficiency, ε_c,

Table 7.1 Doubler

$\gamma = 0.5$

Design Parameter	D=1.0	D=1.3	D=1.6
α	9.95	8.3	8.3
β	0.0227	0.0556	0.0835
$R_{in}\ \omega_1 / S_{max}$	0.080	0.098	0.0977
$R_L\ \omega_1 / S_{max}$	0.1355	0.151	0.151
S_{01} / S_{max}	0.50	0.37	0.28
S_{02} / S_{max}	0.50	0.40	0.34
$V_{dc,\,n}$	0.35	0.28	0.24

Table 7.2 Tripler

(Idler at $2\omega_1$)

$\gamma = 0.5$

Design Parameter	D=1.0	D=1.3	D=1.6
α	11.6	9.4	9.8
β	0.0241	0.0475	0.0700
$R_{in}\ \omega_1 / S_{max}$	0.137	0.168	0.172
$R_L\ \omega_1 / S_{max}$	0.0613	0.0728	0.0722
S_{01} / S_{max}	0.50	0.36	0.26
S_{02} / S_{max}	0.50	0.38	0.31
S_{03} / S_{max}	0.50	0.38	0.30
$V_{dc,\,n}$	0.32	0.24	0.18

Source: C. B. Burkhardt, "Analysis of Varactor Frequency Multipliers for Arbitrary Capacitance Variations and Drive Level," *Bell System Tech. J.*, 44:675. The tables are reprinted with permission from the *Bell System Technical Journal*, Copyright 1965, AT&T.

and output power, P_L, are related to two tabulated parameters, α and β, as follows:

$$\varepsilon_c = \exp(-\alpha/Q_\delta) \tag{7.8}$$

and

$$P_L = \beta\frac{\omega_1(\phi - V_b)^2}{S_{max}} \tag{7.9}$$

where Q_δ is given by (2.69) and is evaluated at the output frequency. S_{max} is the maximum junction elastance (the inverse of the minimum capacitance), V_b is the breakdown voltage, and ω_1 is the input frequency. The tables also include the normalized source and load resistances, R_{in} and R_L, and the average junction elastances at the input and output frequencies, $S_{0,1}$ and $S_{0,n}$, where n is the output harmonic number. These elastances must be resonated by the source and load networks. Table 7.2 includes$S_{0,2}$, the elastance at the idler frequency in the tripler, which must be resonated by the idler. The tables also give the normalized bias voltage, $V_{dc,n}$:

$$V_{dc,n} = \frac{\phi - V_{dc}}{\phi - V_b} \tag{7.10}$$

and V_{dc} is the actual (i.e., not normalized) bias voltage. V_{dc} is easily adjusted empirically to optimize the multiplier's efficiency, so it is not a very important design parameter. When R_{in}, R_L, $S_{0,1}$, and $S_{0,n}$ are known, designing the input and output matching circuits requires matching the simple source and load models shown in Figure 7.2. Reference [7.2] has more extensive tables that include data for multipliers at higher harmonics, with different drive levels and idler configurations, and with values of γ between 0.0 and 0.5. Many of these configurations, however, are practical only at low frequencies or describe types of diodes that no longer are made. The design process will be illustrated by the following example.

7.1.4 Design Example of a Varactor Multiplier

We shall design a 10- to 20-GHz frequency doubler. First we use Burkhardt's data, and then check the design by harmonic-balance analysis. A p^+n varactor would be the logical choice for this application. However,

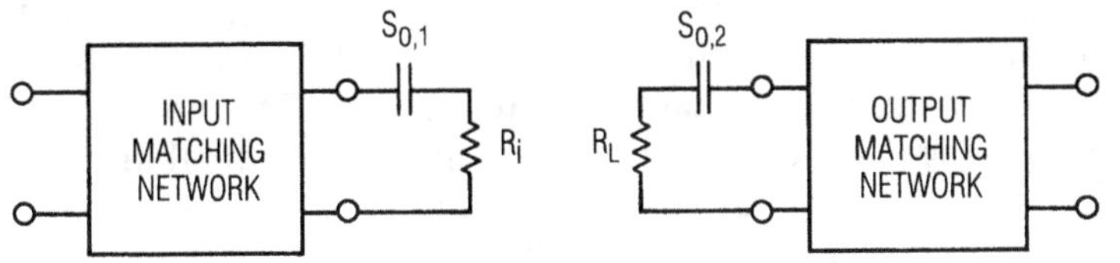

Figure 7.2 Input and output models of the varactor frequency multiplier.

to illustrate some of the problems with such devices, the multiplier uses a Schottky-barrier varactor instead. Because the multiplier is a doubler, no idler is required, but it would still be prudent to short-circuit the diode at high frequencies to prevent power dissipation at the third and higher harmonics.

We begin by selecting a diode that has the parameters $C_{j0} = 0.3$ pF, $\phi = 0.9$V, $V_b = -11.0$V, $\gamma = 0.5$, and $R_s = 4.0\Omega$. The maximum junction voltage that can be achieved without significant conduction is 0.7V. We calculate immediately that the minimum junction capacitance is 0.09 pF and the maximum is 0.61 pF, giving S_{min} and S_{max} equal to $1.64 \cdot 10^{12}$ F^{-1} and $1.11 \cdot 10^{13}$ F^{-1}, respectively, and a dynamic cutoff frequency of 376 GHz or, alternatively, a dynamic Q of 19 at 20 GHz. We may have to suffer some inaccuracy, because this value of Q_δ is lower than the minimum value of 50 required for the Burkhardt analysis. This situation is unavoidable, because a dynamic Q of 50 at 20 GHz would imply a cutoff frequency of 1,000 GHz, a value not beyond the state of the art in varactor diodes, but probably not available in ordinary, inexpensive devices. We also find that the drive level $D = 0.98$ from (7.7), and that the normalizing parameter for the input and output resistances R_{in} and R_L, $\omega_1 / S_{\text{max}}$, is $5.66 \cdot 10^{-3}$.

From Table 7.1, with $D = 1.0$, we find $\alpha = 9.95$ and $\beta = 0.0277$. Substituting these into (7.8) and (7.9), we find the conversion efficiency ε_c to be 0.589, or –2.3 dB, and $P_L = 14.8$ mW, or 11.7 dBm. It is important to note that these quantities include only the loss in the series resistance, and do not include circuit loss, idler loss, or loss in the embedding circuits at unwanted harmonics. Next we find normalized values of R_{in} and R_L from Table 7.1, and quickly determine that $R_{\text{in}} = 14.1\Omega$ and $R_L = 23.9\Omega$. Similarly, we find both $1/S_{0,1}$ and $1/S_{0,2}$ to be 0.18 pF; the source and load impedances are therefore 14.1 + j88 and 23.9 + j44, respectively. Finally, the normalized bias voltage is obtained from the table, and (7.10) gives $V_{\text{dc}} = -2.7$V.

Figure 7.3 shows the circuit. We use transmission-line segments to isolate the input and output; the stubs shown in the figure are one-quarter wavelength long at the fundamental frequency. The open-circuit stub effectively grounds the diode's anode at the fundamental frequency but does not affect the second harmonic. Similarly, the short-circuit stub grounds the cathode at the second harmonic but does not affect the fundamental frequency. The inductors used to tune the capacitive part of the input and output impedance are readily identifiable.

Table 7.3 compares the design and optimized circuit parameters for the multiplier at the design input level of 14 dBm. The agreement is quite good. In tuning the circuit, we find that the most sensitive parameters are the tuning inductors and bias voltage; the source and load impedances are relatively insensitive. Five harmonics were used in the analysis; increasing this to 12 had no measurable effect on the predicted performance. Figure 7.4 shows the junction waveform, which varies from breakdown to ϕ.

7.1.5 Final Details

7.1.5.1 *C*/*V* Characteristic and Modeling

A limitation of this simple design process is that the diode's *C*/*V* characteristic must follow (2.59) at all voltages between breakdown voltage and ϕ. Many modern varactor diodes do not have such ideal *C*/*V* characteristics, especially at high reverse voltages. In an epitaxial Schottky-barrier diode, a high reverse voltage may deplete the epilayer before breakdown occurs, and beyond this voltage the variation in

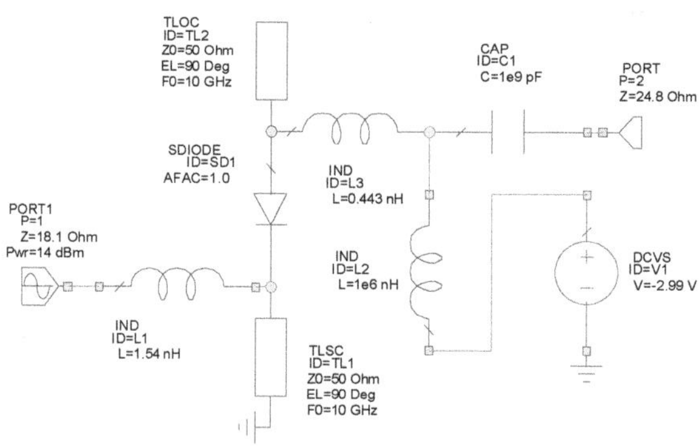

Figure 7.3 Multiplier circuit for the design example, somewhat idealized.

Table 7.3 Multiplier Design vs. Optimized Parameters

Design Parameter	*Design Value*	*Opt. Value*
Conversion loss	2.3 dB	1.8 dB
R_{in}	14.1Ω	18.1Ω
R_L	23.9Ω	24.8Ω
L_{in}	1.40 nH	1.54 nH
L_L	0.35 nH	0.443 nH
V_{dc}	–2.7V	–2.99V

capacitance is minimal. In p^+n *punch-through* or *dual-mode* varactors (Section 2.4.5), the varactor's C/V characteristic is purposely tailored (by adjusting the doping profile) so that the junction-capacitance variation is minimal at high reverse voltages; this characteristic minimizes sensitivity to input level, and may also enhance stability somewhat. When such devices are used in a multiplier, it is best to use the voltage at which the C/V curve begins to limit in place of V_b or Q_B in (7.7) through (7.10).

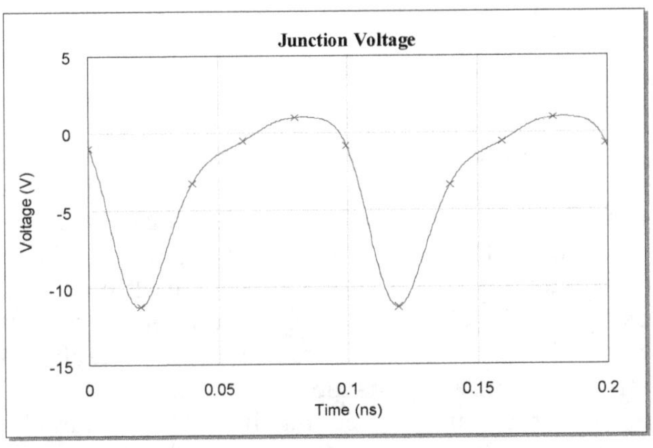

Figure 7.4 Varactor junction voltage waveform at an input level of 14 dBm.

The models used in circuit simulators, primarily versions of the SPICE diode model (Section 2.4.2), are often not well suited for use in varactor multipliers. Uniformly doped Schottky-barrier varactors are reasonably well modeled by the standard diode model, as long as the user is careful to limit the junction voltage to values that do not cause avalanche breakdown or fully deplete the epilayer. *pn*-junction varactors are more of a problem. The diffusion-capacitance model should be adequate for most purposes, and the γ parameter [(2.58), (2.59)] in the depletion part of the model can be adjusted to obtain reasonable accuracy. If these expedients do not produce adequate accuracy, it may be necessary for the user to create his own model.

7.1.5.2 High-Order Multipliers

An important question that often arises in the design of multiplier systems is the following: is it better to realize a high-order multiplier by a single stage or by a cascade of several low-order multipliers? We find from the tables in [7.2] that, in theory, a cascade of low-order multipliers usually has greater efficiency than a single high-order multiplier. However, before concluding that this is true in any particular application, one must consider the additional losses in cascading two multipliers (it is invariably necessary to use an isolator between them) and especially the additional cost of designing, manufacturing, and testing two separate components and their interconnecting hardware. When these practical considerations are included, the answer to this question is not nearly as clear. It must be answered on an ad hoc basis, in view of the requirements of the system in which the multiplier is to be used.

7.1.5.3 Stability

Varactor frequency multipliers are notoriously unstable. Their instability is a kind of chaotic process, not a simple oscillation. The stability of any nonlinear circuit is difficult to assess analytically, but it can be addressed more directly from a practical standpoint. The author has observed that most stability problems encountered in varactor frequency multipliers are the result of practical design deficiencies and are rarely inherent in the nature of the component. Thus, it is best to examine stability from a practical viewpoint, and to note some of the causes of disappointing behavior.

Controlling the broadband embedding impedance characteristic very carefully is the best way to insure good stability. In particular, the input source and output load must be linear and not vary with input or output level. One must not drive a mixer's LO port directly from a multiplier, or the multiplier directly from another multiplier; an isolator should be used. The input and output networks must not have any spurious resonances, especially near harmonics or subharmonics of the input or output frequencies, and the idler resonances must be implemented effectively. In general, the simplest effective matching and idler circuits are least likely to introduce instability. It is also important to have a spectrally clean excitation; the excitation signal must not have significant spurious signals, harmonics, or noise.

7.1.5.4 dc Bias

It is almost always necessary to provide dc bias to the varactor used in a multiplier, although sometimes it is possible to self-bias the device. Even a $p^{+}n$ varactor has some dc current, caused by rectification at high input levels. By introducing a resistor in the diode's dc return path, this current can be used to bias the diode. The resistor also helps to reduce the sensitivity of the output power level to the input power level; as input drive is increased, the resulting increase in dc current further reverse-biases the diode, reducing the multiplier's efficiency and levelling the output power. The design of the bias circuit often has a strong effect on stability. Low-frequency resonances in the bias circuit are a common cause of instability.

7.1.5.5 Noise

Varactor multipliers are low-noise devices. In low-noise applications, even low levels of noise may be a concern. Thus, we need to address the matter of noise in such devices.

Noise in varactor multipliers arises from several sources:

1. The series resistance generates thermal noise. In a well-designed multiplier, this is the dominant noise source.
2. If a Schottky varactor is driven so hard that it rectifies, shot noise is generated. Avalanche breakdown also generates high levels of shot noise.
3. When carriers are accelerated in a strong electric field, the increased energy causes their temperature to be higher than the surrounding lattice. The result is *hot-electron* noise.
4. When GaAs devices have a strong electric field, carriers can be scattered from their normal, high-mobility energy level to a low-

mobility satellite "valley" at a higher level. The resulting change in electron velocity generates *intervalley-scattering* noise.

5. Traps near the junction interface generate low-frequency ("1 /*f*") noise. Although this noise rarely reaches the microwave region, it can modulate the signal applied to the multiplier through the varactor's inherent nonlinearity. This effect can increase the phase noise of a signal beyond what is expected from (7.1).

Accounting for these noise sources in a multiplier design requires a complete noise model and circuit-analysis software capable of performing a complete nonlinear noise analysis. In practice, however, multiplier noise usually can be rendered insignificant by proper design, avoidance of overdrive in Schottky devices, and the use of a high-quality varactor.

7.1.5.6 High-Frequency Considerations

Frequency multipliers using Schottky-barrier varactors are often employed to generate millimeter-wave or submillimeter-wave energy. At such high frequencies, additional phenomena affect the performance and must be considered in the design. Most of these are associated with the diode's series resistance. They include (1) increased series resistance due to skin effect, (2) velocity-saturation effects in the undepleted epilayer, and (3) increases in the substrate's impedance caused by plasma resonances. These phenomena are discussed in Section 2.4.6.

7.2 STEP-RECOVERY DIODE MULTIPLIERS

An SRD can achieve efficient high-order frequency multiplication. The key to its operation is its very strong capacitive nonlinearity, which is realized almost exclusively by charge-storage effects. The SRD multiplier operates by generating a very fast voltage pulse once for each cycle of the input voltage; the pulse is then applied to a filter that converts it to a sinusoidal output voltage. Without the need for idlers, SRD multipliers can achieve conversion efficiency on the order of $1/n$, where n is the harmonic number. They are, however, narrowband components and are limited to output frequencies below approximately 20 GHz.

7.2.1 Multiplier Operation

Because it is consistent with the way SRD multipliers are most often operated, Hamilton and Hall's description of SRD multiplier operation [7.3] is widely accepted. A description of the operation of the SRD in a slightly different circuit is given by Hedderly [7.4]; this paper contains more useful information about the factors that limit efficiency.

We begin by treating the SRD multiplier as a lossless circuit having an ideal diode. First we describe the multiplier circuit as a pulse generator, and then show how the pulse generator is modified to achieve a sinusoidal output. The ideal SRD has the C/V characteristic shown in Figure 7.5; the reverse-bias capacitance is small and independent of voltage, the forward-bias capacitance is infinite, and other parasitics, particularly series resistance, are negligible. We assume that the forward characteristic begins at $V = 0$, although in reality it begins at $V = \phi$; this assumption simplifies the analysis and makes little difference in the results. We also assume that the voltage across the diode never exceeds the reverse breakdown voltage; this requirement limits the output power.

In an ideal SRD, all forward current creates stored charge at the junction without causing a change in voltage. This stored charge must be removed by reverse current before a reverse voltage is possible. If the carrier recombination time of a practical diode is long compared to the inverse of the input frequency, very little of the stored charge recombines (so little charge contributes to resistive conduction) before it is removed, and the diode is nearly ideal in this respect. In the following derivation, we assume that all charge is stored and is recoverable, and that the diode can switch from forward to reverse conduction instantaneously after the stored charge is removed.

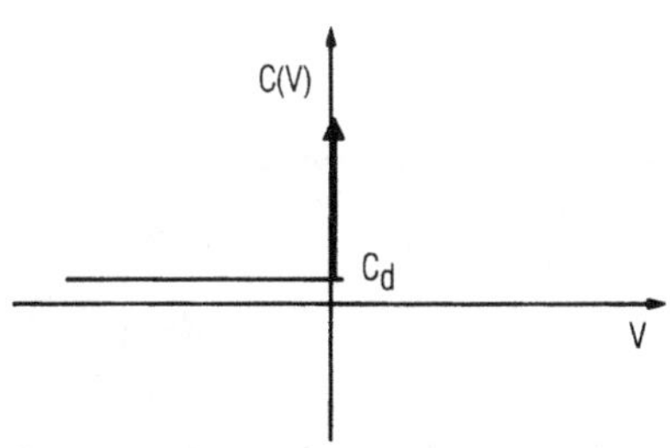

Figure 7.5 C/V characteristic of an ideal step-recovery diode.

Figure 7.6 shows the circuit of the pulse generator. The excitation consists of a sinusoid at frequency ω_1 plus dc bias V_{dc}, and the source impedance $Z_s(\omega)$ is assumed to be zero at dc and all harmonics of ω_1 except, of course, the fundamental; therefore, the input voltage $V_1(t)$ is sinusoidal. The phase of $V_1(t)$ is chosen so that the beginning of the SRD's conduction occurs at $t = 0$. Then

$$V_1(t) = V_1 \sin(\omega_1 t + \alpha) + V_{dc} \tag{7.11}$$

where α is the phase angle of $V_1(t)$ and V_{dc} is the dc component; normally $V_{dc} < 0$. During this interval, the diode is forward-biased, so its capacitance is infinite, and it is effectively a short circuit; the equivalent circuit is shown in Figure 7.7(a). The current is found directly to be

$$I_L(t) = I_L(0) + \frac{V_1}{\omega_1 L}(\cos(\alpha) - \cos(\omega_1 t + \alpha)) + \frac{V_{dc}}{Lt} \tag{7.12}$$

where $I_L(0)$ is the initial current in the inductor at the beginning of the conduction cycle. The second term in (7.12) is the sinusoidal component, and the third term is the linear current ramp generated by the bias source. The voltage waveform $V_1(t)$ and the resulting current $I_L(t)$ are shown in Figure 7.8; when the current is positive, charge is stored in the SRD, and when it is negative, reverse conduction removes this stored charge.

At the end of the conduction interval, the stored charge Q_s is zero:

$$Q_s = \int_0^{T - T_t} I_L(t)\, dt \tag{7.13}$$

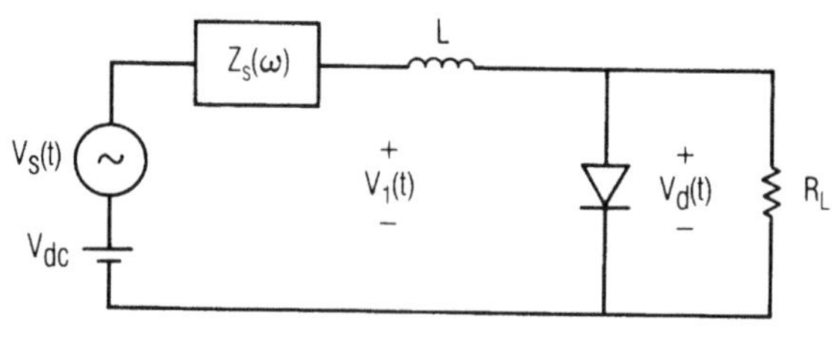

Figure 7.6 Impulse-generator circuit using a step-recovery diode.

where T is the excitation period and T_t is the length of the impulse period. At $t = T - T_t$ all the stored charge has been removed, and the diode switches to its reverse-bias state. At this point the impulse interval begins.

At the instant the diode switches, the current in the inductor is the excitation current for the harmonic-generating impulse. Therefore, we adjust V_{dc} so that the diode switches at the instant when $I_L(t)$ has its maximum negative value. At that instant $dI_L(t) / dt = 0$, so the voltage across the inductor is zero and the diode voltage $V_d(t)$ is zero; $V_1(t)$ is the sum of these voltages, so it must also be zero. Because the SRD switches when $V_1(t) = 0$, the multiplier has the equivalent circuit of Figure 7.7(b), in which the voltage source has been eliminated, and the inductor current $I_L(T - T_t)$ is the only excitation (we call this current I_0 for simplicity; I_0 is a negative quantity). The diode capacitance is now C_d, a relatively small depletion capacitance.

The response $V_d(t)$ of the circuit in Figure 7.7(b) is a damped sinusoid at the resonant frequency of L and C_d:

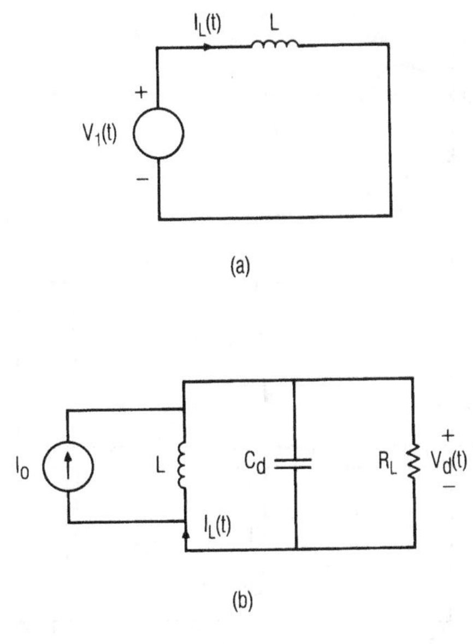

Figure 7.7 Equivalent circuit of the SRD multiplier during (a) the conduction interval, and (b) the impulse interval.

$$V_d(t) = I_0\left(\frac{L}{C_d(1-\varsigma^2)}\right)^{0.5}\exp\left(\frac{-\varsigma\omega_n t'}{(1-\varsigma^2)^{0.5}}\right)\sin(\omega_n t') \tag{7.14}$$

where $t' = t - T + T_t$; that is, t' is time measured from the beginning of the impulse interval. The loaded resonant frequency, ω_n, of the tuned circuit in Figure 7.7(b) is

$$\omega_n = \left(\frac{1-\varsigma^2}{LC_d}\right)^{0.5} \tag{7.15}$$

and the damping factor ς is

$$\varsigma = \frac{1}{2R_L}\sqrt{\frac{L}{C_d}} \tag{7.16}$$

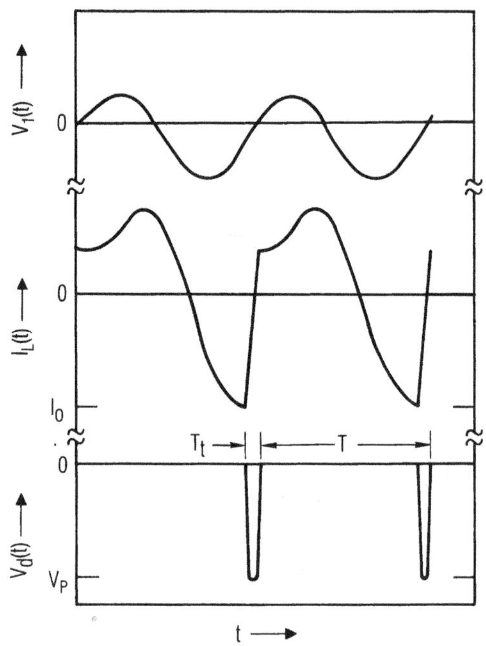

Figure 7.8 Voltage and current waveforms in the SRD impulse generator.

This sinusoid does not last very long; as soon as $V_d(t)$ reaches zero, at the end of one half cycle, the diode again switches to its high-capacitance state and $V_d(t)$ is clamped at zero. Thus, the output voltage consists of a very short-lived pulse, a half-sinusoid at the loaded resonant frequency of the circuit in Figure 7.7(b). The pulse waveform is shown in Figure 7.8; the peak voltage is

$$V_p = -I_0\sqrt{\frac{L}{C_d}}\exp\left(\frac{-\pi\varsigma}{2(1-\varsigma^2)^{0.5}}\right) \tag{7.17}$$

and the pulse width T_t is

$$T_t = \frac{\pi}{\omega_n} \tag{7.18}$$

The current in the inductor during this interval is

$$I_L(t) = I_0 + \frac{1}{L}\int_{T-T_t}^{T} V_d(t)\ dt \tag{7.19}$$

so that

$$I_L(t) = I_0\exp\left(\frac{-\varsigma\omega_n t'}{(1-\varsigma^2)^{0.5}}\right)\left(\cos(\omega_n t') + \frac{\varsigma\sin(\omega_n t')}{(1-\varsigma^2)^{0.5}}\right) \tag{7.20}$$

The power P_o in the pulse train is

$$P_o = \frac{1}{T}\int_{T-T_t}^{T}\frac{V_d^2}{R_L}dt = \frac{\omega_1 V_p^2}{4\omega_n R_L} \tag{7.21}$$

The input impedance of the multiplier circuit $Z_{in}(\omega_1)$, including the inductor L, is the ratio of the fundamental-frequency components of $V_1(t)$ and $I_L(t)$. Because the impulse interval is so short, it is tempting to ignore it in approximating $Z_{in}(\omega_1)$; however, it is only during this interval that power is removed from the circuit, so ignoring the impulse interval gives the trivial result that $Z_{in}(\omega_1) = \omega_1 L$. This is, however, a good approximation of the imaginary part of the input impedance.

The real part of the input impedance can be found from power considerations. Because the diode and inductor are lossless, the power dissipated in the real part of the input impedance must equal the output power. We express the fundamental component of $I_L(t)$ as I_1; then,

$$|I_1|^2 = \frac{V_1^2}{(\omega_1 L)^2 + R^2} \tag{7.22}$$

where $R = \text{Re}\{Z_i(\omega_1)\}$. The input power is P_{in}, and

$$P_{in} = \frac{1}{2}|I_1|^2 R = P_o = \frac{V_p^2 \omega_1}{4 R_L \omega_n} \tag{7.23}$$

(The assumption that $P_{in} = P_o$ is, of course, a stretch. We shall examine it further in the design example in the next section.) We find from other analyses that in most well-designed multipliers $R \approx \omega_1 L$; then substituting (7.22) into (7.23) gives

$$R = \frac{V_p^2 L^2 \omega_1^3}{V_1^2 R_L \omega_n} \tag{7.24}$$

Real diodes, of course, have a series resistance R_s added to R. Finally, the estimate of the input impedance is

$$Z_{in}(\omega_1) \cong j\omega L_1 + R_s + \frac{V_p^2 L^2 \omega_1^3}{V_1^2 R_L \omega_n} \tag{7.25}$$

Equation (7.25) is not very useful for design purposes, because it is difficult to estimate V_p/V_1 without calculating the complete current waveform. Hamilton and Hall give expressions for the real and imaginary parts of the input impedance; however, it appears that they have calculated the inverses of the real and imaginary parts of the input admittance instead. Their tabulated results can be approximated as

$$G_i = \frac{1}{\omega_1 L(1.2 + \varsigma)} \tag{7.26}$$

and

$$B_i = \frac{-1}{\omega_1 L(0.7 + \varsigma)} \tag{7.27}$$

We now have a circuit (Figure 7.6) that generates a pulse train, $V_d(t)$. The spectrum of $V_d(t)$ has components at many harmonics of ω_1, and the envelope of that spectrum has its first zero at $3\,\omega_n$. This circuit can be used effectively to generate a large number of harmonically related tones. However, usually we wish to generate a single output frequency as efficiently as possible. In this case it is not enough just to filter the output; one must use a resonant network that does not dissipate appreciable power at unwanted harmonics, and does not upset the pulse waveform too seriously. Note that the value of R_L in the impulse generator has little effect on the shape of the pulse; even making $R_L \to \infty$ has no effect except to increase V_p [because $\varsigma \to 0$ in (7.17)]. Thus, an open circuit at unwanted harmonics and an appropriate resistance at the desired output frequency are the desired terminations. The resonant network that realizes these terminations is an ideal series LC resonator. The SRD frequency multiplier is shown, in its conceptual form, in Figure 7.9; the box marked f_N is the resonator.

One must be careful to recognize that the circuit in Figure 7.9 is not equivalent to that of Figure 7.6, because the resonator changes the diode's

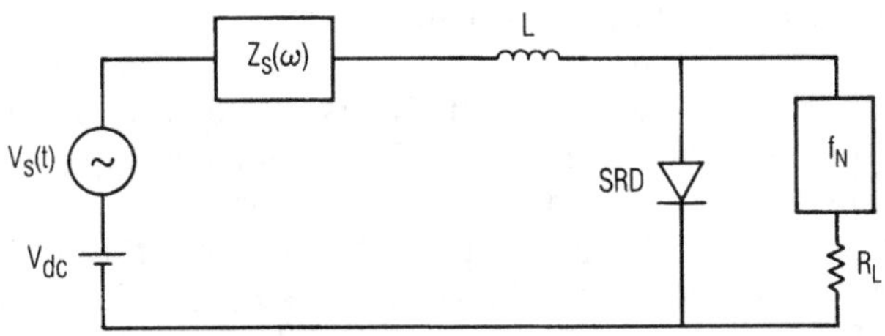

Figure 7.9 Circuit of an SRD frequency multiplier. The block f_N is an ideal series LC resonator tuned to the Nth harmonic of the input frequency.

termination at unwanted harmonics to an open circuit, rather than a finite resistance. Because the diode is a short circuit over most of the excitation period, this change has less effect than one might imagine. The main practical effect is to make the multiplier operate as if it were an impulse generator having a lower damping factor than the value given by (7.16); when terminated in a resonant network, the multiplier is less stable than when terminated in a resistance. Accordingly, it is generally good practice to design an SRD frequency multiplier to have a damping factor of approximately 0.6 to 0.7, rather than the value of 0.4 to 0.5 that would provide stable operation in an impulse generator.

If the multiplier uses an ideal diode and is ideally terminated, all the energy of each impulse is converted to output power at the desired harmonic frequency. Under these conditions the output power is given by (7.16). Unfortunately, because of the distinct paucity of ideal conditions in the world of microwave electronics, the efficiency is considerably lower. The most serious reduction in efficiency comes from the series resistance of the diode. Power is dissipated in the diode's forward resistance not only at the input and output frequencies, but at all the other harmonics as well. Power is also dissipated at these harmonics in the losses in the input matching circuit, the inductor, and the output resonator.

Another important loss mechanism is the recombination current in the diode. Even if the carrier recombination time is long, a fraction of the injected charge recombines and cannot generate output power. This phenomenon has the same effect as adding a resistance in parallel with the diode during the pulse interval. Similarly, the transition time of the diode is always finite and lengthens the pulse interval. The increased pulse length reduces the magnitude of the higher harmonics, and thus reduces the efficiency. The effects of finite pulse length may also limit the SRD frequency multiplier's efficiency.

7.2.2 Design Example of an SRD Multiplier

Designing a step-recovery diode multiplier is a relatively straightforward application of the equations in Section 7.2.1. It is most important to select an appropriate diode and the proper damping factor.

We design an SRD multiplier to generate 20 mW at 4 GHz from a 1-GHz excitation. The diode's recombination time must be long compared to the period of the input excitation, so $\tau >> 10^{-9}$ sec, and in fact 10^{-8} sec would not be too great. The ideal pulse length is one-half period at the output frequency; thus $T_t = 1.25 \cdot 10^{-10}$ sec. The diode's transition time must be considerably shorter than this, no more than approximately 70 to 100 ps. Estimating the optimum value of reverse capacitance C_d is a controversial subject among multiplier designers; this controversy is not unexpected, because the criteria for selecting C_d are mostly empirical. The range of suggested values for the diode's reactance, under reverse bias, varies from 10 or 20 ohms at the output frequency to more than double this value; the best choice is probably an intermediate value that gives a reasonable input impedance without making V_p too great. From (7.26) and (7.27) we see that the input impedance is proportional to $\omega_1 L$, a reactance that must resonate with C_d; increasing C_d decreases L and thus reduces input impedance. We begin by choosing $C_d = 1.0$ pF and $\varsigma = 0.5$, a good compromise between pulse length (low ς) and stability (high ς); from (7.15) and (7.16) we have $L = 1.19$ nH and $R_L = 35\Omega$. We find the input admittance from (7.26) and (7.27) and convert to impedance; the result is $Z_{in}(\omega_1) = 4.2 + j6.0$, a low but reasonable value.

Equation (7.21) can be used to find the peak impulse voltage V_p; V_p must be kept below the diode's reverse breakdown voltage. If the multiplier had 100% efficiency, (7.21) would be directly applicable and could be solved for V_p. However, we expect loss on the order of at least 6 dB; most of this loss is caused by inefficiencies in converting the pulse energy to output power. Accordingly, it would be more realistic, from a design standpoint, to use input power instead of output power in determining V_p. Therefore, to be conservative, we use 80 mW instead of 20 mW in (7.21). This gives $V_p = 6.7$V, considerably below the breakdown voltage of virtually all practical SRDs.

Figure 7.10(a) shows the idealized circuit. The resistive part of the diode is modeled by a conventional Schottky junction; capacitance is a combination of a diffusion capacitance and a linear component representing C_d. A dc bias source is included; dc bias is not necessary, but by controlling the turn-on voltage of the diode, it helps adjust the optimum power level.

Figure 7.10(b) shows the unfiltered output-voltage waveform at an input power level of 23 dBm, which produces an output level of 13.5 dBm at the fourth harmonic. The dc bias is –0.4 V. The pulse width is approximately 0.15 ns, a little wider than intended. No attempt has been made to optimize the inductance or the terminations.

Designing the input matching circuit may be difficult because of the low input impedance; for this reason, a multistage matching network is usually necessary. A low-pass structure consisting of series inductors and parallel capacitors has the required short-circuit output impedance at harmonics of ω_1. To prevent instability, the matching and bias circuits must have no spurious resonances; all capacitors must have series resonant frequencies well above the input frequency, including those in the bias circuit. Because of the matching circuit's low output impedance, the

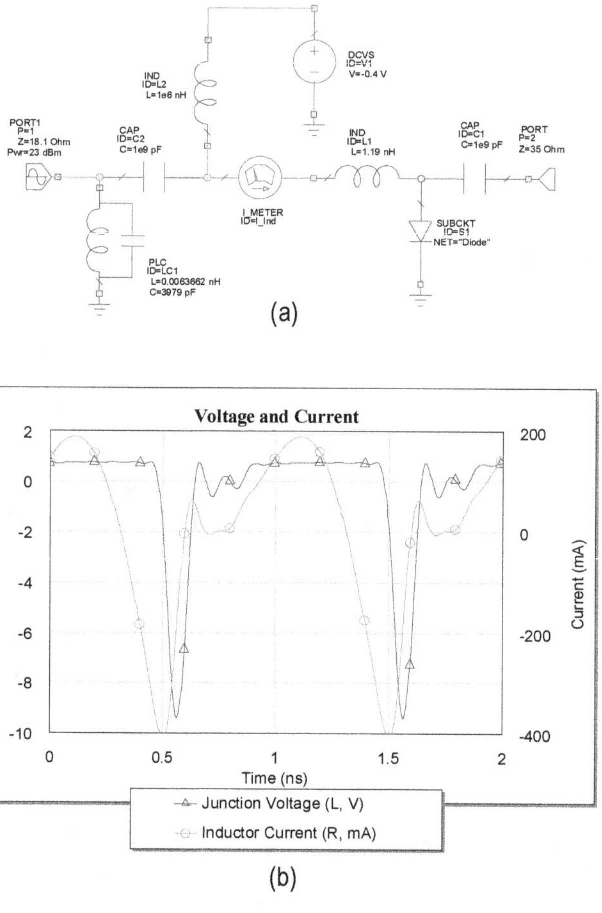

Figure 7.10 (a) Ideal SRD pulse generator and (b) waveforms.

currents in the matching elements are relatively great; these elements must be high-Q parts, or the loss in the matching circuit may be excessive.

There are many ways to design a load network, and in general the simplest designs are best. A lumped-element series resonator is usually not realizable at 4 GHz, so a distributed equivalent network must be used. One possibility is to connect the diode directly to a filter that has the desired out-of-band characteristics; another is to couple it loosely through a capacitor to a narrowband filter. It is wise to design this circuit to provide the impedance transformation between the standard 50Ω coaxial load impedance and R_L. Experienced designers of SRD multipliers report that

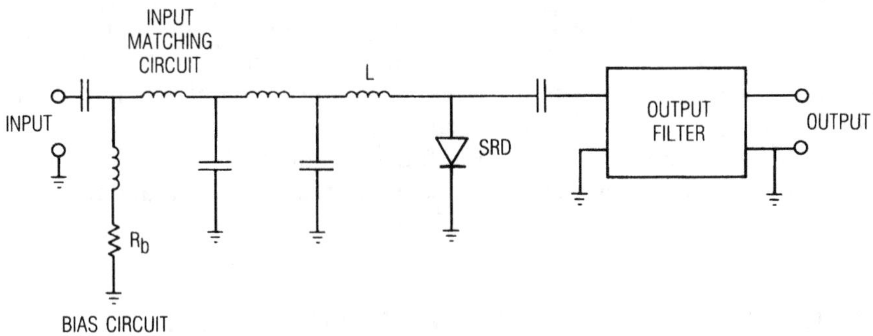

Figure 7.11 The SRD frequency multiplier designed in the example.

some types of resonant networks give better efficiency and stability than others, for reasons that are not always clear. For example, Hamilton and Hall recommend a resonant transmission-line section; this structure, however, can introduce instability if the line impedance is low. Other possibilities are a quarter-wave coupled-line section or a weak capacitive coupling to a quarter-wave coaxial resonator.

Simple resonant structures often have inadequate Q to reject the harmonics closest to the output frequency; in this case the multiplier should be followed by a filter. If the output circuit has been designed to match R_L to 50Ω, the multiplier can be tested easily without this filter in place, and the filter can be tested without the multiplier; this practice significantly eases the testing of both components. The circuit of the multiplier is shown in Figure 7.11.

7.2.3 Harmonic-Balance Simulation of SRD Multipliers

The SRD is fundamentally a diffusion-capacitance device (Section 2.4.7). Such devices are notorious for causing convergence difficulty in harmonic-balance analysis, partly because of their strong capacitive nonlinearity, and partly because they may be unstable. Less obvious is the fact that diffusion and transit time devices can have a Jacobian that is poorly conditioned. We have noted that varactor multipliers are highly sensitive to parameter variations, a property that makes them less "designable" than other types of multipliers. SRD multipliers are no better in this regard; if anything, they are worse.

Harmonic-balance analysis of SRD multipliers designed for high-order operation is especially difficult. The number of harmonics required in the analysis is several times the highest harmonic; if, for example, we design a tenth harmonic multiplier, accurate reproduction of the impulse may require *30* or more harmonics. The results also become quite sensitive to the diode model and to small losses at all the harmonics. Time-domain (SPICE) analysis of the impulse-generator circuit of Figure 7.6 is often successful [7.5]; however, extending time-domain analysis to the complete harmonic generator would be complicated by the need to model distributed circuits.

7.3 RESISTIVE DIODE FREQUENCY MULTIPLIERS

Resistive diode (i.e., Schottky-barrier diode) frequency multipliers have not been employed widely in microwave systems. The reason for their disuse is that they are significantly less efficient than varactor multipliers, and are limited in output power. Furthermore, their efficiency decreases rapidly as harmonic number increases, so resistive diode multipliers are rarely practical for generating harmonics greater than the second. Resistive multipliers, however, have good stability and are capable of wide bandwidths; as such, they complement varactor multipliers nicely, and may be an attractive option in the design of a microwave system.

We saw that reactive multipliers are theoretically capable of achieving 100% efficiency, although, in practice, their efficiency varies approximately as $1/n$. Resistive multipliers are theoretically capable of efficiency no better than $1/n^2$ [7.6]; this is obviously much worse. If AM noise is not a concern, the multiplier's output can be amplified by a FET or HBT amplifier.

7.3.1 Approximate Analysis and Design of Resistive Doublers

Figure 7.12 shows a canonical representation of a resistive multiplier. The diode symbol represents an ideal resistive diode, a Schottky device having no junction capacitance. (We shall see later that the junction capacitance is frequently insignificant in these multipliers.) The series resistance R_s is shown separately from the diode. R_i is the source impedance at f_1, and R_L is the load impedance at $2f_1$. The blocks marked f_1 and $2f_1$ are ideal parallel-resonant filters; that is, they have infinite impedance at frequencies f_1 and $2f_1$, respectively, and zero impedance at all other frequencies. Because of the properties of these resonators, voltage components at only these two frequencies exist across the diode-R_s combination, and only fundamental-frequency and second-harmonic currents circulate in the input and output loops, respectively. V_1 is the magnitude (peak value) of the fundamental component of the diode junction voltage $V_j(t)$, and V_2 is the magnitude of the second-harmonic voltage across R_L. Similarly, I_1 and I_2 are the peak values of the fundamental and second-harmonic components of the diode junction current $I_j(t)$. The source voltage $V_s(t)$ is a sinusoid at frequency f_1; dc bias may also exist.

One can understand the operation of the multiplier by first imagining that the diode is short-circuited at all harmonics except the fundamental, a condition that can be established by letting $R_L = 0$, and that the diode is pumped to a high peak current (≥ 25 mA) by $V_s(t)$. Under these conditions the current waveform, shown in Figure 7.13, is a series of pulses, in phase with the positive excursion of $V_s(t)$ and shaped much like half-cosine pulses. The duty cycle of the pulses is close to 50%. We assume that the current waveform is adequately approximated as a series of half-cosine pulses and, from Fourier analysis, find that the fundamental current component, I_1, is

$$I_1 = 0.5 I_{max} \tag{7.28}$$

where I_{max} is the peak junction current. Similarly, we find the second-harmonic current component, I_2, to be

$$I_2 = \frac{2}{3\pi} I_{max} \approx 0.2 I_{max} \tag{7.29}$$

It is also worth noting that the dc component of the junction current, I_{dc}, is

$$I_{dc} = \frac{1}{\pi} I_{max} \tag{7.30}$$

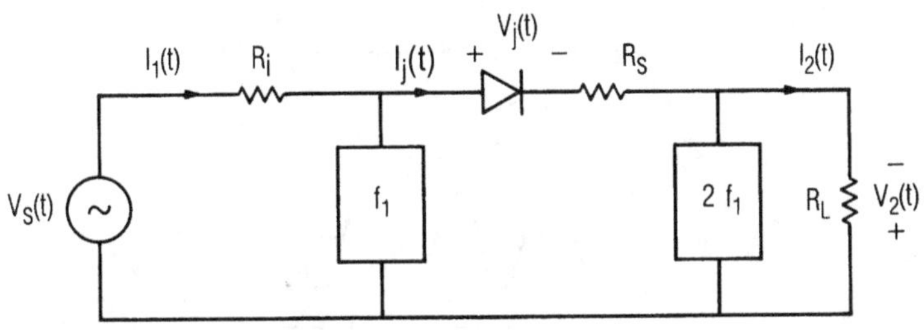

Figure 7.12 Circuit of a resistive frequency doubler. f_1 and $2f_1$ are ideal parallel LC resonators tuned to the fundamental frequency and its second harmonic, respectively.

Because of the source resistance, the junction voltage $V_j(t)$ has more harmonic components than just the first and second. In the time domain $V_j(t)$ is a clipped sinusoid; if $R_s << R_i$, the magnitude of the fundamental component of $V_j(t)$ is

$$V_1 = 0.5(V_s + V_f) \tag{7.31}$$

where V_s is the peak value of $V_s(t)$, and V_f is the forward voltage of the diode, approximately 0.6V for silicon devices, a few tenths of a volt greater for GaAs.

Now imagine that R_L slowly increases from its zero value. I_2 circulates in R_L and generates a voltage $V_2(t)$, the second-harmonic output, shown in Figure 7.13. While R_L is small, I_2 remains approximately constant, so the second-harmonic output power increases with R_L. However, the phase of $V_2(t)$ is such that it reduces the peak positive value of $V_j(t)$, and thus reduces the peak value of $I_j(t)$, I_{max}. This reduction in I_{max} in turn reduces the value of I_2, and eventually a point is reached where the output power levels off and then begins to decrease. If R_L is increased further, V_2 also

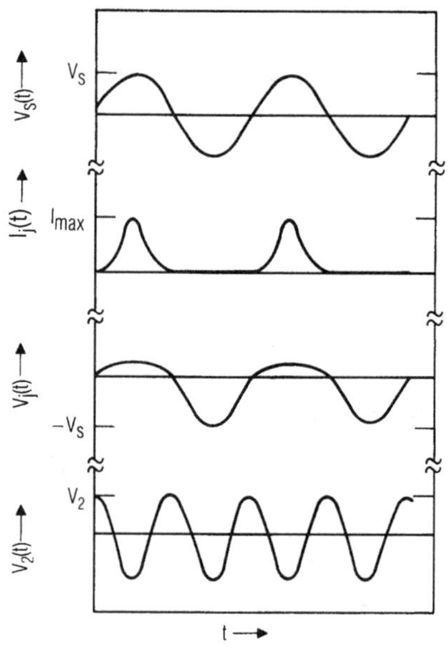

Figure 7.13 Voltage and current waveforms in the resistive doubler.

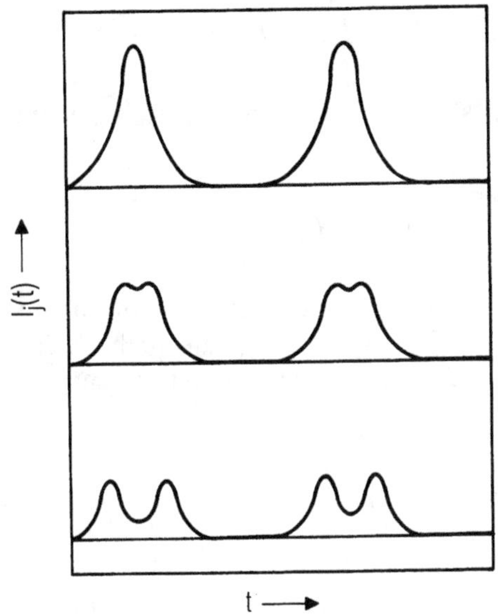

Figure 7.14 Current waveforms in the diode: (a) $R_L = 0$; (b) optimum R_L; (c) R_L greater than optimum. The peak current is greatest in (a), lowest in (c).

increases and eventually the second-harmonic component of the junction current becomes evident as a dip in the peak of the current pulse.

The effect of the magnitude of R_L on the shape of the current pulse is shown in Figure 7.14. It appears at first that the current pulse in Figure 7.14(c) (large R_L) has a strong second-harmonic component; however, this second-harmonic component in fact is relatively weak because the peak current I_{max} is much lower when R_L is large than when R_L is optimum. Harmonic-balance studies of resistive multipliers indicate that optimum efficiency is achieved at the value of R_L where the peak diode current is starting to be compressed by the second harmonic.

In order to design a multiplier, we need to determine the input resistance at f_1, the optimum output load resistance R_L, and the output power as a function of input power. The input quasi-impedance of the junction is the ratio of the fundamental-frequency voltage to current at the junction:

$$R_j = \frac{V_1}{I_1} = \frac{V_s}{I_{max}} \tag{7.32}$$

The input impedance is simply the sum of this impedance and the series resistance:

$$R_{in} = R_j + R_s \tag{7.33}$$

The multiplier's input power equals the sum of the real power of the junction plus the power dissipated in R_s, at all the harmonics, minus the output power. If $R_s << R_j$, the fundamental-frequency power dominates; then

$$P_{in} \cong \frac{1}{2}V_1 I_1 + \frac{1}{2}I_1^2 R_s = \frac{1}{8}I_{max}^2(R_j + R_s) \tag{7.34}$$

We shall see that the efficiency of a resistive multiplier is invariably very low, because most of the input power is dissipated in the diode junction and in the series resistance at the fundamental frequency, and very little is converted to harmonics. When the input is matched, the power available from the source is equal to P_{in}; then

$$P_{av} = P_{in} = \frac{1}{8}I_{max}^2(R_j + R_s) \tag{7.35}$$

We now consider the output. At the peak of the excitation cycle, all the voltage components across the diode must equal the forward voltage. Summing these voltages around the output loop gives

$$V_f = V_1 - V_2 + V_{dc} - I_{max}R_s \tag{7.36}$$

or

$$V_2 = V_1 - I_{max}R_s + V_{dc} - V_f \tag{7.37}$$

where V_{dc} is the dc bias. In (7.36) and (7.37) we have assumed that only the first- and second-harmonic components in $V_j(t)$ are significant. The quantity $V_{dc} - V_f$ is no more than a few tenths of a volt; it can be neglected, and we find V_2 to be

$$V_2 = V_1 - I_{max} R_s \tag{7.38}$$

From (7.28) and (7.32), we can express (7.38) in the more convenient form

$$V_2 = 0.5 I_{max}(R_j - 2R_s) \tag{7.39}$$

We find from harmonic-balance calculations that the value of V_2 given by (7.39) is too great; it results in a value of R_L that is much too high and in a current waveform similar to that shown in Figure 7.14(c). This result occurs because the diode's exponential I/V characteristic causes the current to be very sensitive to junction voltage. The value of V_2 given by (7.39) is not precisely correct for two reasons: first, V_1 is itself approximate; second, it is determined only at a single instant, the peak of the excitation cycle, and does not include effects of high V_2 throughout the period of the junction-voltage waveform. We find empirically that a better value of V_2 is approximately one-third that given by (7.39). Therefore,

$$V_2 \approx 0.167 I_{max}(R_j - 2R_s) \tag{7.40}$$

The load impedance is

$$R_L = \frac{V_2}{I_2} = 0.833(R_j - 2R_s) \tag{7.41}$$

in which we have used (7.29) to express I_2.

The output power is

$$P_L = 0.5 I_2^2 R_L = 0.0167\, I_{max}^2 (R_j - 2R_s) \tag{7.42}$$

The maximum available conversion gain $G_{av,\,max}$ is found from (7.42) and (7.35):

$$G_{av,\,max} = \left.\frac{P_L}{P_{av}}\right|_{\text{input matched}} = 0.133 \frac{(R_j - 2R_s)}{(R_j + R_s)} \tag{7.43}$$

A clear implication of (7.43) is that resistive multipliers suffer from low efficiency. Even if the parasitic series resistance R_s were zero, (7.43) implies that the maximum conversion efficiency of a resistive doubler is only 0.133, or –8.8 dB. This high loss is the unavoidable result of power dissipation in the diode junction. Of course, (7.43) is approximate, so the –8.8 dB limit must also be considered approximate; however, it is difficult to see any way that the efficiency could be more than 1 dB or so greater than this limiting value. Practical resistive diode doublers usually have conversion losses of at least 9 to 10 dB.

7.3.2 Design Example of a Resistive Doubler

We shall design a 20- to 40-GHz frequency doubler. Schottky-barrier diodes are not produced specifically for multiplier use, but good mixer diodes are acceptable and readily available. A typical four-micron chip diode has $R_s = 6.0\Omega$ and $C_{j0} = 0.05$ pF. Initially we shall ignore the junction capacitance; later, we include it in the circuit and design the matching circuit to compensate.

We begin by recognizing that, because of the low conversion efficiency, virtually all the input power is dissipated in the diode. A four-micron Schottky diode has a thermal resistance of approximately 2,000°C per watt. We wish to limit the temperature rise in the junction to approximately 50°C, a prudent limit, so the power dissipation cannot exceed 0.025W, or 14 dBm. We therefore choose the nominal available input power to be 10 dBm, to allow for the effects of input power variation over the input frequency range, changes in environmental temperature, and dc bias power, as well as to maintain a decibel or two of margin.

A second consideration is that the dc junction current must be limited. In order to achieve high output power and efficiency, we wish to have a high value of I_{max}; I_{max}, however, is limited by I_{dc}, which should not exceed approximately 10 mA in a four-micron diode. Using (7.30), we select $I_{max} = 30$ mA. In practice, it may be necessary to provide dc bias in order to achieve this value of I_{max} at the prescribed 10-dBm power level. Equations (7.33) and (7.35) give

$$R_i = R_{in} = R_j + R_s = 89\Omega \tag{7.44}$$

and with $R_s = 6.0\Omega, R_j = 83.0\Omega$. The conversion loss is found from (7.43) to be 9.7 dB, and with 10 dBm of input power, the output power P_L is 0.3 dBm or 1.07 mW. The load resistance R_L is found directly from (7.41) to be 59Ω.

Figure 7.15 shows the circuit of the doubler, which does not include the diode's capacitance. With the design values of R_{in} and R_L, the conversion loss is 8.5 dB, slightly lower than expected. The lower conversion loss is caused by differences between the idealized and calculated shape and peak value of the current pulse; I_2 is relatively sensitive to such differences, especially to the pulse's peak value. The junction voltage and current waveforms are shown in Figure 7.16; the second harmonic is not immediately evident unless the load resistance is decreased to zero.

We now account for the junction capacitance. In a manner analogous to the design of the LO circuit in a diode mixer, we initially assume that the junction capacitance can be approximated as a lumped capacitance equal to C_{j0}, in parallel with the junction. Thus, at the input frequency the diode is equivalent to an 89Ω resistor in parallel with 0.05 pF, and at the output it is equivalent to a 59Ω resistor in parallel with the 0.05 pF capacitor. 0.05 pF is a reactance of 317Ω at 10 GHz and 159Ω at 20 GHz; this is, by itself, low enough that little would be gained by adding a matching circuit to remove its effects. A better approach might be to add transformers to match the source and load impedances to 50Ω; those transformers could be modified to provide tuning. We find that adding a 72Ω line 55 degrees long to the input and a 70Ω line 45 degrees long to the output matches the device to 50Ω source and load impedances, and increases the conversion efficiency insignificantly, a few tenths of 1 dB. Harmonic-balance analysis of the multiplier verifies our approximation that the diode's effective input and output capacitances are quite close to C_{j0}.

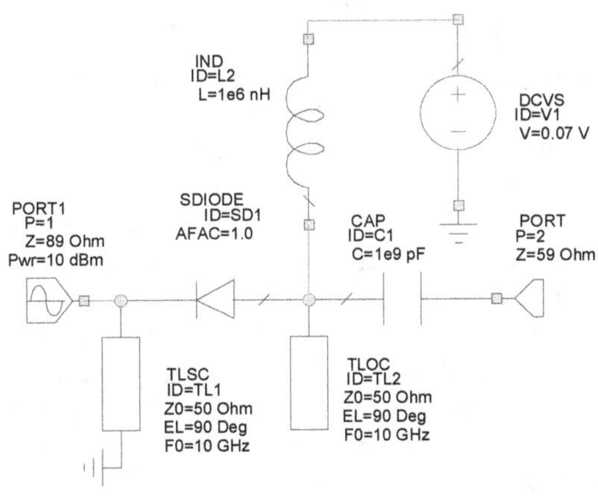

Figure 7.15 Idealized circuit of the resistive frequency doubler.

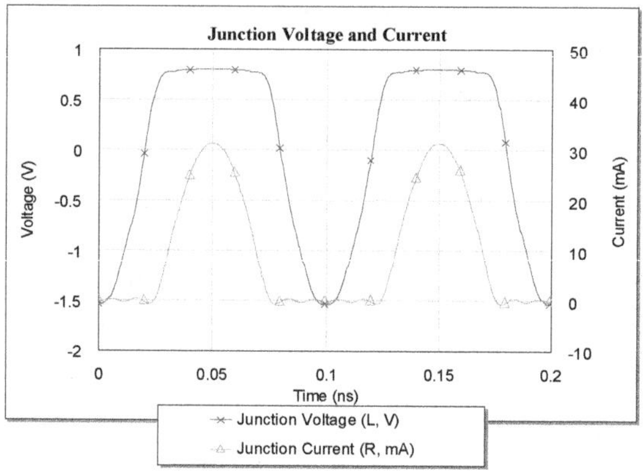

Figure 7.16 Current and voltage waveforms in the junction of the diode used in the resistive doubler.

A few final details should be examined. First, the theory considers only a multiplier having short-circuit embedding impedances, although the embedding impedances of our circuit clearly were not zero at all harmonics. In high-frequency multipliers, the short-circuit case is usually valid, because, regardless of the diode's terminating impedances, the junction capacitance short circuits the resistive junction at the higher harmonics of the input frequency. Second, it may be surprising that we never explicitly considered the diode's I/V characteristic in deriving the expressions for impedance and power. We did, however, account for it implicitly in our assumptions about the shape of the current pulse and $V_j(t)$. The unstated assumption was that the diode does not have an ideal rectifying characteristic (i.e., it is not a short circuit under forward bias), but that it does not have an unusually "soft" I/V characteristic either. [The latter would have rendered invalid the assumptions about $V_j(t)$ and $I_j(t)$.] Third, it may seem cavalier to assume that the desired value of I_{max} is achieved at the desired input level. Of course, P_{av} and I_{max} cannot be selected independently unless dc bias is used; dc bias can be varied to adjust the waveforms to achieve I_{max} and P_{av} simultaneously. Some judgment is necessary here; if one attempts to achieve a value of I_{max} that is unreasonable in view of P_{av}, the $I_j(t)$ and $V_j(t)$ waveforms may not approximate those in Figure 7.13, and the results may be unsatisfactory. Finally, the assumption in (7.35) that all the input power is dissipated in the diode and the conclusion that the efficiency is low may seem like a circular argument. It is not, because this assumption was used only to find an expression for the input power; the output power was determined from other considerations.

7.4 BALANCED MULTIPLIERS

It is a common practice to realize diode frequency multipliers in balanced structures. Balanced multipliers have significant advantages compared to single-ended multipliers; the most important are increased output power and the inherent rejection of certain unwanted harmonics. The input or load impedance of a balanced multiplier in some cases differs by a factor of two from that of a single-diode multiplier; therefore, a balanced multiplier sometimes provides more satisfactory input or load impedance.

Diode multipliers are sometimes interconnected via hybrids, but for economy they are more often used in the antiparallel or series forms described in Section 5.2.1. The antiparallel connection, shown in Figure 5.17, is probably the simplest form of a balanced multiplier; it rejects even harmonics of the input frequency and consequently can be used only as an odd-order multiplier. In an antiparallel-diode multiplier, each diode effectively short circuits the other at the second harmonic, so each diode acts as a type of idler for the other. This circuit does not reject the fundamental frequency, however, so it requires an output filter.

In theory, the antiparallel circuit can be used to realize either resistive or reactive multipliers. However, because the stability of a varactor multiplier is sensitive to slight unbalance between the diodes, varactor multipliers are rarely realized as antiparallel circuits. It is thoroughly practical, however, to realize resistive multipliers this way, although the restriction to third-harmonic operation in the resistive multiplier results in low efficiency.

The bridge rectifier circuit in Figure 7.17 is a practical way to realize resistive frequency doublers. The design of such multipliers is not unlike the design of a diode ring mixer. The diodes are selected to have large junction areas, consistent with a manageable C_{j0}. (We saw from the design example that a capacitive reactance of ~100Ω at the output frequency is usually small enough.) Each transformer is loaded with two sets of two diodes in series; thus, a single diode impedance of $R_s + R_j$. As with balanced diode mixers, high-frequency components require baluns, not transformers; the baluns are designed to match the diode. See Section 6.4.3.3 for an example of balun design as it applies to mixers; balun design for multipliers follows directly from that discussion.

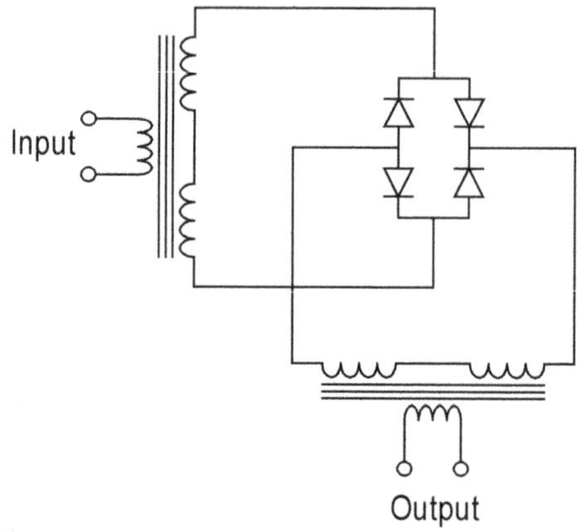

Figure 7.17 A "bridge rectifier" frequency doubler. Note that the diode ring is not identical to that used in a ring mixer.

The voltage and current waveforms in the balanced bridge multiplier are identical to those of a full-wave rectifier in a dc power supply. The current consists of a train of half-sinusoidal pulses, which has no odd-harmonic components. Thus, the multiplier inherently rejects the two most troublesome harmonics, the first and third, and the fourth is usually weak enough to require little or no filtering. Reference [7.7] describes a monolithic realization of such a mixer, covering an output frequency range of 16 to 40 GHz.

It is important to note that the diode "quad" used in the balanced multiplier is not identical to that used in the ring mixer (compare to Figure 6.17). The type of diode required by the multiplier is readily available commercially and uses the same kinds of packages as mixer ring quads. The multiplier version is called a *bridge quad*.

References

[7.1] J. M. Manley and H. E. Rowe, "Some General Properties of Nonlinear Elements," *Proc. IRE*, Vol. 44, 1956, p. 904.

[7.2] C. B. Burkhardt, "Analysis of Varactor Frequency Multipliers for Arbitrary Capacitance Variations and Drive Level," *Bell System Tech. J.*, Vol. 44, 1965, p. 675.

[7.3] S. Hamilton and R. Hall, "Shunt-Mode Harmonic Generation using Step-Recovery Diodes," *Microwave J.*, Vol. 10, No. 4, April 1967, p. 69.

[7.4] D. L. Hedderly, "An Analysis of a Circuit for the Generation of High-Order Harmonics Using an Ideal Nonlinear Capacitor," *IEEE Trans. Electron Devices*, Vol. 9, 1962, p. 484.

[7.5] C. Nguyen, "A 35% Bandwidth Q- to W-Band Frequency Doubler," *Microwave J.*, Vol. 30, No. 9, Sept. 1987, p. 232.

[7.6] C. H. Page, "Frequency Conversion with Positive Nonlinear Resistors," *Journal of Research of the National Bureau of Standards*, Vol. 56, 1956, p. 179.

[7.7] S. A. Maas and Y. Ryu, "A Broadband, Planar, Monolithic Resistive Frequency Doubler," *IEEE Microwave and Millimeter-Wave Monolithic Circuits Symposium Digest*, 1994, p. 443.

Chapter 8

Small-Signal Amplifiers

This chapter is concerned with nonlinear distortion phenomena in small-signal amplifiers. Such amplifiers are designed primarily to have low noise figures or specific values of small-signal gain, and their linearity usually must be optimized within gain and noise constraints. The distortion phenomena of greatest concern are saturation, intermodulation distortion, harmonic distortion, and AM-to-PM conversion; these terms are defined in Section 1.3. We shall see that Volterra-series analysis is applicable to all these phenomena, although harmonic-balance analysis is preferable for determining single-tone saturation effects.

8.1 REVIEW OF LINEAR AMPLIFIER THEORY

8.1.1 Stability Considerations in Linear Amplifier Design

In its simplest form, a small-signal amplifier consists of a transistor, an input-matching network, and an output matching network. Bias circuitry must also be included; however, a well-designed bias network does not affect the RF matching of the device, so we will not consider the bias circuit further. The circuit model of the amplifier is shown in Figure 8.1(a).

The transistor is treated in the design process as a two-port, described by a set of two-port parameters, usually S or Y parameters. When used in a linear amplifier, FET and bipolar devices are usually operated in a common-source or common-emitter configuration, respectively, and the emitter or source is common to the input and output. The S parameters vary with dc bias and therefore must be measured at the bias voltages at which the device will be operated. If the matching networks are lossless (we will assume that they are), they can be represented as lumped impedances or

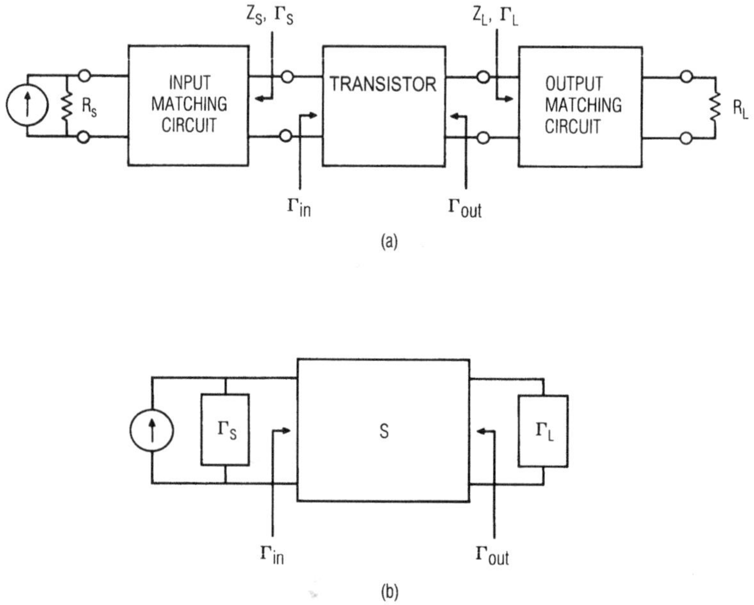

Figure 8.1 (a) Small-signal amplifier consisting of input and output matching circuits and a MESFET; (b) canonical model of the amplifier.

reflection coefficients at the operating frequency, and we can redraw the circuit in Figure 8.1(a) to form the canonical equivalent circuit shown in Figure 8.1(b).

As an alternative to a two-port, a transistor can be represented by an equivalent circuit. This representation can include nonlinear elements for modeling nonlinear phenomena. If the lumped-element model is well conceived, its S parameters can be calculated easily, and they should agree with those measured from the device.

Figure 8.2 shows a widely used small-signal equivalent circuit of a microwave MESFET or HEMT device. Other kinds of FETs (MOSFETs and JFETs), can be represented by an equivalent circuit having a nearly identical structure, but, of course, very different element values; the equivalent circuit of bipolar devices is only slightly different. The following discussion about microwave FETs is at least qualitatively true for bipolars and other types of FETs as well; for this reason, we shall focus first on the microwave FET.

The nonzero value of $S_{1,2}$ implies that the device has feedback, a consequence of elements C_{gd} and R_s in Figure 8.2. A high value of R_s tends to stabilize the device (although it reduces gain and increases noise), but C_{gd} degrades stability. Feedback causes the device's input impedance to be a function of the load impedance, and the output impedance to depend on the source impedance. Occasionally $S_{1,2}$ is small enough to be negligible; a device having $S_{1,2} = 0$ is called *unilateral*[1]; it has no feedback, and therefore the input and output impedances are independent of the load and source impedances.

In some devices, and at some frequencies, it is possible to find a passive source impedance that results in an output impedance having a negative real part or a load impedance that causes the input impedance to have a negative real part. In such cases, it is possible to satisfy the Kurokawa conditions for oscillation (see Section 12.1.3) and, if those conditions are satisfied, oscillation inevitably results. Then, we say that the device is *conditionally stable* or, equivalently, *potentially unstable*; when no such source or load impedances can be found, the device is *unconditionally stable*. Most transistors are conditionally stable in the low-frequency part of their useful range (which, in modern devices, may be several tens of gigahertz) and unconditionally stable at the high end of their useful frequency range.

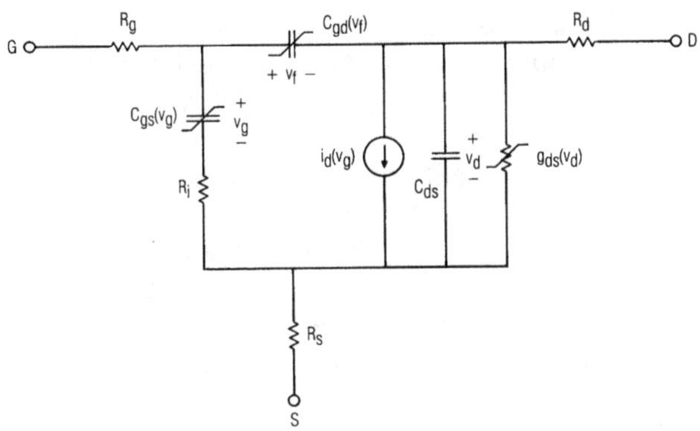

Figure 8.2 Small-signal, nonlinear MESFET equivalent circuit. Four elements—C_{gs}, C_{gd}, i_d, and g_{ds}—are nonlinear, although C_{gd} often can be treated as a linear element.

1. A device having $S_{2,1} = 0$ is called *dead*.

Stability in small-signal amplifiers is important for reasons beyond the natural desire to prevent oscillation; it affects the criteria for which the amplifier can be designed. When a device is unconditionally stable, it is always possible to achieve a conjugate match simultaneously at the input and output; when the amplifier is conditionally stable, it is generally impossible to achieve a simultaneous conjugate match. Furthermore, many of the characteristics of a FET that improve performance—the most important being high transconductance, low C_{gs}, and low R_s—also raise the minimum frequency at which the device is unconditionally stable. Thus, most high-quality MESFETs and HEMTs are only conditionally stable at microwave frequencies.

When a device is unconditionally stable, the design process can be very simple: one calculates the source and load impedances that result in a simultaneous conjugate match (so-called *SCM conditions*) and designs matching networks that present these impedances to the gate and drain of the device over the required bandwidth. The gain that results is the *maximum available gain*, or MAG (Section 1.5). SCM conditions may not be practical, however, for several reasons: (1) the device may be conditionally stable; (2) a value of gain other than the MAG may be desired; (3) SCM conditions do not provide optimum noise figure; or (4) in a broadband amplifier, the designer must mismatch either the source or load at low frequencies to obtain a flat passband. In these cases a unique set of source and load impedances that result in the desired gain generally does not exist. Consequently, our design procedure must allow us to select source and load terminations that result in a specific value of gain and, in some cases, an acceptable noise figure. The design procedure must also prevent us from inadvertently using source or load impedances that cause instability.

The necessary and sufficient conditions for unconditional stability are that the stability factor K be greater than 1.0 and that the magnitude of the determinant of the S matrix, Δ_S, be less than 1.0. If either of these conditions is not met, the device is conditionally stable. SCM conditions can be found if $K > 1$, even if $|\Delta_S| > 1$, although this situation rarely occurs in practical devices. The determinant of the S matrix is

$$\Delta_S = S_{1,1}S_{2,2} - S_{2,1}S_{2,2} \tag{8.1}$$

and the stability factor K is

$$K = \frac{1 - |S_{1,1}|^2 - |S_{2,2}|^2 + |\Delta_S|^2}{2|S_{1,2}S_{2,1}|} \tag{8.2}$$

It is interesting to note that an amplifier having lossless input and output matching circuits has the same value of K as the transistor it uses; that is, K is invariant with lossless, passive matching.

If the device is conditionally stable, we need to know the input and output terminations that can cause oscillation, the source and load reflection coefficients for which

$$|\Gamma_{\text{in}}| > 1.0 \tag{8.3}$$

and

$$|\Gamma_{\text{out}}| > 1.0 \tag{8.4}$$

where Γ_{in} and Γ_{out} are the respective input and output reflection coefficients of the device. These are given by the following relations:

$$\Gamma_{\text{in}} = S_{1,1} + \frac{S_{2,1}S_{1,2}\Gamma_L}{1 - S_{2,2}\Gamma_L} \tag{8.5}$$

$$\Gamma_{\text{out}} = S_{2,2} + \frac{S_{2,1}S_{1,2}\Gamma_s}{1 - S_{1,1}\Gamma_s} \tag{8.6}$$

The solutions of (8.3) and (8.4) are regions in the plane of the load and source reflection coefficients, respectively, and can be plotted conveniently on a Smith chart. The borders of the regions are circles; the values of Γ_L that border the stability region defined by (8.3) and (8.5) is called the *output stability circle*. Its center C_L is

$$C_L = \frac{(S_{2,2} - \Delta_S S_{1,1}^*)^*}{|S_{2,2}|^2 - |\Delta_S|^2} \tag{8.7}$$

and its radius r_L is

$$r_L = \left| \frac{S_{1,2}S_{2,1}}{|S_{2,2}|^2 - |\Delta_S|^2} \right| \tag{8.8}$$

Similarly, the input stability circle defines the boundaries of the region in which Γ_s satisfies (8.4). Its center and radius, C_S and r_S, respectively, are

$$C_s = \frac{(S_{1,1} - \Delta_S S_{2,2}^*)^*}{|S_{1,1}|^2 - |\Delta_S|^2} \tag{8.9}$$

and

$$r_s = \left| \frac{S_{1,2}S_{2,1}}{|S_{1,1}|^2 - |\Delta_S|^2} \right| \tag{8.10}$$

Equations (8.7) through (8.10) identify the boundaries of the stability regions, but they do not indicate whether the region that insures stability is inside or outside the stability circle. The stable region is determined easily from the following considerations: if $|S_{1,1}| < 1.0$, the point $\Gamma_L = 0$, the center of the Smith chart, must be in the stable region; similarly, if $|S_{2,2}| < 1.0$, the point $\Gamma_s = 0$ in the input plane must be within the stable region. In practical devices that do not employ external feedback, the outside of the circle is usually the stable region.

8.1.2 Amplifier Design

Designing a small-signal amplifier involves selecting the appropriate source and load impedances (or reflection coefficients) and designing the input and output matching circuits to present those impedances to the device. If the device is unconditionally stable and maximum gain is desired, the process of determining source and load reflection coefficients is straightforward. The reflection coefficients that provide a simultaneous conjugate match, $\Gamma_{s,m}$ and $\Gamma_{L,m}$ are

$$\Gamma_{s,m} = \frac{B_1 \pm (B_1^2 - 4|C_1|^2)^{1/2}}{2C_1} \tag{8.11}$$

and

$$\Gamma_{L,m} = \frac{B_2 \pm (B_2^2 - 4|C_2|^2)^{1/2}}{2C_2} \tag{8.12}$$

where

$$B_1 = 1 + |S_{1,1}|^2 - |S_{2,2}|^2 - |\Delta_S|^2 \tag{8.13}$$

$$B_2 = 1 + |S_{2,2}|^2 - |S_{1,1}|^2 - |\Delta_S|^2 \tag{8.14}$$

$$C_1 = S_{1,1} - \Delta_S S_{2,2}^* \tag{8.15}$$

and

$$C_2 = S_{2,2} - \Delta_S S_{1,1}^* \tag{8.16}$$

Under SCM conditions, the transducer gain equals the maximum available gain; it is

$$G_t = MAG = \left|\frac{S_{2,1}}{S_{1,2}}\right|[K - (K^2 - 1)^{1/2}] \tag{8.17}$$

If the device is conditionally stable, or if it is unconditionally stable but the desired gain is less than the maximum available gain, the source and load reflection coefficients that give the desired gain are not unique. If the source reflection coefficient is specified, the locus of load reflection coefficients providing a particular value of gain is a circle in the reflection coefficient (Smith chart) plane; conversely, if the load is specified, the source reflection coefficients lie on a circle. Although the desired gain can often be achieved without a conjugate match at either the input or output, it is usually wise to match at least one port; having one port well matched allows stages to be cascaded easily and with minimal gain variation over the amplifier's passband.

An amplifier having one conjugate-matched port can be designed according to its available gain, G_a, or power gain, G_p. These quantities are

$$G_p = \frac{1}{1-|\Gamma_{\text{in}}|^2}|S_{2,1}|^2\frac{1-|\Gamma_L|^2}{|1-S_{2,2}\Gamma_L|^2} \tag{8.18}$$

and

$$G_a = \frac{1-|\Gamma_s|^2}{|1-S_{1,1}\Gamma_s|^2}|S_{2,1}|^2\frac{1}{1-|\Gamma_{\text{out}}|^2} \tag{8.19}$$

where Γ_{in} and Γ_{out} are given by (8.5) and (8.6). From (8.18) we can see that G_p is independent of Γ_s; therefore, designing an amplifier to have a specific value of power gain requires only selecting Γ_L. However, the quantity that we loosely call *gain* is in fact transducer gain, G_t, which is

$$G_t = \frac{1-|\Gamma_s|^2}{|1-S_{1,1}\Gamma_s|^2}|S_{2,1}|^2\frac{1-|\Gamma_L|^2}{|1-\Gamma_{\text{out}}\Gamma_L|^2} \tag{8.20}$$

If the input is conjugate-matched, the power delivered to the network equals the power available from the source, and from (8.8) and (8.20) $G_t = G_p$. Similarly, (8.19) indicates that the available gain is independent of Γ_L; achieving the desired value of G_a requires only selecting Γ_s. If the output is matched, the power delivered to the load equals the power available from the network, and $G_t = G_a$. Thus, one can achieve a specified value of G_t by designing the amplifier to have G_p or G_a equal to the desired value of G_t and then conjugate-matching the input or output, respectively. The design procedure is as follows:

1. Select the desired transducer gain G_t.
2. Decide which port is to be matched.
3. If the input is to be matched, select Γ_L to achieve $G_p = G_t$; then find $\Gamma_s = \Gamma_{\text{in}}^*$ from (8.5).
4. If the output is to be matched, select Γ_s to achieve $G_a = G_t$; then find $\Gamma_L = \Gamma_{\text{out}}^*$ from (8.6).

The remaining problem is to find the values of Γ_L that provide the specified G_p or the values of Γ_s that provide the specified G_a; these quantities lie on circles in the load or source planes, respectively. The center and radius of the Γ_L circle, called the *power gain circle*, are respectively

$$C_p = \frac{g_p(S_{2,2} - \Delta_S S_{1,1}^*)^*}{1 + g_p(|S_{2,2}|^2 - |\Delta_S|^2)} \tag{8.21}$$

and

$$r_p = \frac{(1 - 2Kg_p|S_{2,1}S_{1,2}| + g_p^2|S_{2,1}S_{1,2}|^2)^{1/2}}{1 + g_p(|S_{2,2}|^2 - |\Delta_S|^2)} \tag{8.22}$$

where K is given by (8.2) and $g_p = G_p / |S_{2,1}|^2$. The loci of Γ_s that provide constant available gain are also circles, and their centers and radii are given by the similar relations,

$$C_a = \frac{g_a(S_{1,1} - \Delta_S S_{2,2}^*)^*}{1 + g_a(|S_{1,1}|^2 - |\Delta_S|^2)} \tag{8.23}$$

and

$$r_a = \frac{(1 - 2Kg_a|S_{2,1}S_{1,2}| + g_a^2|S_{2,1}S_{1,2}|^2)^{1/2}}{1 + g_a(|S_{1,1}|^2 - |\Delta_S|^2)} \tag{8.24}$$

where $g_a = G_a / |S_{2,1}|^2$. Comparing (8.23) and (8.9), we see that the centers of the input stability circle and available-gain circle lie on the same line; similarly, the centers of the output stability circle and power-gain circle lie on the same line. Moreover, although it is not obvious from the equations, the circles intersect at the edge of the reflection-coefficient plane.

8.1.2.1 Example: Gain and Stability Circles

A FET has the following S parameters at 10 GHz:

$$\mathbf{S} = \begin{bmatrix} 0.8\angle{-85^\circ} & 0.10\angle 45^\circ \\ 1.7\angle 125^\circ & 0.65\angle{-70^\circ} \end{bmatrix}$$

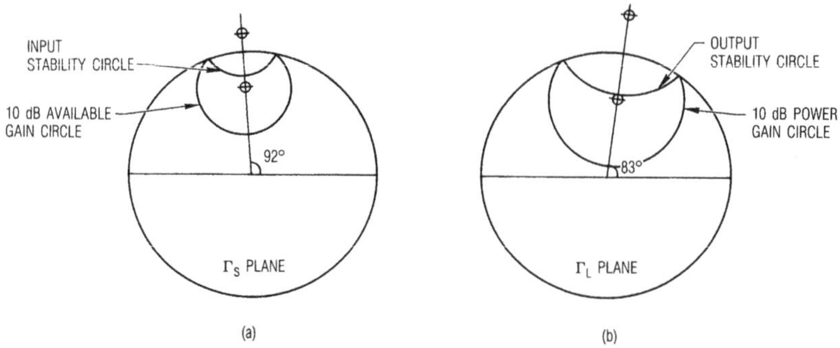

Figure 8.3 Stability and gain circles of the FET in the example: (a) input plane; (b) output plane.

We wish to find the input and output stability circles and gain circles that represent G_p and G_a values of 10 dB. We first use (8.2) to find K, and calculate Δ_S from (8.1). We find that $K = 0.271$ and $\Delta_S = 0.391\angle{-140°}$ so the device is conditionally stable. If the device were unconditionally stable there would be no need to find the stability circles. Equations (8.7) and (8.8) provide the output stability circle; its center and radius are $1.32\angle 83°$ and 0.634, respectively. Similarly, (8.9) and (8.10) give the center and radius of the input stability circle; these are, respectively, $1.45\angle 92°$ and 0.350.

We now calculate the gain circles. First we find $g_p = g_a = 3.16/1.72 = 1.093$. Then, using the K and Δ_S found earlier, (8.21) and (8.22) provide the power-gain circle; its center and radius are $0.636\angle 83°$ and 0.526, respectively. Similarly, we use (8.23) and (8.24) to find the available-gain circle's center and radius, $0.718\angle 92°$ and 0.378, respectively. These circles are plotted on a Smith chart in Figure 8.3.

We have now identified a range of values of Γ_s or Γ_L that provide a specified value of transducer gain; however, we still have no clear rationale for selecting any particular value. Clearly, it is wise to pick a value that is not too close to the stability circle, or a small source or load mismatch may cause oscillation. A consideration in the design of a low-noise amplifier is that Γ_s should be as close as possible to the value that optimizes noise figure; thus, one would pick Γ_s to optimize noise figure and would choose $\Gamma_L = \Gamma_{out}^*$ to match the output. A third criterion (the one we all have been waiting for!) is to pick Γ_s or Γ_L to optimize linearity, perhaps within constraints on gain and noise figure. The latter half of this chapter is devoted to an examination of that criterion.

8.1.3 Characteristics of FETs and Bipolars in Small-Signal Amplifiers

Because the transistor was treated as a general two-port described by S parameters, the design process described in Section 8.1.2 is valid for all types of devices, bipolar as well as FET. Bipolar devices, BJTs and HBTs, are distinctly different, however, and require some special considerations. We describe some of these in this section.

8.1.3.1 Bias

Bipolar transistors are exponential devices: the collector current is an exponential function of base-to-emitter voltage. This characteristic makes it difficult to provide stable bias from a base-to-emitter voltage source. Furthermore, because the current gain is a strong function of temperature, even a dc base-current source provides inadequate stability.

Methods for providing stable dc bias to a bipolar transistor are well known and are standard textbook material. Unfortunately, the methods that provide the best dc stability require an emitter resistor and bypass capacitor. At high frequencies, a bypass capacitor's parasitics may prevent it from working adequately. Methods that do not require an emitter resistor have been developed, but they are not as stable as those that do. Other methods involve active bias, and the use of a current mirror. Reference [8.1] describes a number of such methods.

8.1.3.2 Gain Characteristics

FET devices have relatively low transconductance and low gate-to-source capacitance, while bipolars have high transconductance but high base-to-emitter capacitance. As such, high low-frequency gain (often greater than the in-band gain) is much more likely to occur in bipolar devices than in FETs. Designers of bipolar amplifiers must select matching circuits that suppress low-frequency gain; for example, by using a high-pass circuit structure.

8.1.3.3 Impedances

High-frequency FETs have a high input Q. The input is well approximated by a series RC circuit, which has a large capacitive reactance compared to its resistance, even in the microwave range. In bipolars, however, the large input capacitance short circuits the base-emitter resistance at high frequencies, so the input impedance consists largely of the base resistance.

As a result, the bipolar's input, in common-emitter configuration, is much easier to match over a broad bandwidth.

The output impedance of a BJT or HBT, lacking feedback, would be virtually infinite. Feedback from the base-to-collector capacitance, however, decreases the output impedance dramatically and makes the output impedance much more sensitive to source impedance than in FETs.

Because of the bipolar's high transconductance, its input capacitance, at low frequencies, often consists largely of Miller-effect capacitance. Since the transconductance depends on bias, the input impedance also becomes bias-sensitive. It also can be difficult to model, as it is sensitive to the base-to-collector capacitance, which in turn is quite small and difficult to measure.

8.1.3.4 Distortion

Levels of intermodulation distortion in high-frequency bipolar devices are generally lower than in FETs. HBTs, in particular, often exhibit dramatically lower distortion at signal levels well below the 1-dB compression point. The reason for this characteristic is discussed in detail in Section 8.2.2.

8.1.3.5 Noise Matching

The noise figure of most bipolar transistors is considerably less sensitive to source impedance than in FETs. The noise figures of small-signal bipolars are generally considerably higher than GaAs MESFETs and HEMTs.

In both bipolars and FETs, the source impedance that provides optimum noise figure at low frequencies is higher than the input impedance. As frequency increases, the source impedance decreases; it approaches a conjugate match at the high end of the device's useful frequency range.

8.1.4 Broadband Amplifiers

Section 8.1.2 described the design of amplifiers for a single "spot" frequency. Practical amplifiers must cover a prescribed bandwidth, which, in some cases, may be quite broad. Our design method must address this requirement.

The single-frequency design is usually adequate for amplifier bandwidths up to perhaps 10%. The simplest approach is to design the amplifier for a single frequency, the band center, and use computer analysis to optimize the circuit. For broader bandwidths, however, a new methodology is needed. One simple approach is to make Γ_S the source impedance for optimum noise figure. This defines the output impedance of the device, Γ_{out}. Selecting $\Gamma_L = \Gamma_{out}^*$ matches the output, but unfortunately results in a sloped passband. Thus, it is necessary, in a broadband amplifier, to mismatch the output to achieve flat gain. The equations in Section 8.1.2 can be used to obtain Γ_L values, over the passband, which provide flat gain. This is easiest to do, however, with a computer circuit-analysis program.

Once the values of Γ_s and Γ_L are determined, matching networks can be synthesized, preferably with the aid of a network-synthesis program. If Γ_s or Γ_L are difficult to synthesize, the resulting networks may not provide the desired performance. In this case numerical optimization on the computer may be necessary.

8.1.5 Negative-Image Modeling

The design of broadband amplifiers can become difficult when it involves competing trade-offs between gain, distortion, and noise. An elegant method for resolving those conflicts is called *negative-image modeling* [8.2].

The method is as follows:

1. Create a circuit with "negative-image" source and load networks as shown in Figure 8.4(a); $-C_s$ and $-C_L$ are negative capacitances.
2. Use an appropriate topology for the input and output networks. For best results, they should mirror the structure of the device's equivalent circuit at its respective ports.
3. Optimize the circuit by means of a circuit-analysis program. Use whatever criteria or trade-offs are appropriate. Because of the negative capacitances, the optimization will be surprisingly easy.
4. When satisfactory performance has been achieved, synthesize input- and output-matching networks using the positive versions of the negative-image networks as loads; see Figure 8.4(b).
5. Replace the negative-image circuits with the matching circuits.
6. Do any necessary final optimization.

Why does the method work? If the matching circuits synthesized in Figure 8.4(b) provide a conjugate match to their respective positive-load networks, their output impedances must be equivalent to the negative-image networks. Thus, they provide the same Γ_s and Γ_L that the negative-image networks provided to the FET. Of course, the synthesized networks

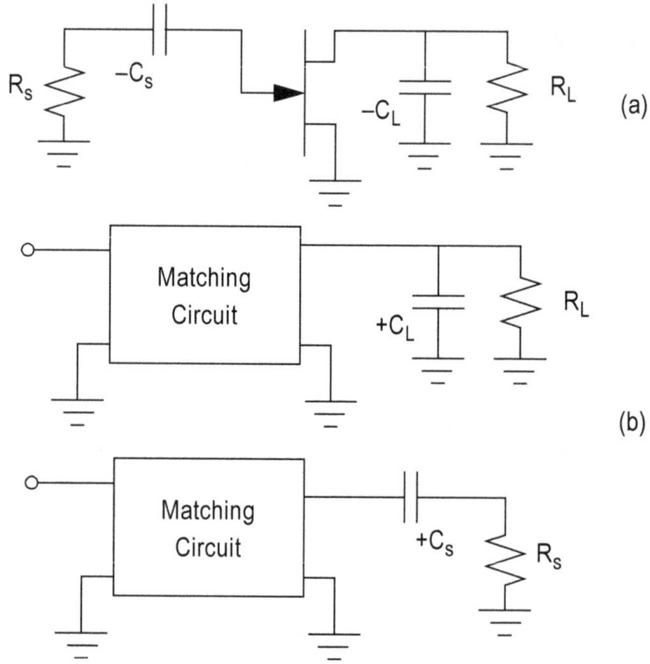

Figure 8.4 Negative-image matching: (a) a FET with negative-image networks; (b) synthesis of equivalent real matching circuits.

may not provide a perfect match to the positive loads, but we expect that they are a good approximation. Some final optimization still may be necessary.

8.1.5.1 Example: Design with Negative-Image Modeling

As a simple example, we use negative-image modeling to design a 7- to 11-GHz amplifier having 10-dB gain. For the example, we use lumped-element matching circuits and do not design a complete, distributed matching circuit. These conditions are, of course, impractical, but they serve to illustrate the method without introducing additional complications.

Figure 8.5(a) shows the amplifier using negative-image matching. The port impedances and capacitor values are adjusted to achieve flat, 10-dB gain over the prescribed band. A simple synthesis program was used to create matching circuits for the positive-image networks, as illustrated in

Figure 8.4(b), and these were attached to the amplifier, as shown in Figure 8.5(b).

Figure 8.6 shows the gain of the amplifier with negative image matching and with the synthesized matching circuits. Without any further optimization, the final circuit's gain is within 1 dB of the negative-image circuit's gain. Optimization can be used to fine-tune the gain, if desired.

8.2 NONLINEAR ANALYSIS

Nonlinear analysis of a small-signal amplifier requires the use of the lumped-element equivalent circuit of Figure 8.2, along with appropriate source and load networks. When the excitation is weak, Volterra methods are the logical means to evaluate such small-signal nonlinear effects such as harmonics, intermodulation, or AM-to-PM conversion. Because the

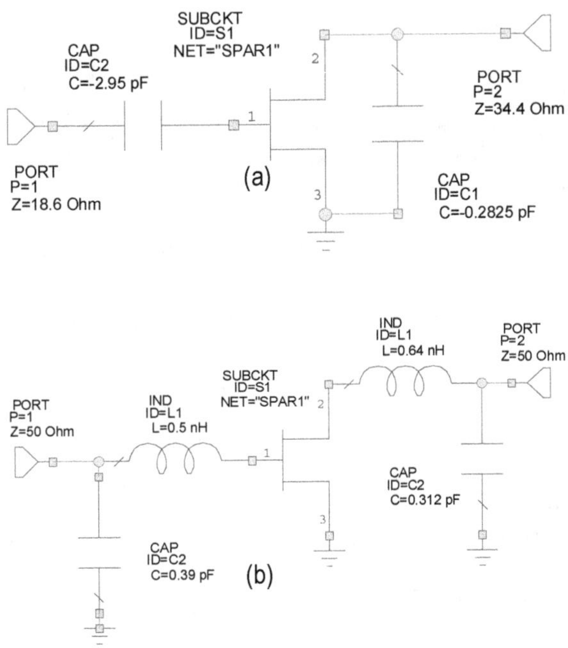

Figure 8.5 (a) Negative-image matching networks connected to a FET; (b) the negative-image networks replaced by equivalent matching circuits having real, positive-valued elements.

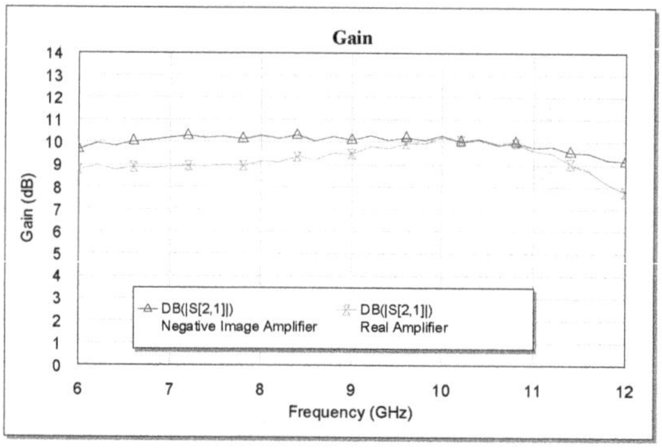

(a)

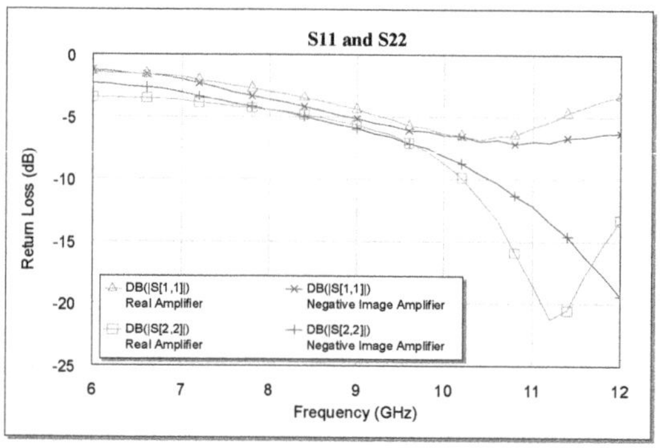

(b)

Figure 8.6 Performance comparison of the real and negative-image amplifiers: (a) gain; (b) input and output return loss.

circuit includes feedback elements and reactive nonlinearities, power-series analysis cannot be used. For Volterra-series analysis, each significant nonlinear element must be characterized by a power series in terms of its small-signal control voltage.

8.2.1 Nonlinearities in FETs

The equivalent circuit in Figure 8.2 shows four nonlinear elements: the gate-to-drain capacitance, C_{gd}, the gate-to-source capacitance, C_{gs}, the controlled current source, i_d, and the drain-to-source conductance, g_{ds}. When used in a small-signal amplifier, a FET is always operated well into its saturation region; the description of the GaAs MESFET's large-signal model in Section 2.5.4 showed that in current saturation C_{gs} depends only weakly upon v_d, and C_{gd} depends so weakly upon v_g and v_d that it often can be treated as a linear element. Thus, C_{gs} is shown in Figure 8.2 as a function of v_g only. If C_{gd} is treated as a nonlinear element, it is a function of only the voltage v_f across it.[2]

The nonlinearities of the capacitances usually do not dominate in the establishment of the small-signal nonlinear performance of the circuit; the dominant element is usually $i_d(v_g)$ or, occasionally, $g_{ds}(v_d)$. Therefore, we can take some reasonable liberties with the nonlinear characterization of these less significant elements. In particular, since C_{gs} is a relatively minor contributor to intermodulation distortion, it usually can be treated as a linear element. Similarly, in small-signal amplifiers where the FET remains in current saturation, C_{gd} also can be treated as a linear element. When the *C*/*V* or *Q*/*V* characteristic of an element has been determined, the incremental power-series representation can be found.

These approximations should be treated with caution. As with all nonlinear capacitances, the significance of the nonlinearities in C_{gs} depends on frequency, bias, and source and load impedance. At low frequencies, the capacitive reactance is high, so it generates little linear or distortion current, regardless of dc bias. As frequency increases, it is possible to find combinations of frequency and bias where the capacitive nonlinearity causes surprisingly high distortion [8.3].

The controlled current source i_d and the gate/drain conductance g_{ds} represent the FET's channel current, a single nonlinearity that has two control voltages, v_g and v_d. Equation (2.10) gives a Taylor-series characterization of a multiply controlled nonlinearity; we can identify v_1 in (2.10) as v_g, v_2 as v_d, and the FET's dc *I*/*V* characteristic $I_d(V_g, V_d)$ as *f*. As in Chapter 2, we let capital letters represent large-signal voltages and currents, while lower-case letters represent incremental ones. If we ignore the cross terms in (2.10) (i.e., those that include the product term $v_1 v_2$), we can treat the nonlinearity as two nonlinear elements in parallel, one depending upon v_g and the other on v_d. The equation can then be split into two parts, one representing the dependence on v_1 and the other representing the dependence on v_2. After substituting and rearranging (2.10), we obtain

2. As before, we view the capacitances in a division-by-capacitance sense (Section 2.5.7), although the assumptions are largely valid for division-by-charge modeling as well.

$$i = \frac{\partial I_d}{\partial V_g} v_g + \frac{1}{2}\frac{\partial^2 I_d}{\partial V_g^2} v_g^2 + \frac{1}{6}\frac{\partial^3 I_d}{\partial V_g^3} v_g^3 + \frac{\partial I_d}{\partial V_d} v_d + \frac{1}{2}\frac{\partial^2 I_d}{\partial V_d^2} v_d^2 + \frac{1}{6}\frac{\partial^3 I_d}{\partial V_d^3} v_d^3 \tag{8.25}$$

where the derivatives are evaluated at the dc bias points V_{g0} and V_{d0}. The terms in (8.25) that contain v_g represent a nonlinear controlled current source, and the ones that contain v_d represent a nonlinear conductance. The former is, of course, the current source $i_d(v_g)$ in Figure 8.2, and the latter is $g_{ds}(v_d)$.

An unfortunate complication of this neat situation is that the drain conductance $g_{ds}(v_d)$ often depends upon frequency, and the value of g_{ds} obtained from dc I/V measurements is usually much lower than the value measured at high frequencies. For this reason, it is best to determine $g_{ds}(v_d)$ by extraction from measured S or Y parameters. It is important to remember that the Volterra-series analysis requires a series expansion of this element's incremental I/V characteristic, not of its G/V characteristic; see Section 2.2.6.

Figure 8.7 shows an example of the measured Taylor-series coefficients of the gate I/V characteristic of a conventional MESFET, as a function of the gate bias voltage. The coefficients are largest near pinch-off, simply because the current changes most rapidly near that voltage. This implies that the distortion is worst near pinch-off as well. It is worth noting that the third derivative has a zero near $V_g = -0.95$V, so we might expect the third-order distortion to be very low at this bias voltage. Unfortunately, this is not the case, because (1) the second derivative is maximum at this point, so the contribution of second-order mixing to third-order distortion is relatively great, and (2) the large variation in the third derivative near the zero implies that the contribution to the $2\omega_2 - \omega_1$ product from higher-order terms (Section 4.1.1) could be relatively large as well. However, the more gradual decrease in the magnitude of the third derivative as $V_g \to 0$ does indeed imply that third-order distortion decreases in that region.

The simplifications described in this section allow modeling of third-order intermodulation distortion intercept points to an accuracy of 1 to 2 dB in modern MESFETs, JFETs, and MOSFETs fabricated in mature technologies. In more advanced technologies, such as short-gate-length pHEMTs and MOSFETs, nonlinearities that were insignificant in mature

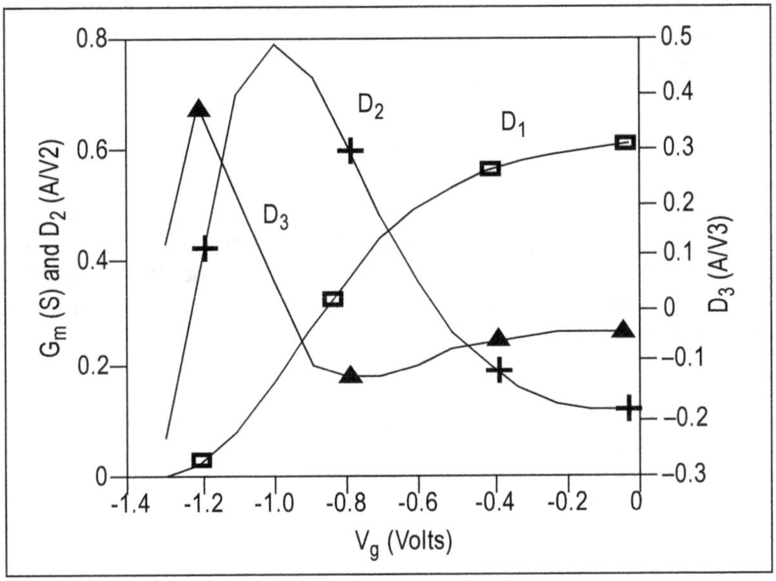

Figure 8.7 The measured derivatives of the gate *I*/*V* characteristic of a conventional GaAs MESFET. The *n*th derivative curve is labeled D_n.

MESFETs may be more important. It may be necessary to examine the device characteristics carefully to determine what must be modeled most accurately. HEMT devices, in particular, have greater g_{ds} than MESFETs, and stronger i_d nonlinearity. Additionally, it may be necessary to model the cross terms in the Taylor series, which have been neglected in (8.25). Reference [8.4] gives some valuable insight into these matters.

8.2.2 Nonlinearities in Bipolar Devices

Bipolar devices have extremely strong, exponential nonlinearities, yet they have relatively low levels of distortion. Two reasons explain this paradox. The first is in the way that the transistor's distortion levels vary with dc bias current. From (2.102), the collector current in a bipolar device, I_c, is

$$I_c = I_s\left(\exp\left(\frac{qV_{be}}{\eta_f KT}\right) - \exp\left(\frac{qV_{bc}}{\eta_r KT}\right)\right) \approx I_s \exp\left(\frac{qV_{be}}{\eta_f KT}\right) \tag{8.26}$$

We saw in Section 4.1.3 that a FET's output third-order IM intercept point, IP_3, is given by

$$IP_3 = 10 \log\left(\frac{2}{3}\frac{a_1^3}{a_3} R_L\right) + 30 \tag{8.27}$$

where a_1 and a_3 are the first- and third-degree Taylor-series coefficients of the I/V characteristic and IP_3 is in dBm. Because of the similarity in the equivalent circuits, this expression is at least qualitatively valid for bipolar devices. Differentiating (8.26), we have

$$\frac{a_1^3}{a_3} = 6I_c^2 \tag{8.28}$$

so

$$IP_3 = 10 \log(4I_c^2 R_L) + 30 \tag{8.29}$$

We see that the intercept point increases dramatically with collector current. A high intercept point can be achieved simply by using a high collector current.

The second reason is a cancellation phenomenon between the components of collector current generated by the resistive and reactive parts of the junction. As a result, there is an *optimum* capacitive nonlinearity, which is that of a classical diffusion capacitance (2.105). This is a surprising result, as it is impossible for the current in a reactance and a resistance to cancel; however, it is possible for the collector current generated by those nonlinearities to cancel. A full derivation of the cancellation phenomenon can be found in [8.5].

The dominant capacitances in a bipolar device are (1) charge stored in the depletion regions around the base-to-emitter and base-to-collector junctions, and (2) diffusion charge stored in the base. The depletion capacitances are well described by the textbook *pn* junction capacitance expression, (2.59), and the diffusion capacitance by (2.105). The base-to-emitter capacitance nonlinearity is quite strong; the diffusion component is, in theory, an exponential function of voltage. In reality, the capacitive nonlinearity is much weaker than (2.105) implies, in part because it is valid only at frequencies well below $1/\tau_f$, and because its nonlinearity is diluted somewhat by the depletion capacitance. The nonlinearity of the base-to-collector capacitance is significant in bipolars, as well; it can be described accurately by (2.59).

8.2.3 Nonlinear Phenomena in Small-Signal Amplifiers

The nonlinear phenomena of greatest concern in amplifiers are AM-to-PM conversion, harmonic generation, intermodulation distortion, and saturation. These phenomena can be analyzed by either Volterra techniques or harmonic-balance analysis. For saturation calculations beyond the 1-dB compression point, harmonic-balance analysis is probably preferable to Volterra-series analysis, because the harmonic-balance approach can include the effects of strong nonlinearities in the device model. These effects are often the dominant ones in establishing saturation characteristics, and are generally not modeled by the Volterra series. Nevertheless, in situations where gain compression effects are dominated by weak nonlinearities, especially a FET's nonlinear transconductance, Volterra-series analysis is an acceptable analytical method.

As in Section 8.1, we view the amplifier as a "black box" (Figure 8.8) having linear and nonlinear transfer functions. The excitation is the signal $v_s(t)$, which consists of Q sinusoidal components,

$$v_s(t) = \frac{1}{2} \sum_{\substack{q=-Q \\ q \neq 0}}^{Q} V_{s,q} \exp(j\omega_q t) \tag{8.30}$$

The response $i(t)$, the output current, is

$$\begin{aligned} i(t) = \sum_{n=1}^{N} \frac{1}{2^n} \sum_{q1=-Q}^{Q} \sum_{q2=-Q}^{Q} \cdots \sum_{qn=-Q}^{Q} V_{s,q1} \\ \cdot V_{s,q2} \cdots V_{s,qn} H_n(\omega_{q1}, \omega_{q2}, \ldots, \omega_{qn}) \\ \cdot \exp[j(\omega_{q1} + \omega_{q2} + \ldots + \omega_{qn})t] \end{aligned} \tag{8.31}$$

The current $i(t)$ is the sum of all the nth-order output currents $i_n(t)$; an nth-order output current is the sum of all current components that arise from mixing between n input frequencies. The function $H_n(\omega_{q1}, \omega_{q2}, \ldots, \omega_{qn})$, called the *nth-order nonlinear transfer function*, relates the output current at the frequency $\omega_{q1} + \omega_{q2} + \ldots + \omega_{qn}$ to the individual components of

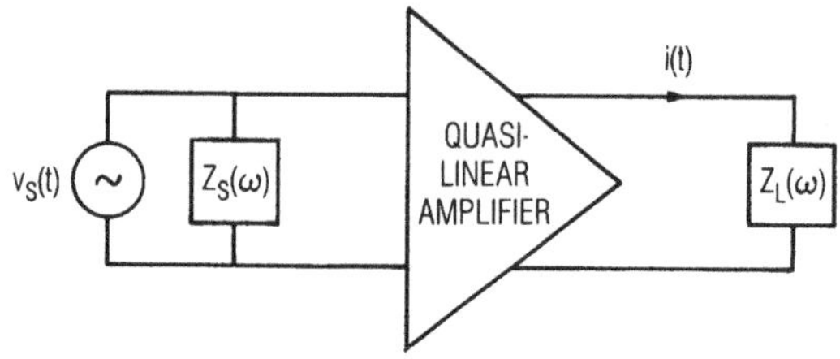

Figure 8.8 Quasilinear amplifier model.

$v_S(t)$ at those frequencies. In this section we assume that the nonlinear transfer functions of the circuit are known, and we show how they can be used to evaluate a circuit's nonlinear behavior. Those transfer functions can be determined by straightforward application of the theory in Chapter 4.

8.2.3.1 Saturation and AM-to-PM Conversion

When the amplifier is driven into saturation by a single sinusoidal signal at frequency ω_1, the output current at ω_1 can be found by evaluating (8.31) under the condition of a single-tone excitation and by retaining only the terms at ω_1. The result is

$$I(\omega_1) = V_{s,1} H_1(\omega_1) + \frac{3}{4} V_{s,1} V_{s,1} V_{s,1}^* H_3(\omega_1, \omega_1, -\omega_1) \tag{8.32}$$

where $I(\omega_1)$ is the component of the output current $i(t)$ at ω. In (8.32) we have considered only the positive-frequency part of $I(\omega_1)$ [so $I(\omega_1)$ is a phasor], and we have limited the summation over n to $N = 3$; components of order greater than three are neglected. The coefficient of 3 in the second term of (8.32), and similar coefficients in the following equations, may be confusing. They arise from the fact that there are multiple identical terms in (8.31) at any particular mixing frequency.

Although it may not be obvious, (8.32) predicts that as $V_{s,1}$ increases, $I(\omega_1)$ saturates and then begins to decrease. Equation (8.32) is valid if $V_{s,1}$ remains small enough that $I(\omega_1)$ does not decrease with an increase in $V_{s,1}$; beyond that point, higher-order terms in the series must be included. The next highest-order component at the fundamental frequency is fifth order; these higher orders become significant as the amplifier is driven more strongly into saturation.

We define the *relative distortion* $D(\omega_1)$ as the ratio of the total output current to the linear (first order) part. $D(\omega_1)$ represents the fractional deviation from linear operation:

$$D(\omega_1) = \frac{I(\omega_1)}{V_{s,1} H_1(\omega_1)} \tag{8.33}$$

and substituting (8.32) into (8.33) gives

$$D(\omega_1) = 1 + \frac{3}{4}|V_{s,1}|^2 \frac{H_3(\omega_1, \omega_1, -\omega_1)}{H_1(\omega_1)} \tag{8.34}$$

Equation (8.34) indicates that $D(\omega_1)$ can be expressed as the sum of two phasors, as shown in Figure 8.9. If $V_{s,1}$ is very small, $D(\omega_1) = 1$, which indicates linear operation. As $V_{s,1}$ increases, however, $D(\omega_1)$ changes in both magnitude and phase; in FET amplifiers the phase of H_3/H_1 is always such that $D(\omega_1)$ decreases, which indicates that the gain decreases, and the output power saturates. The existence of a nonzero phase shift θ shows that, as the device begins to saturate, the phase shift also begins to deviate from its value when $V_{s,1}$ is small; this phenomenon is called *AM-to-PM conversion.*

8.2.3.2 Harmonic Generation

Again we consider a single-tone excitation at ω_1. The positive-frequency component of $I(\omega)$ in (8.31) at the nth harmonic of ω_1 is

$$I(n\omega_1) = 2^{-n+1} V_{s,1}^n H_n(\omega_1, \omega_1, \dots, \omega_1) \tag{8.35}$$

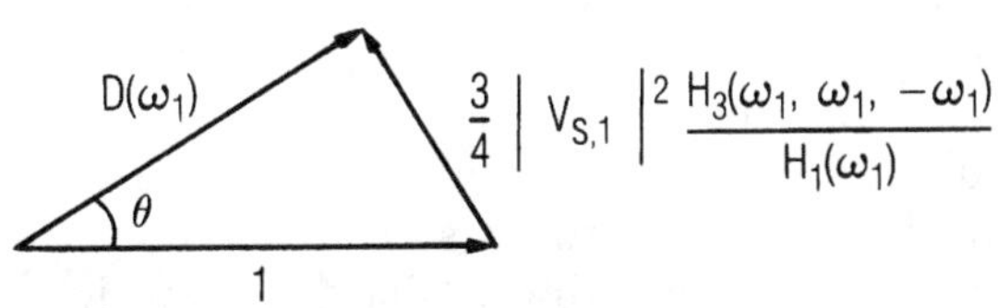

Figure 8.9 Relative distortion vector $D(\omega_1)$ describing saturation and AM-to-PM conversion. $|D(\omega_1)|$ is the gain compression and θ is the phase deviation.

For example, the second harmonic output current is

$$I(2\omega_1) = \frac{1}{2} V_{s,1}^2 H_2(\omega_1, \omega_1) \tag{8.36}$$

and the third harmonic is

$$I(3\omega_1) = \frac{1}{4} V_{s,1}^3 H_3(\omega_1, \omega_1, \omega_1) \tag{8.37}$$

A harmonic can also have a component at a higher order; for example, the second harmonic can include a fourth-order component,

$$I(2\omega_1) = \frac{1}{2} V_{s,1}^2 H_2(\omega_1, \omega_1) + \frac{1}{2} V_{s,1}^3 V_{s,1}^* H_4(\omega_1, \omega_1, \omega_1, -\omega_1) \tag{8.38}$$

Note that there are four identical terms in (8.31) that contribute to the second term in (8.38). We can, of course, pick the phase of $V_{s,1}$ arbitrarily without losing generality, so the conjugate quantity is not significant. In general, an even harmonic can have components at all even orders, and an odd harmonic can have components at all odd orders. The components at orders greater than the lowest, however, are only significant when $V_{s,1}$ approaches saturation.

The relative distortion of the lowest-order component at the nth harmonic is, from (8.33),

$$D(n\omega_1) = 2^{-n+1} V_{s,1}^{n-1} \frac{H_n(\omega_1, \omega_1, \ldots, \omega_1)}{H_1(\omega_1)} \tag{8.39}$$

8.2.3.3 Intermodulation Distortion

Intermodulation involves the effects of mixing between the fundamental frequencies and harmonics when two or more excitation frequencies exist. If the excitation contains the frequencies $\omega_1, \omega_2, \omega_3, \ldots$, the output may contain the frequencies $m\omega_1 + n\omega_2 + p\omega_3 + \ldots$, where m, n, and p are integers. Many of these mixing products are potentially troublesome, but the case that is universally annoying is the one in which two excitation frequencies exist, ω_1 and ω_2, and the intermodulation distortion product has the frequency $2\omega_1 - \omega_2$ or $2\omega_2 - \omega_1$. Then,

$$I(2\omega_1 - \omega_2) = \frac{3}{4} V_{s,1}^2 V_{s,2}^* H_3(\omega_1, \omega_1, -\omega_2) \tag{8.40}$$

Higher-order current components can contribute to a mixing product at $2\omega_1 - \omega_2$. Thus,

$$\begin{aligned} I(2\omega_1 - \omega_2) &= \frac{3}{4} V_{s,1}^2 V_{s,2}^* H_3(\omega_1, \omega_1, -\omega_2) \\ &+ 5 V_{s,1}^3 V_{s,1}^* V_{s,2}^* H_5(\omega_1, \omega_1, \omega_1, -\omega_1, -\omega_2) \\ &+ \frac{15}{2} V_{s,1}^2 V_{s,2} V_{s,2}^{*2} H_5(\omega_1, \omega_1, \omega_2, -\omega_2, -\omega_2) \end{aligned} \tag{8.41}$$

As with the other distortion products, the components of order greater than three represent saturation effects and are not significant at very small $V_{s,1}$ and $V_{s,2}$. The relative distortion, when $V_{s,1}$ and $V_{s,2}$ are small, is

$$D(2\omega_1 - \omega_2) = V_{s,1} V_{s,2}^* \frac{H_3(\omega_1, \omega_1, -\omega_2)}{H_1(\omega_1)} \tag{8.42}$$

The relative distortion, as it is defined for intermodulation and harmonic generation, is an important quantity. Its magnitude squared is the ratio of the power in the distortion component to the linear power, or, more colloquially, the signal-to-distortion ratio. This is an important quantity in specifying a system, and can be used to define the intermodulation intercept point of a system or component (Section 4.1.3).

8.2.3.4 IMD, Saturation, and the 10-dB Rule

A commonly used rule, throughout the industry, is to estimate the output third-order intercept point (for the $2\omega_1 - \omega_2$ product) as 10 dB greater than the 1-dB gain compression point. This rule seems to hold remarkably well in a wide range of devices. Although often viewed as an empirical observation, the 10-dB rule has some basis in theory: we see the third-order nonlinear transfer function in both the expression for gain compression, (8.34), and for intermodulation distortion (IMD), (8.42), so there should be no surprise that the two are linked. In fact, a simple analysis gives a

10.6-dB difference between the compression point and the third-order intercept point.[3]

For better or worse, the 10-dB rule often collapses dramatically. For example, the IP_3 of an HBT amplifier is often much more than 10 dB greater than the compression point. This puzzle can be resolved by noting that even a perfectly linear amplifier (in the sense that $H_n(\omega_1, \dots, \omega_n) = 0$, $n > 1$) must still compress at some point, as it has only limited dc bias power available to create RF output power. Thus, if the amplifier compresses because of weak nonlinearity, the 10-dB rule holds, but if it compresses because of dc limitations, the rule may not apply.

8.2.3.5 Spectral Regrowth

Spectral regrowth, which has been examined in Section 4.2.8, is a manifestation of intermodulation distortion in components or systems involving modulated waveforms. When a bandlimited signal is distorted, the odd-order distortion components appear as an increase in spectral power adjacent to the linear spectrum. Figure 4.13 shows the spectrum when the signal is subjected only to third-order distortion; however, higher-order distortion can increase the bandwidth even further.

In many communication systems, users are assigned contiguous channels. Thus, the distortion components fall into adjacent channels and cause interference. The *adjacent-channel power ratio* can be defined as

$$ACPR = \frac{\int_{f_2}^{f_3} S(f)\, df}{\int_{f_1}^{f_2} S(f)\, df} \tag{8.43}$$

where f_1 and f_2 are the boundaries of the adjacent channel and f_2 and f_3 are those of the prescribed channel. It is important to note that many wireless and cellular-telephone standards define this quantity in different ways.

3. Some sources give a figure of 9.6 dB, which comes from ignoring the 1 dB of gain compression.

8.2.4 Calculating the Nonlinear Transfer Functions

Nonlinear transfer functions of small-signal amplifiers are best calculated by the method of nonlinear currents, described in Section 4.2.5. Section 4.2.6 describes the application of this method to large circuits. The output current as a function of excitation voltage can be used to determine the transfer function; for example, from (8.40) we obtain

$$H_3(\omega_1, \omega_1, -\omega_2) = \frac{4}{3} \frac{I(2\omega_1 - \omega_2)}{V_{s,1}^2 V_{s,2}^*} \tag{8.44}$$

Currently available commercial circuit-analysis software can perform this analysis.

8.3 LINEARITY OPTIMIZATION

In this section, we examine ways to optimize the distortion in small-signal amplifiers using both FET and bipolar devices. We assume throughout that the dominant nonlinearity is the weak nonlinearity of the *I*/*V* and *Q*/*V* characteristics, so Volterra-series analysis is applicable. In particular, we assume that the device is not driven hard enough so that its strong nonlinearities, such as a FET's gate pinch-off and the knee of its drain *I*/*V* characteristic, have any significance. In effect, we assume that the device is biased in the ordinary manner, and that RF voltages are small relative to the dc bias voltages. When these conditions are not met, harmonic-balance analysis should be used instead of Volterra methods.

8.3.1 Linearity Criteria

One of the first problems in optimizing linearity is to select a quantity to optimize. An immediate choice is the output intermodulation intercept point, IP_n. Closer inspection shows, however, that IP_n is not a very good candidate as a figure of merit for linearity. For example, if the dominant nonlinearity is located near the input of a two-port (or cascade of two-ports), IP_n can be increased arbitrarily by increasing the linear gain, without improving the linearity of the nonlinear part of the circuit. Nevertheless, if the output power is controlled to a specific value (e.g., by an AGC loop), the output IP_n may well be the most important quantity. Conversely, if the input power is the controlled quantity, the input intercept point, IP_{ni}, may be most important.

Another possibility is the *dynamic range* of the system. Dynamic range is defined as the difference between the maximum and minimum signal levels that the system can accommodate. The maximum and minimum levels are sometimes defined rather arbitrarily; the minimum signal level is often defined as the noise level, and maximum as the point where intermodulation-distortion components exceed the noise level. For third-order distortion,

$$P_{min} = KTB \tag{8.45}$$

where K is Boltzmann's constant and T is the noise temperature. In dBm,

$$P_{min} = -174 + 10\log(B) + 10\log\left(\frac{T}{T_0}\right) \tag{8.46}$$

where $T_0 = 290\text{K}$ by definition. From (4.38),

$$P_{min} = P_{IM} = 3P_{max} - 2IP_{3i} \tag{8.47}$$

where P_{min} and P_{max} are input powers, and IP_{3i} is the *input* third-order intercept point. A little algebra gives the dynamic range in decibels:

$$P_{max} - P_{min} = \frac{2}{3}\left[IP_{3i} + 174 - 10\log(B) - 10\log\left(\frac{T}{T_0}\right)\right] \tag{8.48}$$

Equation (8.48) is probably the most generally valid criterion for optimization, as it describes the quantity that system designers usually need to optimize.

Another problem is that linearity—however defined—may not be the most important characteristic of an amplifier, and other characteristics, usually noise figure and gain, may be more important. Unfortunately, the conditions that optimize linearity may not satisfy constraints on noise figure or gain. When this situation arises, the designer must make a prudent trade-off between the conflicting requirements.

8.3.2 MESFETs and HEMTs

8.3.2.1 Bias Effects

It is a general rule that a dc drain current of approximately 0.5 I_{dss} maximizes a FET's gain and IM intercept points. The gain and intercept points achieved at this bias level are 1 to 2 dB greater than those obtained at the bias that optimizes noise figure, approximately 0.1 I_{dss} to 0.2 I_{dss}. In fact, the gain and intercept point increase further at higher drain current, but the higher gate bias voltage introduces the possibility of large-signal distortion from rectification in the gate-to-channel Schottky junction. In HEMTs, distortion is minimal near the peak transconductance.

In Chapter 4 we saw that intermodulation distortion can be related directly to the coefficients of the Taylor-series expansion in the vicinity of a dc bias point. Those coefficients are derivatives of the *I*/*V* characteristic, so it should be no surprise that distortion is greatest where the gate-to-drain *I*/*V* characteristic is most strongly curved. In MESFETs, JFETs, and MOSFETs, curvature is greatest near the pinch-off or threshold voltage, so distortion is worst at low current, when the device is biased near threshold. In HEMTs, the situation is more complex, as HEMTs' transconductance often peaks at a gate voltage well above pinch-off, and there exist multiple minima and maxima of all derivatives.

For this reason, the primary means to improve a FET amplifier's linearity has always been to increase its dc drain current, although this approach inevitably compromises the amplifier's noise figure.

dc bias has another effect on the linearity of the amplifier. Even if increasing the drain current increases the transconductance without changing the linearity of the curve, the change still increases the output intercept point. This situation would exist even if increasing the drain current multiplied all the Taylor coefficients in by the same factor. The effect of such a change would be to "scale up" the output power of the device by a decibel or two, so that all linear and IM powers would be increased by the same factor. Because the intercept point is a point on the extrapolated linear and IM output power curves, it would be increased by that same factor of 1 or 2 dB.

A final consideration is large-signal distortion. Intermodulation distortion is generated not only by the small-signal nonlinearity of the device, as manifested by the curvature of the *I*/*V* characteristic, but also by large-signal nonlinearities. Examples of the latter are the turn-on voltage, and rectification in the gate-to-channel Schottky junction. Biasing the device at 0.5 I_{dss} approximately centers the RF voltage between these limits; the clipping that results from attempts to exceed these limits generates large-signal intermodulation.

8.3.2.2 Effect of Source and Load Impedances

The selection of source and load impedances that optimize linearity has always been a major concern in the design of FET amplifiers. Various researchers have shown both theoretically and empirically that the appropriate selection of these impedances, particularly the load impedance, strongly affects the output intercept point of a microwave amplifier [8.6–8.11]. The power-series analysis of a simplified FET equivalent circuit in Section 4.1.3 is consistent with this idea; in particular, (4.42) and (4.43) imply that the IM intercept point of the simplified FET equivalent circuit is a function solely of the load resistance and the power-series coefficients of the controlled current source. In the case of a real FET, the situation is more complicated, but the idea that the load impedance and the linearity of the controlled current source primarily establish the FET amplifier's intercept point is still entirely valid.

Optimization of source and load impedances depends largely on the parameter to be optimized. It is quite clear that the load impedance has a strong effect on the *output* intercept point, but the source impedance has little effect. Conversely, the source impedance has a much stronger effect on the *input* intercept point. In cases where the output intercept point is the important quantity, the load impedance can be selected to optimize distortion, while the input can be adjusted to achieve minimal noise (within the constraints of optimizing bias for low distortion) or flat passband. In the latter case, however, there is a clear trade-off between distortion and noise figure.

Some researchers [8.6] have suggested that selecting $\Gamma_L = S_{2,2}^*$ optimizes the output intercept point. This rule has become "conventional wisdom," and is actually fairly accurate in most cases. Similarly, conjugate matching (which often is not much different from selecting $\Gamma_L = S_{2,2}^*$) has been suggested for IM optimization. Reference [8.12] presents a criterion, using an available-gain design approach, for optimizing distortion under constraints of gain and even noise figure. It shows that the optimum load impedance lies on an available-gain circle that is far from the stability circle and usually closest to the Smith chart's real axis.

8.3.2.3 Effect of Constraints on Gain, Match, and Noise Figure

In Section 8.1 we saw that we could design a FET amplifier to have a specific value of transducer gain by first designing it to have that same value of power gain or available gain and then matching one port. If the output port is to be matched, the source impedance (or equivalently the source reflection coefficient) of the amplifier is chosen to achieve the desired available gain; conversely, if a conjugate match at the input port is desired, the load impedance is selected to achieve the desired power gain. The values of source or load impedance that result in a specific value of available or power gain lie on a circle in the Γ_s or Γ_L plane.

Although all the values of Γ_s or Γ_L on one of these circles provide the same gain, they do not provide the same intermodulation intercept point. This fact should be clear in the case of power-gain design, in which the input is matched and Γ_L is selected from the power-gain circle; a wide range of Γ_L values can be used, but there is no guarantee that the optimum value lies along the constant-gain circle. However, Γ_L is not fixed in available-gain design either; because of the requirement that the output port be matched, Γ_L varies as Γ_s is varied. Consequently, neither the available-gain nor the power-gain design processes guarantee that the optimum value of Γ_L can be used.

Nevertheless, intermodulation performance still can be optimized within the constraints of one matched port and a specified value of gain. Because there is considerable variation in intercept point with values of Γ_s or Γ_L that lie along the gain circles, it is important to select the source or load reflection coefficient optimally. This selection can be made by drawing the gain circle and then calculating the amplifier's intercept point at a range of Γ_s or Γ_L values along the circle.

This kind of plot is shown in Figure 8.10, which presents available-gain circles of a conventional MESFET representing gains of 6 to 11 dB. Output intercept values are plotted along the gain circles at various values of Γ_s. It is clear from this plot that the variation in IP values is relatively small, as long as the values of Γ_s are well removed from the stability circle. The optimum values are those closest to the real axis of the Smith chart, especially on the high-impedance side (i.e., $\angle \Gamma_s \cong 0$).

Performing a trade-off between noise figure, gain, and intermodulation in this design process is straightforward. Matching the input of a MESFET amplifier invariably results in noise figure that is much greater than the minimum value; in order to minimize the noise figure, we must be free to vary the amplifier's source impedance, so the available-gain design process must be employed. We first draw the gain circles as in Figure 8.10, and then draw circles of constant noise figure on the same chart (noise figure and noise figure circles are not within the scope of this book; see [8.1]). Finally, we add the values of the third-order intercept point periodically along the gain circles. Having this information, we can determine imme-

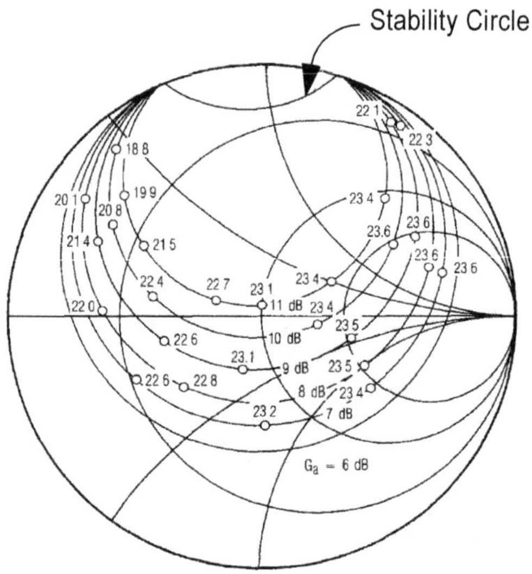

Figure 8.10 Available gain circles plotted on the Γ_s plane, with corresponding values of the IM intercept point.

diately the gain, noise figure, and intermodulation intercept point of the amplifier that results from any proposed value of Γ_s.

It is important to recognize that these results represent only one third-order IM product, the one at $2f_2 - f_1$, and apply strictly to only one device at only one frequency. The situation may change somewhat at different frequencies or in different MESFETs.

8.3.2.4 Effect of Source and Load Terminations at Low-Order Mixing Frequencies

In Chapter 4 we saw that the second-order nonlinear transfer function $H_2(\omega_1, \omega_2)$ often is a part of the third-order transfer function $H_3(\omega_1, \omega_2, \omega_3)$. This situation occurs because mixing between the second-order voltages at $f_2 - f_1$ and $2f_2$ contribute to the nonlinear source currents at $2f_2 - f_1$. Therefore, it seems possible that the termination of the MESFET's input or output at the second-order mixing frequencies might affect the intermodulation performance at the third-order IM frequency. FET amplifiers are normally not designed to have some particular termination at their second-order frequencies; any sensitivity of third-order IM levels to those terminations could partially explain any variation in the intercept point in different amplifiers using the same device.

Measurements of FET amplifiers often show asymmetry in the third-order IM products. Normally, the levels of the mixing products at $2f_2 - f_1$ and $2f_2 - f_1$ are identical, but in some cases, especially power amplifiers, they differ. The difference, often several decibels, makes IM characterization difficult. Carvalho and Pedro [8.13] have attributed this phenomenon to the existence of a reactive part in $H_2(-\omega_1, \omega_2)$, in conjunction with a complex $H_3(\omega_1, \omega_2, \omega_2)$. These requirements imply that the amplifier must have some kind of difference-frequency feedback and a significant reactive nonlinear element in the input. The RF circuitry of FET small-signal amplifiers rarely satisfies either of these requirements, but it is not unusual to have significant difference-frequency feedback in the bias circuits. In bipolar devices, however, the large base-to-emitter capacitive nonlinearities, combined with modest feedback effects, are enough to cause such asymmetry.

8.3.2.5 Effects of Individual Nonlinear Elements

The significance of the individual nonlinear elements in the MESFET's equivalent circuit can be found by replacing each of the nonlinear elements with a linear one, and by recalculating the IM level. These changes affect only the IM performance; they have no effect on such linear parameters as the small-signal gain. Table 8.1 shows the results of one such study. It involves a conventional GaAs MESFET fabricated in a mature technology, and probably is typical of such devices. It may not be applicable to HEMTs or MOSFETs, however.

Table 8.1 Change in IM Output Level Due to Linearization of Certain Elements

Case No.	C_{gs}	g_{ds}	i_d	ΔP_{IM} *(dB)*
1	NL	NL	NL	0.00
2	lin	NL	NL	–0.29
3	NL	lin	NL	–1.34
4	NL	NL	lin	–8.66
5	NL	lin	lin	–7.60
6	lin	NL	lin	–12.22
7	lin	lin	NL	–2.52

In Table 8.1, *lin* means that only the linear part of the element's *C*/*V* or *I*/*V* expansion is used in the calculation; *NL* means that the first three terms of its Taylor-series expansion were used. The nonlinearity of $i_d(v_g)$ is clearly the dominant one in this MESFET; in cases 4, 5, and 6, where $i_d(v_g)$ is linear, the amplifier has significantly lower IM levels than in those in which $i_d(v_g)$ is nonlinear. Most studies of intermodulation in MESFETs have drawn the same conclusion, although in at least one study [8.10], g_{ds} was found to be dominant, and others [8.3] have shown that the normally insignificant nonlinearities can sometimes become significant.

It is important to be cautious with such generalizations, because many devices don't obey the rules. HEMTs, for example, often have a higher and more strongly nonlinear $g_{ds}(v_d)$ than MESFETs. By adjusting the doping profile, it is possible to make a FET's transconductance approximately constant with gate voltage, thus linearizing the device, or even for g_{ds} and g_m nonlinearities to cancel [8.14]. Such forms of linearization inevitably require making the device more strongly nonlinear in some operating region; for example, if $g_m(v_g)$ is flat above pinch-off, yet zero below pinch-off, there must be a region of relatively strong nonlinearity near the pinch-off voltage. In practice, such conditions cause the IM level to be low at low excitation levels, but to increase suddenly when the excitation level exceeds some threshold.

8.3.2.6 Conclusions

It is most important to note that optimized values of source and load impedance can be selected to minimize distortion in a small-signal amplifier. Selection of terminating impedances and dc bias are the designer's main degrees of freedom in minimizing distortion in small-signal amplifiers. The fact that the linearity of the $i_d(v_g)$ characteristic usually dominates the amplifier's IM performance is also important, because that characteristic can be measured relatively easily (Section 2.8.2.2). Thus, a designer can select FETs having good IM performance on the basis of relatively simple screening.

8.3.3 HBTs and BJTs

Both HBTs and homojunction BJTs exhibit low levels of intermodulation distortion as small-signal amplifiers. The reason, as we have noted, is a cancellation phenomenon between collector currents generated by the resistive and reactive parts of the base-to-emitter junction. This phenomenon is evident only above a critical frequency, ω_c; in a conjugate-matched device, ω_c is approximately

$$\omega_c \approx \frac{1}{2R_b C_{be}} \tag{8.49}$$

where R_b is the base resistance and C_{be} is the total base-to-emitter capacitance.

As with FETs, the designer's tools for distortion minimization are (1) device selection, (2) dc bias, and (3) source and load optimization. Because all bipolar devices are fundamentally exponential (Section 2.6.3), one cannot say that any particular device is inherently more linear than another. Apparent differences in linearity are probably caused by such things as feedback from emitter resistance, which may appear to reduce IM at the cost of noise figure and gain, not inherent linearity of the *I*/*V* or *Q*/*V* characteristics of the intrinsic device.

Equation (8.28) shows that the intermodulation intercept point increases rapidly with an increase in collector bias current. Although this expression does not account for IM cancellation, the cancellation terms increase in largely the same manner as collector current, so the conclusion is largely valid. Empirical evidence shows that the reduction of distortion in bipolar devices, with increased bias current, is greater than in FETs. Both the current gain-band width product, f_t, and the maximum available gain, $f_{\max}$, increase with current; therefore, when noise is not a consideration, bipolar devices are operated at their maximum practical current. As with FETs, noise figure is optimum at a particular bias current; however, it is usually less sensitive, at least for small current variations, than in MESFETs or HEMTs. The noise figure of bipolars also is generally less sensitive to source impedance; this characteristic allows the source impedance to be used more freely to optimize gain and, in some cases, IM.

Because of their large values and strong nonlinearities, nonlinear capacitances are more significant in bipolars than in FETs. This fact is of great concern because accurate separation of the diffusion and depletion components of the capacitance is critical to accurate IM analysis. It may be best simply to treat the base-to-emitter capacitance as a single nonlinear element, and to find its Taylor coefficients by other means. The base-to-collector capacitance has a significant effect on IM analysis; fortunately, it is a relatively easy element to characterize. The most important resistive element is, unsurprisingly, the collector current as a function of base voltage, but the other base-to-emitter diodes (which model current gain) can also be significant.

References

[8.1] G. Gonzalez, *Microwave Transistor Amplifiers*, Englewood Cliffs, NJ: Prentice Hall, 1984.

[8.2] M. Medley and J. L. Allen, "Broad-Band GaAs FET Amplifier Design Using Negative-Image Device Models," *IEEE Trans. Microwave Theory Tech.*, Vol. 27, 1979, p. 784.

[8.3] J. A. Garcia, A. M. Sanchez, and J. C. Pedro, "Characterizing the Gate-to-Source Nonlinear Capacitor Role on GaAs FET IMD Performance," *IEEE Trans. Microwave Theory Tech.*, Vol. MTT-46, 1998, p. 2344.

[8.4] J. C. Pedro and J. Perez, "Accurate Simulation of GaAs MESFET's Intermodulation Distortion Using a New Drain-Source Current Model," *IEEE Trans Microwave Theory Tech.*, Vol. 42, 1994, p. 25.

[8.5] S. A. Maas, B. L. Nelson, and D. L. Tait, "Intermodulation in Heterojunction Bipolar Transistors," *IEEE Trans Microwave Theory Tech.*, Vol. 40, 1992, p. 442.

[8.6] C. Y. Ho and D. Burgess, "Practical Design of 2–4 GHz Low Intermodulation Distortion GaAs FET Amplifiers with Flat Gain Response and Low Noise Figure," *Microwave J.*, Vol. 26, Feb. 1983, p. 91.

[8.7] R. A. Minasian, "Intermodulation Distortion Analysis of MESFET Amplifiers Using the Volterra Series Representation," *IEEE Trans. Microwave Theory Tech.*, Vol. MTT-28, 1980, p. 1.

[8.8] R. S. Tucker, "Third-Order Intermodulation Distortion and Gain Compression in GaAs FETs," *IEEE Trans. Microwave Theory Tech.*, Vol. MTT-27, 1979, p. 400.

[8.9] F. N. Sechi, "Design Procedure for High-Efficiency Linear Microwave Power Amplifiers," *IEEE Trans. Microwave Theory Tech.*, Vol. MTT-28, 1980, p. 1157.

[8.10] G. M. Lambrianou and C. S. Aitchison, "Optimization of Third-Order Intermodulation Product and Output Power from an X-Band MESFET Amplifier Using Volterra Series Analysis," *IEEE Trans. Microwave Theory Tech.*, Vol. MTT-33, 1985, p. 1395.

[8.11] J. A. Higgins and R. L. Kuvas, "Analysis and Improvement of Intermodulation Distortion in GaAs Power FETs," *IEEE Trans. Microwave Theory Tech.*, Vol. MTT-28, 1980, p. 9.

[8.12] A. M. Crosmun and S. Maas, "Minimization of Intermodulation Distortion in GaAs MESFET Small-Signal Amplifiers," *IEEE Trans. Microwave Theory Tech.*, Vol. MTT-37, 1989, p. 1411.

[8.13] N. B. Carvalho and J. C. Pedro, "Two-Tone IMD Asymmetry in Microwave Power Amplifiers," *IEEE International Microwave Symposium Digest*, (CD ROM), 2000.

[8.14] P. K. Ikalainen, L. C. Witkowski, and Y. C. Kao, "Low-Noise, Low DC Power Linear FET," *European Microwave Conf. Proc.*, 1992, p. 570.

Chapter 9

Power Amplifiers

Transistor power amplifiers can be realized with either FETs or bipolar devices. For many years, BJTs have been used in high-power amplifiers at frequencies up to a few gigahertz. HBTs are rapidly supplanting BJTs in such applications, as they offer improved gain and efficiency, and require only a positive dc power supply. This is especially important in such portable systems as cellular telephones. Similarly, new MOSFET technologies, such as laterally-diffused MOS (LDMOS) devices, have found application in power amplifiers, especially for fixed base stations. MESFETs and HEMTs are used as power amplifiers in the higher microwave and millimeter-wave frequency ranges.

As with small-signal amplifiers, our fundamental concern is for the single-tone properties of power amplifiers—gain, output power, and impedance. Although linear theory has some use in the design of power amplifiers, linear theory by itself is usually inadequate for determining all the properties of a power amplifier that we need to know; it is necessary to take into account the device's nonlinearities as well. For this reason harmonic-balance techniques are the logical method for analyzing power amplifiers.

9.1 FET AND BIPOLAR DEVICES FOR POWER AMPLIFIERS

9.1.1 Device Structure

Power devices must be designed to survive much greater electrical stresses than small-signal devices. A power device must support high current, survive high drain-to-gate or collector-to-base voltages, endure high temperatures, and dissipate a large amount of heat. Furthermore, like a small-signal device, a power device must provide good gain, linearity, and efficiency, and often must be useful at high frequencies.

A transistor's output-power capability is established primarily by three factors: (1) its breakdown voltage, (2) its maximum current, and (3) its thermal properties. Obtaining high power involves maximizing breakdown voltage and channel current, as well as maintaining good heat-dissipation properties, while avoiding the introduction of excessive resistive or capacitive parasitics. In FETs, channel current can be increased arbitrarily by simply increasing the gate width; however, increasing gate width exacerbates many device parasitics, especially the gate-to-source capacitance and, unless measures are employed to keep it low, the gate resistance. For a FET's gain to remain constant with changes in gate width, the gate resistance must decrease in proportion to the change in gate width. Although it is possible to decrease the gate resistance by modifying the FET's geometry, it usually cannot be reduced in proportion to the increase in gate width, so gain decreases as gate width increases. Consequently, power FETs usually have low gain, compared to small-signal devices. A power FET's gain is often marginal at high frequencies, and its maximum operating frequency decreases with gate width.

The channel current in a FET cannot exceed a value slightly above its I_{dss}. Bipolar devices do not have such a well defined limit, but for reliability, and to maximize their gain-bandwidth products, maximum current and power dissipation must be constrained. Power HBTs are usually biased to approximately 20 to 30 kA/cm^2 of emitter area; peak current is, in most types of amplifiers, approximately twice this value. Bipolar devices have base resistance and base-to-emitter capacitance that are analogous to the gate parasitics of a FET, but the base resistance scales inversely with emitter area, while the gate resistance of FETs generally does not.

In order to allow for adequate current, and to obtain good thermal properties, a power device is designed as a number of *cells*—individual, small devices—connected in parallel. In FETs, the gates of the individual cells may have multiple feed points, or they may be arranged as a number of small sections. This cell structure has a price, however: the cell interconnections introduce additional inductive and capacitive parasitics. In modern devices, the cells are often interconnected by air bridges, which minimize stray capacitance.

The use of multiple cells and multiple short gate segments places difficult requirements upon the manufacturing process. Because even one flaw in one gate segment can ruin the entire device, each power FET must have a perfect gate, sometimes several millimeters wide. Because the difficulty of fabricating flawless gates decreases with increasing gate length, the gates of power FETs usually are longer than those of small-signal devices. Transconductance decreases and capacitance increases with gate length, so a long gate results in low gain.

Because it establishes a fundamental limit to the power capability of the device, the gate-to-drain breakdown voltage of a power device must be much greater than that of a small-signal device. A device designer can maximize a FET's breakdown voltage by optimizing the ohmic contact technology, using a recessed-gate structure, and leaving adequate space between the gate and drain. The gate-to-drain spacing cannot be increased arbitrarily, however, because it increases the drain series resistance. The full drain current passes through that resistance; if the resistance is too great, the resulting I^2R loss can degrade the gain, efficiency, and output power.

Inductance in series with a FET's source or a bipolar's emitter can reduce the gain of a power amplifier, especially in devices having high transconductance. The series inductance, L_s, adds a frequency-independent, resistive component R_{Ls} having an approximate value of $R_{Ls} = g_m L_s / C_\pi$, where C_π represents either the gate-to-source or base-to-emitter capacitance. It also creates an inductive component of value L_s in series with the gate or base. These additional elements reduce the FET's maximum available gain and make impedance matching more difficult. If L_s is fixed, R_{Ls} remains approximately constant with changes in gate width; most of the other resistive parasitics in the input decrease with gate width, however, so source inductance becomes more significant as gate width increases. Furthermore, mutual inductance between bond wires prevents the series inductance from decreasing in proportion to the number of wires. Therefore, source/emitter inductance has a particularly strong effect on the gain in power devices, so a low-inductance ground connection is critical to the performance of a power device. One highly effective way to reduce the inductance is to include *via holes*—metallized holes connecting the source or emitter metallization to the underside of the chip—in the design of the device. Virtually all modern high-power FETs and HBTs use via-hole grounding.

The third factor that limits output power is the chip's ability to dissipate heat. Thermal properties of GaAs devices are especially worrisome because GaAs has poor thermal conductivity, significantly lower than silicon. Furthermore, a power transistor must dissipate quite a lot of heat; the dc-to-RF efficiency of a power amplifier is rarely above 50%, and some types of amplifiers dissipate more power in the absence of RF output than in operation. Consequently a power device must dissipate, in the form of heat, 1.5 to four times its RF output power, and often must do so in the presence of one or more other chips dissipating equal amounts of heat. Because the heat dissipation can be so great, the chip must be designed carefully to minimize its thermal resistance: the cells must not be placed too close together, the chip must be made quite thin (some large chips are thinned to 50 μm or even 25 μm), and often a thick gold layer must be plated onto the chip's underside. The resulting thermal resistance between the channel and the mounting surface may be from one to two C/W (in the case of a large chip) to 50 C/W or more (for single-cell, medium-power devices). The resulting increase in channel temperature may be several tens of degrees Celsius at full power.

Bipolar devices, but not FETs, are subject to thermal instability. Thermal runaway in silicon BJTs is a well-known phenomenon. HBTs exhibit *thermal collapse*, in which the central cells in a large device become hotter than the outer cells, and the resulting decrease in base-to-emitter voltage causes them to draw a disproportionately large base current. The current in the central devices becomes much greater than the outer devices, causing them to become even hotter and their gain to decrease. Meanwhile the outer cells conduct less current, causing the total collector current to decrease and the device gain likewise to decrease. The use of series resistors, called *ballast resistors*, in the base, emitter, or both can reduce this effect. See Section 9.5.8 for further information.

9.1.2 Modeling Power Devices

Large-signal modeling of FETs and bipolar devices is covered in Chapter 2, primarily Sections 2.5 and 2.6. That material is fairly general, but makes the point that models should be designed for their intended use. In this section, we address the special requirements of device models for nonlinear analysis of power amplifiers.

9.1.2.1 Considerations in Power-Device Modeling

Thermal Effects and Self Heating

Power devices get hot. The large amounts of power dissipated in such devices can raise their temperatures to well over 100°C. The characteristics of a device at a high temperature are certain to be very different from its characteristics at room temperature, so temperature must be a parameter of a power-device model.

There are two ways to approach the problem of thermal modeling. The first is to allow for *thermal scaling*, in which the user estimates the temperature of the device and enters it as a model parameter. In this case, the user's temperature estimate must be reasonably accurate. Developing a sufficiently accurate temperature estimate is not difficult for a single device, but in an integrated circuit, which may have many tens or even hundreds of devices, estimating the temperature of each device may not be practical. The second method is to use a *self-heating model*, in which the model monitors its power dissipation and, through the use of an appropriate thermal network model, calculates the temperature of the device. The thermal network model is usually a simple electrical analog of the thermal circuit, consisting of a simple thermal resistance and capacitance. It may also be a fairly complex characterization, which models the nonlinear thermal conductivity of the semiconductor and thermal coupling between cells. For more information on thermal modeling, see Section 2.7.

Traditionally, the self-heating analysis has been built into the device model. It is also possible to make it part of the simulator; then, any thermal-scaling model can be used in a self-heating analysis, models would be simplified considerably, and simulator convergence would be much more robust. This capability has not been implemented in commercial harmonic-balance software, however, as of this writing.

Geometrical Scaling

A power device consists of a number of cells connected in parallel to form a larger device. The equivalent circuits in Chapter 2 should be viewed as models of individual cells. When N cells are connected in parallel, generating an equivalent circuit of the combination merely requires dividing all the resistances of a single-cell model by N and multiplying capacitances and current-source currents by N. In large power devices, however, the interconnection parasitics are rarely negligible, and they prevent such a simple expedient. Furthermore, cells in the center of the device run hotter than those at the outer ends, so some accommodation must be made for temperature differences.

Conversely, it is usually not practical to describe each cell by its own equivalent circuit; since a power device may have tens or hundreds of cells, such a description would be prohibitively complex. Generally, the designer must treat the device as a number of groups of cells, where each group can be modeled by a simple parallel interconnection. These groups are then interconnected, with appropriate parasitics, to form the complete model.

Devices usually scale approximately, but not precisely, in proportion to a FET's total gate width or a bipolar's emitter area. When improved accuracy is needed, nonlinear scaling rules (i.e., something other than a direct or inverse proportionality to N) may be used. A FET's gate resistance is an example of a parameter where such special rules are needed. As a FET is made wider, the gate resistance increases in proportion to width. To prevent the resistance of power devices from becoming too great, the gate is broken into multiple segments. These resistances are in parallel. Thus, the resistance scales as

$$R_g \to R_g \frac{A_W}{A_F} \tag{9.1}$$

where $A_W = W_g/W_{g0}$ is the ratio of the scaled width of each gate finger to the original, and $A_F = N_F/N_{F0}$ is the ratio of the number of gate fingers. When W_g is defined as the total gate width, $A_W = (W_g/N_F)\ /\ (W_{g0}/N_{F0})$; then,

$$R_g \to R_g \frac{A_W}{A_F^2} \tag{9.2}$$

Avalanche Breakdown

Power devices experience avalanche breakdown. Unless avalanche breakdown is included in a device model, an analysis may predict much greater output power and efficiency than the device can really supply. Many kinds of FETs experience "soft" breakdown, which has a more gradual onset than classical avalanche breakdown.

Breakdown is often modeled as a resistive phenomenon; however, significant time delays may be associated with avalanche multiplication.

"Four Quadrant" Operation

Most early FET models were designed to operate only with $V_{ds} > 0$ and $I_d > 0$. In fact, in many power amplifiers, reactive elements in the output matching circuit may cause the drain voltage to drop below zero momentarily. In other kinds of circuits, operation at $V_{ds} < 0$ and $I_d < 0$ may be the norm; for example, FET resistive mixers and switches are biased at $V_{ds} = 0$. Thus, it is clearly necessary for models to be valid at or below zero drain voltage.

One method to create a model that works at $V_{ds} < 0$ is to exploit the symmetry of the FET. Then, when the voltage drops below zero, the model exchanges V_{gs} with V_{gd}. Although seemingly an effective solution, this practice can create a discontinuity at $V_{ds} = 0$, leading to convergence failure in harmonic-balance analysis. In Volterra-series analysis using such models, the derivatives at $V_{ds} = 0$ are indistinct, so large errors result.

Fortunately, the SPICE Gummel-Poon BJT model, the mainstay of BJT and HBT circuit design, is well defined for inverse operation. The same is true of virtually all advanced bipolar models.

Parasitics

Interconnecting a large number of cells introduces parasitic capacitance, inductance, and resistance. These parasitics can be difficult to estimate. Additional parasitics are associated with the long conductors needed to connect the large device to its matching circuits.

When the width of a multicell power device approaches a significant fraction of a wavelength (which, to be more concrete, we might define as approximately 0.1λ), it becomes difficult to guarantee that all cells are driven equally by the source. Electromagnetic simulation may be necessary to design structures that provide uniform drive to all cells.

9.1.2.2 MOSFET Models

For many years, the SPICE MOSFET models have been the dominant ones in the industry. In particular, the Berkeley SPICE level 3 model has been used for most MOSFET design of all kinds. This model has a number of limitations, which have been well documented in the technical literature. Of great concern for harmonic-balance analysis is the existence of multiple discontinuities in the model's functions and their derivatives.

The limitations of the level 3 model have motivated the development of new models. At this writing, more than 50 such models exist. There is certainly no shortage of MOSFET models, but a great shortage of consensus on which model to use. Recently, the University of California at Berkeley was contracted to develop an industry-standard MOSFET model. The result was BSIM, an extremely complex model that underwent multiple revisions. The current (as of late 2002) and probably final revision of that model is BSIM3 version 3.22 [2.15]. Currently, BSIM4 is under development.

BSIM3 has not been received with unqualified admiration. The model's complexity is daunting, and parameter extraction requires considerable expertise. It has not, on the whole, resulted in better circuit simulations than much simpler models (see Section 2.3.12). Because the model exists in so many forms, it has not solved the problem of support for multiple models; instead of multiple models, we have multiple implementations and multiple versions of a single model. That is not much of an improvement. Finally, BSIM3 is a "general-purpose" model; it is not specifically designed for power amplifiers, and, for all its complexity, lacks such features as self-heating that are essential for the design of power-amplifier ICs.

A promising model for power LDMOS devices is the Motorola Electrothermal Model (MET) developed by Curtice et al. [9.1]. Unfortunately, a complete description of this model has not been published in the open literature, but a good description is available in an unnumbered report from Motorola [9.2]. Another such model, whimsically called ELMO,[1] uses BSIM3 as its core [9.3].

For more insight on the dominant MOSFET models and their applicability to power devices, see [2.14] and [2.19].

9.1.2.3 MESFET and HEMT Models

One of the earliest compact MESFET models, from Curtice [9.4], was originally devised for power amplifier use. Since then, dozens of FET and HEMT models have been produced. Many of the simpler models are quite serviceable for straightforward amplifier, mixer, and frequency multiplier calculations, but may not be adequate for accurate power amplifier design. Missing from them are thermal scaling or self-heating, breakdown phenomena, lack of correct operation at $V_{ds} < 0$, and inconsistent capacitance formulation. More modern, advanced models have solved many of these problems. Examples of the latter are those of Parker and Skellern [9.5], Angelov [9.6], and Cojocaru [9.7].

9.1.2.4 BJT and HBT Models

Like the SPICE MOSFET models, the SPICE Gummel-Poon model has been the industry workhorse since the early 1970s. This model is an extension of the model described in Gummel and Poon's original paper [2.17].

The limitations of this model are well known. Among the most serious are the lack of self heating, avalanche breakdown, poor thermal scaling, and poor scaling of transit time with current and temperature. The model is designed for silicon homojunction devices, but it can be modified acceptably, although clumsily, for use with HBTs. As we might expect, this situation has engendered the development of advanced BJT and HBT models. The resulting models are more complex than the SPICE Gummel-Poon model, but not so daunting as BSIM3. Three important advanced BJT models are VBIC [2.18], MEXTRAM [2.19], and HICUM [2.20]. Of these,

---

1. For *E*ricsson *L*DMOS *Mo*del. The author suggested this name as a humorous remark, and somehow it stuck. See how technology develops?

only VBIC is designed specifically as a power-amplifier model, but their designers have claimed that the latter models are perfectly adequate for power amplifier use as well. VBIC, conversely, includes complex substrate modeling, which is unnecessary for discrete devices and for devices fabricated on insulating substrates, such as GaAs and InP HBTs.

Two models developed specifically for HBTs are the Anholt [2.21] and UCSD [2.23] models. The Anholt model, like VBIC, uses much of the SPICE Gummel-Poon model, changing only the parts necessary for modeling HBTs. It also includes self-heating. The UCSD model is a more extensive revision. Neither of these models are specifically designed for power amplifier use, but they do include appropriate features for power-amplifier modeling.

A new model by Angelov [2.22] may also prove useful for HBT power-amplifier design.

9.2 POWER-AMPLIFIER DESIGN

9.2.1 Class-A Amplifiers

Figure 9.1 shows a simplified circuit of a FET power amplifier. We will derive some of the fundamental properties and limitations of power amplifiers from this circuit. As in other chapters, we use a FET simply to keep our examples concrete; the conclusions apply equally to bipolar power amplifiers.

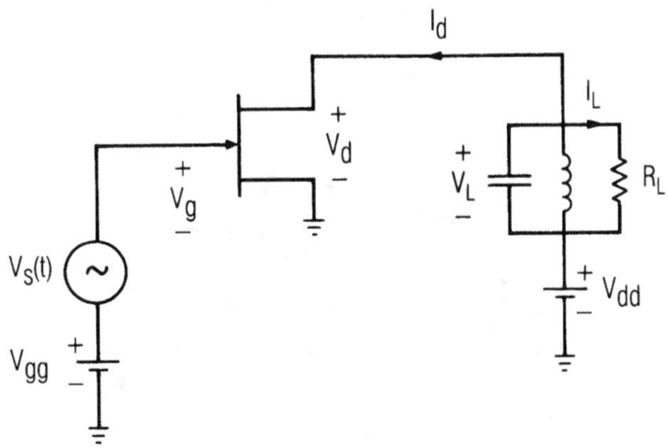

Figure 9.1 Equivalent circuit of an ideal FET power amplifier.

The circuit consists of a FET, excitation and gate-bias sources, a tuned circuit, and a load, R_L. The drain-bias voltage is V_{dd}, and the gate bias is adjusted so that, in the absence of excitation, the dc drain current is I_{dd}. Initially we shall assume that the FET is an ideal transconductance amplifier; that is, it has no resistive or reactive parasitics, so the external and internal voltages are identical ($V_{gs} = V_g$ and $V_{ds} = V_d$). The tuned circuit in the figure is resonant at the excitation frequency.

The application of a sinusoidal excitation $V_s(t)$ to the gate generates an RF component of drain current, $\Delta I_d(t)$. If the tuned circuit is resonant at the RF frequency, that current must pass entirely through R_L. The RF component of the drain voltage, $\Delta V_d(t)$, is equal to the voltage drop across R_L:

$$V_L(t) = \Delta V_d(t) = -\Delta I_d(t) R_L \tag{9.3}$$

Each curve in the FET's drain I/V characteristic, shown in Figure 9.2, represents a range of values of V_d and I_d that can exist when the gate voltage V_g has a specified value; (9.3) expresses an additional constraint on V_d and I_d. Thus, the drain voltage and current must satisfy both (9.3) and the I/V curve for V_g simultaneously; these values of V_d and I_d are found at the point where the I/V curve and (9.3) intersect. Figure 9.2 shows (9.3) plotted on top of the FET's drain I/V curves; when the FET is excited by $V_s(t)$, $V_d(t)$ and $I_d(t)$ must always lie along the straight line. That line is called a *load line*.

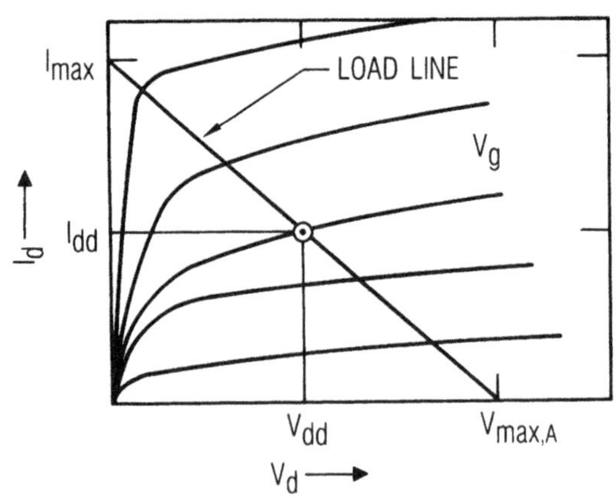

Figure 9.2 Drain I/V characteristics and the load line of the FET in Figure 9.1.

In a power amplifier, we wish to maximize the power delivered to R_L. This power is clearly maximum when both $V_L(t) = \Delta V_d(t)$ and $I_L(t) = -\Delta I_d(t)$ have their maximum excursions. If we recognize that V_d and I_d can not be less than zero, these maximum excursions occur when $|V_L(t)| = V_{dd}$ and $|I_L(t)| = I_{dd}$; the geometry of the load line dictates that these conditions are met when $R_L = V_{max,\,A} / I_{max} = V_{dd} / I_{dd}$. Then, if $V_s(t)$ and V_{gg} are chosen appropriately, the drain voltage varies from zero to $V_{max,\,A} = 2\,V_{dd}$, and the drain current varies from zero to $I_{max} = 2\,I_{dd}$. The $V_d(t)$ and $I_d(t)$ waveforms in this case are shown in Figure 9.3.

The output power P_L under these conditions is

$$P_L = 0.5|V_L(t)||I_L(t)| = 0.5 V_{dd} I_{dd} \tag{9.4}$$

Usually we wish to maximize the output power of a specified transistor. In that case $V_{max,\,A}$ and I_{max} are the device's maximum drain voltage and current, and the maximum output power is

$$\begin{aligned} P_L &= \frac{1}{2}\left(\frac{1}{2}V_{max,\,A}\right)\left(\frac{1}{2}I_{max}\right) \\ &= \frac{1}{8}V_{max,\,A}\,I_{max} \end{aligned} \tag{9.5}$$

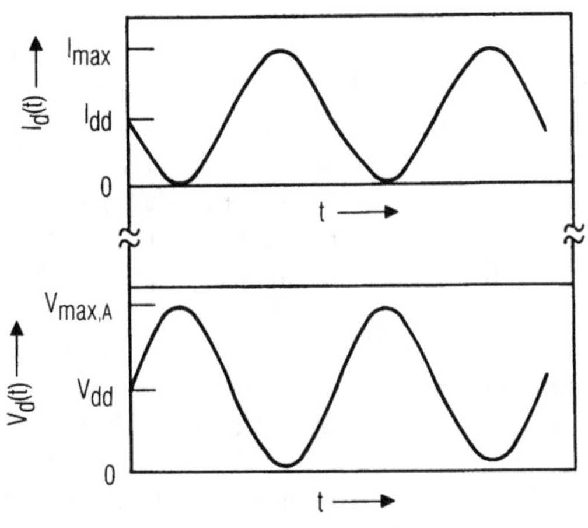

Figure 9.3 Drain voltage and current waveforms in the ideal class-A FET power amplifier; the bias voltages, excitation, and load resistance are chosen optimally, causing both $V_d(t)$ and $I_d(t)$ to vary between zero and their maximum values.

Ideally, the dc current remains constant at I_{dd} at all excitation levels; therefore, the dc power $P_{dc} = V_{dd} I_{dd}$, and the dc-to-RF conversion efficiency is

$$\eta_{dc} = \frac{P_L}{P_{dc}} = \frac{0.5 V_{dd} I_{dd}}{V_{dd} I_{dd}} = 50\% \tag{9.6}$$

An amplifier operated in this manner is called a *class-A* amplifier (although this arcane terminology was originally used to describe vacuum-tube amplifiers, it has been transferred with little modification from vacuum tubes to bipolar transistors and finally to FETs). In theory, the maximum efficiency of such an amplifier is 50%, so the transistor in a class-A amplifier dissipates at least as much power in the form of heat as it delivers to the RF load. In theory, it uses the same dc power at all excitation levels, and that power is divided between output power and heat dissipation in the device. At full output, a class-A amplifier transistor has minimum power dissipation.

Two factors complicate this simple reasoning. First, it is not possible, in practice, to vary the drain voltage and current all the way to the peak of the load line, where $I_d = I_{max}$ and $V_d = 0$, because of the knee in the uppermost I/V curve in Figure 9.2. As a result, $|V_L(t)|$ cannot quite equal V_{dd}, and $|I_L(t)|$ must be less than I_{dd}, so both the output power and efficiency are somewhat lower than the values given by (9.5) and (9.6). Second, the FET is nonlinear, so the $I_d(t)$ waveform is generally not sinusoidal. The tuned circuit constrains $I_L(t)$ to be sinusoidal, however, so the assumption that $I_L(t) = -\Delta I_d(t)$ is not precisely correct, and in fact $|I_L(t)| < |\Delta I_d(t)|$, which further limits output power and efficiency. Nevertheless, because the purpose of this derivation is to illustrate fundamental properties of power amplifiers, we shall continue to assume that $I_d(t)$ can reach I_{max} and that the FET is linear. We will modify these assumptions when we face the problem of accurately designing practical power amplifiers.

Two undesirable characteristics of the class-A amplifier are its relatively low efficiency and its dissipation of a great amount of power even when it is not excited; in fact, class-A amplifiers dissipate more power under quiescent (i.e., unexcited) conditions than when they are operating. Thus, a class-A amplifier must be designed either to dissipate safely its quiescent power, or to be turned off when not in use. Both alternatives are unacceptable in many applications.

9.2.2 Class-B Amplifiers

Many of the disadvantages of class-A operation are circumvented by class-B operation. The gate-bias voltage of an ideal class-B amplifier is set at the turn-on (or *threshold*) voltage, V_t; therefore, the FET's quiescent drain current is zero, so the FET dissipates no power in the absence of excitation. The bias point is thus V_{dd} on the voltage axis of the FET's I/V curves. It is not possible to draw a true load line describing the single-device amplifier in Figure 9.1 when the amplifier is biased to achieve class-B operation because the harmonic components of I_d, which are substantial in a class-B amplifier, do not circulate in R_L; therefore, (9.3) is not valid here.

During the half cycle when $V_s(t)$ is positive, $V_g(t) > V_t$ and the drain conducts; during the other half cycle, $V_g(t) < V_t$ so the drain current is zero. The drain current $I_d(t)$ is therefore a pulse train, and each pulse has the half-cosine shape shown in Figure 9.4. The dc drain current is the average value of the half-cosine waveform; from Fourier analysis, we find that, under full excitation, $I_{dc} = I_{max} / \pi$, and the amplifier's dc power is

$$P_{dc} = V_{dd}\frac{I_{max}}{\pi} \tag{9.7}$$

Because the tuned circuit allows only the fundamental and dc components of drain voltage to exist, the ac part of $V_d(t)$, which is equal to

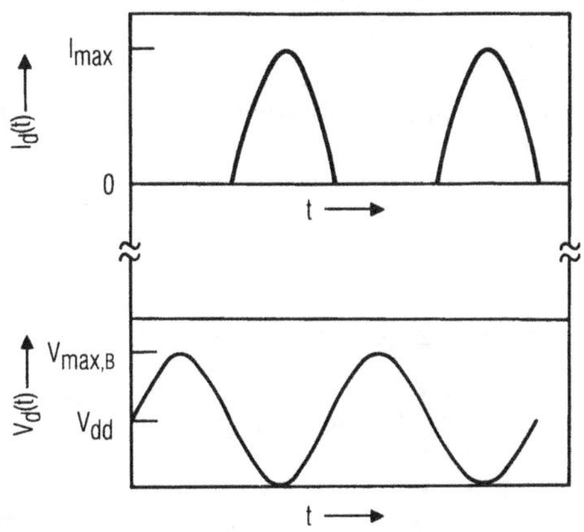

Figure 9.4 Drain voltage and current waveforms in the ideal class-B amplifier. The drain conducts in sinusoidal pulses because the gate is biased at V_t.

$V_L(t)$, is a continuous sinusoid. The tuned circuit also allows only the fundamental component of $I_d(t)$ to pass through R_L. The power delivered to the load is

$$P_L = 0.5 I_1 |V_L(t)| \tag{9.8}$$

where $I_1 = |I_L(t)|$ is the magnitude of the fundamental component of $I_d(t)$. From Figure 9.4, $|V_L(t)| = |\Delta V_d(t)| = |V_{dd}|$, and from Fourier analysis, $I_1 = 0.5\, I_{max}$. Then

$$P_L = \frac{1}{2}\left(\frac{1}{2} I_{max}\right) V_{dd} = \frac{1}{4} I_{max} V_{dd} \tag{9.9}$$

and the dc-to-RF efficiency is

$$\eta_{dc} = \frac{P_L}{P_{dc}} = \frac{\pi}{4} = 78\% \tag{9.10}$$

Theoretically, a class-B amplifier has a maximum efficiency of 78%, much better than the 50% limit of the class-A amplifier. It has achieved this improvement by allowing the channel to conduct during only half the period of the excitation; during the time that the FET is turned off, it dissipates no power. However, the peak value of the class-B amplifier's drain current is twice the peak value of $\Delta I_d(t)$ in the class-A amplifier, so the fundamental-frequency component of the output current is the same in both types of amplifiers.

To find the maximum output power in terms of the device's limitations, we let the maximum drain voltage be $V_{max,\,B}$ and note that $V_{max,\,B} = 2\, V_{dd} = 2\,|V_L(t)|$. Then,

$$\begin{aligned} P_L &= \frac{1}{2}\left(\frac{1}{2} V_{max,\,B}\right)\left(\frac{1}{2} I_{max}\right) \\ &= \frac{1}{8} V_{max,\,B}\, I_{max} \end{aligned} \tag{9.11}$$

which is the same as that of the class-A amplifier if $V_{max,\,A} = V_{max,\,B}$.

To achieve the maximum output power, the load resistance R_L must be such that

$$I_1 R_L = 0.5 I_{\max} R_L = |V_L(t)| = V_{dd} \tag{9.12}$$

so

$$R_L = \frac{2V_{dd}}{I_{\max}} = \frac{V_{\max,B}}{I_{\max}} \tag{9.13}$$

and we see that the load resistance of the class-B amplifier is the same as that of the class-A, again, if $V_{\max,A} = V_{\max,B}$. Furthermore, because the load resistance and the fundamental component of the load current are the same in both amplifiers, the output power must also be the same.

Because the maximum drain voltage is limited by gate-to-drain avalanche breakdown, $V_{\max,A}$ is generally greater than $V_{\max,B}$. In a class-A amplifier, the maximum drain-to-gate voltage occurs when $V_d = V_{\max,A}$ and $V_g = V_t$. Thus, if V_a is the drain-to-gate avalanche breakdown voltage,

$$V_{\max,A} = V_a - |V_t| \tag{9.14}$$

The class-B amplifier is biased at $V_{gg} = V_t$, so the maximum negative excursion of V_g is 2 V_t. Then,

$$V_{\max,B} = V_a - 2|V_t| \tag{9.15}$$

so $V_{\max,B}$ is less than $V_{\max,A}$ by an amount equal to $|V_t|$. Accordingly, the maximum output power of a class-B amplifier is slightly lower than that of a class-A amplifier using the same device.

The difference in maximum output power between class-A and class-B amplifiers is not the most significant one; there is a much greater difference in their gains. The gate voltage of a class-A amplifier varies between zero and V_t; in a class-B amplifier the gate voltage varies between zero and 2 V_t. More input power is required to achieve the class-B amplifier's wider gate-voltage variation, but the output power is nearly the same; thus, class-B amplifiers have inherently lower gain than class-A.

Another disadvantage of the class-B amplifier is that it generates a high level of harmonics in the drain current by switching the FET on and off during each excitation cycle. If the device is terminated in the same impedance at the fundamental and second-harmonic frequencies, the second-harmonic output of an ideal class-B amplifier is only 7.5 dB below the fundamental output (for reasons that will be examined in Section 9.3, the second-harmonic output of a practical amplifier is usually somewhat lower). One solution to the problem of harmonics is to use a "push-pull" configuration, in which the excitation is applied out of phase to the inputs of two class-B amplifiers, and the outputs are combined out of phase. The phase shift of the output combiner must be 180 degrees at the harmonic frequencies as well as the fundamental. This configuration, in conjunction with an appropriate design of the output matching network, can reduce significantly the levels of even harmonics.

In order to avoid the class B amplifier's inherently low gain, and because the turn-off characteristics of power FETs are often very "soft," power FETs are rarely operated in a true class-B mode. So-called class-B microwave amplifiers are usually biased near 0.1 $I_{\max}$, and are actually operated in a mode somewhere between class B and class A. Conversely, class-A amplifiers are often not operated in a classical class-A mode; they are sometimes biased to a minimal current level and driven well into saturation. Both types of operation are called *class AB*, and both represent a compromise between the extremes of either class. Class-AB amplifiers usually have better efficiency than class-A amplifiers and better gain than class-B amplifiers.

Power-added efficiency is used more often than dc-to-RF efficiency as a figure of merit for power amplifiers. It is defined as the ratio of the additional RF power provided by the amplifier to the dc power:

$$\eta_a = \frac{P_L - P_{\text{in}}}{P_{\text{dc}}} \tag{9.16}$$

where P_{in} is the RF input power. One can show easily that

$$\eta_a = \eta_{\text{dc}}\left(1 - \frac{1}{G_p}\right) \tag{9.17}$$

where G_p is the power gain; $G_p = P_L/P_{\text{in}}$. Equation (9.17) implies that the low gain of the class-B amplifier somewhat offsets the advantage of high dc-to-RF efficiency; practical class-B amplifiers usually have, at best, only slightly better power-added efficiencies than class-A amplifiers. Class-B amplifiers are most valuable for amplifying pulsed signals having low duty cycles, where their low average current requirements are a distinct advantage.

9.2.3 Other Modes of Operation

Other classes of operation are possible, but they are used less often in microwave and RF circuits. We mention a few of them here for completeness.

9.2.3.1 Class C

We saw that decreasing the *operating angle* of an amplifier (the fraction of the excitation cycle, expressed in degrees of phase, over which it conducts) increased the efficiency of the amplifier, at the cost of distorting the current waveform. The efficiency comes from the absence of drain (or collector) current over half the excitation cycle. The power dissipated in the device is

$$P_d = \frac{1}{T}\int_T V_d(t) I_d(t)\,dt \tag{9.18}$$

where T is a period of the excitation waveform. If $I_d(t) = 0$ over half the cycle, the integral is zero during this period and power dissipation decreases.

In a class-B amplifier, the distortion was acceptable, even for linear applications, as it (theoretically, at least) generated only even-order products. In many applications, such as FM or phase-modulated communications, linearity is not necessary, so trading off even greater distortion for efficiency is acceptable. By decreasing the operating angle further, so it is less than 180 degrees, efficiency can approach 100% in theory, although rarely greater than 75% in practice. Such amplifiers are called *class-C amplifiers*.

Unfortunately, decreasing the operating angle, while keeping the peak current constant, decreases the magnitude of the fundamental component of current. The peak current must increase, as operating angle decreases, to maintain practical levels of output power. In FETs, increasing the drain current beyond I_{dss} is impossible, but in bipolar devices a high peak collector current is possible. The problem of decreased gain, however, which was evident in class-B amplifiers, is more severe in class C. Thus, class-C amplifiers are practical only at relatively low frequencies, where device gain is high. Class-C bipolar amplifiers are also subject to instability if the collector is not effectively shorted at harmonics of the excitation frequency.

9.2.3.2 Class D

Figure 9.5 illustrates the idea behind the class-D amplifier. In the figure, L_2 is an RF choke and C_2 is a large capacitor, which keeps the voltage at point A equal to V_{cc}. L_1 and C_1 are resonant at the output frequency. The switch, which is implemented by a pair of transistors, creates a square wave of voltage across the resonant circuit. The resonator forces the current in the load to be sinusoidal at the fundamental frequency. Since the transistors conduct only when they are saturated, at all times either the collector/drain voltage or current are zero, so the power dissipation, from (9.18), is also zero and the theoretical efficiency is 100%. In practice, efficiency is limited by parasitic resistances in the devices and their switching time.

Class-D amplifiers are not used frequently. They have been used occasionally in high-power, low frequency applications such as AM and short-wave broadcast transmitters.

9.2.3.3 Class E

Like class D, class E is a switching mode method of amplification, using approximately 50% duty cycle and achieving a theoretical 100% efficiency [9.8]. Unlike class D, however, only a single device is needed.

Figure 9.6 shows a class-E amplifier. The transistor operates as a switch, and L_1 is an RFC, which maintains a constant dc current I_{dc}. When the transistor turns on, the switch is closed, $V_{ce} = 0$, and $I_c = I_{dc}$. When the switch is opened, $I_c = 0$ and a pulse of current is applied to the C_1, C_2, and L_2 combination. The current pulse excites a damped, second-order system with $V_{ce} = 0$ as an initial condition. During the half cycle while $I_c = 0$, a

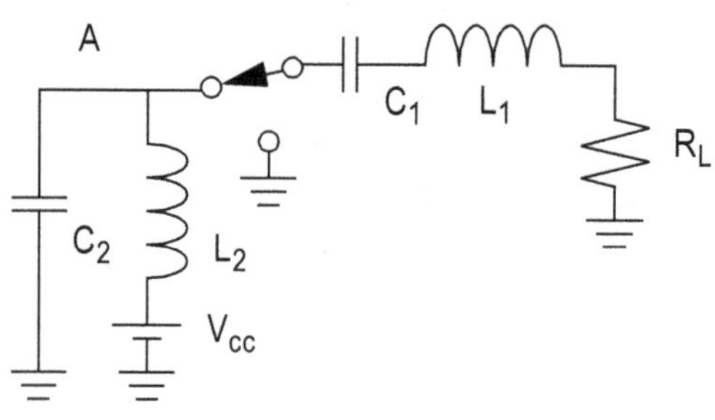

Figure 9.5 Conceptual circuit of a class-D amplifier. C_2 is charged through L_2, an RF choke, providing a constant voltage at point A. The switching operation creates a square wave of voltage at the series LC resonant circuit.

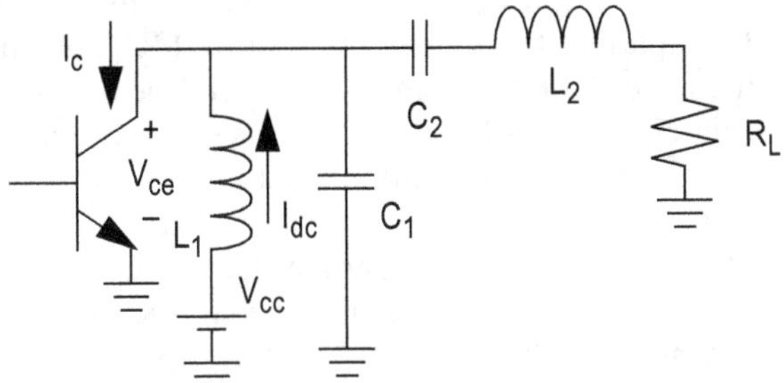

Figure 9.6 The output circuit of a class-E amplifier. L_1 is an RFC, and C_1, C_2, and L_2 provide waveform shaping.

pulse of voltage is generated. At the end of that half cycle, the switch closes again, setting $V_{ce} = 0$. If the circuit is designed properly, the overlap between the current and voltage across the transistor is virtually zero. Additionally, the efficiency does not depend as critically on switching time as in the class-D amplifier.

Class E is a strongly nonlinear mode of amplification and therefore is practical only in applications where high levels of distortion are tolerable. Nevertheless, class-E amplifiers are thoroughly practical for many applications, usually (but not exclusively) in the VHF to UHF frequency ranges.

9.3 DESIGN OF SOLID-STATE POWER AMPLIFIERS

In designing power amplifiers, we follow the general procedure used in the previous three chapters: we employ the usual components of approximation and engineering judgment to generate an initial design, then optimize that design via numerical techniques. The numerical process we use to optimize the power amplifier is harmonic balance. Because a class-A amplifier is ideally a linear component, its initial design can employ linear circuit theory, usually very successfully. This is not the case with the class-B amplifier, however, so we must be more careful with its design.

9.3.1 Approximate Design of Class-A FET Amplifiers

The first step in the design of a power amplifier is to select an appropriate device. Most manufacturers of power devices know their output-power capabilities, and this information is listed prominently on the specification sheets along with other traditionally optimistic claims. Most importantly, the device must be capable of handling the required RF current and voltage, and these quantities are derived from the required power and available dc supply voltage, which we shall calculate presently. Finally, the device's thermal resistance must be low enough so that the channel temperature remains within prescribed limits.

In designing the amplifier, we recognize that an ideal class-A amplifier is, after all, a linear component. Therefore, we should be able to rely fairly heavily on linear-amplifier theory in the initial, approximate design. The fundamental task in designing a class-A amplifier, as in designing a small-signal linear amplifier, is to pick the appropriate source and load impedances and to bias the device appropriately. In a power amplifier, the load impedance must be selected to achieve the desired output power, and the source impedance must provide a conjugate input match. Additionally, we must select a bias point that results in both adequate power and good efficiency.

We use the load-line approach described in Section 9.2 to select the real part of the load admittance. However, in order to select the load conductance properly, we must take into account the limits on the drain voltage and current as explained in Section 9.2. Figure 9.7 shows the terminal *I*/*V* characteristics of a power MESFET (i.e., with I_d expressed as a function of the terminal voltages V_{gs} and V_{ds}, instead of a function of the internal voltages V_g and V_d); we would prefer to have a plot of the *internal* *I*/*V* characteristics, the function $I_d(V_g, V_d)$, which does not include the voltage drops across the drain and source resistances. However, such curves are difficult to generate, and recognizing that this initial design is, after all, approximate, we shall accept a plot of the MESFET's terminal *I*/*V* characteristics as an approximation of the internal ones.

V_{min}, the minimum drain-to-source voltage, is limited to approximately 1.5V by the knee of the *I*/*V* curve at $V_g = 0.6$V; I_{max} is similarly limited. Because of subthreshold conduction (or, if you prefer, the variation in V_t with V_d) and the gate-to-drain avalanche limitation, V_d usually cannot be driven to the point where $I_d = 0$. Thus, there is a finite drain current I_{min} at V_{max}, the maximum value of V_d. V_{dd}, the dc drain-to-source voltage, is selected precisely halfway between V_{max} and V_{min}; I_{dd}, the quiescent dc drain current, is halfway between I_{max} and I_{min}. The gate-bias voltage that establishes this bias point is read directly from the *I*/*V* curves. We draw the load line superimposed on the *I*/*V* curves so that it connects these points; the load conductance is equal to the slope of the load line:

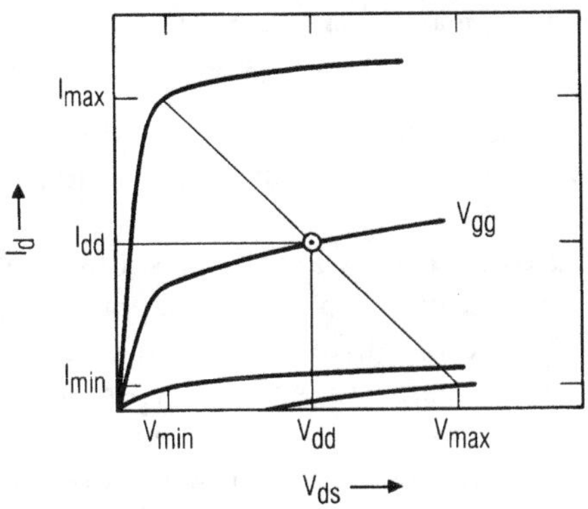

Figure 9.7 Drain *I*/*V* characteristics of a MESFET and the amplifiers load line. Because of the knee of the uppermost *I*/*V* characteristic, the minimum voltage is greater than zero. The optimum bias points are halfway between the maximum and minimum values of both voltage and current.

$$G_L = \frac{V_{max} - V_{min}}{I_{max} - I_{min}} \tag{9.19}$$

When an unpackaged MESFET is biased in its saturation region, the dominant component of its output admittance is the drain-to-source capacitance, C_{ds}. Because we wish to present a real load of conductance G_L to the terminals of the controlled current source I_d, the susceptance of the load must resonate with C_{ds}. Thus, the initial estimate of the load admittance is

$$Y_L = G_L - j\omega C_{ds} \tag{9.20}$$

If a packaged FET is used, determining the load impedance is complicated somewhat by the presence of the package parasitics, but the underlying principle—presenting a real conductance of value G_L to the terminals of the current source—remains the same.

Because the load impedance at the terminals of the current source is real, the ac part of the drain voltage $\Delta V_d(t)$ [which equals the load voltage $V_L(t)$] and the load current $I_L(t) = -\Delta I_d(t)$ are in phase. The output power is their product:

$$P_L = \frac{1}{2}\left[\frac{1}{2}(V_{max} - V_{min})\right]\left[\frac{1}{2}(I_{max} - I_{min})\right] \tag{9.21}$$

Some of this power is dissipated in the drain and source resistances, so for this reason, as well as the others discussed in Section 9.2, (9.21) represents a slightly optimistic estimate.

The drain current of a class-A amplifier should remain constant at the dc value under all excitation levels up to approximately the 1-dB gain compression point. As the amplifier is driven further into saturation, the $I_d(t)$ waveform becomes distorted and its average current may change. Below the compression point, the dc power equals the product of V_{dd} and I_{dd}; above the compression point, the dc power is usually greater, but much of it is converted to RF output power. Therefore, the quiescent dc power can be considered an upper limit to the power dissipated by the device. If the amplifier has high gain and is to be operated only under excitation, the power dissipated by the device is approximately the difference between the output power and dc power. Designating the power dissipation P_d and the thermal resistance of the device from the channel to the mounting surface θ_{jc}, we find the temperature of the channel T_{ch} to be

$$T_{ch} = T_a + P_d\theta_{jc} \tag{9.22}$$

where T_a is the temperature of the mounting surface. Equation (9.22) presupposes that the junction between the device and the mounting surface is thermally perfect; flaws in that junction, such as solder voids, can change the thermal resistance significantly or can cause "hot spots" on the surface of a large chip. In high-reliability circuits, chips are sometimes X-rayed to find such flaws.

The input of the power FET amplifier is designed to be conjugate matched, so we need to know the input impedance of the terminated device. We can estimate this impedance by using small-signal S parameters and (8.5). Finally, the small-signal gain can be found from (8.20), and stability factors and circles can be found from the appropriate equations, (8.2) and (8.7) through (8.10); the load impedance that optimizes output power is usually well within the stable region. Harmonic-balance analyses show that the input impedance varies only slightly with power level up to the point where the FET's gate begins to rectify the input signal significantly. Furthermore, in a well-designed amplifier, a good margin of small-signal stability is usually adequate to guarantee large-signal stability.

When the approximate source and load impedances are known, we can turn to the computer and a harmonic-balance program for optimization. First, we terminate the device with an ideal load, and optimize the output power, bias, and load impedance. Optimum tuning of the output can be determined by plotting the *internal* drain voltage and current, $V_d(t)$ and $I_d(t)$. When their phase difference is precisely 180 degrees and they vary from V_{max} to V_{min} and I_{max} to I_{min}, respectively, the circuit is optimized. The input need not be perfectly matched for this operation.

Once the optimum load impedance is determined, an output matching circuit can be designed, and the FET's large-signal input impedance determined from the harmonic-balance analysis. If all is well, it should not be very different from the value determined from S parameters. Finally, knowing the input impedance, we can design an input matching circuit and connect it to the FET. When the entire combination of input matching, FET, and output matching is simulated, it should be very close to the optimum.

Designing the matching networks is complicated by the low source and load impedances and the need to short-circuit the drain at the harmonics of the excitation frequency. The latter requirement is not very important for class-A amplifiers, because the second and higher harmonic currents are not great, but it is much more important in class-B amplifiers. However, the combination of low impedances and high current densities requires careful consideration. The gate and drain currents in a power amplifier can be on the order of a few amperes, so even very small resistances can cause significant power dissipation. Capacitors—even those used for such prosaic purposes as dc blocking—must have high Qs, and inductors should not be made from narrow microstrips or fine wire (gold ribbon is a good material for inductors that must carry high currents). The topology of the matching circuit can often be selected to minimize the currents in relatively lossy components.

9.3.2 Approximate Design of Class-A Bipolar Amplifiers

Design of bipolar amplifiers—both BJT and HBT—follows the same pattern as with FETs. The device is biased at half its maximum collector current, the output power is found from (9.21), and the load conductance from (9.19). The output susceptance of a bipolar device depends strongly on feedback (collector-to-base capacitance) and the source impedance, so it may be necessary to determine the imaginary part of Y_L empirically.

As with FETs, the input impedance can be estimated by linear analysis. Because of the high base-to-emitter capacitance and pronounced Miller effect in bipolar devices, the impedance of a power bipolar amplifier can be extraordinarily low, and therefore difficult to match. Packaged discrete devices often employ *prematching*: LC elements, within the package, that increase the input impedance to a manageable value. In ICs, similar techniques can be used on-chip to raise the input impedance. In some cases, it is not possible to match a power device in any practical manner; then, power combining with power dividers or similar structures may be necessary.

Increasing the drive level of a class-A bipolar amplifier can increase the rectified current in the base, increasing the collector current. To avoid this phenomenon, bipolar amplifiers can use current-source biasing. When the base is biased by a current source, the collector current is forced to remain approximately constant at all drive levels; as drive is increased, the dc base current source causes the base-to-emitter voltage to decrease, keeping the base and collector currents from increasing.

9.3.3 Approximate Design of Class-B Amplifiers

The design of the class-B amplifier parallels that of the class-A amplifier. The load impedance of an ideal class-B amplifier is the same as that of an ideal class-A amplifier having the same output power, and it is determined identically. In general, however, it is not possible to estimate the linear gain or input impedance of a class-B amplifier from small-signal S parameters; instead we must use nonlinear analysis to determine gain and input impedance.

The maximum value of V_d allowable in a class-B amplifier is somewhat lower than that of a class-A amplifier. In FET amplifiers that are limited by gate-to-drain avalanching, the output power in class-B operation is lower than that in class-A operation. However, if the amplifiers are not limited by avalanche breakdown, the output powers of both classes are nearly identical. Thus, one can use the same procedure to select the load impedance of a class-B amplifier as is used for a class-A amplifier, as long as V_{max} is chosen to have its class-B value.

The dc drain current of a class-A amplifier under full excitation can be estimated as I_{max} / π. The dc power dissipation is

$$P_d = V_{dd}\frac{I_{max}}{\pi} \tag{9.23}$$

This estimate of the dc drain current is reasonable up to the 1-dB compression point; however, because the drain-current waveform distorts with drive level, it is not valid at other levels. Furthermore, because of the inherently low gain of the class-B amplifier, the RF input power may be relatively high, and therefore may contribute significantly to power dissipation. Equation (9.22) is a valid expression for the channel temperature of a class-B amplifier as well as a class-A amplifier.

In an ideal FET class-A amplifier, the gate-to-source voltage V_g varies between V_t and the threshold of gate conduction, approximately 0.5 V. In a class-B amplifier, V_g varies between approximately 2 V_t and the same maximum voltage. Therefore, in order to deliver the same output power, the class-B amplifier requires approximately twice the voltage across the input capacitance as the class-A. Accordingly, one might conclude that the class-B input power must be 6 dB greater, so the gain must be 6 dB lower. This conclusion is troubling, because many microwave power devices do not provide high gain, and 6-dB gain decrease is not tolerable. Fortunately, the situation is not quite that bad, for several reasons: first, even in the ideal case, the differences in voltage is usually slightly less than a factor of two; second, a class-B amplifier is often biased slightly above V_t, in class-AB operation, so it has a small quiescent drain current, which reduces the difference in the variation of V_g even further; and third, because the gate-bias voltage is more negative, the gate-to-source capacitance in a FET or the depletion component of the base-to-emitter capacitance in a bipolar device is lower in class B than in class A. As a result, the difference in gain between class-B and class-A amplifiers using the same FET is usually from 3 to 5 dB, still significant, but less than 6 dB.

A workable approach to the design of a class-B amplifier, either FET or bipolar, is as follows:

1. Determine the load conductance for maximum output power from

$$G_L = \left(\frac{V_{max, B} - V_{min, B}}{I_{max} - I_{min}}\right)^{-1} \tag{9.24}$$

2. Add a shunt reactance and optimize the power and efficiency using harmonic-balance analysis. Bias should allow moderate drain current when there is no excitation. Do not be concerned about input matching at this point.

3. Once the output is designed, calculate the input impedance, defined as $V_i(\omega)/I_i(\omega)$, where $V_i(\omega)$ and $I_i(\omega)$ are the fundamental-frequency components of the input voltage and current, respectively.

4. Design the input and output matching networks.

5. Replace the ideal output impedance with the matching network and verify that the circuit is still optimized.
6. Add the input matching network and check the input VSWR. Minor tweaking may be necessary. If major changes are needed, there is a significant design error.

9.3.4 Push-Pull Class-B Amplifiers

Figure 9.8 shows a push-pull amplifier. It consists of two transistors biased as class-B amplifiers, connected by 180-degree transformers or, for high-frequency circuits, hybrids. (Matching circuits, not shown in the figure, can be included as well.) In this configuration, one transistor conducts when the excitation cycle is positive, and the other when it is negative. Thus, a linear amplifier results.

A push-pull amplifier is a practical implementation of a pair of 180-degree hybrid-coupled components, discussed in Section 5.1.3 and shown in Figure 5.11. We noted in Section 5.2.1 that this configuration rejects even harmonics of the excitation frequency. The class-B amplifier generates only even harmonics, so rejecting these effectively turns a class-B amplifier into a linear amplifier. Additionally, the circuit inherently provides a short-circuit termination to the transistors' collectors at even harmonics; this is the ideal termination for such devices.

9.3.5 Harmonic Terminations

An early paper by Snider [9.9] identified optimum terminations for transistor power amplifiers. He concluded that the optimum output

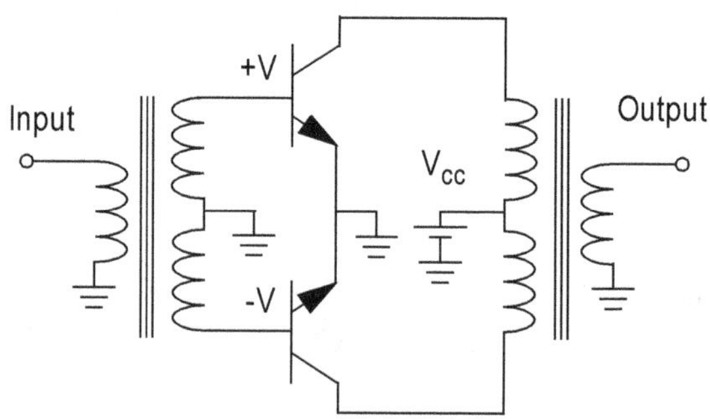

Figure 9.8 A push-pull amplifier consisting of two class-B stages interconnected by 180-degree hybrids.

terminations are short circuit at even harmonics and open circuit at odd harmonics other than, of course, the fundamental. These terminations ideally must be realized at the internal collector-to-emitter or drain-to source junctions. These terminations create a square wave of voltage, resulting in theoretically zero power dissipation in the device.

Achieving such terminations, at high frequencies and with large devices is difficult. The large size of power devices often prevents the placement of a stub close enough to the junction to realize the required short circuits, and device parasitics make an open circuit, at microwave frequencies, almost impossible to achieve. In some lower-frequency amplifiers, however, it may be possible to approximate these terminations at the first few harmonics.

9.3.6 Design Example: HBT Power Amplifier

We wish to design a single-stage HBT amplifier integrated circuit. The amplifier must cover 1.6 to 2.1 GHz, have at least 10-dB gain at full output, and operate at a power-supply voltage of 3.4V. To save chip area and minimize output loss, the output matching circuit is off-chip. A conventional foundry process will be used; the foundry offers InGaP HBT technology having an f_{max} of approximately 50 GHz. The DC bias regulator also will be off-chip, probably a CMOS IC. Thus, the output matching network and DC bias need not be part of the design.

To design the amplifier, we start at the output and work our way toward the input. The process is as follows:

1. Evaluate the device;
2. Determine the device size, bias point, and optimum load;
3. Determine the device input impedance;
4. Synthesize an input matching network;
5. Connect the input network to the device and make sure the combination works properly;
6. Design the output matching network.

We address each step in order.

Evaluate the Device

Before beginning the design, it is essential to perform a sanity check on the device models that the foundry provides. Often, the model's parameters may seem questionable, and such problems should be resolved before the design begins. Most foundries use the SPICE Gummel-Poon (SGP) model to characterize their devices. SGP is an old model and has many limitations for HBT design. More advanced models, such as the UCSD HBT model, would be preferable, but such advanced models are not uniformly implemented in circuit simulators, while SGP is supported on virtually all.

It is a simple matter to evaluate the device. We use two simple test circuits in the nonlinear simulator, one to calculate S parameters and power performance, the other to create *I*/*V* characteristics. We calculate the device's small-signal current gain and maximum available gain to make sure they are reasonably close to the expected f_t and $f_{\max}$ advertised for the device. We find that the current gain, H_{21}, and the maximum available gain, $G_{\max}$, indicate that f_t and $f_{\max}$ are both 45 GHz, in good agreement with the expected ~50 GHz. To make certain that we do not have any potential instability problems, we compute stability circles using conventional, linear analysis. Finally, we sweep the HBT's *I*/*V* characteristic to make certain that the DC part of the model is reasonable, and to determine the base bias current that provides the proper collector current.

Determine the Device Size, Bias Point, and Optimum Load

We begin with the load impedance. The resistive part is given by the well-known relation,

$$R_L = \frac{V_{cc} - V_{\min}}{I_{cc} - I_{\min}} \tag{9.25}$$

and the output power, P_L, is

$$P_L = 0.5(V_{cc} - V_{\min})(I_{cc} - I_{\min}) \tag{9.26}$$

The device's *I*/*V* curves show that $V_{\min} \sim 0.5\text{V}$, and we estimate $I_{\min} \sim 0.05\text{A}$. Noting that $V_{cc} = 3.4$, and experimenting a little with (9.26), we find that $I_{cc} = 0.5\text{A}$. This results in $P_{\text{out}} = 0.65\text{W}$ and $R_L = 6.4\Omega$. These are starting values, which may have to be modified somewhat. According to the foundry, we must limit the current density in the devices, under bias conditions, to 25 kA/cm². The individual cells have areas of 50 μm², so we need a device having 40 cells. The foundry offers a 20-cell device, so we can use two of these in parallel.

In most power amplifier designs, we must provide a shunt inductance to resonate the device's output capacitance. However, from linear analysis, we find that the output capacitance is negligible, so no reactive tuning is needed. The load is purely resistive.

We now use the harmonic-balance simulator to optimize the bias and load impedance, using the evaluation circuit of Figure 9.9. We make no attempt to match the input at this time; we simply increase the excitation until we achieve maximum output power. We adjust the load impedance while monitoring the collector waveforms and adjusting the power. It is a simple matter to do this with the simulator's *tune* mode; numerical optimization is not necessary. The optimum condition is achieved when both the voltage and current minima are near zero, but not clipping; if the resulting output power is not right, we adjust the bias current and load resistance until the correct power is achieved. Note that we allow for an extra fraction of 1 dB in output power, to compensate for losses in the

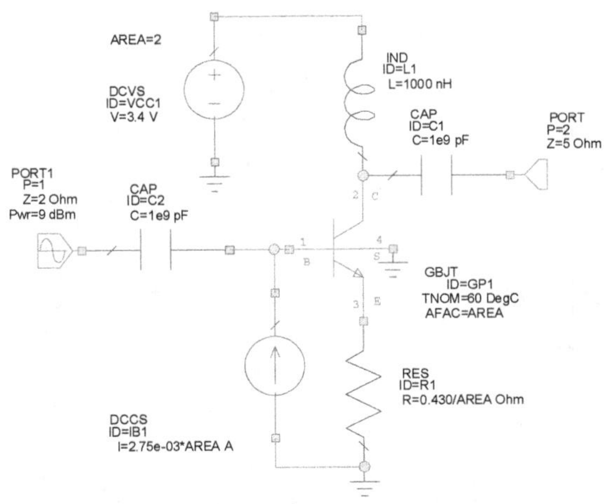

Figure 9.9 Evaluation circuit for the half-watt, 2-GHz HBT power amplifier design. This circuit can be used to determine power performance, the optimum load, and the input impedance. Note that emitter ballast resistance has been included.

output matching network. The final collector current is 0.47A and load resistance is 5.0Ω.

Figure 9.10 shows the collector voltage and current waveforms at bandcenter. These are *internal* quantities; that is, they are the current in, and voltage across, the collector-to-emitter controlled source. The internal voltage and current exhibit a precise 180-degree phase difference, showing that the output reactance is negligible (or if it were not negligible, proper output tuning) and no saturation or clipping.

Determine the Input Impedance

To design an input matching circuit, we must first calculate the large-signal input impedance, $Z_{in}(\omega)$, defined as

$$Z_{in}(\omega) = \frac{V_{in}(\omega)}{I_{in}(\omega)} \tag{9.27}$$

where $V_{in}(\omega)$ and $I_{in}(\omega)$ are the input voltage and current Fourier components, respectively, at the excitation frequency. This is the device-input impedance that should be used for designing a matching circuit.

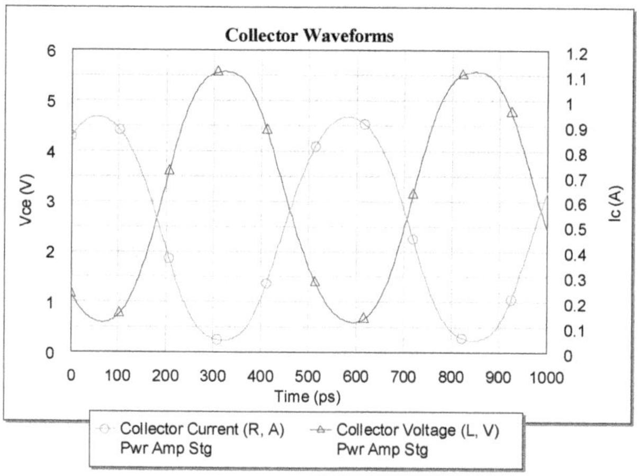

Figure 9.10 The internal collector-to-emitter voltage and current are precisely 180 degrees out of phase and vary from V_{min} to V_{max} and I_{min} to I_{max}, while providing the desired output power with minimal waveform distortion. These conditions show that the output circuit is optimized.

To design a matching circuit, it is helpful to have a lumped-element model of the HBT's input. The dominant input elements in the HBT model are the base resistance and collector-to-emitter capacitance, so it is no surprise that a series RC network models the input impedance quite well. By plotting the input impedance of the HBT and the model on the same graph, we can easily adjust the model to fit the input impedance. Again, optimization could be used for this task, but it is a simple task with the simulator's tuner. The input model consists of 16.9 pF capacitance and 2.2 ohms resistance.

Synthesize the Input Matching Network

Several considerations drive the design of the input network. To eliminate low-frequency gain, it should have a high-pass structure, and it should allow for easy biasing and DC blocking. Because of the high Q of the load, and the need to transform from a very low impedance to 50Ω, the design of the network is not simple.

To meet these requirements, we use a series-L, shunt-C design. We employ a "constant-Q" approach, in which elements are selected by moving along a contour of constant Q on the Smith chart. This is an entirely graphical process, which can be performed with the circuit simulator in the tune mode. Additionally, we use resistive loading to optimize the input match over the relatively wide bandwidth of 1.6 to 2.1 GHz. The loading introduces loss, of course, but the gain of modern HBTs is so great that it is acceptable. It also reduces the sensitivity of input return loss to uncertainties in the device model. Square spiral inductors, characterized by EM simulation, are used in the matching circuit. Because of the high current in these inductors, it is essential to include their losses. The input return loss of the complete circuit is better than 20 dB across the 1.6- to 2.1-GHz band.

Connect the Matching Circuit to the HBT

We now connect the input matching circuit to the HBT and analyze the combination. We find that no further tuning or optimization of the circuit is needed.

Design the Output Matching Circuit

Because of the low load impedance required by the amplifier, an output matching circuit is unavoidably lossy. Most of the loss is generated where the currents are greatest, in the elements closest to the chip. Ideally, these should use capacitive microstrip stubs, but size constraints may dictate the use of chip capacitors instead. In this case, the main problem is a trade-off between capacitor cost and Q. A second problem is the large impedance transformation between ~5Ω at the chip and the invariably 50Ω outside world, which creates a direct trade-off between bandwidth and loss.

A matching circuit consisting of series transmission lines and shunt capacitors represents a good trade-off between loss and size. High-quality RF ceramic chip capacitors must be used. The chip must also be designed to allow the use of multiple bond wires, as even bond-wire loss can be significant. The output matching circuit includes the bias circuit.

Performance

Figure 9.11 shows the final circuit and the calculated performance of the amplifier. It provides a minimum of 27 dBm over the band with 14-dB minimum gain. The output matching circuit is not shown.

9.4 HARMONIC-BALANCE ANALYSIS OF POWER AMPLIFIERS

9.4.1 Single-Tone Analysis

Harmonic-balance analysis of power amplifiers is generally straightforward, especially when only single-tone analysis is required. Still, a few caveats are necessary.

Class-A amplifier analysis usually does not require a large number of harmonics; five or six is usually adequate. Analysis of class-B and other types of switching mode amplifiers may require more harmonics, but rarely is it necessary to use more than 8 or 10 harmonics. The more strongly driven circuits require the largest number of harmonics.

Because power amplifiers have large current components, termination criteria should not be too tight. Excessively stringent termination criteria can result in apparent nonconvergence of the analysis. There is no need, for example, to force current components to converge to error levels below 10^{-6} A when the current itself is on the order of amperes. This is especially true of bipolar amplifiers, where tight termination criteria in terms of current imply even more severe tolerances on the voltage components.

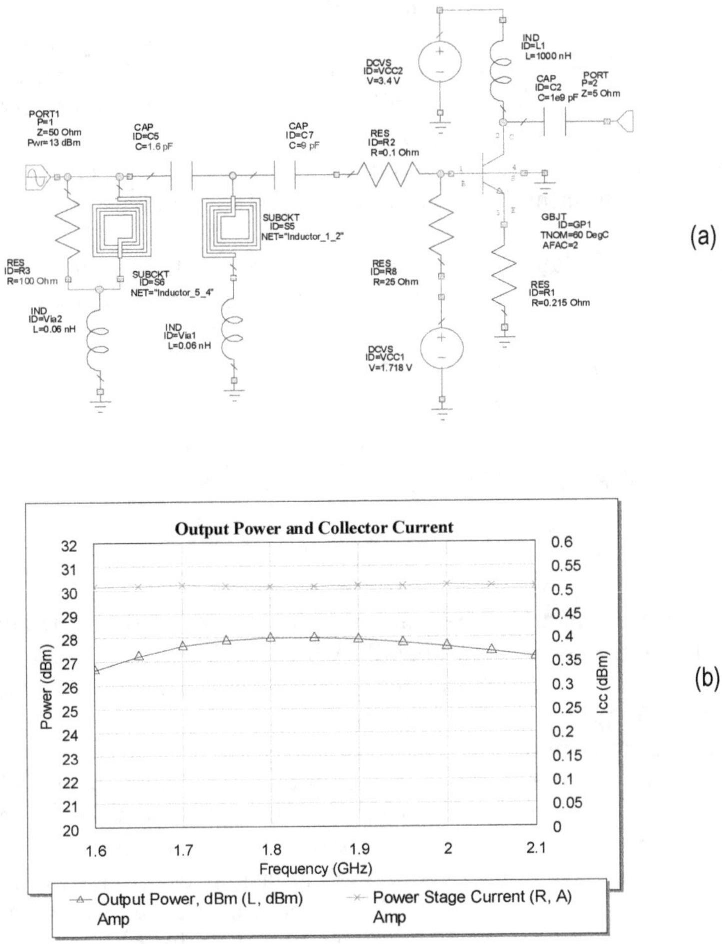

Figure 9.11 (a) The amplifier circuit and (b) performance. The output matching circuit is not included, so the output circuit is idealized.

9.4.2 Multitone Analysis

Multitone problems involve such phenomena as intermodulation distortion in power amplifiers, spectral regrowth, and adjacent-channel interference. These are largely manifestations of intermodulation distortion. Thus, the analysis of intermodulation distortion can be treated as a fundamental requirement of the more complex analyses.

Analysis of intermodulation distortion in strongly driven power amplifiers places severe requirements on a harmonic-balance software. First, the analysis of intermodulation distortion is a multitone problem; that is, it requires at least two noncommensurate excitation frequencies. The resulting number of frequency components is quite large; moreover, it is difficult to determine how many frequency components, and which components, must be retained.

Certain simulators give the user more control over frequency set selection than others. Most use a so-called *diamond truncation*, defined as

$$\omega_{m,n} = m\omega_1 + n\omega_2 \qquad |m| + |n| < Q \tag{9.28}$$

and the user specifies the maximum order of the product. A second method, called a *rectangular* or *box truncation*, is to select

$$\omega_{m,n} = m\omega_1 + n\omega_2 \qquad |m| < M \qquad |n| < N \tag{9.29}$$

where the user specifies the maximum harmonic numbers, M and N. This method is more versatile, especially when combined with the constraint in (9.28). Note that $M = N = Q$ in (9.29) gives approximately double the number of frequency components as (9.28).

Second, intermodulation distortion arises largely from the clipping of the envelope of the composite, multitone waveform near voltage and current minima and maxima. These, in turn, require that the device be well modeled at the voltage and current extremes of its operation, a requirement that goes well beyond the usual modeling task. Most device models are not as accurate in these regions as in the central region; for example, FET capacitances become strongly nonlinear at low drain-to-source voltage, and these must be modeled well to handle clipping phenomena adequately.

Third, harmonic-balance analysis is based on an assumption that all signals are periodic, but multitone waveforms are not. Thus, we must use an imprecise time-to-frequency transform, which may be less accurate than a classical, single-tone fast Fourier transform (FFT). The reduced accuracy greatly affects weak distortion components. Finally, the distorted waveforms contain a mix of large frequency components (the fundamental excitation frequencies and their lower harmonics) and very weak components (the distortion components). The latter are the products of most interest. Therefore, we must use a termination criterion that guarantees convergence of the weak components while not making the criterion too strong for the stronger components. These problems all are solvable, but few simulators have implemented effective solutions to all of them. They are addressed in detail in Sections 3.6 and 3.7.

Three approaches are possible in dealing with modulated signals in power amplifiers. One is to force the modulating function to be deterministic and periodic, allowing the signal to be expressed as a Fourier series. Although it has a large number of frequency components, the excitation has only two independent, noncommensurate basis frequencies, the carrier and the fundamental modulating frequency. This approach is no different, in principle, from the use of single-tone deterministic signals to analyze and test all kinds of circuits that are eventually used with more complex signals.

A second technique is behavioral analysis. In this case, the amplifier's nonlinear AM-to-AM and AM-to-PM responses are determined, and the amplifier is modeled as a simple, memoryless two-port having these responses. A random signal can then be applied to the amplifier, and statistics of the amplified signal (e.g., bit error rate) can be measured. This method is frequently used in system simulators. It is valid as long as the signal is narrowband and memory effects in the amplifier are negligible.

A third method is called *envelope analysis*, which is discussed in Section 3.7. In this approach, the modulated waveform is sampled at a rate based on the envelope frequency, not the carrier frequency. A single-tone harmonic-balance analysis is performed at each sample point, using the carrier frequency as the fundamental. Considerable effort must be expended to characterize the linear circuit properly at the carrier frequency and its harmonics; otherwise, envelope analysis devolves to a complicated form of behavioral analysis.

9.5 PRACTICAL CONSIDERATIONS IN POWER-AMPLIFIER DESIGN

9.5.1 Low Impedance and High Current

Transistor power amplifiers operate at low voltages. Amplifiers used in cellular telephones, for example, typically operate from a dc supply of only 3.4V, which decreases as the battery power is expended. At such low voltages, high currents are necessary to provide even a watt or two of power. To provide such high currents, large devices are needed, and the resulting base-to-emitter or gate-to-collector capacitances may be large. Low-frequency amplifiers using bipolar devices have high voltage gain, which increases the input capacitance because of Miller effect. The input and load impedances of a 1W HBT RF amplifier, in the 1- to 2-GHz range, are on the order of only 1 ohm.

When impedance levels are so low, currents are high, and I^2R losses in circuit elements can be significant. In ICs, the loss is highest in such components as transmission-line segments and spiral inductors, especially those closest to the device in the input matching circuit, where the current is highest. Even capacitors in off-chip output matching circuits can introduce significant losses; microstrip stubs have lower loss, and should be used wherever size allows. Similarly, small parasitics, such as a 0.05 nH via-hole inductance, or the series inductance of a short bond wire, can have a surprisingly great effect on the circuit. For a design to be accurate, all such parasitics must be included in the circuit model.

When impedances are so low that it is impossible to match them in any practical manner, or resistive losses are simply too great, power-combining a number of lower-power amplifiers may be necessary. The simplest combiner is probably a "tree" of power dividers; such structures are practical up to 8- or 16-way combining before imbalance and the effect of imperfect interface VSWR make the approach impractical. Other kinds of power-combining structures, such as radial combiners [9.10], have been employed. The ubiquitous quadrature-coupled amplifier, discussed in Section 5.1.3.2, is arguably a simple type of power combining structure. It is practical for power amplifiers as well as small-signal ones.

9.5.2 Uniform Excitation of Multicell Devices

The large dimensions of power chips introduce several practical difficulties. Because good output power and efficiency requires that all the cells operate at full power, it is important that all cells in the device have equal excitation. If a discrete device is very wide (large chips can be several millimeters in width), the bond wires from the cells near the center of the chip to the microstrip line are often shorter than those from the cells that are close to the chip's ends, causing the outer cells and inner cells to be driven unequally. Even if the chip is no wider than the microstrip to which it is connected and the bond wires have equal lengths, the connections to the outer edges of the microstrip have source impedances different from those close to the center, and these unequal impedances may cause the cells' drive levels to be nonuniform. Even in ICs, driving all the cells equally can be difficult, especially at high frequencies. A symptom of unequal drive in FETs is the existence of dc gate current at power levels well below saturation.

A simple way to avoid the problem of unequal drive is to use a tree structure in the input microstrip. The input microstrip is split into two branches, those are then split in two, and the process continues until there are enough branches to provide a separate microstrip line to each cell. In such structures, it is essential to avoid the possibility of odd-mode oscillation (Section 9.5.3). Of course, power-combining lower-power amplifiers, although an expensive solution, avoids this problem completely.

9.5.3 Odd-Mode Oscillation

Figure 9.12 shows two FETs connected in parallel by a tree of transmission lines, as suggested in Section 9.5.2, to equalize drive to the two devices. It is possible for such transistors to oscillate in an odd mode; that is, where the currents and voltages in the two devices have a phase difference of 180 degrees. In that case, the connection points of the transmission lines are virtual grounds, and each device is effectively terminated in a shorted stub. The impedance of the stub can satisfy oscillation conditions for the device at some frequency. Such oscillation can be puzzling, because the virtual ground isolates the output port, so oscillatory output power may not be evident on a spectrum analyzer.

Figure 9.12 shows the simple solution to the problem: add stabilizing resistors between the devices. In normal oscillation, there is no voltage across either of the resistors.[2] If the devices were oscillating in an odd mode, however, they would each be terminated in half the resistance. As long as the resistors' values are selected appropriately, such oscillation is impossible.

The situation becomes much more complex when many devices are interconnected with a tree feed structure. Then, it is possible to have multiple modes of oscillation, and the simple approach shown in Figure 9.12 is not optimum. In practice, however, we find that the simple resistor network invariably provides adequate stability.

9.5.4 Efficiency and Load Optimization

Optimizing efficiency and output power of a particular device is largely a process of optimizing the dc bias and load. When a class-A amplifier is optimized, the load, as seen from the intrinsic junction, is resistive, and the dynamic load line is a straight line extending from the knee of the uppermost drain I/V curve to the $I = 0$ axis. Ellipticity in the dynamic load

2. We assume that the resistors can be viewed as lumped-element components. However, at high frequencies, the resistors may not be short relative to a wavelength, and thus behave as lossy, open-circuit stubs with the open-circuit point at their centers. The resistors then can dissipate power. It is therefore essential that the stabilizing resistors be kept small.

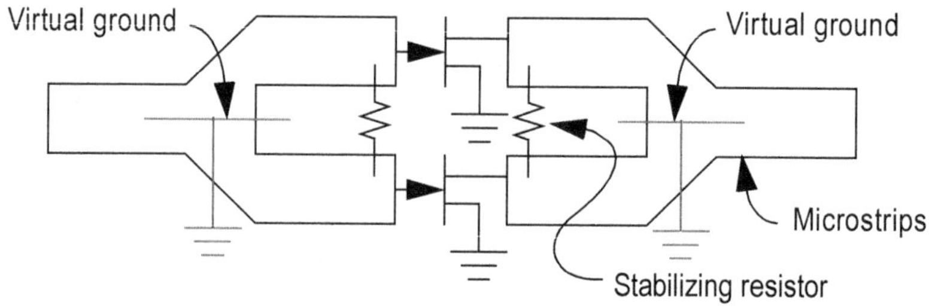

Figure 9.12 When odd-mode oscillation occurs, the connection point between the devices is a virtual ground, effectively terminating each device in a stub.

line indicates the presence of a reactive component at the fundamental or harmonic frequencies. Figure 9.13 shows a nearly ideal set of curves, calculated as part of an 850-MHz, class-A HBT amplifier design.

In assessing whether the amplifier is optimized, it is essential to view the voltage at the intrinsic junction (the voltage and current at the controlled source representing the FET channel or the bipolar collector-to-emitter current), not at the device's terminals. The latter includes the reactive current in parasitic drain-to-source or collector-to-emitter capacitance, and the voltage across the parasitic drain/source or collector/emitter resistances. The current and voltage at the terminals may well have a phase difference other than 180 degrees.

The slope of the load line is the inverse of the load resistance. This must be adjusted, along with the bias voltage and current, to locate the load line properly.

9.5.5 Back-off and Linearity

Linearity in power amplifiers can be specified in several ways. In some cases, a classical intercept point is the most meaningful characterization. More often, however, adjacent-channel power or the two-tone intermodulation level, at full output power, are more meaningful.

Distortion in power amplifiers can arise from two different phenomena. At low levels, in class-A amplifiers, distortion is caused by the same nonlinearities that affect small-signal amplifiers. These phenomena are discussed in Sections 4.2 and 8.3. As the amplifier is driven into saturation, however, distortion caused by clipping the amplitude peaks of the modulated carrier waveform becomes the dominant phenomenon, and the distortion generated in this manner is much greater than the small-signal

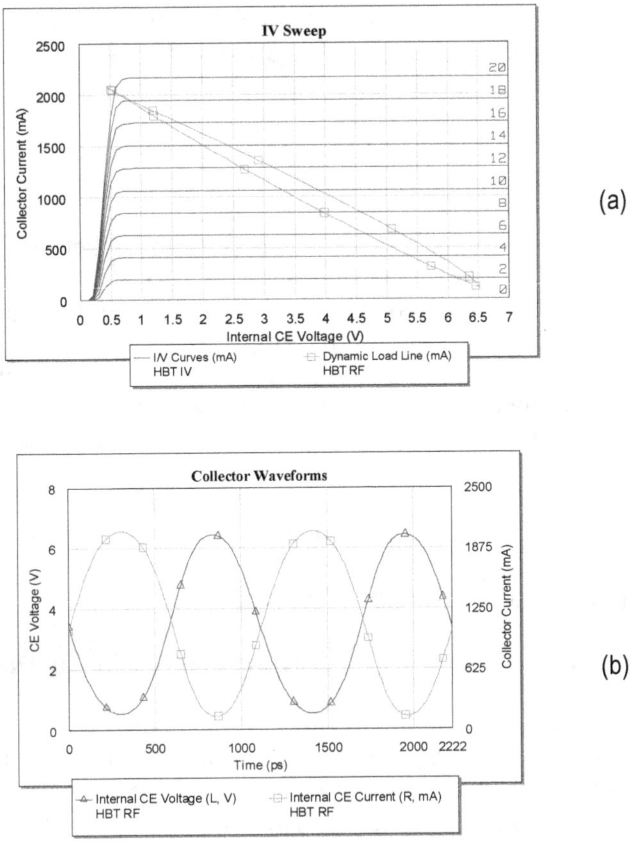

Figure 9.13 Waveforms in an optimized 1.5W class-A amplifier: (a) dynamic load line, superimposed on a set of collector *I*/*V* curves; (b) voltage and current waveforms. Note that we view internal, not external, voltages and currents. Ideally, the current minimum should be zero, but this would introduce distortion that would degrade the amplifier's linearity.

distortion. As a result, distortion increases significantly—much more rapidly than small-signal intermodulation levels—as the amplifier is driven into saturation, and it is not possible to define a meaningful intercept point.

Minimizing clipping distortion requires optimizing the load impedance. It is frequently noted that the load impedance providing optimum output power and efficiency is significantly different from the impedance that minimizes distortion. Figure 9.13 illustrates why this is so: an increase in the load resistance flattens the dynamic load line, reducing the clipping, especially at minimum current. The range of the drain-voltage variation must be reduced to compensate, reducing the output power, but the reduction in distortion is great.

In many applications, the distortion at full output power is unacceptable, so the amplifier's power is reduced to decrease the distortion level. This *back-off* may be several decibels below full sinusoidal output power. Efficiency decreases as output power decreases, so backed-off amplifiers are usually inefficient. For this reason, other linearization schemes are sometimes employed. One is *predistortion*, in which the input signal is distorted in a manner that compensates for the amplifier's distortion [9.11]. Predistortion linearizers can reduce distortion at levels close to the amplifier's full output power; however, they cannot decrease distortion in hard saturation. Predistortion linearizers are not easy to design; the design depends strongly on the type of amplifier with which they are used, and they are notoriously temperature sensitive. Another technique is *feedforward linearization*, a type of distortion cancellation scheme. Feedforward linearization is used extensively in base-station amplifiers [9.12]. It is not applicable to low-power, mobile applications such as cellular telephones.

9.5.6 Voltage Biasing and Current Biasing in Bipolar Devices

The base bias of a bipolar device used in a class-A amplifier can be provided by either a voltage or a current source. When a voltage source is used, increasing the input power results in an increase in the rectified dc base-to-emitter junction current. This allows the collector current to increase with power as well, causing the gain and output power to saturate gradually.

When a current source provides base bias, increased input power cannot increase the dc base current. As the RF drive increases, the dc base voltage decreases to maintain constant current. Consequently, the dc collector current remains constant as the RF input level increases, and the amplifier is driven into saturation. The result is a "hard" saturation characteristic in which the transition from linear to saturated operation occurs rapidly with increased input power.

Other biasing schemes, including certain forms of active bias, can provide saturation characteristics that are somewhere between these two extremes. For example, biasing the base from a voltage source and series resistor allows the dc current to increase with increased drive, but the dc base voltage also decreases. The saturation characteristics then depend on the value of the base resistance.

Because the average dc collector current must increase when the device is driven, class-AB or B amplifiers cannot be current-biased.

9.5.7 Prematching

For many reasons, a designer may prefer to use packaged devices. Packaged devices are usually practical at low frequencies (below a few gigahertz, depending upon power level), but at higher frequencies, the package parasitics may complicate an already difficult matching problem. One solution is for the device manufacturer to place some of the matching components inside the package. These raise the input and optimum load impedances to a level that can be matched easily by the external circuit. Such devices are called *internally matched* or *prematched*; some internally matched power devices are so carefully designed that it is not necessary to use any external matching circuits at all. A disadvantage of internal matching is that the internal circuit has a specific, limited frequency range that cannot be adjusted by the user.

9.5.8 Thermal Considerations

The thermal design of the power amplifier must be performed carefully so that the device's channel or junction temperature is minimized. Temperature affects both the performance and the reliability of an amplifier. In FETs, the transconductance varies approximately in inverse proportion to temperature. Furthermore, a transistor's mean time to failure increases exponentially with temperature; bipolars are subject to thermal runaway, while HBTs exhibit thermal collapse. Although the circuit designer can do little to change the thermal design of the device itself, he can do much to minimize the temperature increase caused by factors under his control. If an unpackaged chip is used, the chip must be soldered effectively to the mounting surface; a packaged device must be screwed or soldered in place, according to its design. The housing in which the device is mounted must provide good heat transfer to its outer surface, and in many cases a separate heat dissipator, sometimes including forced-air cooling, must be used.

Hot areas on the surface of the device can be caused by solder voids under the chip. These "hot spots" can lead to early failure of the device, so devices used in high-reliability applications must be free of them. The most commonly used method of identifying such problems is to perform an infrared scan across the surface of the device. Another popular method is to use a liquid crystal material that can be deposited directly on the chip and observed under polarized light. Finally, devices can be X-rayed to view any voids directly.

High-performance discrete power devices often are fabricated on very thin substrates. ICs sometimes also can be thinned to 50 μm, instead of the standard 100 μm, to decrease thermal resistance. Although effective in reducing temperature, such thin substrates are fragile and difficult to handle. Spreading out the cells in a power device can improve cooling as well, but this expedient may increase interconnection parasitics, size, and cost.

In large, multicell bipolar transistors, the center cells have the highest temperature and therefore the lowest dc base-to-emitter voltage, V_{be}. The base current in the center cells is therefore significantly greater than in the outer ones, as is the collector current and power dissipation. Eventually, as the device heats, the hottest cells carry all the collector current and the cooler cells conduct very little. In silicon BJTs, the current gain increases with temperature, causing thermal runaway and eventual destruction of the device. In HBTs, the current gain decreases with temperature, causing thermal collapse of the I/V characteristic.

To prevent thermal runaway or collapse, *ballast resistors* can be placed in series with the base or emitter [9.13, 9.14]. Emitter ballast can be used in either HBTs or silicon homojunction devices, while base ballast is best used only in HBTs [9.14]. Ballast resistors provide negative dc feedback, which helps to stabilize the device thermally and insure uniform dc bias in the individual cells. Ballast also provides a more uniform input impedance at all the cells, which helps to make the RF drive more uniform as well. Unfortunately, the ballast resistors also introduce considerable loss, which decreases gain, output power, and efficiency.

In some cases, in HBT amplifiers, it is possible to use ballast resistors in series with the base of each cell, and to bypass them with capacitors. This way, RF performance is not degraded. This approach is most practical in low-frequency amplifiers, where the parasitics that are introduced and the increased layout size are tolerable. It is usually not workable with large devices at high frequencies.

Ballast-resistor design requires knowledge of device characteristics, such as dV_{be}/dT, that cannot be estimated a priori, but must be measured from test devices. The value of the ballast resistors is very sensitive to these quantities; the thermal scaling equations of the device model (particularly those of the SPICE Gummel-Poon model) are rarely accurate enough to estimate them.

In any power device, the instantaneous power dissipation varies with time. When the device amplifies a high-frequency sinusoid, the period of the temperature variation is essentially that of the RF waveform. That period is short compared to the thermal time constant (which is on the order of tens of microseconds to, at most, milliseconds), so the temperature does not fluctuate. However, when the device amplifies a modulated waveform having a time-varying amplitude, the power dissipation varies on the time scale of the modulating waveform. The latter may be on the same order as the thermal time constant, so the device temperature varies with time as well. In effect, the amplifier is modulated by a new waveform, the device temperature. Such phenomena are called *memory effects*, and are evident when the thermal time constant has the same order as the envelope period.

References

[9.1] W. Curtice et al., "A New Dynamic Electro-Thermal Nonlinear Model for Silicon RF LDMOS FETs," *IEEE MTT-S International Microwave Symposium Digest*, 1999, p. 419.

[9.2] http://e-www.motorola.com/collateral/MET_LDMOS_MODEL_DOCUMENT_0502.pdf

[9.3] O. Tornblad and C. Blair, "An Electrothermal BSIM3 Model for Large-Signal Operation of RF Power LDMOS Devices," *IEEE MTT-S International Microwave Symposium Digest*, 2002.

[9.4] W. R. Curtice and M. Ettenberg, "A Nonlinear GaAs FET Model for Use in the Design of Output Circuits for Power Amplifiers," *IEEE Trans. Microwave Theory Tech.*, Vol. MTT-33, 1985, p. 1383.

[9.5] A. E. Parker and D. J. Skellern, "A Realistic Large-Signal MESFET Model for SPICE," *IEEE Trans. Microwave Theory Tech.*, Vol. MTT-45, 1997, p. 1563.

[9.6] I. Angelov, "A New Empirical Nonlinear Model for HEMT and MESFET Devices," *IEEE Trans. Microwave Theory Tech.*, Vol. MTT-40, 1992, p. 2258.

[9.7] V. I. Cojocaru and T. J. Brazil, "A Scalable General-Purpose Model for Microwave FETs Including DC/AC Dispersion Effects," *IEEE Trans. Microwave Theory Tech.*, Vol. MTT-45, 1997, p. 2248.

[9.8] N. Sokal and A. D. Sokal, "Class E—A New Class of High-Efficiency Tuned Single-Ended Switching power Amplifiers," *IEEE J. Solid-State Circuits*, June 1975, p. 168.

[9.9] D. Snider, "A Theoretical Analysis and Experimental Confirmation of the Optimally Loaded and Overdriven Power Amplifier," *IEEE Trans Electron Dev.*, Vol. ED-14, 1967, p. 851.

[9.10] I. Stones, J. Goel, and G. Oransky. "An 18 GHz 8-Way Radial Combiner," *IEEE MTT-S International Microwave Symposium Digest*, 1983, p. 163.

[9.11] S. Cripps, *RF Power Amplifiers for Wireless Communication*, Norwood, MA: Artech House, 1999.

[9.12] N. Pothecary, *Feedforward Linear Power Amplifiers*, Norwood, MA: Artech House, 1999.

[9.13] G.-B. Gao et al., "Emitter Ballasting Resistor Design for, and Current Handling Capability of AlGaAs/GaAs Power Heterojunction Bipolar Transistors," *IEEE Trans. Electron Devices*, Vol. 38, 1991, p. 185.

[9.14] W. Liu, J. Sweder, and H.-F. Chau, "The Use of Base Ballasting to Prevent the Collapse of Current Gain in AlGaAs/GaAs Heterojunction Bipolar Transistors," *IEEE Trans. Electron Devices*, Vol. 43, 1996, p. 245.

Chapter 10

Active Frequency Multipliers

Active frequency multipliers have significant advantages over diode multipliers. While passive resistive diode multipliers are broadband and inefficient, and varactors are narrowband and efficient, active multipliers can have broad bandwidths and conversion gain. They can realize efficient multipliers; a high-frequency FET or bipolar multiplier chain usually consumes little dc power and dissipates little heat; this is an important advantage in space systems. In contrast, receiver LO chains using multipliers often require high-power, high-gain driver amplifiers; such amplifiers often are a dominant drain on dc power.

This chapter is primarily concerned with low-power "class-B" multipliers, which operate in a manner analogous to a class-B power amplifier. Such multipliers are very stable and have good gain, efficiency, and output power, and they are usually the most practical form for an active frequency multiplier.

10.1 DESIGN PHILOSOPHY

In the past, frequency multipliers were often used to generate high levels of microwave RF power. High-power multipliers were important components because microwave solid-state power amplifiers did not exist; power amplification at microwave frequencies could be provided only by vacuum devices, which were expensive, unreliable, and had high dc power requirements. Accordingly, a "high-power" multiplier chain (which rarely had an output power greater than a fraction of 1W) consisted of a power amplifier (often a UHF bipolar amplifier) that delivered several watts to a cascade of varactor or SRD multiplier stages.

Today, solid-state power amplification at microwave frequencies is possible, so high-power multiplier chains are rarely needed. Instead, the functions of power amplification and signal generation are usually separated; signals at the required frequencies are generated at relatively low powers, and if greater power is needed, those signals are amplified. Keeping these functions separate has two important advantages: first, it minimizes the consumption of dc power and the generation of heat, and allows the components that dissipate the most heat to be separate from others that may be temperature-sensitive. Second, because the multipliers operate at low power, the levels of spurious signals and harmonics are reduced. Furthermore, many systems do not require high-power signals. The majority of frequency-multiplier chains are used in low-power systems, as mixer local oscillators (LOs), in test instruments, in frequency synthesizers, or as low-power drivers for transmitters. The output power of such chains is usually on the order of 10 dBm.

When used as frequency multipliers, small-signal FETs and bipolar transistors can achieve conversion gain over broad bandwidths while maintaining good dc-to-RF efficiency. In contrast, diode multipliers always exhibit loss. Varactor multipliers are lossy, narrowband components that operate best at moderate to high power levels; resistive (Schottky-diode) multipliers are more broadband but have even greater loss and limited power-handling ability. Thus, the medium- to high-power driver amplifiers required by such multipliers generate RF power that is eventually dissipated in the diodes and matching circuits. It is not unusual for a driver amplifier and diode multiplier chain to require several watts of dc power to generate a few milliwatts of RF power. The dc power advantage of active multipliers is essential for RF and wireless applications.

The low-power, class-B multipliers we examine in this chapter generate low-level RF output power (normally below 10 dBm) at low harmonics, have at least unity gain, and may have high output frequencies, sometimes in the millimeter-wave region. The design approach we shall develop is, of course, applicable to FET or bipolar multipliers operating at higher powers and lower frequencies; designing a high-power multiplier requires only using a larger device and providing greater dc and RF input power. Like the class-B power amplifier discussed in Chapter 9, the gate or base of a frequency-multiplier device is biased near the turn-on point, the channel conducts in pulses having a duty cycle near 50%, and the device's terminals are short-circuited at all unwanted harmonics of the excitation frequency.

10.2 DESIGN OF FET FREQUENCY MULTIPLIERS

Following the pattern of the previous chapters, we begin with an approximate design procedure, use it to generate an initial design, and then optimize that design via harmonic balance. To keep the discussion concrete, we focus on FET frequency multipliers; the extension to bipolars is straightforward. We begin by examining the properties of a large-signal multiplier circuit that uses an ideal FET, then modify the circuit to account for parasitic elements.

10.2.1 Design Theory

Figure 10.1 shows the circuit of a frequency multiplier that uses an ideal FET. The output resonator is tuned to the *n*th harmonic of the excitation frequency, so it short circuits the FET's drain at all other frequencies, especially the excitation frequency, ω_p. We assume throughout this section that a short-circuit termination is optimum; in Section 10.4.1 we examine this assumption further.

For reasons that will be clear shortly, the gate-bias voltage in an efficient FET multiplier must be equal to or less than (more negative than) the threshold voltage, V_t. Thus, the FET's channel conducts only during the positive half of the excitation cycle, and the drain conducts in pulses; the shape of the pulses is approximately a rectified cosine. In this derivation we assume that the drain-current waveform can be modeled as a train of half-cosine pulses, an assumption that is justified by the results of harmonic-balance analyses. The duty cycle of the pulses varies with the dc gate bias V_{gg}; if $V_{gg} = V_t$, the duty cycle is 50%, but if $V_{gg} < V_t$ (the usual

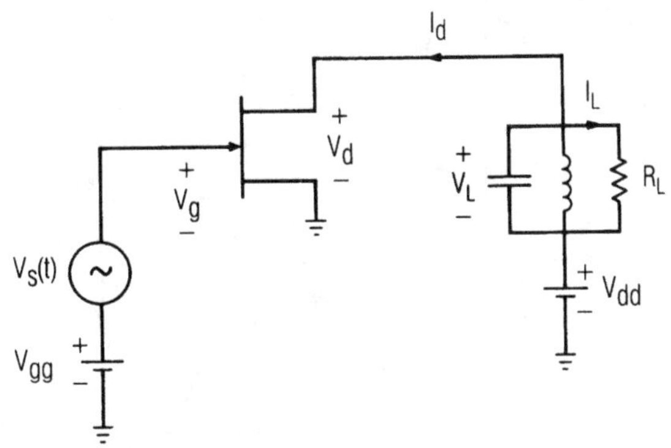

Figure 10.1 Circuit of an ideal FET frequency multiplier.

situation), the FET is turned off over most of the excitation cycle. The duty cycle then is less than 50%.

Figure 10.2 shows the voltage and current waveforms of an ideal FET used as a frequency doubler. Because the output resonator eliminates all voltage components except the one at the nth harmonic, the drain voltage $V_d(t)$ is a sinusoid at radian frequency $n\omega_p$. For best efficiency and output power, the drain voltage must vary between V_{max} and V_{min}; V_{min} is the value of drain voltage at the knee of the drain I/V curve when the gate voltage has its maximum value $V_{g,\,max}$. V_{max} and V_{min} are established by the same considerations as those used in power amplifiers; V_{dd}, the dc drain voltage, is halfway between V_{max} and V_{min}. The gate voltage varies between $V_{g,\,max}$, the peak gate voltage (limited to approximately 0.5V by rectification in the gate/channel Schottky junction), and $2V_{gg} - V_{g,\,max}$, a relatively high reverse voltage. The drain current peaks at the value I_{max}, and the current pulses have the time duration t_0; $t_0 < T/2$, where T is the period of the excitation. If we define $t = 0$ as the point where the current is maximum, the Fourier-series representation of the current has only cosine components:

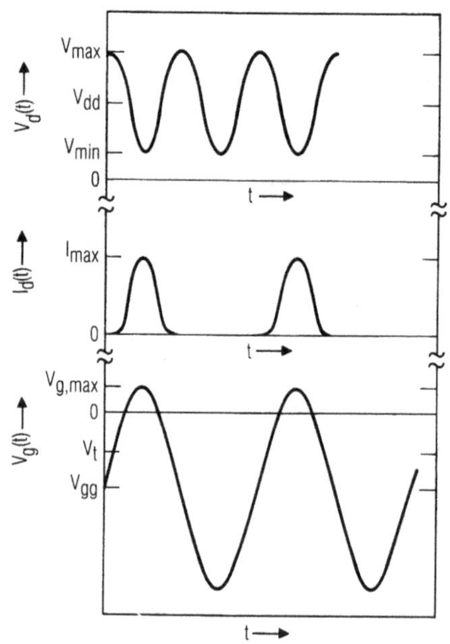

Figure 10.2 Voltage and current waveforms in an ideal FET frequency multiplier.

$$I_d(t) = I_0 + I_1\cos(\omega_p t) + I_2\cos(2\omega_p t) + \dots \tag{10.1}$$

When $n \geq 1$ the coefficients are

$$I_n = I_{\max}\frac{4t_0}{\pi T}\left|\frac{\cos(n\pi t_0/T)}{1-(2n\pi t_0/T)^2}\right| \tag{10.2}$$

and when $n = 0$,

$$I_n = I_{\max}\frac{2t_0}{\pi T} \tag{10.3}$$

When $t_0/T = 0.5/n$, $n \neq 0$, (10.2) is indeterminate. Then, I_n is

$$I_n = I_{\max}\frac{t_0}{T} \tag{10.4}$$

Because the tuned circuit in Figure 10.1 is an open circuit at the output frequency $n\omega_p$, all of the nth-harmonic current I_n circulates in R_L and contributes to output power. Accordingly, in order for the FET multiplier to achieve maximum output power and efficiency, we must maximize I_n. Equation (10.2) shows that we have only one means to do so, adjusting t_0/T. Figure 10.3 shows a plot of $I_n/I_{\max}$ as a function of t_0/T when $n = 2$ through $n = 4$; each of these curves has a clear maximum below $t_0/T = 0.5$. It appears that, in order to achieve the optimum value of I_n, we need only adjust V_{gg} so that $I_d(t)$ has the desired period of conduction, t_0.

Unfortunately, two problems arise in this attempt to achieve a short conduction period. First, we would have to make $V_{gg} \ll V_t$, and this large bias voltage would make the magnitude of the peak reverse voltage, which is approximately 2 V_{gg}, a very great value. Ideally, the peak reverse gate voltage occurs at the minimum drain voltage, but because of phase shifts in practical multipliers and the more rapid variation of $V_d(t)$ than $V_g(t)$, the peak drain-to-gate voltage can be nearly $V_{\max} - 2\,V_{gg}$. If V_{gg} is adjusted to make t_0/T very small, the peak drain-to-gate voltage may be much greater than the breakdown voltage of the FET. The second problem is that, even if the device could survive this high voltage, the input power required to achieve such a wide gate-voltage variation would be so great that the multiplier's conversion gain would be poor. Thus, it is necessary in most cases (especially in a multiplier having an output harmonic greater than the

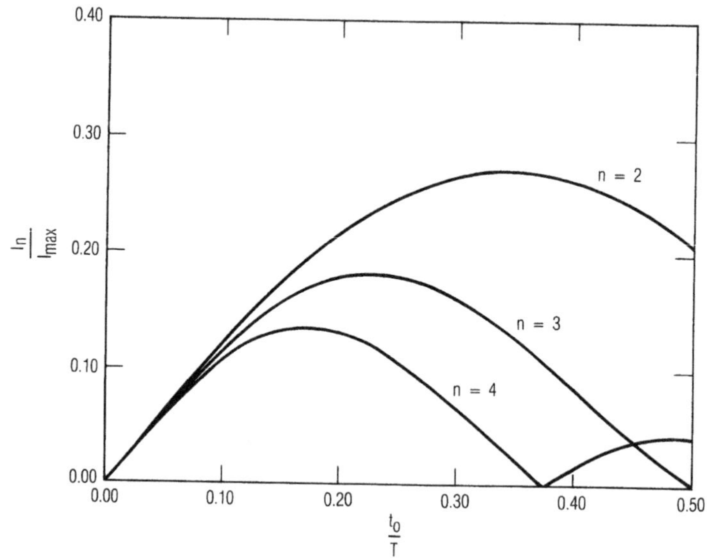

Figure 10.3 Harmonic drain-current components as a function of t_0/T when the drain-current waveform is a half-sinusoidal pulse train.

second) to use a value of t_0/T that is greater than the optimum. Selecting t_0/T to achieve an acceptable trade-off between gain and output power is an important part of the design process.

The maximum reverse gate voltage that the FET can tolerate establishes one limit on t_0/T. If the gate voltage varies between $V_{g,\,\max}$ and the peak reverse voltage $V_{g,\,\min}$, the phase angle, θ_t, over which $V_g(t) > V_t$ is

$$\theta_t = 2\,\mathrm{acos}\left(\frac{2V_t - V_{g,\,\max} - V_{g,\,\min}}{V_{g,\,\max} - V_{g,\,\min}}\right) \tag{10.5}$$

The bias voltage that achieves this value of θ_t is

$$V_{gg} = \frac{V_{g,\,\max} + V_{g,\,\min}}{2} \tag{10.6}$$

θ_t is sometimes called the *conduction angle* of the device. Equation (10.5) shows that a large negative value of $V_{g,\,\min}$ decreases the conduction angle. It also shows that decreasing $V_{g,\,\max}$ has the same effect and, by decreasing the range of $V_g(t)$, reduces input power. However, decreasing $V_{g,\max}$ is not a good way to achieve a low value of t_0/T; decreasing $V_{g,\max}$ decreases $I_{\max}$, and thus reduces output power. Furthermore, when $V_{g,\max}$ is not as great as possible, the multiplier may not be operated in gain saturation, and the output power may vary appreciably with input power (in most practical applications, multipliers are operated in gain saturation in order to stabilize their gains).

The difficulty of achieving a low value of t_0/T can be illustrated by an example. Suppose that a FET has the parameters $V_t = -1.5$V, $V_{g,\min} = -7.0$V, and $V_{g,\max} = 0.5$. Equation (10.5) indicates that $\theta_t = 2.183$ (125 degrees), and therefore $t_0/T = 0.35$. This is the minimum t_0/T that can be achieved with this device if $V_{g,\max}$ is not reduced. Figure 10.3 shows that this value of t_0/T is nearly optimum for a doubler, and is not too far from the optimum value for a tripler (although $I_3 < I_2/2$, so a tripler's output power would be approximately 6 dB below a doubler's). However, $t_0/T = 0.35$ is near the zero of I_4, so a fourth-harmonic multiplier having this value of t_0/T would have very low output power and efficiency. If a fourth-harmonic multiplier were desired, it would be better to increase t_0/T to 0.5, although even then the output power would be at least 16 dB below that of the doubler. It is easy to see from this example why the published research shows that successful FET frequency multipliers have most frequently been doublers.

The current in the load resistance R_L is I_n. For the voltage V_L across the load to vary between $V_{\max}$ and $V_{\min}$,

$$|V_L(t)| = I_n R_L = \frac{V_{\max} - V_{\min}}{2} \tag{10.7}$$

The optimum load resistance is

$$R_L = \frac{V_{\max} - V_{\min}}{2I_n} \tag{10.8}$$

Because I_n is relatively small compared to I_1 in a class-B amplifier, R_L in a multiplier is usually much greater. The output power at the nth harmonic $P_{L,n}$ is

$$P_{L,n} = \frac{1}{2} I_n^2 R_L = \frac{1}{2} I_n \frac{V_{\max} - V_{\min}}{2} \tag{10.9}$$

As with a power amplifier, the dc drain bias voltage is halfway between V_{max} and V_{min}:

$$V_{dd} = \frac{V_{max} + V_{min}}{2} \tag{10.10}$$

The dc power is

$$P_{dc} = V_{dd} I_{dc} = V_{dd} I_0 \tag{10.11}$$

Substituting I_0 from (10.2) into (10.11) gives

$$P_{dc} = \frac{2t_0}{\pi T} I_{max} V_{dd} \tag{10.12}$$

The dc-to-RF efficiency is

$$\eta_{dc} = \frac{P_{L,n}}{P_{dc}} \tag{10.13}$$

Because the harmonic output current in a multiplier is usually much less than the fundamental-frequency current in an amplifier, η_{dc} is usually much lower in a FET multiplier than in a FET amplifier.

We can approximate the RF input power by employing the same set of assumptions that is used to approximate the LO power in a FET mixer (Section 11.1.2). Because the drain is short-circuited at the fundamental frequency, the input of the FET can be modeled as a series connection of $R_s + R_i + R_g$ and $C_{gs}(V_{gg})$. The excitation source must generate an RF voltage having the peak value $V_{g,max} - V_{gg}$ across C_{gs}; if the source is matched, the power available from the source must equal P_{in}:

$$P_{av} = P_{in} = \frac{1}{2}(V_{g,max} - V_{gg})^2 \omega_p^2 C_{gs}^2 (R_g + R_i + R_s) \tag{10.14}$$

The expression shows that the required input power is proportional to ω_p^2, so the required input power increases 6 dB per octave; or, in other terms, the available gain decreases by 6 dB per octave. If the input is well matched across a broad bandwidth, a gain slope inevitably results. A broadband multiplier requires low-frequency input mismatching to have a flat response.

The transducer conversion gain is simply $P_{L,n}/P_{av}$. The power-added efficiency of a FET multiplier is

$$\eta_a = \frac{P_{L,n} - P_{in}}{P_{dc}} \tag{10.15}$$

or

$$\eta_a = \eta_{dc}\left(1 - \frac{1}{G_p}\right) \tag{10.16}$$

where G_p is the power gain of the multiplier ($G_p = P_{L,n}/P_{in}$).

A final consideration is the trade-off between V_{max} and $V_{g,min}$. Neither of these parameters can be established independently in any FET; V_{max} and $V_{g,min}$ must be chosen so that the drain-to-gate avalanche voltage is not exceeded. The maximum drain-to-gate voltage is approximately $V_{max} - V_{g,min}$, so we have the limitation

$$V_{max} - V_{g,min} < V_a \tag{10.17}$$

where V_a is the drain-to-gate avalanche voltage. Thus, we can increase $|V_{g,min}|$ by decreasing V_{max}. Decreasing V_{max} decreases the optimum value of R_L, not an undesirable result in view of the fact that R_L is often too great to be realized in practice. It is usually not possible to decrease V_{min} when V_{max} is reduced, however, so from (10.9) we see that decreasing V_{max} reduces $P_{L,n}$. The design process is illustrated by the following example.

10.2.2 Design Example: A Simple FET Multiplier

We wish to design a 10 to 20-GHz MESFET frequency doubler. The FET has the following parameters:

$V_a = 12.0\text{V}$	$V_t = -2.0\text{V}$
$L_s = 0.005$ nH	$C_{ds} = 0.10$ pF
$C_{gs} = 0.25$ pF (at $V_{gs} = V_{gg}$)	$C_{gd} = 0.08$ pF
$R_s = 2.0\Omega$	$R_i = 2.0\Omega$
$R_g = 1.0\Omega$	$R_d = 2.0\Omega$
$I_{dss} = 80$ mA (at $V_{ds} = 3.0\text{V}$)	

We use a Curtice model [9.4] to describe the FET. From the I/V curves we estimate $I_{max} = 80$ mA and $V_{min} = 1.0$V. $V_{g,min}$ and V_{max} must obey the constraint expressed by (10.17), so we choose $V_{g,min} = -7.0$V and $V_{max} = 5.0$V; we also choose $V_{g,max} = 0.2$V, slightly below the lowest value that allows rectification. Equations (10.10) and (10.6), respectively, give $V_{dd} = 3.0$V (a convenient value) and $V_{gg} = -3.4$V; substituting these values into (10.5) gives $\theta_t = 2.36$ (135 degrees), or $t_0/T = 0.37$. Figure 10.3 shows that this value of t_0/T is close to the optimum for a doubler, and that $I_2 = 0.27\, I_{max}$, or 21.6 mA.

Equation (10.14) can now be used to find the input power; (10.16) implies that $P_{in} = 8.0$ mW, or 9.0 dBm. If the input is conjugate matched, the input power is equal to the power available from the excitation source. The output power $P_{L,2}$ is given by (10.9); it is 21.6 mW or 13.3 dBm, making the conversion gain 4.3 dB. The dc drain current from (10.11) and (10.2) is 19.9 mA, which gives 59.7 mW dc power and 36% dc-to-RF efficiency. Finally, R_L is found from (10.8) to be 92.6Ω, and in order to resonate the output capacitance, C_{ds}, there must be a susceptance in parallel with R_L of $-2\omega_p C_{ds}$, or -12.5 mS. Converting this load to an impedance gives $Z_L(2\omega_p) = 39.4 + j45.8\Omega$. The estimated input impedance is simply $R_s + R_i + R_g + 1 / j\omega_p C_{gs}(V_{gg})$, or $5 - j63\Omega$.

The rest of the design involves realizing the input and output matching networks. The output matching network is relatively easy to design; it consists of a filter, to short-circuit the drain at the fundamental frequency and unwanted harmonics, followed by matching elements. A half-wave filter is ideal for the output; it consists of a cascade of alternating high- and low-impedance transmission-line sections, each $\lambda/4$ long at ω_p; these sections are $\lambda/2$ long at $2\omega_p$ and $3\lambda/4$ long at $3\omega_p$. Thus, the frequencies of maximum rejection occur at ω_p and $3\omega_p$, but the filter has no rejection at the output frequency $2\omega_p$. From Figure 10.3 we see that $I_4 \sim 0$ at $t_0/T = 0.37$, so the fourth-harmonic output should be very low. The gate's short circuit at the second-harmonic frequency is less critical; a shorted stub $\lambda/4$ long at ω_p is adequate to provide the termination. This stub has no effect on the excitation, but is $\lambda/2$ long at $2\omega_p$ and thus short-circuits the gate at this frequency. Figure 10.4 shows the circuit of the multiplier.

A quarter-wavelength open-circuit stub could also be used to short-circuit the drain at the fundamental frequency. This would also short-circuit the third harmonic and require less space, an advantage in an IC. Because of the limited Q of a microstrip stub, however, rejection would not be as good, and bandwidth would be less.

The validity of this design was tested by a harmonic-balance calculation. To insure validity, we compare the performance when the approximate design and the harmonic-balance calculation have the same gate-voltage variation. By normalizing the gate voltage instead of the available input power, we can separate the effects of the input- and output-circuit designs more easily. Accordingly, the input power and bias were adjusted in the harmonic-balance calculation until the estimated peak-to-peak voltage of 7.2V across C_{gs} was achieved.

The multiplier's operating parameters found by the harmonic-balance analysis are compared in Table 10.1 to those from the approximate design. The two sets of data agree reasonably well, although the output power calculated by harmonic balance is 1.6 dB lower than the estimated output power. The main reason for the difference is that the current pulse is not precisely a half-sinusoid; the pulse is somewhat distorted, so that its shape appears to be something between a cosine and a triangle. This distortion reduces the magnitude of I_2, and thus decreases the output power at $2\omega_p$. The second-harmonic peak-to-peak voltage across R_L of 3.5V, instead of 4.0V, is evidence that I_2 is lower than intended; this difference in voltage alone accounts for 1.2 dB of the difference in output power. The calculated value of t_0/T, 0.44, is slightly greater than the estimated value, 0.37; this difference further reduces I_2.

It is also possible that the load impedance is not precisely optimum; certainly R_L could be increased to achieve the full peak-to-peak output voltage of 4.0V; this change would increase the output power approximately 0.6 dB. A plot of the output current and voltage shows that C_{ds} is effectively resonated, because the peak of the drain current pulse $I_d(t)$ occurs almost exactly at the minimum of the drain voltage, $V_d(t)$; this condition implies that the impedance presented to the terminals of the controlled source I_d is entirely real.

It is a worthwhile exercise for the reader to compare this design process to those of the varactor and resistive multipliers in Chapter 7. Although the

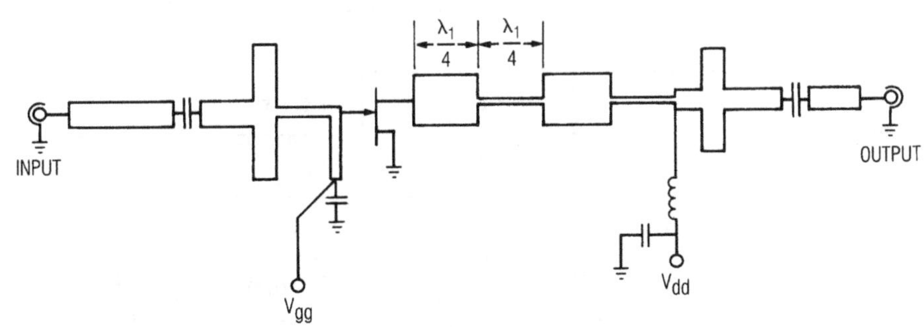

Figure 10.4 Circuit of the FET frequency doubler designed in the example.

latter are no more difficult to implement, the approximate design of the FET frequency multiplier is much "cleaner" than those of the diode multipliers: the design is more intuitive, the approximations are not as severe, less empiricism is required, and there is a better initial agreement between the approximate and harmonic-balance analyses. Indeed, after performing harmonic-balance analyses of a FET multiplier and a varactor multiplier, we can see immediately that the performance of the FET multiplier is far less sensitive to virtually every circuit parameter than is the varactor. This property—that the FET multiplier is more "designable"—is difficult to

Table 10.1 10- to 20-GHz FET Frequency Doubler Design

Parameter	*Approximate Analysis*	*Harmonic-Balance Analysis*
$V_g(t)$ Range	–7.0 to –0.2V	–6.9 to –0.1V
$V_d(t)$ Range	1.0 to 5.0V	1.1 to 4.6V
$I_{\max}$	80.0 mA	91.0 mA
I_{dc}	19.9 mA	22.7 mA
V_{gg}	–3.4V	–3.3V
V_{dd}	3.0V	3.0V
t_0 / T	0.37	0.44
Z_L	$39 + j46$	$39 + j46$ (not optimized)
Z_{in}	$5 - j63$	$3.4 - j56$
P_{av}	9.0 dBm	11.0 dBm
$P_{L,2}$	13.3 dBm	11.7
G_t	4.3 dB	0.7 dB
P_{dc}	59.7 mW	68.1 mW

quantify but is nevertheless one of the multiplier's most important characteristics.

10.2.3 Design Example: A Broadband Frequency Multiplier

Now that we know what we're doing, we can illustrate how the design process can be simplified, as we did in Chapter 9 with power amplifier circuits.

Here we consider the design of a frequency doubler having an input frequency of 8.6 to 9.8 GHz. We will not go through the entire process of making the initial, approximate design, as we did in the first example, but instead we will start with an ideal circuit, use it to determine the optimum load and source impedances, and finally synthesize them on the computer.

The ideal circuit is shown in Figure 10.5. The circuit is ideal in the sense that it uses elemental source and load networks and ideal bias sources, but the FET is described by its complete Curtice model. It is a 0.25-μm MESFET, having an I_{dss} of 80 mA and threshold voltage of –1.5V. The gate-to-source capacitance is 0.35 pF at zero gate bias. An open-circuit, quarter-wavelength stub short-circuits the drain at the fundamental frequency and odd harmonics.

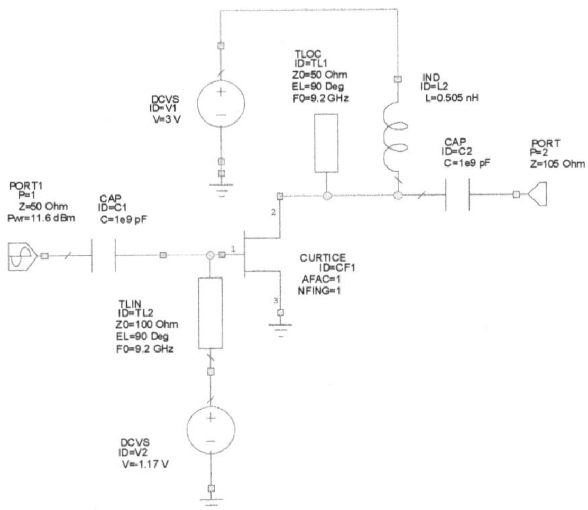

Figure 10.5 Ideal circuit of the frequency multiplier in the example of Section 10.2.3. This circuit is used to obtain basic operating parameters—input and load impedances, input power, and bias conditions—which will be used in the complete design.

The ideal circuit is optimized by adjusting the input power, gate bias, and load impedance. The latter consists of the port resistance, which is freely adjustable, and the drain-bias inductance, which provides the reactive part of the load. The circuit simulator's *tune* mode is adequate for this purpose; there is no need for numerical optimization. When the circuit is optimized, we find that $P_{L,2}$ = 12.1 dBm, $V_{gg} = -1.17$, $R_L = 105\Omega$, L_L = 0.505 nH, and $Z_{in} = 9.7 - j58.9$.

Figure 10.6 shows the voltage and current waveforms in the device. The pulse of drain current coincides with the gate-to-source voltage (some delay is evident) and the second harmonic in the drain voltage is also clearly evident.

Figure 10.7(a) shows the final circuit. Knowing the optimum load impedance, we can approximate the 0.5-nH drain inductance by a transmission line. The drain-bias line serves nicely for this purpose. The length of the bias line is adjusted to optimize the output power. To realize the 105Ω resistive load, we use a quarter-wave transformer, 72Ω, approximately 30 μm wide. The quarter-wave stub is also realized as a real microstrip line. Finally, when the output is optimized, the input matching circuit is designed and optimized. To achieve flat frequency response, we tune the circuit empirically on the computer, mismatching the input at the low end of the band. The circuit does not include discontinuities such as tee and step junctions; these should be included before the circuit is fabricated.

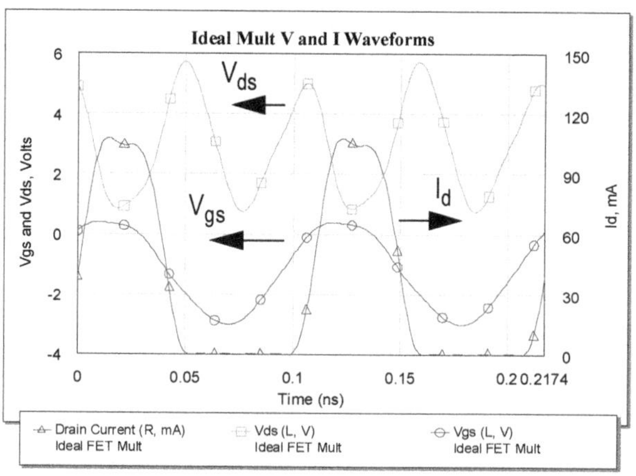

Figure 10.6 Voltage and current waveforms in the ideal multiplier at bandcenter, 9.2 GHz.

Figure 10.7(b) shows the output power at the fundamental frequency and at the second, third, and fourth harmonics. The fourth harmonic is actually greater than the fundamental and third harmonics because the stub does not attenuate it.

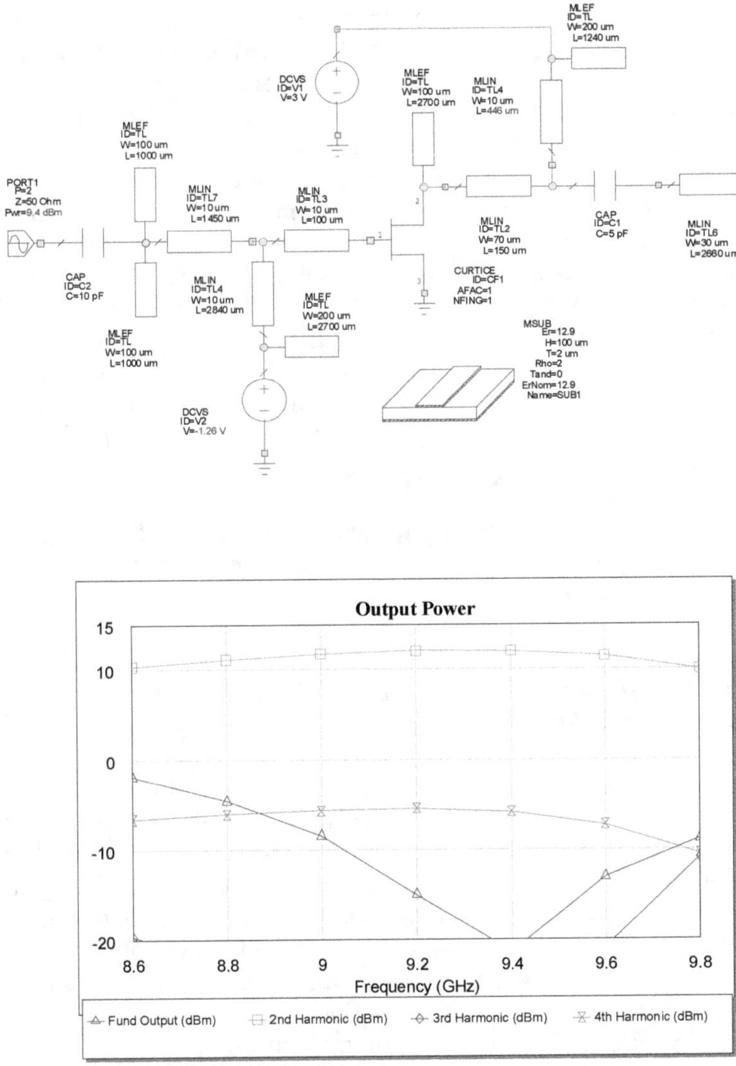

Figure 10.7 (a) Final circuit of the 8.6 to 9.8 GHz multiplier, and (b) output levels at the first four harmonics.

10.2.4 Bipolar Frequency Multipliers

The theory of bipolar multipliers is essentially the same as that of FET multipliers. A few notes on the differences, however, are in order.

Unlike FETs, whose channel currents are limited to a little over I_{dss}, bipolar devices do not have such a strict limit. Silicon BJTs experience high-level injection effects, which tend to limit the peak current and reduce transconductance at high collector current. HBTs do not exhibit such effects but still should be limited in peak current for reliability reasons. The peak current depends strongly on the HBT technology, but in most devices, the peak current should be kept below approximately 40 to 50 kA/cm^2 of emitter area. Of course, collector current must also be restricted to minimize heat dissipation.

Bipolar devices have a large, strongly nonlinear base-to-emitter capacitance. Because of that capacitance, bipolar multipliers are susceptible to modes of oscillation that are not unlike those of p^+n junction varactor multipliers. As with varactors, the best (and simplest) way to avoid such instability is to short-circuit the base and drain at all unwanted harmonics. Similarly, the designer must make certain that active dc bias supplies do not exhibit negative resistance or couple the collector to the base at low frequencies.

Because multiplying devices are turned off under quiescent conditions, bipolar multipliers should not be current-biased; they must be biased from a voltage source, perhaps with a series resistance.

10.3 HARMONIC-BALANCE ANALYSIS OF ACTIVE FREQUENCY MULTIPLIERS

The harmonic-balance analysis of frequency multipliers is virtually identical to the analysis of class-B power amplifiers described in Section 9.4. The main differences between the power-amplifier and multiplier analyses are that frequency multipliers usually employ small-signal or medium-power devices, not high-power transistors, and the output is taken at a harmonic of the excitation frequency, not at the fundamental frequency.

Because of the stronger nonlinearities, convergence may be slower in frequency multipliers than in other nonlinear active circuits. In bipolar multipliers, especially, the strongly pumped base-to-emitter capacitance can be unstable; in that case, convergence failure is virtually assured. The best way to avoid convergence failure caused by an unstable circuit is to make certain that the circuit is stable, by effectively short-circuiting unwanted harmonics at the collector and base.

10.4 PRACTICAL CONSIDERATIONS

10.4.1 Effect of Gate and Drain Terminations at Unwanted Harmonics

The previous sections were based on the hypothesis that the optimum gate and drain terminations at unwanted harmonics are both short circuits. Empirical evidence shows that short-circuit terminations at these frequencies results in good performance, but we have accepted on faith, not proven, that short-circuit terminations are in some sense optimum. In fact, there have been reports that the use of other terminations, especially an open-circuit drain termination at the fundamental frequency, has advantages over a short circuit. The primary advantage of using other terminations is that greater gain can be achieved, although the increase in gain usually is the result of undesirable feedback.

Short-circuit terminations for unwanted harmonics are optimum, in a practical sense, because most solid-state devices operate as voltage-controlled current sources, and their capacitive parasitics are in shunt with their terminals. Open-circuiting the drain in a multiplier implies that the second harmonic current is sinusoidal, while the voltage can have some arbitrary waveform determined by the characteristics of the device. This condition violates the gate-to-drain I/V characteristic of the device and is therefore impossible; trying to enforce it can only result in a circuit that has poor performance in virtually all respects.

A study by Rauscher [10.1] describes the performance of a small-signal MESFET operating as a frequency doubler between 15 and 30 GHz. Because of the multiplier's high output frequency and the relatively large value of C_{gd} in this device, feedback effects are very significant, and the effect of the drain-terminating impedance on gain and stability is pronounced. The conversion gain is approximately 2 dB when the drain is shorted, but rises monotonically with the reactance of an inductive fundamental-frequency termination. When the terminating reactance is only 45Ω, the multiplier oscillates; when the termination is an open circuit, it has low conversion efficiency, –4 dB. One is forced to conclude that using an open-circuit drain termination at the fundamental frequency may have unpredictable results, and its effect on stability in particular is likely to be deleterious.

10.4.2 Balanced Frequency Multipliers

By far the most practical and the one most commonly used balanced multiplier is the antiseries or, less elegantly, the "push-push" multiplier, a circuit that has existed since the days of vacuum tubes [10.2]. An especially nice property of an antiseries active doubler is that the connection between the two drains or collectors is a virtual ground; it therefore eliminates the problem of achieving a broadband fundamental-frequency short circuit at the drains or collectors. Figure 10.8 illustrates the antiseries multiplier circuit.

The general properties of the push-push circuit are described in detail in Section 5.2.1. Although that section describes circuits consisting of two-terminal nonlinear elements, the circuit in Figure 10.8 is conceptually identical. The gates of two FETs are connected, by individual matching circuits, to two mutually isolated ports of a 180-degree hybrid. The delta port of the hybrid is used as the input, and the sigma port is terminated. The gates of the two FETs are driven by signals having a 180-degree phase difference; therefore, the fundamental-frequency components of the drain currents are out of phase, so each FET effectively short-circuits the other at the fundamental frequency and all odd harmonics, creating a virtual ground at the drain. The even harmonics of the drain currents in the two FETs have no phase difference, however, so the drain-current components at those frequencies combine in phase at the output.

This configuration has several advantages over a single-device circuit. First, the output matching circuit can be located close to the drains of the FETs; it need not be separated from them, as in the single-device multiplier, by the intervening filter. Eliminating the parasitic effects of this filter allows the balanced multiplier to have greater bandwidth than would a single-device multiplier. Second, like other balanced circuits, a balanced multiplier has 3-dB greater output power than an equivalent single-device circuit. This can be a significant advantage when used at high frequencies, with small devices, which have low output power. Third, it is often easier

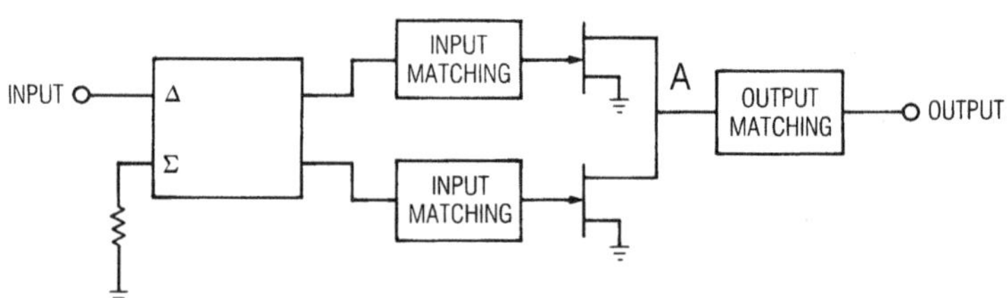

Figure 10.8 Antiseries or "push-push" frequency multiplier. The devices are driven out of phase from a 180-degree hybrid or balun, but the outputs are in parallel. Point A is a virtual ground at the fundamental frequency and all odd harmonics, but even-harmonic components combine at that point.

to realize the load impedance of a balanced multiplier than that of a single-device multiplier. As we saw in Section 5.2.1, the effective load impedance presented to each device in an antiseries circuit is twice the actual load impedance of the balanced circuit. The load impedance required by a single-device FET multiplier often is relatively high; however, the load impedance of the balanced multiplier need only be half that of a single-device circuit. This property significantly eases the task of matching the output at the second harmonic.

In practical balanced multipliers, it is important that the drains (or collectors, in the case of bipolar multipliers) be connected together as close as possible to the device. In particular, hybrids or power dividers should not be used to combine the outputs, and a single matching circuit, as shown in Figure 10.8, should be employed. Any length of conductor between the drain or collector and the common node, point *A* in the figure, becomes an inductance in series with the drain, and the drain no longer has its ideal, virtual ground at odd harmonics.

10.4.3 Noise

We saw in Chapter 7 that their very low noise levels was one of the most attractive properties of varactor frequency multipliers, especially in such applications as receiver LO sources. Since their gain, bandwidth, and efficiency make FET multipliers attractive for generating LO signals in communications receivers, and phase noise and AM noise are important properties of the receiver's LO, it seems wise to examine the noise properties of FET frequency multipliers. This is particularly true of receivers used in phase-modulated communications systems, because the phase noise of the receiver LO is transferred degree-for-degree to the received signal.

GaAs MESFETs are known to have relatively high levels of $1/f$ noise, and this noise can modulate the phase of a signal applied to the FET. Bipolar devices are somewhat better, but operate at lower frequencies. This phenomenon is responsible for most of the phase noise in oscillators, and it can also increase the noise in FET frequency multipliers beyond the inevitable 20 log(n) dB, the minimum carrier-to-noise ratio degradation in any frequency multiplier (Section 7.1.1). Other noise sources, such as noise from the multiplier's bias circuits, can also introduce low-frequency phase noise.

Like other active devices, active frequency multipliers can generate amplitude (AM) noise as well as phase noise. When a multiplier is used in the LO chain of a receiver, the AM noise can be coupled into the mixer with the LO signal, increasing the receiver's noise figure. This phenomenon is analyzed in detail in [2.5].

10.4.4 Harmonic Rejection

Another important concern is the rejection of the fundamental-frequency output and unwanted harmonics. Transistors are, after all, amplifying devices, so unless special effort is made to prevent it, the FET or bipolar device amplifies the input signal, creating a large fundamental-frequency output. Viewed another way, the fundamental component of the drain current in an active multiplier is much greater than the harmonic components, so the fundamental output power may not be much lower than the desired harmonic.

In balanced multipliers, the balance of the hybrid and the individual multipliers can provide approximately 20 dB rejection of the fundamental output; greater rejection is possible in narrowband circuits, ICs, and in other circuits where good balance is relatively easy to obtain. Nevertheless, because most circuits require even more rejection, some degree of high-pass filtering at the output may be needed. Because the rejection band is invariably far below the output passband, a simple filter is usually adequate. The third-harmonic output of a FET doubler is usually very weak, so minimal filtering is needed at this frequency. Many types of microwave filters (e.g., half-wave filters) have frequencies of maximum rejection near 0.5 and 1.5 times the passband frequency; this property makes such filters ideal for use in FET multipliers. The fourth harmonic of a well-designed FET frequency doubler is often virtually nonexistent; this fact is evident from the zero of I_4 / I_{max} near the peak of I_2 / I_{max} in Figure 10.3. Thus the fourth and higher harmonics are rarely of concern.

A single microstrip quarter-wavelength open-circuit stub provides an adequate drain of collector termination for optimizing conversion efficiency but may not provide adequate harmonic rejection. At high frequencies, the Q of a stub is limited not only by resistive losses, but by radiative losses as well, and those can be difficult to quantify. Thus, the harmonic rejection of a stub is difficult to predict.

10.4.5 Stability

Reference [10.1] indicates that the gain of the doubler studied in that research increases dramatically when the fundamental-frequency load susceptance resonates the output capacitance, creating a fundamental-frequency open circuit at the drain. Although that paper does not say so explicitly, this is an obvious indication of instability. Instability in active frequency multipliers usually results from a reactive drain termination at frequencies where the drain should be short-circuited. Such nonoptimum terminations often result from an attempt to increase conversion gain. Increasing conversion gain in this manner is similar to increasing amplifier gain by introducing positive feedback; gain can indeed be increased, but only at the cost of decreasing stability margins.

10.4.6 High-Order Multiplication

Figure 10.3 and the discussion in Section 10.2.1 lead to the inevitable conclusion that FETs and bipolar transistors do not make very good frequency multipliers beyond second order.[1] It is inevitable that the conversion gain and output power of a third- or higher-order multiplier are lower than those of a doubler, but such multipliers may still be practical.

An important difficulty in triplers is the need to short circuit the drain or collector at the unwanted harmonics. In a doubler, this is easy to do; a quarter-wave stub, for example, effectively shorts the first and third harmonic, while the fourth and higher harmonics are weak enough to neglect. There is no such elegant solution for terminating the drain or collector in a tripler. The output network can be maddeningly difficult to design; the inevitable result is a suboptimum termination, which results in suboptimum efficiency and a genuine risk of instability.

Because the magnitude of the harmonic current, I_n, decreases with n, the load resistance (10.8) increases, and quickly becomes unrealizable in practice. Output power and gain suffer, not simply because of the decrease in I_n, but also because of the lower-than-optimum value of R_L.

If even-harmonic frequency multiplication is needed, a designer should consider using a cascade of doublers instead of a single multiplier. If this is not possible, the best gain is achieved when high-harmonic multipliers are closer to the input. It may be necessary to follow the high-harmonic stages with amplifier stages to increase power and to provide isolation.

References

[10.1] C. Rauscher, "High-Frequency Doubler Operation of GaAs Field-Effect Transistors," *IEEE Trans. Microwave Theory Tech.*, Vol. MTT-31, 1983, p. 462.

[10.2] F. E. Terman, *Radio Engineers' Handbook*, New York: McGraw-Hill, 1943.

---

1. Without being entirely facetious, we could make the point that nothing else does, either.

Chapter 11

Active Mixers and FET Resistive Mixers

Although diode mixers are used more frequently than active mixers, it is possible to design active FET and bipolar mixers that are in many respects superior to them. X-band mixers having 4- to 5-dB single-sideband (SSB) noise figures, 6- to 10-dB gain, and 20-dBm third-order intermodulation intercept points are regularly produced, and this performance can be achieved at lower LO power levels than would be required for diode mixers [11.1, 11.2]. MESFETs and HEMTs can produce conversion gain well into the millimeter-wave region. Dual-gate MESFETs reduce the problem of obtaining adequate LO-to-RF isolation in single-device FET mixers; the RF and LO are applied to separate gates, and the low capacitance between the gates provides approximately 20 dB of isolation without the need for filters or hybrids. Balanced FET mixers reject spurious responses and LO noise in a manner similar to balanced diode mixers.

11.1 DESIGN OF SINGLE-GATE FET MIXERS

11.1.1 Design Philosophy

In the design of diode mixers, we often wish to minimize conversion loss, because low conversion loss generally guarantees low-noise operation. In microwave FET mixers, high gain is usually relatively easy to obtain, but it does not automatically insure that other aspects of performance will be good. Indeed, high mixer gain is often undesirable in receivers because it tends to increase the distortion of the entire receiver. Therefore, in most receiver applications, an active mixer is designed not to achieve the maximum possible conversion gain, but to achieve a low noise figure and modest gain, unity or only slightly greater.

As in earlier chapters, to keep our discussion concrete, we begin with FET mixers and later discuss mixers using bipolar devices. Indeed, most active microwave mixers use FETs, although occasionally bipolar devices are used in the lower microwave region. Most bipolar mixers use the Gilbert-cell configuration, a type of doubly balanced mixer, which we discuss in Section 11.3.5.

Figure 11.1 shows a diagram of a single-device FET mixer. The mixer consists of a FET and RF, LO, and IF matching circuits (bias circuits, not shown in the figure, are also required). The matching circuits provide filtering as well as matching; they terminate the FET's gate and drain at unwanted frequencies (mixing products and LO harmonics) and provide port-to-port isolation.

Although other types of mixers have been proposed, most FET mixers have the LO and RF signals applied to the gate and the IF filtered from the drain. The time-varying transconductance is the dominant contributor to frequency conversion. These are sometimes called *transconductance mixers* or *transconductance downconverters*. In such mixers, the effects of harmonically varying gate-to-drain capacitance, gate-to-source capacitance, and drain-to-source resistance are often deleterious and must be minimized.

Because the time-varying transconductance is the primary contributor to mixing, it is important to maximize the range of the FET's transconductance variation. In simple downconverters, we are most concerned with the magnitude of the fundamental-frequency component of the transconductance. To maximize the fundamental-frequency component of the transconductance variation, the FET must be biased close to its threshold voltage, V_t, and must remain in its current-saturation region

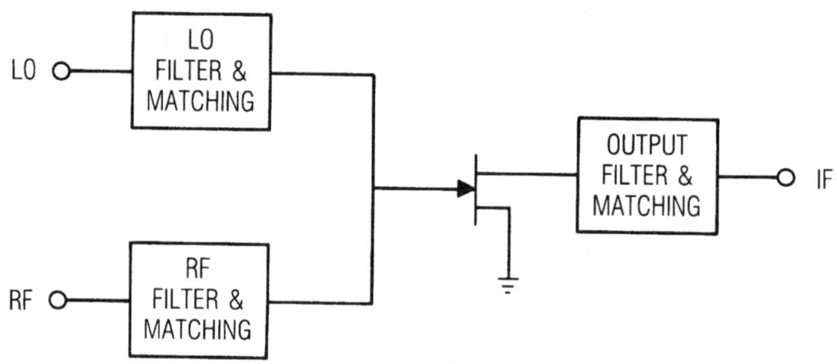

Figure 11.1 Single-gate, single-device FET mixer.

throughout the LO cycle. Full saturation can be achieved by ensuring that the drain voltage $V_d(t)$ under LO pumping remains at its dc value, V_{dd}. This condition is achieved by short circuiting the drain at the fundamental LO frequency and all LO harmonics. If the drain is effectively shorted, the drain LO current, which may have a fairly high peak value, cannot cause any drain-to-source voltage variation; then, the LO voltage across the gate-to-drain capacitance is minimal, so feedback is minimal and the mixer is stable. In this case, the drain current has the same half-sinusoidal pulse waveform as a class-B power amplifier, and the transconductance waveform is similar.

If the drain is not effectively shorted, the drain voltage varies with the LO excitation. Then, the voltage is likely to drop, at the current peaks, as it does in a class-B amplifier. If the voltage dips enough that the FET drops into its linear region, the peak transconductance also decreases, so the fundamental-frequency component of the transconductance is not maximized. Similarly, the peak drain-to-source conductance increases, increasing the average output conductance, creating an additional loss mechanism.

It is usually best to bias the FET at the same drain voltage it would require when used in an amplifier. Although the optimum gate bias is usually near V_t, minor adjustment of the gate voltage must be made empirically as part of the circuit tuning. A well-designed mixer is usually insensitive to small changes in dc drain voltage, but may be moderately sensitive to dc gate voltage.

FET mixers are often conditionally stable, so it is impossible to find source and load impedances that simultaneously match the RF input and IF output ports. Even when the mixer is unconditionally stable, the output impedance of a FET downconverter having an IF frequency below X-band is very high. The resistive part is on the order of several hundred ohms, and there may be a small shunt capacitive reactance. The resistive part is much greater than the drain-to-source resistance of an unpumped, dc-biased FET. Except at low frequencies and over very narrow bandwidths, it is nearly impossible, in practice, to obtain a conjugate match to such a high impedance; therefore, it is usually impossible to match the IF output of an active FET mixer. A better choice is to use a resistive load at the IF, its value selected to obtain the desired conversion gain. In this case, the mixer's output VSWR is, of course, high; however, theoretical and practical limitations of impedance matching dictate that the high output VSWR is unavoidable, regardless of the philosophy employed in designing the IF circuit. Nevertheless, a resistive load, if properly implemented, provides stable operation, flat frequency response, and the desired gain.

The high IF output impedance is a consequence of pumping the FET with the LO. It exists, in principle, in all gate-driven FET transconductance mixers, whether used as upconverters or downconverters. In mixers having high IF frequencies, however, including most microwave upconverters, capacitive parasitics may lower the output impedance somewhat. Nonlinear analysis may be necessary to determine the IF output impedance and an appropriate matching circuit.

Ordinary small-signal HEMTs and MESFETs are used to realize single-gate FET mixers. A FET designed to be used in low-noise amplifiers within some specific frequency range usually works well as a mixer within the same range. Special situations often affect the choice of a device; for example, it is generally easier to obtain a high intermodulation intercept point from a device having a relatively wide gate, and there is some experimental evidence that good noise figures are more readily obtained by using narrow devices. Most millimeter-wave devices are optimized for amplifier use, and therefore have very narrow gates. It may be difficult to obtain conversion gain at high frequencies from such devices.

When the FET is pumped strongly by the LO, its transconductance waveform is approximately a rectified sinusoid. That waveform has an average (dc) value, which allows the FET to amplify as well as mix. Amplification must be minimized to achieve good stability and to prevent spurious effects. In particular, the mixer must not have appreciable linear gain at the IF frequency, or spurious inputs at the IF frequency (especially noise from the gate-bias circuit or other sources) can be amplified and appear in the output.[1] Similarly, RF and LO amplification can result in instability and spurious responses. The only way to minimize unwanted amplification is to mismatch the device at either the gate or drain at these frequencies; therefore, one should design the mixer to have a short circuit at the gate and drain at, ideally, all unwanted mixing frequencies and LO harmonics, especially the IF. This precaution also helps to prevent large-signal instability that might be caused by the pumped nonlinear gate-to-source capacitance, and by ordinary feedback effects.

The effect of parametric instability caused by pumping the gate-to-source capacitance can be insidious, as it can mimic intermodulation distortion in two-tone IMD measurements. Minimizing the source-lead inductance can help enormously in preventing such oscillation, as can short-circuiting the gate at LO harmonics.

Achieving adequate LO-to-IF isolation can be difficult in active mixers. (Of course, if the LO is really short-circuited at the drain, there can be no LO leakage. The short circuit is never perfect, however, so some degree of leakage is inevitable.) The LO current in the FET's drain is very

1. In the author's experience, amplification at the IF frequency is the most common cause of high noise figure in active mixers.

great; its peak value is somewhat above I_{dss}, which, even in small-signal devices, may be over 100 mA. Consequently, the LO-frequency output power is potentially very high. Unfortunately, it is difficult to design an IF matching circuit that provides high LO isolation and still meets all the other requirements placed upon it; therefore, even in well-designed mixers, the level of the LO leakage from the IF port often is high, sometimes even higher than the applied LO power. This LO leakage can saturate the IF amplifier or generate spurious signals. Accordingly, it is important that the IF output circuit include sufficient filtering to provide adequate LO-to-IF isolation. The required rejection depends upon the FET's output power capability and the level of LO leakage that the IF amplifier can tolerate. For example, most small-signal FETs have saturated output levels of at most 10 to 16 dBm. If the leakage is to be kept to –30 dBm or lower, 40 to 46 dB of rejection may be necessary. This large amount of rejection may dictate that a separate LO-rejection filter be used.

11.1.2 Approximate Mixer Analysis

We now perform an approximate analysis of a FET mixer. The results of this exercise can be used for an approximate design, which is optimized by means of nonlinear analysis, or for assessing the performance capabilities of some particular FET. The analysis in this section is valid only for gate-driven transconductance downconverters, but with a little careful thought, it can be modified to include upconverters or other types of mixers.

The design of a FET mixer must optimize the large-signal LO pumping (i.e., it must vary the transconductance over the widest range possible while using as little power as possible) as well as the small-signal operation. We begin with the LO design, recalling that the FET must be short-circuited at the drain at all LO harmonics, and at the gate at all harmonics except the fundamental frequency. If the gate and drain are well shorted at unwanted mixing frequencies and LO harmonics, it is possible to simplify the FET equivalent circuit to obtain the approximate unilateral equivalent circuit shown in Figure 11.2(a). In generating this circuit we assumed that the source inductance, L_s, is negligible; included the source resistance R_s in the input loop; and recognized that when the drain is shorted, C_{gd} is effectively in parallel with C_{gs}. Usually $C_{gd} \ll C_{gs}$, so C_{gd} can be neglected. The parallel-tuned circuit at the output is tuned to the IF frequency, and the input tuned circuit is assumed to be broadband enough to include both the RF and LO frequencies. These resonators short-circuit the drain and gate at all other frequencies.

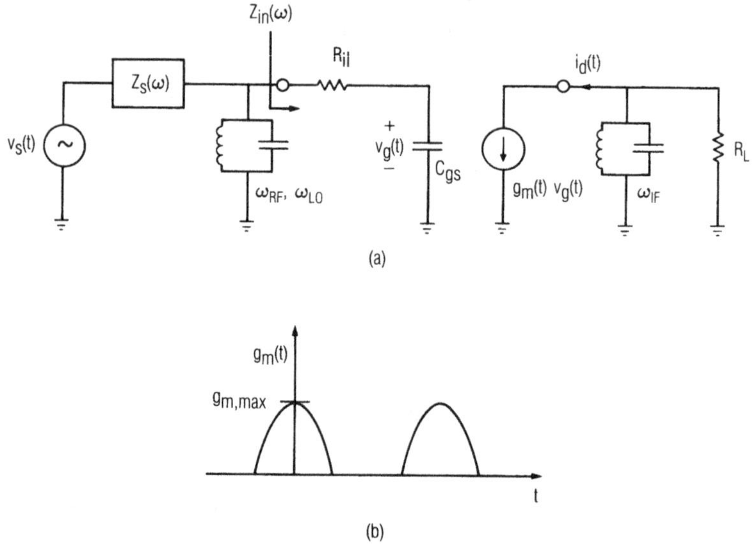

Figure 11.2 (a) Simplified equivalent circuit of the single-gate FET mixer; (b) the MESFET's transconductance waveform when $V_{gg} = V_t$.

The input impedance is found from Figure 11.2(a) to be

$$Z_{\text{in}}(\omega) = R_{il} + \frac{1}{j\omega C_{gs}} \tag{11.1}$$

where C_{gs} is the gate-to-source capacitance at the bias voltage $V_{gg} = V_t$; $\omega = \omega_p$, the LO frequency; and R_{il} is the resistance in the input loop. In a MESFET or HEMT, $R_{il} = R_g + R_s + R_i$, the sum of the gate, source, and intrinsic resistances.

Ideally, the input matching circuit should match the input impedance of the FET at both the RF and the LO frequencies. In many cases, however, the LO and RF frequencies are significantly different, and it is impossible to match the device successfully at both frequencies. When this conflict exists, it is better to match the device at the RF frequency and to accept a mismatch at the LO frequency. A poor RF match degrades conversion performance, but the only consequence of a poor LO match is to waste a little LO power.

The minimum required LO power can be estimated from Figure 11.2(a), under the assumption that the input is conjugate matched. We assume that the gate is biased at V_t, and that the LO voltage at the gate varies between $V_{g,\max}$ (the maximum forward gate voltage, limited by gate-to-channel rectification in MESFETs; $V_{g,\max} \sim 0.5\text{V}$) and the maximum reverse voltage, $2V_t - V_{g,\max}$. The LO power is

$$P_{LO,\min} = \frac{1}{2}(V_{g,\max} - V_t)^2 \omega_p^2 C_{gs}^2 R_{il} \tag{11.2}$$

If the gate is not conjugate matched at the LO frequency, reflection losses must be included.

If we make the reasonable assumption that the transconductance waveform can be approximated by the pulse train of half-sinusoids shown in Figure 11.2(b), the circuit in Figure 11.2(a) can be analyzed relatively easily to determine its conversion gain. Because the input impedance of a FET is not highly sensitive to signal level (as long as the gate is not driven to the point of rectification), the expression for the input impedance of a FET mixer at the RF frequency is the same as the LO input impedance. Therefore, (11.1) is a valid expression for RF input impedance when the RF frequency is substituted for ω_p. The FET's RF input is usually conjugate matched; although it is likely that the noise figure could be improved by input mismatching, as is done with FET amplifiers, it is not clear that similar techniques improve the noise figure of a FET mixer.

The RF excitation $v_s(t)$ in Figure 11.2(a) is

$$v_s(t) = V_s \cos(\omega_1 t) \tag{11.3}$$

where ω_1 is the RF frequency; we use the notation shown in Figure 6.4. If the source is matched, $Z_s(\omega_1) = Z_{\text{in}}^*(\omega_1)$ and the small-signal gate voltage is

$$v_g(t) = \frac{V_s \cos(\omega_1 t + \phi)}{2\omega_1 C_{gs} R_{il}} \tag{11.4}$$

The phase shift ϕ will not be evaluated, because it does not affect the conversion gain. The fundamental-frequency component of $g_m(t)$ in Figure 11.2(b) is

$$g_{m,\,1}(t) = \frac{1}{2} g_{m,\,\max} \cos(\omega_p t) \tag{11.5}$$

where $g_{m,\max}$ is the peak value of $g_m(t)$. The small-signal drain current $i_d(t)$ is

$$i_d(t) = g_m(t) v_g(t) \tag{11.6}$$

The current $i_d(t)$ includes components at the RF and IF frequencies, and at all other mixing frequencies shown in Figure 6.4. Substituting (11.4) and (11.5) into (11.6), employing the usual trigonometric identities, and retaining only the terms at the IF frequency gives the IF component of $i_d(t)$, $i_{\mathrm{IF}}(t)$. Note that only the fundamental component $g_{m,\,1}(t)$ of $g_m(t)$ contributes to frequency conversion:

$$i_{\mathrm{IF}}(t) = \frac{g_{m,\,\max} V_s \cos(\omega_0 t)}{8\omega_1 C_{gs} R_{il}} \tag{11.7}$$

where ω_0 is the IF frequency. The IF output power is

$$\begin{aligned} P_L(\omega_0) &= \frac{1}{2} |i_{\mathrm{IF}}(t)|^2 R_L \\ &= \frac{g_{m,\,\max}^2 V_s^2 R_L}{128\omega_1^2 C_{gs}^2 R_{il}^2} \end{aligned} \tag{11.8}$$

The available power from the conjugate-matched source is

$$P_{\mathrm{av}}(\omega_1) = \frac{V_s^2}{8\mathrm{Re}\{Z_s(\omega_1)\}} = \frac{V_s^2}{8R_{il}} \tag{11.9}$$

and the transducer conversion gain, G_t, is the ratio of (11.8) and (11.9):

$$G_t = \frac{P_L(\omega_0)}{P_{\mathrm{av}}(\omega_1)} = \frac{g_{m,\,\max}^2 R_L}{16\omega_1^2 C_{gs}^2 R_{il}} \tag{11.10}$$

Equation (11.10) is remarkably accurate in practice, as long as the optimum short-circuit embedding impedances are achieved and the gate is optimally biased, near the FET's turn-on voltage, V_t.

Equation (11.10) seems to imply that it is possible to make the conversion gain arbitrarily high by increasing the IF load impedance, R_L, or by increasing the device's width, thus increasing $g_{m,\max}$. These implications are generally valid; however, practical difficulties limit the conversion gain. Problems involving stability and realizability limit R_L to 100Ω to 200Ω, and the FET's output capacitance limits the bandwidth if R_L is made too great. If device width is increased too far, the resulting decrease in input impedance introduces matching difficulties. Furthermore, as we noted earlier, it may not be desirable to have high gain in a mixer. It is possible, however, to achieve remarkably high gain (above 10 dB) at X-band in mixers using medium-power devices (having gate widths around 0.5 mm) and high load impedances. Because of the mixer's high output impedance, it is even possible in some cases for a MESFET to achieve greater gain as a mixer than as an amplifier.

An active FET mixer's input intermodulation intercept point is largely constant with R_L. Thus, a valid approach to designing a mixer for low distortion is to use a large device, pump it adequately, and use a relatively low value of R_L to keep the gain reasonable and to provide stability. It is better to use a smaller device, strongly pumped, than a larger device with inadequate LO power.

The design process is relatively simple. The first task is to estimate the important parameters of the FET, $g_{m,\max}$, R_{il}, and $C_{gs}(V_t)$. The peak transconductance, $g_{m,\max}$, can be found from dc measurements, as can the resistances; C_{gs} can be estimated with adequate accuracy from the FET's S parameters. One should then select a value of R_L that is achievable in practice and satisfies the gain requirements, as indicated by (11.10), and then estimate the input impedance by (11.1). If the input Q of the device is so high that it cannot be matched over the required bandwidth, reflection losses must also be included in the gain estimate. The final step is to design the input and output networks to conjugate match the input, to present R_L to the drain at the IF frequency, and to short-circuit the gate and drain at all other significant frequencies.

11.1.3 Bipolar Mixers

The requirements for the design of bipolar mixers, both conventional homojunction BJTs and HBTs, are essentially the same as in FETs. As with FETs, providing an LO short circuit at the drain is probably the most important requirement. The general design goals—conjugate matching the input and terminating the output in an appropriate value of R_L—are likewise identical.

The above analysis is generally applicable to bipolar-transistor mixers, both homojunction and heterojunction, when C_{be} is substituted for C_{gs} and R_b for R_{il}. Bipolar transistors have an additional base-to-emitter junction resistance, R_{je}, which is in parallel with C_{be}; the analysis can be modified easily to include it when necessary. In many cases, however, R_{je} can be neglected, because the large, parallel C_{be} has a much lower impedance at RF and microwave frequencies.

In bipolar mixers, the peak transconductance is usually much greater than in FET mixers. This allows much greater conversion gain at lower frequencies, which may not be desirable; it may exacerbate IF gain and stability problems. At high frequencies, however, the high C_{be} may create difficulties in achieving adequate conversion gain.

Bipolar mixers are rarely implemented as single-device mixers, and only occasionally as singly balanced mixers. The most common implementation is a *Gilbert cell*, a type of doubly balanced structure. We examine Gilbert-cell mixers in Section 11.3.5. For an example of a singly balanced HBT mixer, see [11.3].

11.1.4 Matching Circuits in Active Mixers

The input and output matching circuits in active mixers have unique requirements, so designing them requires special care. The input matching circuit must not only match the RF source to the FET's gate or BJT's base-to-emitter junction, but it must also provide an IF short circuit to the device. If the IF frequency is much lower than the RF, this short can be realized via the bias-circuit elements. As long as the IF short is realized effectively, the only other critical function of the input matching network is impedance matching at the RF frequency and, if possible, at the LO frequency.

Because of its limited Q, a quarter-wave stub may not be adequate to short-circuit the drain at the RF and LO frequencies; it is better to realize the IF matching circuit as a low-pass filter connected directly to the FET's drain, and to include additional elements to provide the desired IF terminating impedance. The IF matching network is a critical part of a FET mixer, and applying a little creative thought to its design can do much to ensure that the mixer's performance will be good. A standard, textbook filter design is often not a good choice for the IF filter, because a filter having even very high rejection may present a reactive termination, rather than a short circuit, to the drain. In many cases it is possible for the IF circuit to provide both impedance transformation and filtering functions from a single structure; this approach minimizes circuit loss and complexity.

In many FET mixers, especially those having RF and LO frequencies below a few gigahertz, it may be impossible, in any practical way, to match the input. It is easy to see why. The input Q of the FET, Q_i, is

$$Q_i = \frac{X}{R} = \frac{1}{(\mathrm{Re}\{Z_s(\omega_1)\} + R_{il})C_{gs}\omega} \tag{11.11}$$

As an illustration of the problem, suppose we are designing a 5-GHz mixer. If the input is conjugate matched, $\mathrm{Re}\{Z_s(\omega_1)\} = R_{il} = 5\Omega$, $C_{gs} = 0.25$ pF, and $Q_i = 12.7$. Even with a complex matching circuit, which may be difficult, in practice, to realize, the bandwidth cannot exceed approximately 10%. Even if a broadband conjugate match were possible, the conversion gain would have a slope, since conjugate matching does not guarantee that the voltage across C_{gs} will be flat with frequency.

In such cases, a different approach is needed. The goal is to achieve a voltage across C_{gs} that is adequately flat over the frequency range of interest, not to achieve a good input VSWR. Since (11.6) shows that the IF output current is proportional to that voltage, this approach should result in a flat frequency response. We select a gate inductance that approximately resonates C_{gs} and adjust the real part of the source impedance to reduce Q_i to a reasonable level. Finally, the values of these elements are adjusted, on the computer, to achieve a flat voltage response across C_{gs}.

Unlike FET mixers, which usually have a high input Q, bipolar mixers usually have a low input Q. Bipolar devices—both conventional homojunction devices and HBTs—have a large base-to-emitter capacitance in series with a moderate base resistance; at high frequencies, the reactance of the capacitance is very low, often negligible, making the input impedance equal to the base resistance. In homojunction bipolars, the base resistance is on the order of tens or hundreds of ohms; in heterojunction devices, it is on the order of a few tens of ohms at most.[2] Real impedances in these ranges are usually not difficult to match; therefore, input matching in bipolar mixers is usually much easier than in FETs.

2. These broad generalities are offered in a desperate attempt to be quantitative. Because so many different types and sizes of transistors exist, these generalizations may not be valid in some cases.

11.1.5 Nonlinear Analysis of Active Mixers

Nonlinear analysis of active mixers, both FET and bipolar, is a relatively straightforward application of harmonic-balance analysis. Two methods are possible: multitone harmonic-balance analysis and large-signal/small-signal analysis. The latter is more efficient and is best used to calculate conversion loss and input/output impedances. Multitone harmonic-balance analysis is necessary when distortion, compression, or other nonlinear effects are of interest.

Certain mixer calculations are easier than others. It is usually easy to calculate conversion loss and port impedances accurately, as long as the device model is adequate and the passive circuit elements are well modeled. Isolation is always difficult to calculate because it depends strongly on circuit-element Q and value tolerance. When isolation is high, it may be dominated by coupling outside the circuit, for which the circuit simulator obviously cannot account.

More complex phenomena, such as spurious responses and intermodulation distortion, involve mixing between harmonics of the RF and LO. Often they result in a mix of strong and weak frequency components, so the concerns regarding criteria for terminating the analysis, described in Section 3.3.9.7, are especially relevant. Multitone harmonic-balance analysis requires a multitone Fourier transform, which inevitably has less numerical range than a classical fast Fourier transform. It can sometimes be difficult to determine the accuracy of a weak intermodulation component; in some cases, the mixing product can be lost in numerical noise, but the simulator still produces a value for it, however invalid.

Device modeling also requires special care. In Section 2.3.2, we noted that accurate analysis of nth-order distortion requires a device model whose first n derivatives are accurate. In a small-signal amplifier, the derivatives must be accurate only at the bias point, but in a mixer they must be accurate over the entire range of the LO voltage. This is a difficult requirement to meet.

11.1.6 Design Example: Simple, Active FET Mixer

Given a model of the device and the I/V characteristics, (11.1), (11.2), and (11.10) can be used to estimate the input impedance, conversion efficiency, and minimum LO power. As discussed in Section 11.1.4, however, it may not be possible to match the input over both the LO and RF bands, so the estimated conversion efficiency may be high and LO power requirement low. Nonlinear analysis can increase the accuracy of those estimates.

We design a mixer operating from 7.9 to 8.4 GHz with a 7.4-GHz LO frequency and 0.5- to 1.0-GHz IF. A conventional Ku-band MESFET is available; its I_{dss} is 100 mA, $V_t = -2.5$V, and it is characterized by a Curtice model. The mixer will be realized as a hybrid circuit on a 0.635-mm thick alumina substrate ($\varepsilon_r = 9.8$).

We use much the same approach as for the single-diode mixer design in Section 6.4.2: begin with an ideal circuit and replace the ideal parts with real ones to create the complete design. The ideal circuit is shown in Figure 11.3. The LO and RF are combined by an ideal combiner, which eventually will be replaced by an appropriate diplexer. Since the LO frequency is constant, the ring resonator shown in Figure 6.6 might be a good choice. The FET's drain is shorted by a stub, and the FET itself is modeled by the complete Curtice model. We have included a high-impedance series line to tune the gate, and the source resistance is selected according to the controlled-Q matching approach described in Section 11.1.4. In the simulator, the source impedance can be adjusted directly at the port; it is not necessary to include a transformer or other such structure. Being unimaginative, however, we start with 50Ω, and see how well that works.

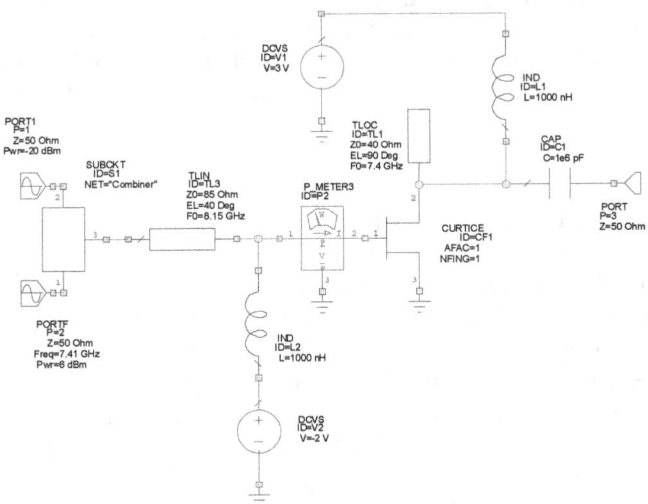

Figure 11.3 Idealized single-device active FET mixer. This circuit is used to optimize the LO power, dc bias, and load resistance; by optimizing the pieces of the mixer individually, little or no numerical optimization should be necessary.

The load impedance is similarly adjustable, and for precisely the same reasons, we start with a 50Ω load.

The bias, LO power, and input tuning are adjusted to optimize the design. The result is a flat response over the required bandwidth and, with 50Ω source and load, 5.6-dB gain. The input return loss is quite low, only about 1 dB; decreasing the source impedance can improve it at the cost of more passband gain variation and higher gain, neither of which is desirable. A better approach in dealing with the input mismatch is to use isolators or to make sure that the source return loss at each port is high.

We next create the real circuit. We replace the ideal stubs by microstrip ones and replace the ideal bias circuits with real ones. The circuit is shown in Figure 11.4, along with a plot of the conversion gain. The gain is virtually identical to that of the ideal circuit.

Because of the low IF frequency, it is easy to minimize amplifier-mode gain. A bypass capacitor in the gate-bias circuit should be adequate. The 2KΩ resistor, although intended primarily for gate protection, also helps to isolate the gate from power-supply noise.

To complete the circuit, we still must design a diplexer and add discontinuity parasitics. The discontinuities and the diplexer's output impedance (which is never precisely equal to 50Ω) may detune the circuit somewhat. Lengths of the tuning elements can be adjusted to account for these additions. It is especially important to make certain that the drain stub, once the discontinuity elements are added, still provides the required short circuit. This can be assured by adjusting the length to minimize LO leakage at the IF.

11.2 DUAL-GATE FET MIXERS

Dual-gate FETs have an important advantage over single-gate FETs when used as mixers: the LO and RF can be applied to separate gates. Because the capacitance between the gates is low, the mixer has good LO-to-RF isolation. Because of its high isolation, a single-device, dual-gate FET mixer often can be used in applications where a balanced mixer would otherwise be needed. Dual-gate FETs are also used in integrated circuits, where filters and distributed-element hybrids may be impractical, and good LO-to-RF isolation may otherwise be difficult to achieve.

Dual-gate MOSFET mixers have been used successfully in many kinds of portable and fixed radio receivers for many years. Because of this success, it was originally expected that dual-gate FET mixers would become the devices of choice for most receiver applications. Unfortunately, the reported performance of dual-gate FET mixers has not been very good,

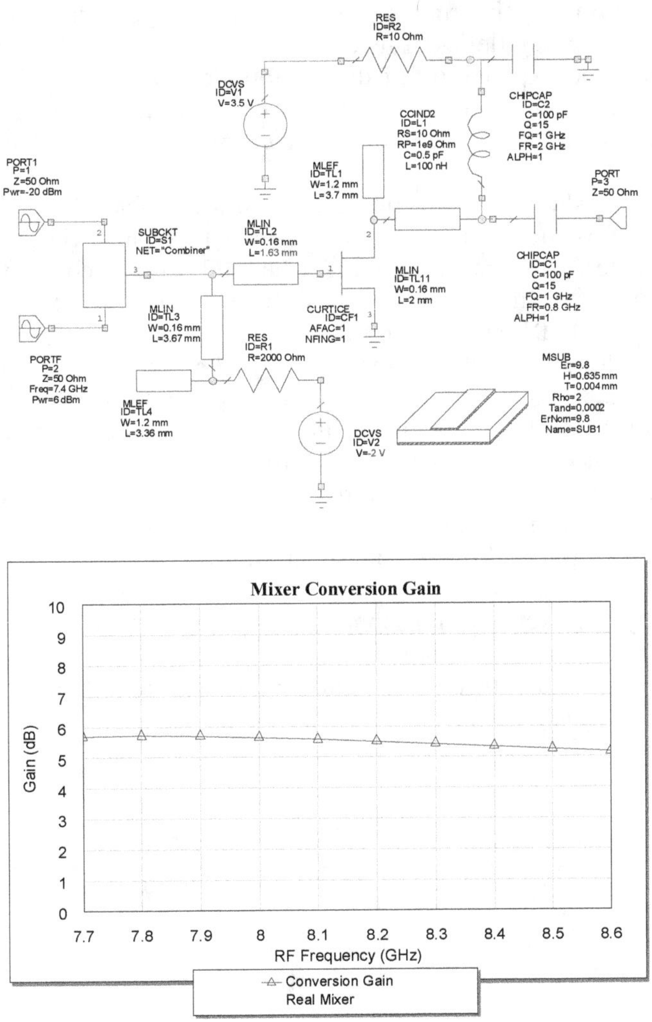

(a)

(b)

Figure 11.4 (a) Circuit and (b) conversion gain of the single-device mixer. The RF-LO combiner still must be designed and discontinuity parasitics added to complete the circuit.

and after some initial enthusiasm in the mid 1980s, they are now not produced very often. Although dual-gate mixers usually exhibit reasonably good gain, their noise figures have been disappointing, considerably worse than those of single-gate mixers. One reason is that dual-gate FET mixers have inherent disadvantages compared to single-gate mixers; another is that a dual-gate mixer is a much more complex component than a single-gate mixer, and the subtleties of its operation are not always appreciated by designers (a good explanation of these subtleties can be found in [11.4]). Still, dual-gate mixers have their place: most important is their use in ICs to obtain many of the advantages of balanced mixers without the need for hybrids.

Figure 11.5 shows a simplified circuit of a dual-gate FET mixer. The dual-gate FET is modeled as two single-gate FETs in series. The LO is applied to gate 2, the gate of FET 2 (the FET that is connected to the external drain terminal), and varies V_{gs2}; the RF signal is applied to gate 1, the gate of FET 1. RF and LO sources are connected to gate 1 and gate 2 through matching circuits, represented by the embedding impedances $Z_{s,\,\mathrm{RF}}(\omega)$ and $Z_{s,\,\mathrm{LO}}(\omega)$, respectively; a series-resonant element (which can be an LC tuned circuit, a stub, or simply a bypass inductor) is used to ground gate 2 at the IF frequency. As with the single-gate FET mixer, the load impedance $Z_L(\omega)$ is a short-circuit at all LO harmonics and mixing frequencies except the IF; this termination guarantees that the LO power is not dissipated in the IF load, and that the drain voltage V_{ds} remains constant over the LO cycle.

A dual-gate mixer is a transconductance mixer, so mixing must occur by variation of the transconductance between V_{gs1} and I_d. The transconductance variation must come from varying the drain voltage of FET 1.

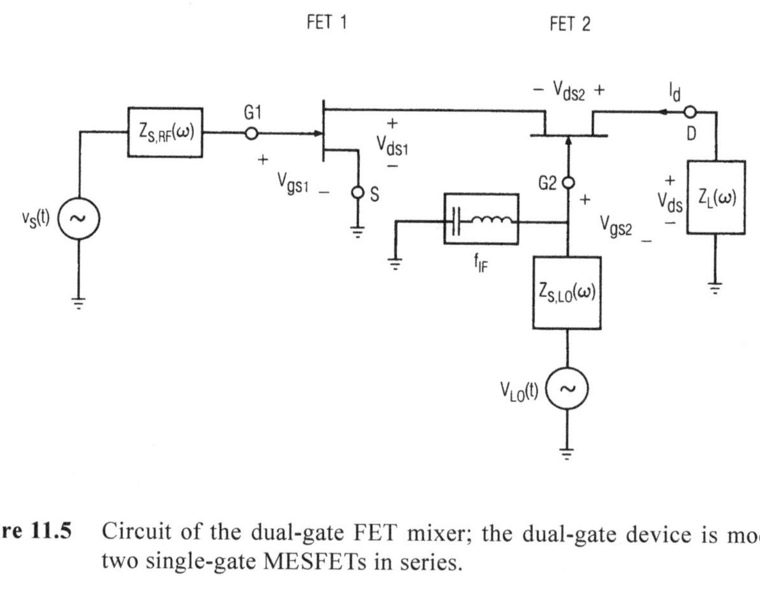

Figure 11.5 Circuit of the dual-gate FET mixer; the dual-gate device is modeled as two single-gate MESFETs in series.

Figure 11.6 shows the dc drain *I*/*V* characteristic of FET 1 in Figure 11.5 as a function of the gate voltages V_{gs1} and V_{gs2} when V_{ds} is fixed at 5.0V. V_{ds} must be divided between the channels of the two single-gate MESFETs; $V_{ds1} + V_{ds2} = V_{ds}$. When two FETs are connected in series, it is impossible to have a stable operating point if both devices are in current saturation, because in this case the FETs' channels are equivalent to two current sources in series. Inevitably, one device must be saturated, and the other must operate in its linear region; most of V_{ds} is dropped across the saturated FET.

If FET 2 is linear and FET 1 is saturated (i.e., the operating point is close to the right side of the set of curves in Figure 11.6), varying V_{gs2} with the LO voltage, while V_{gs1} is constant, does not vary the transconductance between V_{gs1} and the drain current I_d; therefore, no mixing can occur. Significant transconductance variation occurs only when the gate voltages lie within the shaded region of Figure 11.6, the region in which FET 2 is saturated and FET 1 is linear. In this case, the V_{gs1}-to-I_d transconductance variation occurs primarily because the drain voltage of FET 1 is varied from nearly zero—a value that forces the FET to be in its linear region, and its channel to be a low-value resistance—almost to the point of current saturation.

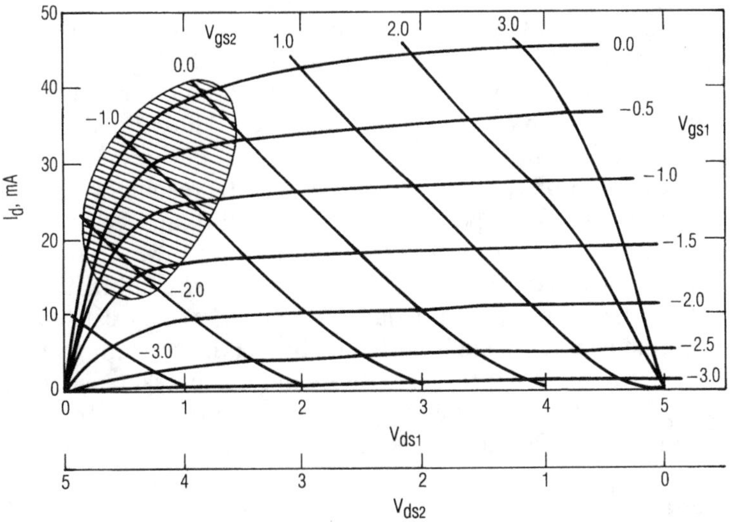

Figure 11.6 *I*/*V* characteristics of the dual-gate FET when V_{ds} = 5.0V.

In a dual-gate mixer, mixing occurs primarily in FET 1; its transconductance and drain-to-source resistance vary with time while the device is in its linear region. In this mode of operation, the peak transconductance of FET 1 is relatively low, and its low drain-to-source resistance shunts the IF output, further reducing conversion gain. In contrast, a single-gate device is in current saturation throughout the LO cycle, so its transconductance is greater and its drain-to-source resistance is very high. For this reason, the single-gate FET is a more efficient mixer than a dual-gate FET.

In the dual-gate mixer, FET 2 remains in its saturation region throughout the LO cycle, and its high transconductance varies only moderately. Consequently this FET provides some mixing between the RF drain current of FET 1 and the LO, but its primary effect is to amplify FET 1's IF output. The series resonator grounds the gate of FET 2, so that FET operates as a common-gate amplifier at the IF frequency. The input impedance of this amplifier is approximately $1 / \langle g_m(t) \rangle$, where $\langle g_m(t) \rangle$ is the average transconductance of FET 2; this impedance is usually a mismatch to the IF output impedance of the mixing FET, so the amplifier's input coupling is not optimum. As a result, its gain is not great. The mixing FET's poor conversion transconductance and the poor current coupling to the input of the amplifying FET cause the dual-gate mixer's gain and noise performance to be poorer than that of a single-gate FET.

The procedure for designing a dual-gate mixer is much the same as that for designing a single-gate mixer. The dual-gate mixer requires both a carefully designed RF-LO filter at its drain and a resistive IF load. As with a single-gate mixer, the IF output impedance of a dual-gate FET mixer is relatively high, although for a different reason: the high output impedance is a property of a common-base FET amplifier. Thus, good gain can be achieved, in spite of the inherent limitations of the device, by using a relatively high value of IF load resistance. The IF resonator connected to gate 2 has a critical effect upon the mixer's stability and LO efficiency. If the resonator's reactance at the LO frequency is too low, the LO matching may be poor; however, at some frequency, the combination of the resonator's reactance and the impedance of $Z_{s,\,\mathrm{LO}}(\omega)$ may cause the mixer to oscillate. One can avoid such problems by making certain that $Z_{s,\,\mathrm{LO}}(\omega)$ and the resonator do not present a high inductive reactance to Gate 2 outside the LO frequency range. As with a single-gate mixer, source and load impedances $Z_{s,\,\mathrm{RF}}(\omega)$ and $Z_L(\omega)$ should be short circuits at unwanted mixing frequencies.

11.3 BALANCED ACTIVE MIXERS

11.3.1 Singly Balanced Mixers

A pair of single-gate FET or BJT mixers can be combined by quadrature or 180-degree hybrids to create a singly balanced mixer. The properties of balanced transistor mixers—LO isolation, spurious-response rejection, and LO noise rejection—are essentially the same as in balanced diode mixers. However, FETs and bipolars cannot be "reversed," as can diodes, so the structures of singly balanced active mixers are not entirely analogous to those of singly balanced diode mixers. A balanced active mixer employs the same type of hybrid and input structure as a diode mixer, but because the IF currents in the individual devices are out of phase, an active mixer always requires an IF hybrid to subtract them. Because the output hybrid complicates both the circuit and its layout, the need for an output hybrid is a disadvantage of single-gate FET balanced mixers.

Both the 180-degree and 90-degree (quadrature) mixers shown in Figure 11.7(a) and 11.7(b), respectively, require 180-degree output hybrids, and in both mixers the IF output is derived from the delta port. In Figure 11.7(a), the RF and LO are applied to the sum (sigma) and difference (delta) ports, respectively; if the ports are reversed, the conversion gain and noise figure are the same, but the spurious-response characteristics are not. Because the IF currents are subtracted instead of added, the spurious-response characteristics of a singly balanced active mixer are precisely the opposite of those of a singly balanced diode mixer, described in Section 6.4.1. Pumping the devices out of phase [the case shown in Figure 11.7(a)] rejects spurious responses arising from odd harmonics of the RF mixing with even harmonics of the LO. If the LO were applied to the sigma port of the input hybrid, the devices would be pumped in phase and the opposite would occur: the mixer would reject mixing products between odd harmonics of the LO and even harmonics of the RF. In both cases, however, the mixer would reject all responses that arise from even harmonics of both the RF and LO.

There are other valid reasons for applying the LO or the RF to a particular port. If the LO is applied to the sigma port, it may be possible to achieve LO rejection via the output hybrid. This property is particularly valuable when the LO frequency is close to the IF frequency, as might occur in an upconverter, and it may not be possible to separate the frequencies by filters. If the LO and IF are both within the output hybrid's bandwidth, the hybrid combines the IF but rejects the LO. The rejection level depends upon the amplitude and phase balances of both the mixer and the hybrid, but well designed hybrids and mixers should have LO rejection

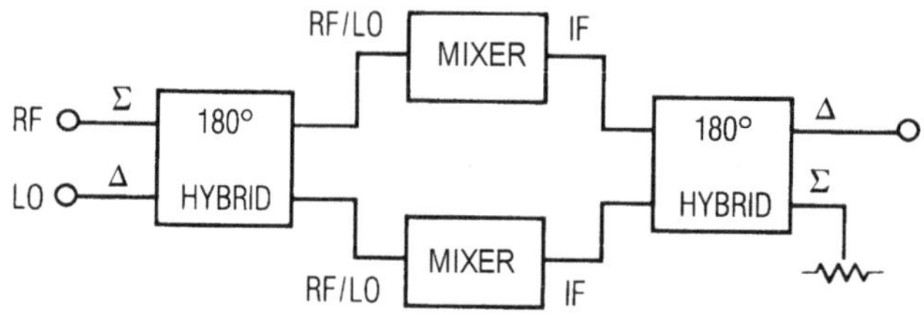

(a)

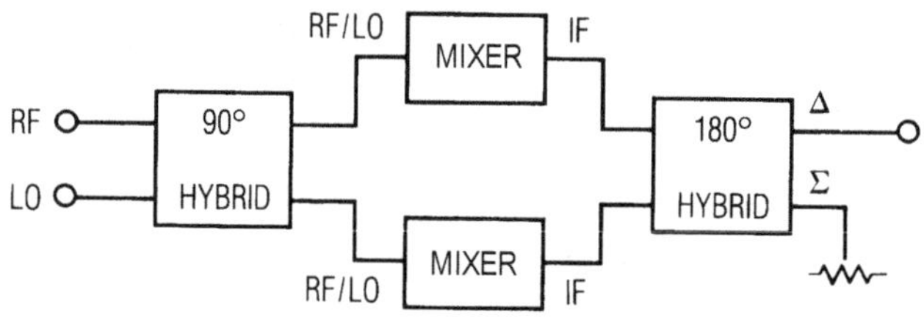

(b)

Figure 11.7 Active (a) 180-degree and (b) quadrature singly balanced FET mixers. The block marked *mixer* can be either a FET or bipolar single-device mixer.

on the order of 20 dB. Conversely, in conventional downconverters, where the LO and RF frequencies are high compared to the IF, there may be an advantage to using the delta port for the LO. The drains or collectors of the two devices can then be connected by a small-value capacitor, which connects the drains or collectors together at the LO frequency but leaves them separate at the IF. Because the LO currents in the devices are 180 degrees out of phase at the fundamental frequency and all odd harmonics, each device effectively short-circuits the other, reducing LO leakage significantly.

The singly balanced quadrature mixer, shown in Figure 11.7(b), has a 90-degree hybrid at the input, and the RF and LO are applied to one pair of mutually isolated ports; the other pair of isolated ports is connected to the inputs of two single-device mixers. This configuration has the same properties as a quadrature diode mixer, specifically that the isolation between the RF and LO ports is good only if the inputs of the FETs or bipolar devices are well matched at both the LO and RF frequencies. This mixer has the same spurious response properties as a quadrature diode mixer: it rejects only spurious responses associated with even harmonics of both the RF and LO frequencies.

A singly balanced FET mixer can be realized with the differential structure shown in Figure 11.8(a). This approach is often more practical than the one described above, as it requires only a simple LO balun. The two FETs connected directly to the LO balun operate as switches, while the lower FET, whose gate is connected to the RF port, operates as a transconductance element. The node connecting the sources of the upper devices (point *A* in the figure) is a virtual ground, so there is no LO voltage on the lower FET's drain, and the upper devices operate as if their sources were grounded. Consequently, the RF and LO input impedances are simply those of a common-source FET.

As with all mixers, the drains of the upper devices must be shorted at the LO fundamental frequency. If the IF frequency is well below the LO frequency, the short can be provided by simply connecting the drains together with a capacitor. This expedient does not short the even LO harmonics; however, the FETs are not operated in their active regions, so little LO harmonic energy is generated.

The design of this mixer is straightforward. Like other active FET mixers, the IF output impedance is likely to be high, so the load is resistive and selected for appropriate gain. The RF and LO input matching is essentially the same as in any other common-source circuit. Because of the FET's high input *Q*, a conjugate match over a wide bandwidth may not be

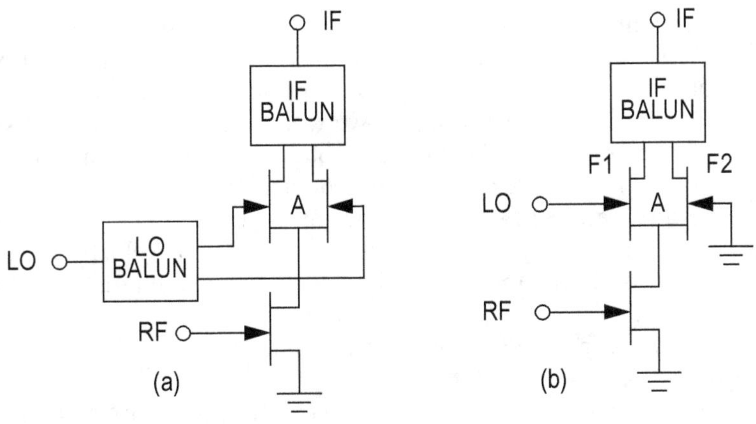

Figure 11.8 Two simple singly balanced FET mixers: (a) conventional and (b) configured so that a balun is not needed.

possible or even desirable; it may be necessary to use the methods described in Section 11.1.4. Determination of the gain and optimization of the matching circuits should be done on the computer. The LO and RF matching networks should be tuned to achieve a flat frequency response of $V_{gs}(t)$ at their respective FETs; then, flat gain should be achieved easily.

Figure 11.8(b) shows a version of the mixer that uses no balun. It takes advantage of the fact that, in a differential amplifier, the applied voltage is divided between the gate-to-source junctions of the two devices. This circuit has two serious problems: first, the source node (point A) is no longer a virtual ground, so there is significant LO voltage at the drain of the RF FET. This LO voltage component can pump the drain voltage of the RF FET, causing a decrease in conversion gain. The second problem is that the drain-to-source resistance of the RF FET, which is never particularly high, is in parallel with the gate-to-source junction of the device marked F1, but not in parallel with the same junction of F2. The lack of symmetry causes unequal pumping of the two FETs, and consequent imbalance in the mixer. Isolations suffer, especially RF-to-IF, as does spurious-response rejection. This circuit does not offer high performance, but it may still be useful in cases where a balun cannot be used and moderate performance is adequate.

11.3.2 Design Example: Computer-Oriented Design Approach

As an example, we design another 7.9- to 8.4-GHz mixer of the type illustrated in Figure 11.8(a), using the same FET as in the example in Section 11.1.6. Because the performance is difficult to approximate analytically, we use a fully computer-based design approach.

First, we must select the bias for the devices. The upper pair of devices is biased in their linear region, while the lower device is biased into saturation. For a decent noise figure, we bias the lower device well below 0.5 I_{dss}, but not as low as 0.15 I_{dss}, the approximate bias for minimum noise figure in amplifier operation. (If we were to bias the device at such a low current, we might not be able to obtain any conversion gain.) We therefore select $I_d = 35$ mA for the lower device, or 17.5 mA for the upper devices. We also select $V_{ds} = 3$V for the lower device and $V_{ds} = 0.5$V for the upper ones, giving 3.5V for the dc supply. These selections are not critical; we can optimize them later. From the I/V characteristic in Figure 11.9(a), we see that $V_{gs} = -1.3$V for the lower device and $V_{gs} = -1.5$V for the upper ones. Figure 11.9(b) shows that we must apply –1.5V to the gate of the lower device and 1.7V to the upper ones.

Next, we assemble the circuit. We use ideal bias circuits, an ideal hybrid to serve as the LO balun, and an ideal transformer for the IF output. We use the complete FET model, and include quarter-wave open-circuit

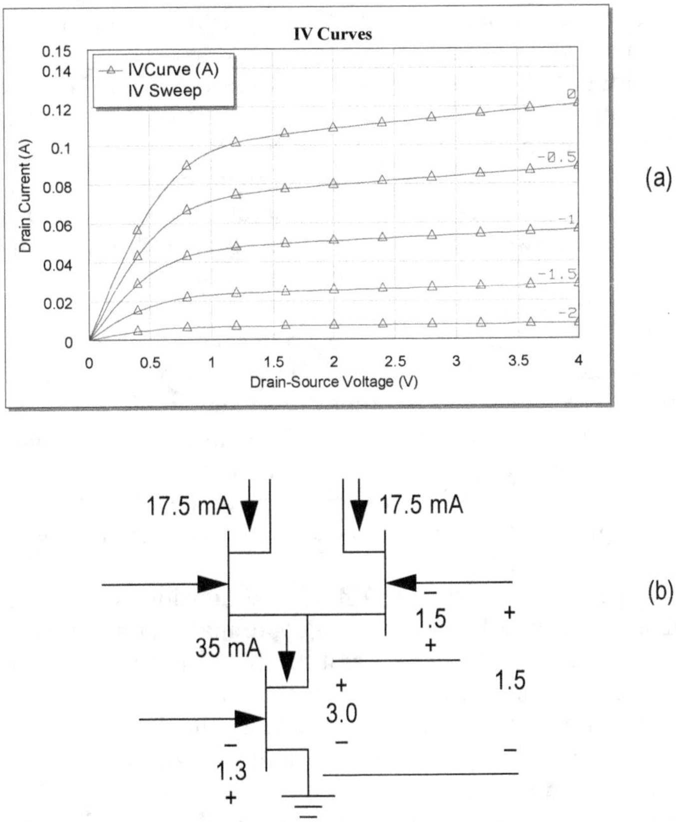

Figure 11.9 (a) Drain I/V characteristics of the FETs used in the singly balanced mixer design example; (b) quiescent bias voltages and currents.

stubs to ground the drains at the LO frequency. Finally, we include high-impedance lines in series with the gate, as in the example of Section 11.1.6, to resonate the gate capacitance.

Optimization requires adjustment of the LO level, dc bias, and gate tuning elements to achieve maximum conversion efficiency. This may seem like a paradox; after all, we stated earlier that high conversion gain is not necessarily desirable. Nevertheless, we need to distinguish between optimizing the circuit, for which conversion gain is an indicator, and designing it to achieve high gain. If our optimization results in gain that is too high, we can easily reduce it by decreasing the load impedance or the bias current in the lower device. In optimizing the bias, the dc gate voltage of the lower FET should not be varied appreciably; doing so would increase the bias current in that FET. Instead, we concentrate on the bias to the upper devices. After a few minutes of work, we have reduced the bias on the upper FETs to 0.5V and set the LO level to 4 dBm. The drain-to-source voltage of the lower FETs is now approximately 2.3V and on the upper FETs, 1.1V. The upper FETs bias must be adjusted, to maintain the correct current in the circuit, because pumping them causes a change in their dc current. With $V_{gs} = -1.9$V, they are operating on the edge of the linear region, which allows them to switch quickly and with minimal LO power. Conversion gain, at this point, is approximately 6 to 7 dB with a 1-dB gain slope across the band.

Finally, we replace the ideal elements with real ones. The hybrid, a simple rat-race design, is realized in microstrip with appropriate discontinuity elements. By treating it as a subcircuit, we can easily assess and optimize its performance outside of the mixer circuit. Ideal dc blocking capacitors are replaced by chip-capacitor models. Finally, the microstrip bias line on the lower FET is used as a tuning element and adjusted to flatten the gain. The resulting circuit and conversion performance are shown in Figure 11.10.

Several minor modifications might be considered. Because the LO is narrowband, it would be easy to conjugate match it and further reduce the LO power requirement. The separation of the LO and RF circuits allows this; it is an important advantage over the mixers in Figure 11.7, where the input matching circuit must encompass the combined RF and LO passbands. Second, the gain of 7 dB is a little high for some applications. It might be worthwhile to reduce the current in the lower FET, to reduce the conversion gain to a 3 to 5 dB while reducing the mixer's dc input power.

11.3.3 Doubly Balanced FET Mixers

Doubly balanced FET mixers have most of the same beneficial characteristics as doubly balanced diode mixers: good port isolation, broad bandwidth, rejection of LO AM noise, and rejection of all spurious responses that include an even harmonic of either, or both, of the RF or LO frequencies. Doubly balanced mixers need baluns at all ports, including the IF, but those baluns can sometimes be implemented as active circuits. This makes the mixers practical for monolithic integration.

Figure 11.11 shows a doubly balanced FET mixer without its baluns. It can be viewed as a balanced connection of two of the mixers shown in Figure 11.8(a). Alternatively, it can be viewed as a FET version of the Gilbert-cell mixer, discussed in Section 11.3.5.

The LO is usually applied to the upper devices, and the RF to the lower ones. As with the related singly balanced mixer, the upper devices operate as switches. The upper devices are biased in their linear region, while the lower ones are biased into current saturation. A current-source device can be used instead of grounding the sources of the RF FETs directly; this may provide some improvement in balance, but requires that V_{dd} be increased by 2V or so to bias the transistor. Including the current-source device, this

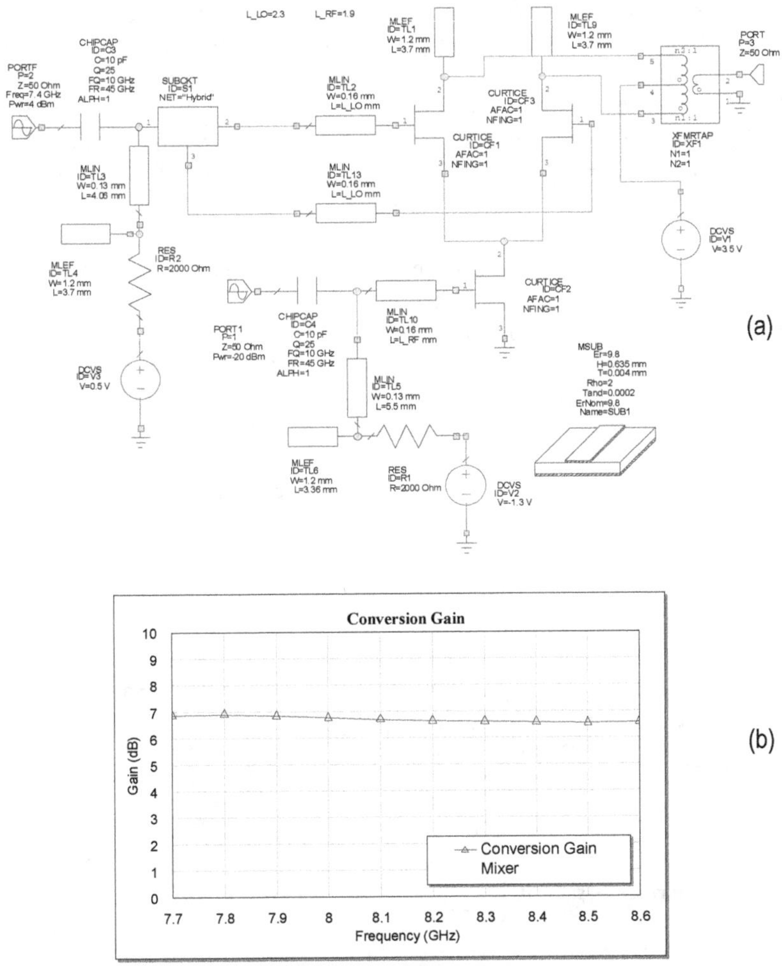

Figure 11.10 (a) Complete mixer circuit; (b) conversion gain.

circuit requires a minimum of 5V to operate, and usually considerably more. As such, it may not be suitable in RF circuits designed for low-voltage portable operation, which must operate from a dc supply as low as 3V. For such applications, a FET resistive mixer may be more suitable.

11.3.4 Active Baluns

Active baluns are, in fact, linear amplifiers having two outputs that have equal amplitudes but differ in phase by 180 degrees. They can provide the phase split necessary for balanced mixers. Such baluns are much smaller than their distributed counterparts, and therefore may be more useful in applications, such as ICs, where space must be minimized.

It is difficult to make a good active balun. The fundamental problem is that FETs' low drain-to-source resistance prevents them from making good current sources. Active baluns suffer from a number of additional problems:

- An active balun must often be designed primarily to achieve broad bandwidth in combination with good phase and amplitude balance. It is often not possible to optimize its noise figure or linearity within such

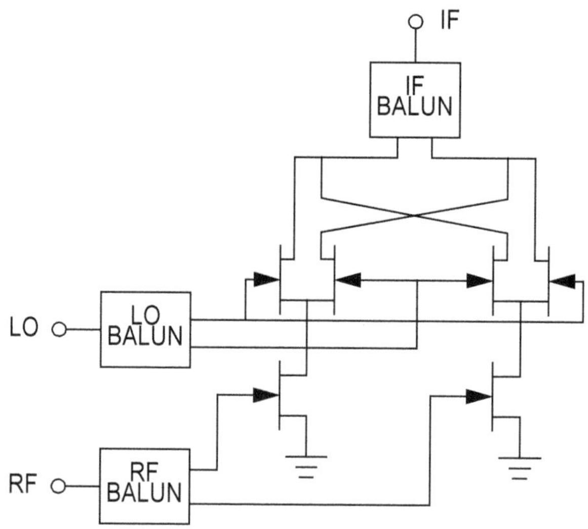

Figure 11.11 A doubly balanced FET mixer. This circuit can be viewed as a balanced interconnection of the singly balanced circuit in Figure 11.8(a).

constraints. Therefore, an active balun may introduces a substantial amount of noise and distortion.

- Amplitude and phase balance of the balun are often poor.
- The balun's impedances and frequency response are often different at its two outputs.
- The bandwidth and gain flatness of the balun may limit that of the entire mixer.

The greatest advantage of an active balun is its small size. Active baluns are much smaller than distributed ones, making them practical for integrated circuits.

Figure 11.12 shows two types of active baluns. The first, in Figure 11.12(a), uses the well-known property of a transistor amplifier in which the signal at its drain and source have, ideally, a 180-degree phase difference. In practice, this property exists only at low frequencies, and the large number of mid- and high-frequency poles in its equivalent circuit introduce substantial phase shifts. As a result, the voltage gain between the input and the two outputs is generally unequal, and the difference varies with frequency. In brief, it is difficult to achieve good phase and amplitude balance with this circuit.

Figure 11.12(b) shows a more common approach: the use of a differential amplifier. This circuit suffers from the same problems as the singly balanced mixer in Figure 11.8(b) and described in Section 11.3.1. The low drain-to-source resistance of the current-source device is in

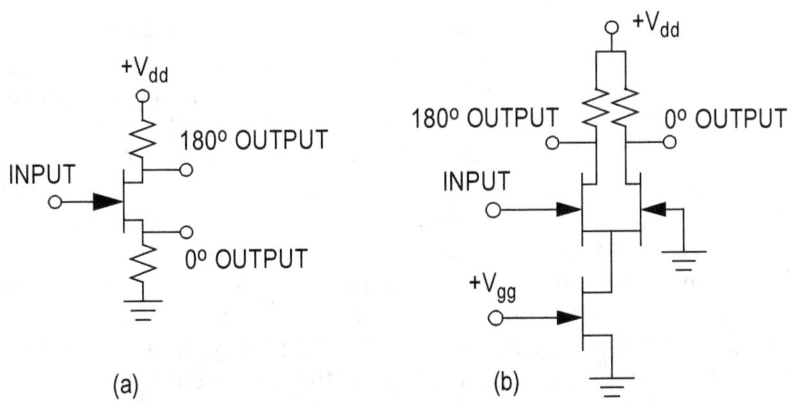

Figure 11.12 Two active balun circuits: (a) a classical phase splitter and (b) a differential amplifier.

parallel with one gate-to-source junction but not the other, so it unbalances the balun.

11.3.5 Gilbert-Cell Mixers

The Gilbert multiplier [11.5] was originally conceived to be a bilinear, four-quadrant analog multiplier. It is a BJT circuit using a diode-connected transistor as a linearizer; the logarithmic $V(I)$ response of the diode cancels the transistor's exponential I/V characteristic.

For a Gilbert multiplier to operate as a bilinear multiplier, it must operate at frequencies where the transistors' capacitances are negligible. Even then, the noise figure may be high. In RF and microwave circuits, the devices' capacitance is not negligible, and high noise figure is not tolerable, so some other mode of operation is needed. Still, except for the linearizing devices, the circuit is essentially the same as the original, and thus bears the same name.

A Gilbert multiplier is a doubly balanced mixer. As such, it is similar to the FET mixer in Figure 11.11, and operates in much the same manner. In RF applications, the LO is applied to the upper devices, which operate as commutating switches. The lower transistors realize a differential amplifier, whose outputs are modulated by the switching devices. The virtual-ground conditions described in Section 11.3.3 apply to a bipolar Gilbert mixer as well as to a doubly balanced FET mixer.

Figure 11.13 shows a Gilbert-cell mixer. The dc current source is helpful for setting the bias of the mixer devices, but if the RF signals are applied as shown, it is not essential. In many cases, however, the RF is applied with no balun [in a manner similar to Figure 11.12(b)]. The current source, when needed, is designed as in any differential amplifier. Usually, it is realized by a single BJT in a Wilkinson connection.

Unlike the drain-to-source impedance of FETs, bipolar devices usually have a high low-frequency collector-to-emitter impedance. At high frequencies, the collector-to-base feedback reduces the collector-to-emitter impedance, but the effect is usually not as severe as in FETs, so practical current sources are possible. This characteristic allows Gilbert multipliers to be operated, at low to moderate frequencies, without baluns.

The design of a Gilbert multiplier parallels that of a doubly balanced FET mixer, which parallels that of a singly balanced mixer. The latter is described in Sections 11.3.1 and 11.3.2. The key to the design is to recognize that the ungrounded terminal of the dc current source is a virtual ground for the RF emitters, and that the RF collectors are virtual grounds for the LO devices. Then, to design the individual parts of the circuit, the RF and LO devices can be treated as simple, common-emitter stages. The

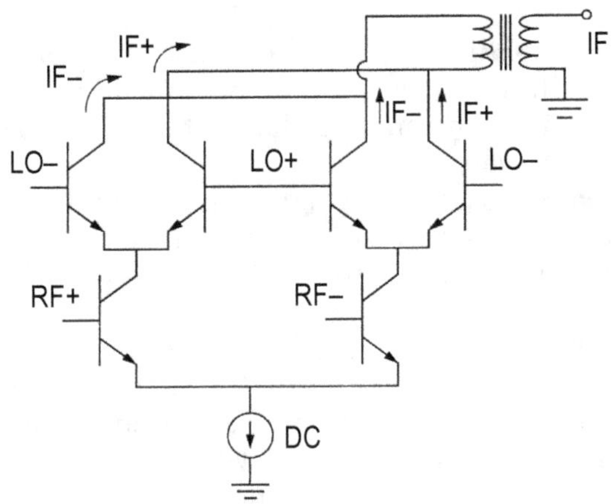

Figure 11.13 Gilbert mixer circuit. The baluns are not shown, but the polarity of the applied RF and LO signals, and the IF polarity, are as indicated.

RF devices are designed as a differential amplifier, to have constant gain over the RF frequency range. This requires, in turn, that the base-to-emitter voltage be constant with frequency. Matching to the LO devices is designed similarly, even though they operate primarily as switches.

It is often helpful to add emitter resistance (emitter degeneration) to the RF and LO devices, to obtain flat gain. Usually only a few ohms are necessary; too much feedback of this type reduces the gain-bandwidth product and can cause instability.

11.4 FET RESISTIVE MIXERS

The FET resistive mixer is a relatively new idea. It was first described in [11.6], and a balanced version was described in [11.7]. Since then, many such mixers have been reported, occasionally in the form of commercial products. The advantages of such mixers are very low distortion, low $1/f$ noise, and no shot noise; the conversion loss of such mixers is comparable to diode mixers, around 6 dB, and, since the high-frequency noise is virtually entirely thermal, the noise figure equals the conversion loss.

11.4.1 Fundamentals

Although it uses a nonlinear device to realize it, a mixer is fundamentally a linear device; shifting a signal from one frequency to another obeys superposition, and is therefore a linear operation. This operation is performed by a time-varying, linear circuit element such as a time-varying resistor. We create this time-varying linear element by applying a large-signal LO to a nonlinear element, but there is no fundamental, theoretical reason why this is necessary. Any time-varying *linear* circuit is capable of mixing.

As long as we use a nonlinear device to perform the mixing operation, mixers have relatively high levels of intermodulation distortion, spurious responses, and other undesirable nonlinear phenomena. However, if we could obtain this time-varying element without nonlinearity, we could use it to realize a mixer having no distortion. All we need is some way to modulate a resistance (or some other linear parameter) at the LO frequency.

The channel of a FET, at low drain-to-source voltages, is a good approximation of a linear resistor. It becomes significantly nonlinear only above some minimum drain voltage. In most FETs, this occurs at a few tenths of a volt to 1V, depending on the gate voltage. At normal, small-signal voltages (a few millivolts), the FET's resistive channel is very linear.

The resistance of this linear channel can be modulated by applying an LO voltage to the gate. That voltage changes the depth of the depletion region under the gate and therefore the resistance of the channel. When the gate voltage drops below V_t, the FET's turn-on voltage, the channel becomes an open circuit; when the gate voltage reaches its maximum value (just below the value that causes gate-to-channel rectification, about 0.5V), the channel resistance drops to a few ohms. This range of resistances is entirely adequate to achieve good conversion performance in a resistive mixer; it is, in fact, not very different from the junction resistance of a diode mixer.

FET resistive mixers are based on this principle. To realize such a mixer, we must do the following:

- Apply the LO to the gate; dc gate bias is usually necessary as well;
- Apply the RF to the drain;
- Filter the IF from the drain;
- Short circuit the LO at the drain;
- Especially, *apply no dc bias to the drain*!

Of course, appropriate filtering is required to separate the RF from the IF. Filtering is also necessary to prevent LO leakage from being coupled to the drain, through the gate-to-drain capacitance, C_{gd}, and pumping the drain conductance. This latter point is important; because the drain is unbiased, the gate-to-drain capacitance, C_{gd}, is much greater than it would be in a more conventional application, such as an amplifier. When the FET's dc drain voltage is zero, the gate-to-channel capacitance is approximately equally divided between C_{gd} and the gate-to-source capacitance, C_{gs}. Therefore, the mixer's matching circuits must be designed to short-circuit the drain at the LO frequency. Similarly, the gate should be short-circuited at the RF frequency to prevent the RF voltage from being coupled to the gate and introducing nonlinearity by varying the channel conductance. This latter requirement is less important than the former, and therefore sometimes can be ignored.

FET resistive mixers can achieve low conversion loss with surprisingly low LO power. At low LO levels, the RF input and IF output impedances are usually relatively high and dc gate bias must be adjusted carefully. At low LO levels, distortion performance is poor, often worse than that of a diode mixer. As LO level is increased, distortion performance improves, the optimum dc bias becomes more negative, and the conversion loss becomes less sensitive to bias. Minimum distortion in MESFETs and HEMTs occurs when the LO drive is just short of causing breakdown on negative peaks, or rectification on positive peaks. In MOSFETs, the optimum level is less pronounced; as LO level is increased, a point of diminishing returns is gradually reached, and further increases in LO power provide little improvement in performance.

11.4.2 Single-FET Resistive Mixers

Figure 11.14 illustrates the basic, single-device circuit. The LO, RF, and IF are applied as specified above, and provision is included for dc bias at the gate. The gate bias voltage is usually somewhat lower (i.e., more negative) than V_t; the result is a pulsed conductance waveform little different from that of a diode. The dc drain voltage must be 0V; to guarantee this, it may be necessary to create a path to ground with an RF choke, stub, or even a resistor.

Fortuitously, the channel's RF input and IF output impedances are usually surprisingly practical; when a conventional, 250-μm-wide MESFET or HEMT is used, the RF input impedance is usually close to 50Ω, and the IF output impedance is often the same or a little higher, perhaps 50Ω to 100Ω. Because FET's parasitic capacitances are substantial, greater than those of a diode, these impedances usually have a significant

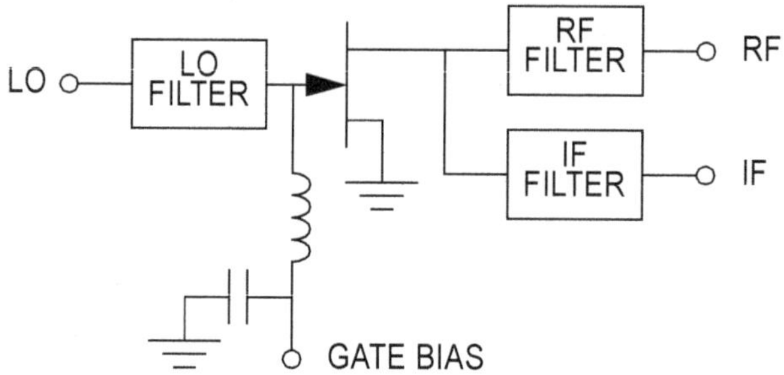

Figure 11.14 Single-FET resistive mixer.

imaginary part as well. The values of input and output impedance depend somewhat on LO level and dc bias, although the latter are best adjusted to achieve low distortion, and should not be used as tuning adjustments to minimize port VSWR.

The LO input impedance at the gate is largely the same as in any common-source FET. See Section 11.1.4 for ways to handle the high input Q.

Because the FET's channel is purely resistive, its noise is almost entirely thermal. Therefore, in terms of its noise, a FET resistive mixer should behave as a simple passive attenuator, with an effective temperature equal to the mixer's physical temperature. This is somewhat lower than that of a diode mixer, which includes both shot and thermal noise.

11.4.3 Design of Single-FET Resistive Mixers

The design of single-FET resistive mixers is straightforward. LO input matching is essentially the same as for any common-source FET; the only addition is the need to short-circuit the gate at the RF frequency. Although the gate RF short is theoretically optimum, we have found, in practice, it is not essential; the mixer works almost as well without it.

Drain matching is a more complex problem. The LO short circuit at the drain is essential at frequencies where significant gate-to-drain coupling, through the large gate-to-drain capacitance, can be expected. At the same time, the circuit must provide a conjugate match to the drain. Finally, it must separate the IF signal from the RF with appropriate isolation.

If the IF frequency is not very high (say, less than 10% of the RF), the LO frequency is close to the RF, and it may be difficult to provide the necessary short circuit at the drain without affecting RF matching. In such cases, a balanced mixer is probably the best solution.

LO input and RF output impedances can be found in the same manner as with other circuits: create an ideal circuit, one having an ideal RF-IF diplexer, ideal dc bias circuits, and drain-shorting structure, but the complete FET model. Then, optimize its bias voltage and LO level, and calculate the FET's RF and LO input impedances and IF output impedance. Finally, design the matching circuits and replace the ideal circuits, one at a time, with the real ones; when each circuit is replaced, do essential tuning to make sure that the performance of the ideal circuit is retained.

11.4.4 Design Example: FET Resistive Mixer

As an example, we design a resistive FET mixer having an RF frequency of 14 to 16 GHz, IF from 2.5 to 4.5 GHz, and a fixed LO frequency of 11.5 GHz. We use the same FET as in the active-mixer example. It is described by a Curtice model, which we assume to be accurate near $V_{ds} = 0$. The circuit will be fabricated in hybrid form on an alumina substrate.

The "ideal" mixer (which is not entirely ideal, as it includes the full nonlinear FET model) is shown in Figure 11.15. The circuit is used to determine the optimum dc bias, LO level (as indicated by the voltage variation at the gate) and port impedances. It uses an ideal combiner to separate the IF and RF signals and an ideal, high-Q, series-resonant circuit

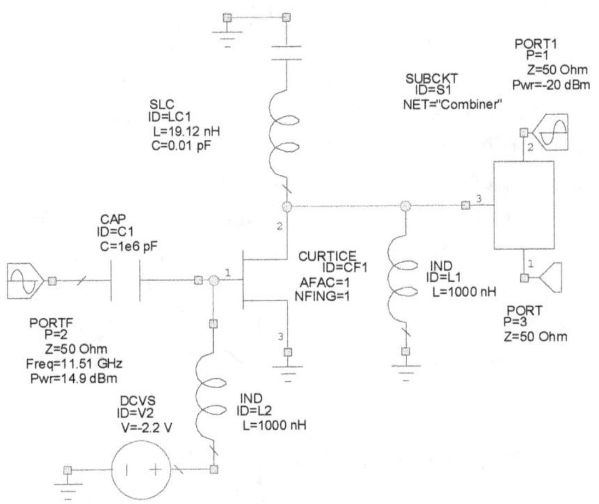

Figure 11.15 Ideal FET mixer circuit used to obtain port impedances, gate-bias voltage, and the optimum LO level.

to short-circuit the drain at the LO frequency. We quickly determine that the optimum bias is –2.2V and the LO voltage at the gate varies from almost –5V to a few tenths of a volt positive. The LO input impedance is $6 - j\,58$ and the RF input impedance is $49 + j\,18$. The slightly inductive RF port impedance is no great surprise; it is a consequence of the mixing phenomena. The IF output impedance is $49 + j\,4$. The RF and LO port impedances are particularly convenient. They are not unusual, however, for conventional, 250 μm × 0.25 μm Ku-band MESFETs at this frequency. The ideal circuit has 7-dB conversion loss.

Converting this ideal circuit to a real one requires matching the LO, changing the LO level to retain the correct voltage variation at the gate, and designing a practical diplexer to replace the combiner. The diplexer must provide the drain short circuit as well.

We begin at the gate. Since the LO is fixed frequency, a straightforward stub-matching circuit is adequate. By monitoring the gate voltage in the time domain, we see that 5 dBm of LO power is adequate to achieve the desired gate voltage variation. Because of the FET's high input Q, the stub-matching circuit is a sensitive element; it probably will require manual tuning if a low LO input VSWR is necessary.

The diplexer is a much more difficult design. For the RF filter, we select a quarter-wave, coupled-line structure. A series line is adjusted to make the filter's output impedance equal to zero at the LO frequency, and, as with all such filters, the low-frequency output impedance is very high. The IF filter is a simple stub structure. It includes an RFC, at the input end, to guarantee that $V_{ds} = 0$ at dc. The last section, at its output (the drain end of the filter) is adjusted to present an open circuit at the RF frequency. Finally, the two filters are connected to form the diplexer, and the performance is adjusted by a little additional tuning. Figure 11.16 shows the filters and the diplexer's passbands.

Finally, we replace the drain circuitry of the ideal mixer with the diplexer and calculate the performance. With no further tuning or optimization, the conversion loss is 7 to 8 dB across the band, and the LO and RF port VSWRs are less than 1.5. The circuit is shown in Figure 11.17.

As with previous examples, to maintain lucidity, we have left out the microstrip discontinuity parasitics. These should be included in a final design.

11.4.5 Balanced FET Resistive Mixers

The advantages and disadvantages of balanced FET resistive mixers are essentially the same as those of other types of balanced mixers. There is, however, one important consideration in the design of such mixers. We

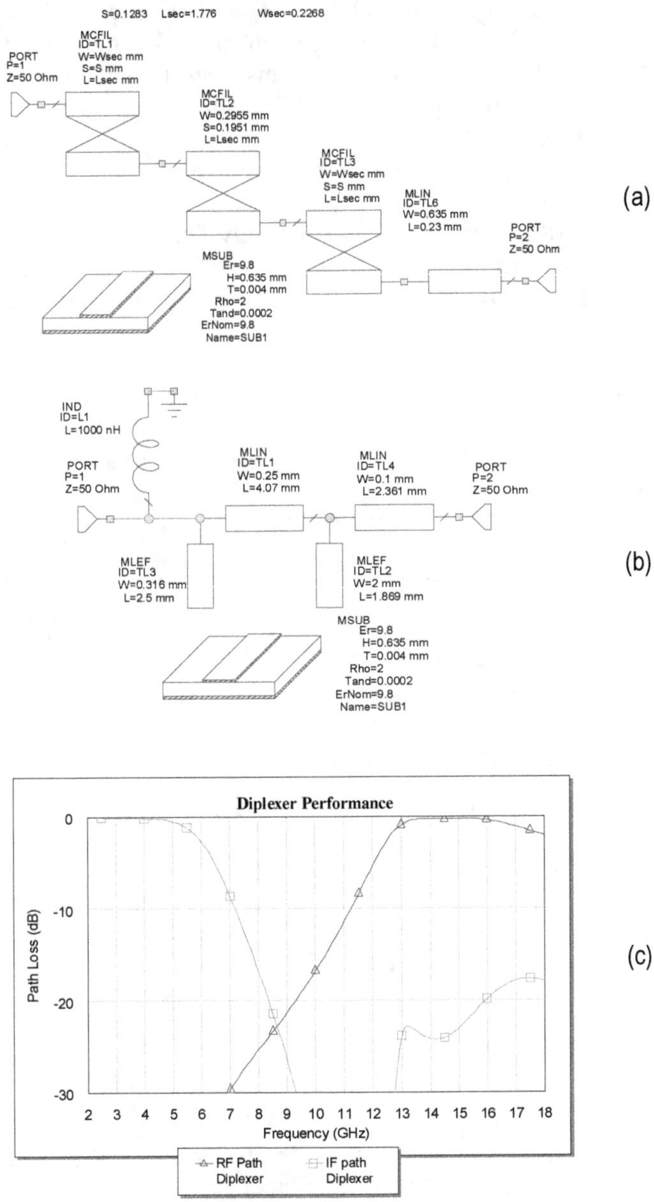

Figure 11.16 Diplexer filters and performance: (a) RF filter; (b) IF filter; (c) transmission through both ports.

noted earlier that the mixer's matching circuits must be designed to short-circuit the LO at the drain and the RF at the gate. Unfortunately, most microwave baluns are driven in an even mode by the waveform they are required to short-circuit, and the baluns present an *open* circuit to such excitation. A very few types of baluns present a short circuit to even-mode excitation; for example, a simple transformer provides the appropriate termination. So does the half-wave "hairpin" balun, which we shall describe presently.

Figure 11.18 shows a microstrip singly balanced FET resistive mixer. The LO is applied to the gates through a balun, and the RF is applied to the drains in phase. It is best, in this circuit, to use baluns and direct connection

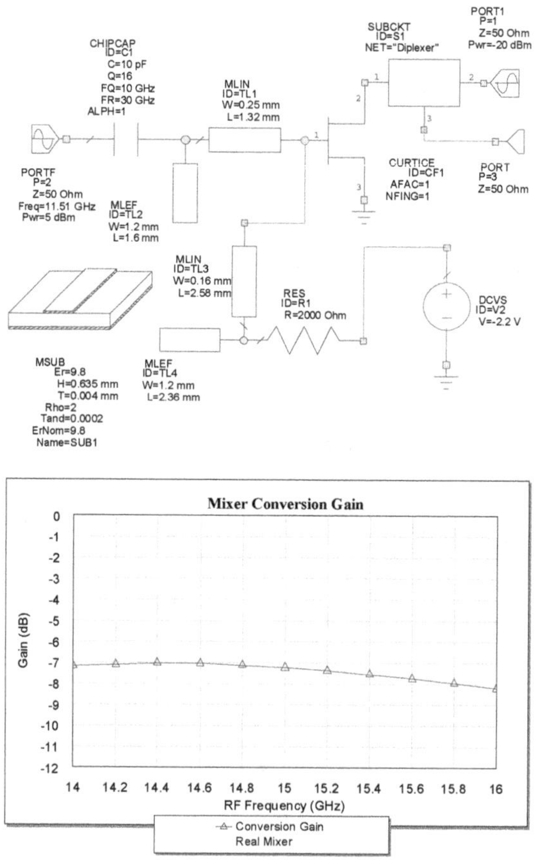

(a)

(b)

Figure 11.17 (a) Complete mixer circuit, and (b) conversion loss.

of the RF; this configuration provides optimum drain and gate terminations. In particular, hybrids or power dividers should not be used; these do not terminate the gates or drains optimally.

The drains are connected together at all frequencies except the IF by capacitors, thus providing an LO virtual ground at the drains. The LO balun, consists of a *u*-shaped half-wavelength "hairpin" transmission line. Although the simple balun shown in Figure 11.18 does not have very wide bandwidth, it presents a short circuit to even-mode excitation, in this case, RF leakage. The balun's bandwidth can be increased by the use of a multisection structure. As with active balanced mixers, the IF currents in the drains are out of phase; thus, an output balun or hybrid must be used to combine them. RF tuning can be applied to the line from the capacitors to the RF terminal. LO tuning elements should not be located on the hairpin; they should be placed between the hairpin and the LO terminal.

It is possible to have a singly balanced mixer structure in which the gates are driven in phase by the LO and the drains out of phase by the RF. In this case, the gates are a virtual ground for the RF, which is certainly desirable, but the drains are no longer LO virtual grounds. In this case, some type of filter must be used to short-circuit the drains, and this complicates the design somewhat.

Figure 11.19 shows a doubly balanced FET resistive mixer. Such mixers are very practical for RF applications; with microwave baluns instead of transformers, they can be used at high frequencies as well. This

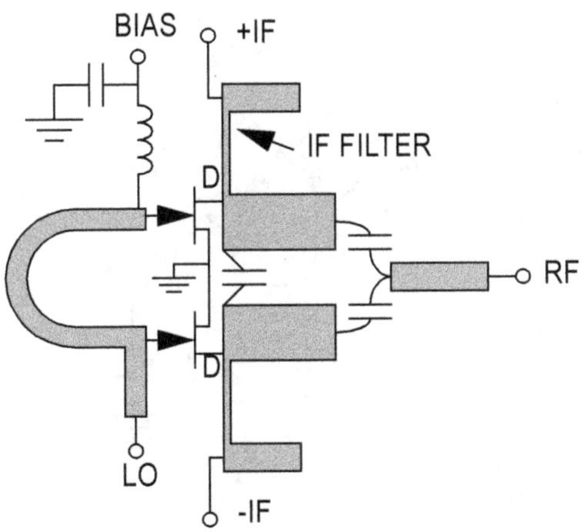

Figure 11.18 A singly balanced FET resistive mixer using a half-wavelength "hairpin" LO balun.

is a type of commutating mixer, and it operates very much like the ring diode mixer described in Section 6.4.3. The FET mixer, however, requires three hybrids instead of the diode mixer's two. The RF, LO, and IF are connected to the ring by these hybrids. All four corners of the ring are virtual grounds for the LO; the IF connection points are virtual grounds for the RF, the RF connection points are virtual grounds for the IF, and the gates are virtual grounds for both. The existence of these virtual grounds implies that the RF, LO, and IF are inherently isolated.

The circuit of this mixer includes tuning inductors in the RF and LO paths. At low frequencies, these may not be necessary; conversely, if the IF frequency is high, some IF tuning may be needed. As with diode mixers, it is best to minimize tuning (to preserve balance) and to achieve matching by adjusting the device sizes and, when practical, the output impedances of the baluns or transformers. As shown in Figure 11.19, the mixer can operate with either positive or negative dc bias; negative bias is applied directly to the gates, while positive bias is applied equally to the four source and drain connection points, preserving the $V_{ds} = 0$ condition but creating a negative gate-to-source voltage. The unused bias terminal should be grounded.

The virtual-ground properties listed above can be used to make the design of such mixers entirely straightforward. Note that, when the virtual grounds are considered, each balun is terminated in two parallel sets of two series impedances. Figure 11.20 illustrates this situation for the LO circuit; the RF and IF follow directly. In Figure 11.20(a), we see that the LO balun

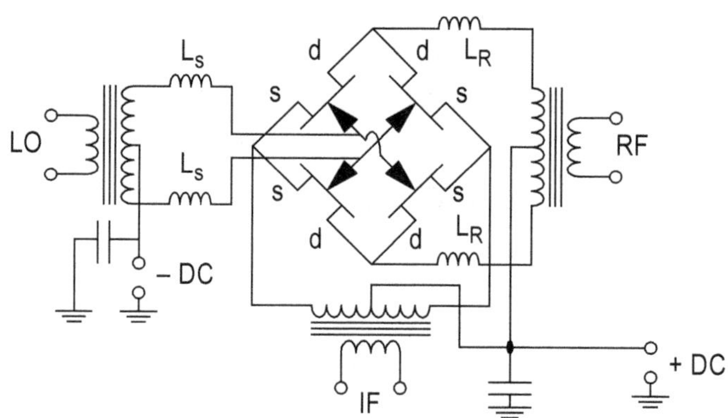

Figure 11.19 Commutating ring mixer using resistive FETs. The circuit is configured so that either positive or negative bias can be used. The operation of this mixer is similar to the diode ring mixer.

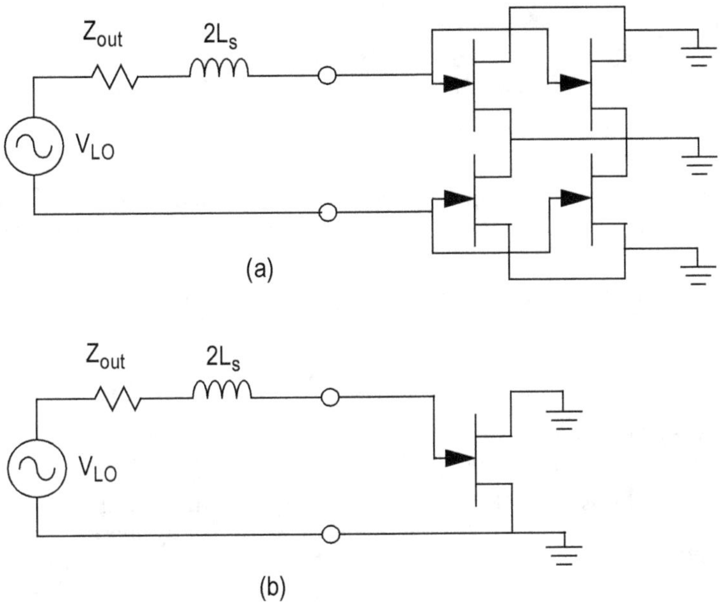

Figure 11.20 (a) LO equivalent circuit of the ring-FET resistive mixer; (b) the single-device equivalent circuit.

drives four gates, two in series and two in parallel. The load is equivalent to a single device, shown in Figure 11.20(b).

The FET ring mixer has the same intermodulation and spurious-response rejection properties as a diode-ring mixer: all even-order products are rejected. Good rejection requires careful balance, a condition not difficult to achieve at RF frequencies. At high frequencies, however, the balance can be upset easily by the large number of parasitics introduced by the inevitably complex layout. Good odd-order distortion performance requires hard pumping of the devices; the FETs must be driven as hard as possible, but the gate-to-channel junction must not be allowed either to rectify the LO or to break down. In this respect, silicon MOS devices are ideal, because of the linearity of their channel resistance and lack of gate rectification. MOS devices are somewhat limited in frequency, however, so MESFETs or HEMTs may be necessary for microwave realizations.

References

[11.1] R. A. Pucel, D. Masse, and R. Bera, "Performance of GaAs MESFET Mixers at X-Band," *IEEE Trans. Microwave Theory Tech.*, Vol. MTT-24, 1976, p. 351.

[11.2] S. A. Maas, *Theory and Analysis of GaAs MESFET Mixers*, Ph.D. Diss., University of California, Los Angeles, 1984.

[11.3] M. Case et al., "An X-Band Monolithic Active Mixer in SiGe HBT Technology," *IEEE MTT-S International Microwave Symposium Digest*, 1996, p. 655.

[11.4] C. Tsironis, R. Meierer, and R. Stahlmann, "Dual-Gate MESFET Mixers," *IEEE Trans. Microwave Theory Tech.*, Vol. MTT-32, 1984, p. 248.

[11.5] B. Gilbert, "A Precise Four-Quadrant Multiplier with Subnanosecond Response," *IEEE J. Solid-State Circuits*, Vol. SC-3, 1968, p. 365.

[11.6] S. A. Maas, "A GaAs MESFET Mixer with Very Low Intermodulation," *IEEE Trans. Microwave Theory Tech.*, Vol. MTT-35, 1987, p. 425.

[11.7] S. A. Maas, "A GaAs MESFET Balanced Mixer with Very Low Intermodulation," *IEEE MTT-S International Microwave Symposium Digest*, 1987, p. 895.

Chapter 12

Transistor Oscillators

This chapter is concerned with the design and nonlinear analysis of high-frequency FET and bipolar oscillators. We begin with the classical approach to both feedback and negative-resistance oscillators, and then reexamine these classical concepts in view of our understanding of nonlinear circuits. Finally, we examine some practical circuits and design techniques.

12.1 CLASSICAL OSCILLATOR THEORY

12.1.1 Feedback Oscillator Theory

An amplifier circuit can be made to oscillate by feeding some of its output energy back to the input. The oscillation conditions for such a circuit are well known. A feedback oscillator is shown schematically in Figure 12.1; the gain of the feedback circuit, A_v, can be found easily to be

$$A_v = \frac{V_o}{V_i} = \frac{A}{1 - AF} \tag{12.1}$$

where A is the voltage gain of the amplifier and F is the voltage gain of the feedback network; these are generally complex. Clearly, as $AF \rightarrow 1$, $A_v \rightarrow \infty$, implying that an output is possible with a vanishingly small input. The condition $AF = 1$ shows that oscillation occurs when (1) the loop gain, AF, is unity, and (2) the loop phase is zero. If these conditions are established at some particular frequency, the circuit can oscillate at that frequency.

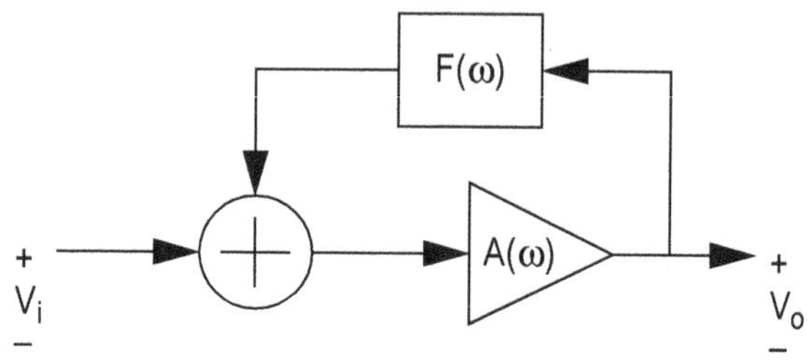

Figure 12.1 A model of a feedback oscillator.

This conclusion comes from linear circuit theory. In effect, it guarantees that the transfer function has a pole on the real axis. This doesn't tell us much about how the oscillator actually operates; for example, the output level is indeterminate. To find out what we really need to know, we must examine the circuit's nonlinear behavior. In a real oscillator, the circuit is unstable in the linear sense; that is, its transfer function has a pole in the right half plane, near the $j\omega$ axis. This allows any small perturbation, such as noise or the turn-on transient, to create a sinusoidal output whose magnitude increases exponentially with time. Eventually, the amplifier saturates, limiting the output, and decreasing the gain to the level where (12.1) is satisfied.

Since the amplifier is operated in saturation, nonlinear analysis is necessary to determine oscillation frequency and output level. Nevertheless, linear analysis can be used for an approximate, initial design, as long as we recognize the limitations of the linear analysis and modify it appropriately. Most importantly, we must modify the oscillation condition to $AF > 1$ to put the transfer-function pole in the right half plane, to allow the oscillation to commence. Then, as the oscillation builds, the amplifier saturates, the gain decreases, and the oscillation stabilizes at $AF = 1$. We shall examine this paradoxical idea of a "stable oscillation" in Section 12.1.3.

To illustrate the parts of a feedback oscillator, we use the Colpitts circuit in Figure 12.2. Figure 12.2(a) shows the oscillator, and Figure 12.2(b) shows its simplified equivalent circuit. We can readily see that the controlled source represents the amplifier portion of the circuit, and the *pi* network represents the feedback portion. In this case, we identify

$$\begin{aligned} A &= -g_m \\ F &= Z_{2,1} \end{aligned} \tag{12.2}$$

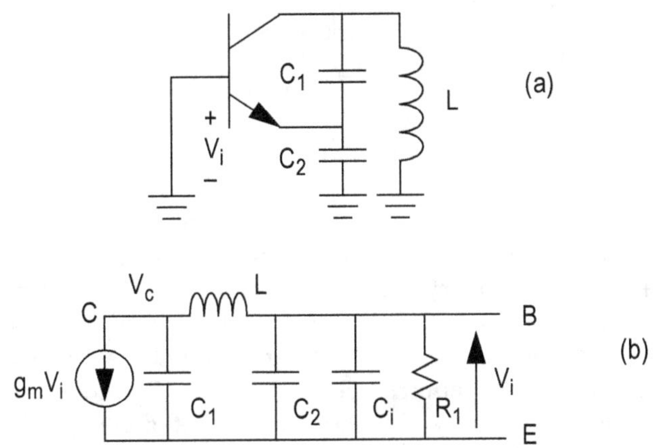

Figure 12.2 (a) A transistor Colpitts oscillator, and (b) its equivalent circuit. C_i and R_1 are the input (base-to-emitter) capacitance and resistance of the transistor.

where $Z_{2,1}$ is a Z parameter of the *pi* network. The oscillator could be designed by deriving $Z_{2,1}$ and finding the conditions for which $AF = 1$.

A more elegant approach is to recognize that the nodal equations of the network are

$$\begin{bmatrix} g_m - \dfrac{1}{j\omega L} & \dfrac{1}{j\omega L} + j\omega C_1 \\ j\omega C_t + \dfrac{1}{R_1} + \dfrac{1}{j\omega L} & -\dfrac{1}{j\omega L} \end{bmatrix} \begin{bmatrix} V_i \\ V_c \end{bmatrix} = \begin{bmatrix} 0 \\ 0 \end{bmatrix} \tag{12.3}$$

where $C_t = C_2 + C_i$, C_i is the input capacitance, and we have switched the ground node to the emitter. We note that the input resistance $R_1 = \beta/g_m$ and that the determinant must be zero for this system of equations to have a nontrivial solution. A little algebra gives

$$\beta = \frac{C_1}{C_t} \tag{12.4}$$

(which means, in practice, that $\beta > C_1/C_t$) and the resonant frequency,

$$f_0 = \frac{1}{2\pi}\sqrt{\frac{C_1 + C_t}{LC_1C_t}} \tag{12.5}$$

12.1.2 Feedback Oscillator Design

More generally, we have the case shown in Figure 12.3 [12.1]. The oscillator consists of a transistor and some type of transmission resonator. The resonator can be a crystal, an LC circuit, a surface acoustic wave (SAW) device, an electromagnetic resonator coupled to a pair of ports, a ceramic piezoelectric device, or anything else that resonates at the desired frequency and has other required characteristics. Z_s and Z_L are not used in the circuit; they exist only for the purpose of analysis. Z_T is the load connected to the oscillator's output port.

The circuit is adjusted until the following conditions are obtained:

$$\begin{aligned} |S_{2,1}| &> 1.0 \\ \angle S_{2,1} &= 0 \\ Z_s = Z_L &= Z_{\text{in}} \end{aligned} \tag{12.6}$$

These conditions are equivalent to $|AF| > 1$ and $\angle AF = 0$. When these conditions are obtained, we need only connect the collector to the input of the resonator to complete the design.

The resonator is a critical part of the design. If it is coupled very weakly to the circuit, its loss is high, but so is its Q. A high Q, as we shall see, results in low noise and makes the resonator, not the transistor, dominant in setting the oscillator frequency. This is a desirable situation, because, with proper care in its design, the resonator is thermally more

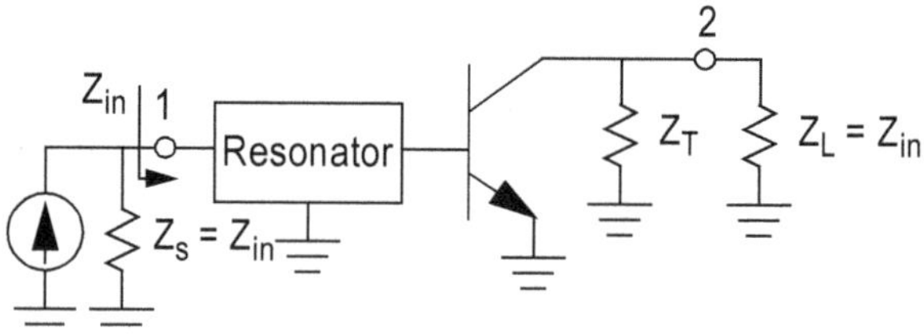

Figure 12.3 A circuit for calculating the open-loop gain of an oscillator. This circuit can be analyzed in terms of S parameters, making it useful for design by a microwave circuit simulator.

stable than the transistor. High resonator loss, however, makes it more difficult to satisfy the gain condition, $|S_{2,1}| > 1$.

Figure 12.4 shows an example of this approach to oscillator design. The circuit shows a 900-MHz voltage-controlled oscillator (VCO) using a bipolar transistor. The transistor is described by scattering parameters, so the circuit includes no bias source, but because of their effect on the gain, the bias resistors must be included. The resonator consists of the inductor and capacitor L_2 and C_4; C_4 represents a varactor, and L_1 is its bias RF choke. The 14-pF capacitors C_3 and C_6 adjust the coupling to the resonator. R_2 is the 50Ω output port. The value of the source and load resistance used to calculate the gain is treated as a variable quantity.

In adjusting the circuit, we try to achieve a linear gain of at least 6 dB, and preferably 10 dB. This allows margin for circuit losses and ensures

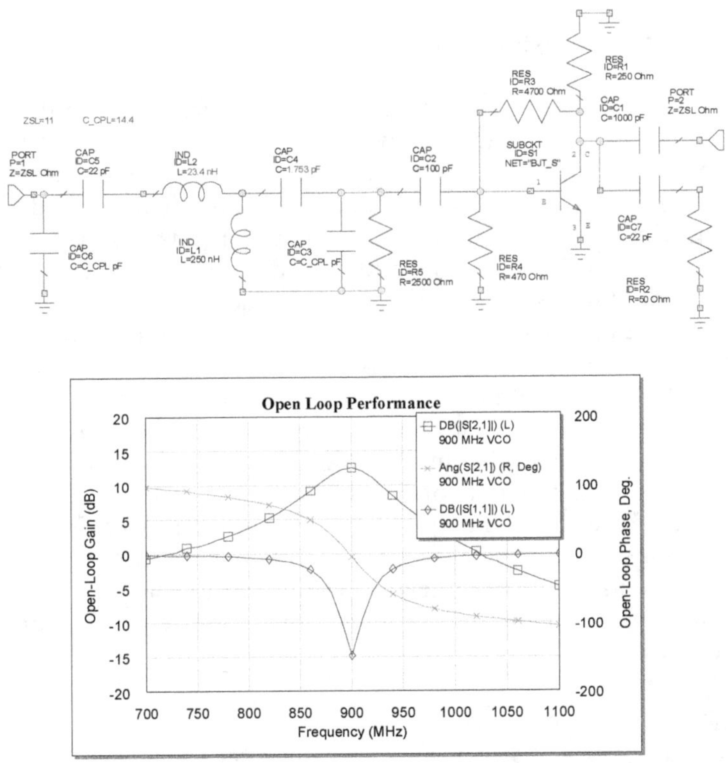

Figure 12.4 The open-loop model and performance of a 900-MHz VCO. C_4 and L_2 are the resonator, while C_6 and C_3 adjust the coupling. R_2 is the load. Other resistors provide bias and limit the low-frequency gain.

reliable start-up. The plot of $|S_{2,1}|$ shows a peak of 12.5 dB and zero phase at the desired frequency of 900 MHz, and the plot of $|S_{1,1}|$ indicates that the input impedance is also close to the source and load values.

In a feedback oscillator, it is relatively easy to avoid spurious resonances, which could cause the oscillator to oscillate at an undesired frequency. As long as the resonator has transmission only at the resonant frequency, a condition not difficult to establish, the oscillator can oscillate only at the desired frequency. Feedback oscillators, unfortunately, can be difficult to design at high frequencies, because of phase shift in the long connection from the amplifier output to the resonator, so high-frequency oscillators are usually designed by means of a negative-resistance theory. In Section 12.1.5 we shall see some examples of high-frequency negative-resistance oscillators; because they use feedback to establish the negative resistance, they also can be considered feedback oscillators.

12.1.3 Negative-Resistance Oscillation

A general understanding of the operation of electronic oscillators has existed almost as long as active devices. However, more recent work by Kurokawa [12.2] is the basis for the design of modern negative-resistance microwave oscillators. In this work, a microwave oscillator is modeled as a one-port in which the real part of the port impedance is negative. The one-port can represent a two-terminal solid-state device, such as a Gunn device or tunnel diode, that exhibits "negative resistance," meaning that its port impedance has a negative real part. It can also represent one port of a two-port that includes appropriate feedback.

An oscillator modeled in this manner is shown in Figure 12.5. The load impedance $Z_L(\omega)$ is linear, but the source impedance $Z_s(I_0, \omega)$ (the output impedance of the oscillator) is modeled in an unusual fashion: it is a linear impedance that is a function of I_0, the magnitude of the fundamental-frequency component of the output current. The real part of Z_s is negative and decreases with an increase in I_0. Although no linear impedance behaves in this manner, a nonlinear impedance can behave this way if the current and voltage harmonics are ignored. Precisely, we define the impedance as

$$Z_s(I_0, \omega) = \begin{cases} \dfrac{V(\omega)}{I(\omega)} & \omega = \omega_p \\ 0 & \omega = n\omega_p \end{cases} \tag{12.7}$$

so the voltage across the device is zero at all harmonics. Additionally, we assume that the harmonic components of $Z_L(\omega)$ are zero, so any harmonic currents that may exist are of no consequence. The small-signal source $v(t)$ in Figure 12.5 represents a perturbation in the voltage across the combined impedances; in practical circuits it represents noise, an injection-locking signal, or the turn-on transient of the circuit.

Kurokawa proved that the conditions for oscillation are

$$Z_s(I_0, \omega) + Z_L(\omega) = 0 \tag{12.8}$$

that is, the real parts of the impedances cancel and the imaginary parts resonate. Then, if an infinitesimal perturbation $v(t)$ exists, the magnitude of the response $i(t)$ increases exponentially with time and becomes sinusoidal at some frequency ω_p where $\mathrm{Im}\{Z_s(\omega_p)\} = -\mathrm{Im}\{Z_L(\omega_p)\}$.

In real oscillators, the condition is slightly different. We must have, at start-up, $\mathrm{Re}\{Z_s\} + \mathrm{Re}\{Z_L\} < 0$. Then, as the amplitude of the oscillation, I_0, increases, $|\mathrm{Re}\{Z_s\}|$ decreases and eventually stabilizes at the point where (12.8) is satisfied. If I_0 were to increase beyond the point at which (12.8) is satisfied, I_0 would decrease, and eventually $|\mathrm{Re}\{Z_s\}|$ would rise to the point where (12.8) would again be valid. Thus, the value of I_0 that satisfies (12.8) is stable, so I_0 remains at that level and oscillation continues at a constant amplitude. The decrease in $|\mathrm{Re}\{Z_s\}|$ with increasing I_0 is an inevitable consequence of the fact that the amplitude of $i(t)$ cannot, in practice, become infinite.

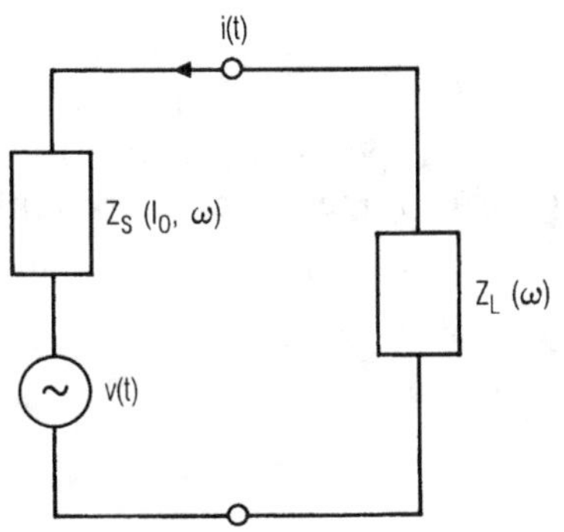

Figure 12.5 The classical model of a negative-resistance oscillator. The voltage source $v(t)$ provides a perturbation necessary to start oscillation in the unstable circuit.

The source could also be described by a nonlinear conductance $Y_s(V_0, \omega)$; then, the oscillation condition is

$$Y_s(V_0, \omega) + Y_L(\omega) = 0 \tag{12.9}$$

where, analogous to (12.7),

$$Y_s(V_0, \omega) = \begin{cases} \dfrac{I(\omega)}{V(\omega)} & \omega = \omega_p \\ 0 & \omega = n\omega_p \end{cases} \tag{12.10}$$

This is the case of a parallel resonance having a total negative conductance, in which the transient perturbation comes from a shunt small-signal current source, and V_0 is the magnitude of the shunt voltage. The oscillation begins when the real part of the shunt conductance is negative and $\mathrm{Re}\{Y_s\}$ decreases as the oscillation increases until (12.9) is satisfied.

In practice, Z_s or Y_s is realized by a solid-state device, which inevitably includes nonlinear capacitances. The average values of those capacitances—and thus $\mathrm{Im}\{Y_s\}$ or $\mathrm{Im}\{Z_s\}$—vary at least slightly with V_0 or I_0. Thus, the frequency at which oscillations begin (when V_0 or I_0 is small) is not necessarily the same as that for which (12.8) or (12.9) is satisfied (and V_0 or I_0 are large). Nevertheless, if a transistor oscillator circuit includes a high-Q resonator, that resonator, not the reactances of the solid-state device, will dominate in establishing the frequency. In a high-Q resonator, $\mathrm{Im}\{Y_L\}$ varies rapidly with frequency close to resonance, so changes in $\mathrm{Im}\{Y_s\}$ do not cause much frequency deviation.

The oscillation is stable if the sinusoidal voltage or current returns to its steady-state value after it is perturbed. Kurokawa derived a condition for stable oscillation; in terms of impedance, the condition is[1]

$$\frac{\partial R_s}{\partial I}\frac{\partial X_L}{\partial \omega} - \frac{\partial X_s}{\partial I}\frac{\partial R_L}{\partial \omega} > 0 \tag{12.11}$$

1. Some texts give an expression that appears to disagree with this one. The cause is a difference in sign convention. In [12.2], the device impedance was written as $Z_S = -R_s + jX_S$, where $R_s > 0$. More conventional notation, today, is $Z_S = R_S + jX_S$, where $R_s < 0$. We use the latter.

where $R_s = \mathrm{Re}\{Z_s\}$, $X_s = \mathrm{Im}\{Z_s\}$, $R_L = \mathrm{Re}\{Z_L\}$, $X_L = \mathrm{Im}\{Z_L\}$, and the derivatives are evaluated at $I = I_0$ and $\omega = \omega_p$. Note that, in a simple case where X_s is independent of I and the load is a simple series RL or RC circuit, (12.11) is always satisfied.

The idea of a "stable oscillation" seems, at first, to be a contradiction: the device has to be unstable to oscillate. In fact, we can define many types of stability. The classical concept of stability from linear circuit theory, which requires that all network poles remain in the right half plane, is only one. Stability factors, such as the K factor in small-signal amplifiers, is another type, which is not precisely the same as classical linear-network stability. In the present case, we seek a type of bounded stability, in which the magnitude of the oscillation is limited and returns to its steady state if perturbed. Such operation can occur only in a nonlinear circuit.

12.1.4 Negative Resistance in Transistors

We have already noted that negative resistance can occur from physical processes in certain two-terminal devices, including tunnel diodes and Gunn devices. It is also possible to obtain negative resistance at one port of an amplifier by introducing feedback. Our "amplifier" is usually just a transistor, and one port, usually the output, is terminated; the other port becomes, effectively, a two-terminal, negative-resistance device. Such circuits are arguably feedback oscillators, but we can view them equivalently as negative-resistance oscillators.

We saw in Chapter 8 that a two-port could oscillate if its source and load impedances were chosen appropriately. For such oscillation to occur, it must be possible to obtain an input or output impedance having a negative real part or, equivalently, an input or output-reflection coefficient greater than unity. This condition can occur only if both $S_{2,1}$ and $S_{1,2}$ are nonzero, which requires that the two-port have forward gain and feedback. In designing small-signal amplifiers, we usually wish to minimize the effects of feedback; however, in oscillators, we do our best to enhance it, even to the point of introducing additional feedback, to cause the device to oscillate.

We now examine negative-resistance or negative-conductance phenomena in transistor circuits heuristically by means of a very simple model, and we apply our understanding of large-signal and small-signal properties of nonlinear circuits to show how $\mathrm{Re}\{Z_s\}$ or $\mathrm{Re}\{Y_s\}$ changes with I_0 or V_0. Figure 12.6(a) shows an ideal FET and a feedback network F; we assume that the magnitude of the voltage gain of F is A_F, its phase shift is 180 degrees, and the port impedances of F are infinite. V_d and I_d are the static or instantaneous voltage and current at the output terminals; the time-waveforms are $V_d(t)$ and $I_d(t)$. The FET has no capacitive or resistive parasitics (it does have an ideal Schottky-barrier gate-to-source junction), its transconductance is g_m, and it is biased at $V_d = V_{dc}$ and $I_d = I_{dc}$. Its transfer function thus has a 180-degree phase shift, making AF real. Because we are using this example to illustrate only some of the properties of negative resistance in transistor circuits, we have not included a resonator or any other reactive elements; these would be necessary in practice to establish a sinusoidal oscillation at some particular frequency.

By a simple small-signal linear analysis, one can show that the port conductance of the circuit G_s is

$$G_s = -g_m A_F \tag{12.12}$$

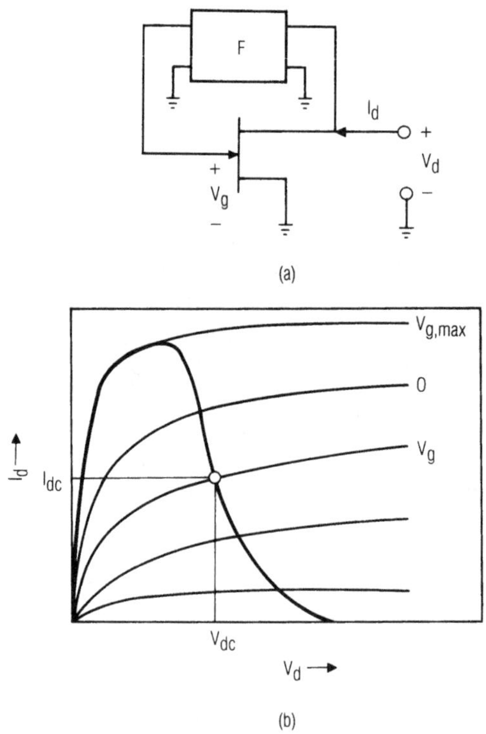

Figure 12.6 (a) An ideal FET and a phase-reversing network; (b) the resulting I/V characteristic at the terminals.

and G_s is negative for all positive values of g_m and A_F. G_s is the circuit's incremental port conductance in the vicinity of the bias point (V_{dc}, I_{dc}). The large-signal terminal I/V characteristic $I_d(V_d)$ is graphed on top of the FET's I/V curves. The slope of $I_d(V_d)$ is negative at the bias point, indicating negative conductance.

The negative conductance exists only over a limited range of V_d. Because of the clamping action of the gate-to-channel Schottky junction, the gate voltage V_g cannot increase beyond $V_{g,\,max}$, approximately 0.6V; accordingly, $I_d(V_d)$ follows the curve of constant $V_{g,\,max}$ at low values of V_d. Similarly, beyond the point where $-A_F V_d = V_t$, where V_t is the threshold (or pinch-off) voltage, I_d is zero. If V_d is increased further, I_d can increase only through avalanching or other second-order effects (e.g., changes in V_t with V_d). Thus, the device has an incremental conductance that is positive at low V_d, is negative over a region of V_d, and then becomes positive (or zero) again at even greater voltages.

Figure 12.7 shows how the ac part of $I_d(t)$, ΔI_d, behaves as the amplitude of the ac part of $V_d(t)$, ΔV_d, increases. When ΔV_d is very small, ΔI_d is also small, is nearly sinusoidal, and is 180 degrees out of phase with ΔV_d. The current waveform is shown in Figure 12.7(a): as V_d increases, I_d decreases, and the circuit exhibits negative conductance over the entire range of $V_d(t)$. The magnitude of the large-signal negative conductance is defined in a manner identical to that of the large-signal impedance that we encountered in the analysis of diode circuits:

$$G_s = \frac{I_{d,\,1}}{V_{d,\,1}} \tag{12.13}$$

where $V_{d,\,1}$ and $I_{d,\,1}$ are the fundamental-frequency components of $V_d(t)$ and $I_d(t)$, respectively (note that these are not precisely the same as ΔV_d and ΔI_d, which can contain harmonic components). The slope of the terminal I/V characteristic, Figure 12.6(b), is relatively constant in the vicinity of the bias point, so as long as V_d is small, G_s does not vary much. If V_d increases, however, it encounters the lower part of the I/V curve, which has a more positive slope. Then $I_{d,1}$ does not increase as fast as $V_{d,\,1}$, and $|G_s|$ decreases.

If the amplitude of ΔV_d increases further, the peaks of $V_d(t)$ eventually exceed the range of the negative-resistance region, and $I_d(t)$ has the waveform shown in Figure 12.7(b). The waveform shows two "dips" at its peaks; these occur when V_d enters the positive-resistance range. At this point the increase in $I_{d,\,1}$ with $V_{d,\,1}$ virtually ceases, and accordingly $|G_s|$ decreases rapidly, although G_s still remains negative. If V_d increases

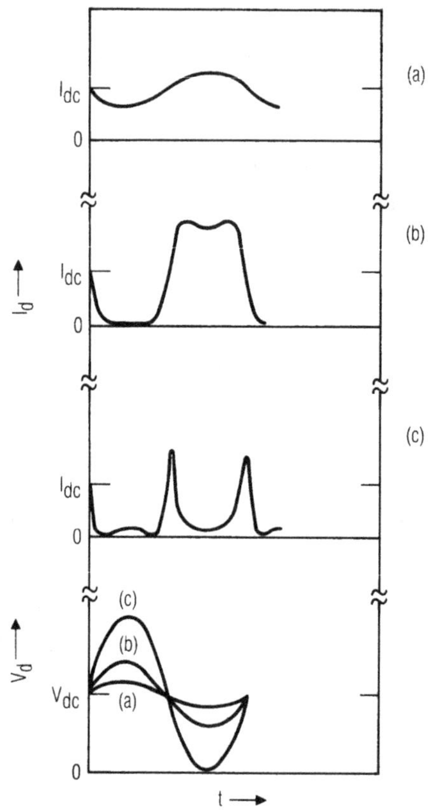

Figure 12.7 The voltage and current waveforms in the circuit of Figure 12.6(a): (a) $V_d(t)$ is entirely within the negative-resistance region; (b) $V_d(t)$ peaks at the edge of the region; (c) $V_d(t)$ peaks well outside the region.

further, these dips become deeper, and eventually a point is reached where $I_{d,1} = 0$ and therefore $G_s = 0$.

If ΔV_d is increased even further, the current waveform becomes as shown in Figure 12.7(c). In this case, ΔV_d is so great that $V_d(t)$ remains within the positive-resistance range over most of its period, remaining in the negative-resistance region only briefly while $V_d(t) \approx V_{dc}$. The only part of the $I_d(t)$ waveform that implies negative resistance is the rising part of the peak that occurs when $V_d \approx V_{dc}$; over the rest of the period, the variations in $I_d(t)$ are in phase with $V_d(t)$. Thus, throughout most of the period, $I_{d,1}$ is in phase with $V_{d,1}$, and consequently, the resulting positive resistance dominates, making $G_S > 0$. We see that as the amplitude of the oscillation increases, the large-signal conductance G_S increases from its incremental value, which is negative, to zero, and finally becomes positive.

In this example, we have "assumed away" all the reactive parts of the circuit. In a real oscillator, reactive parasitics would exist, and some type of resonator would be used to set the oscillation frequency; these elements would add an imaginary part to Y_S, with which the load would have to resonate. If the resonator were to have a high Q, its reactance would dominate the imaginary part of Y_S making $\mathrm{Im}\{Y_S\}$ very frequency-sensitive, and keeping the oscillation frequency close to the resonant frequency. Thus, the temperature stability of the oscillator would be essentially that of the resonator, and the resonator's narrow bandwidth would act as a filter to minimize phase and amplitude (AM) noise.

12.1.5 Oscillator Design by the Classical Approach

Although an oscillator is in reality a large-signal, nonlinear component, small-signal linear considerations are usually sufficient to ensure that oscillation conditions are met, and to approximate the operating frequency. A design based on linear theory is valid because the oscillator, at the onset of oscillation, is in fact a linear, small-signal component. If the frequency does not change appreciably as the amplitude of the oscillation increases, and if precise knowledge of the output power is not needed, small-signal design may be adequate by itself. By using nonlinear analysis, however, one can predict output power precisely and determine the voltage waveforms across critical components (such as a tuning varactor) in the circuit. The latter may be very valuable in maximizing power and efficiency or minimizing noise in voltage-controlled oscillators (VCOs).

The classical approach to the design of oscillators involves four steps:

1. Select a circuit structure and method of obtaining feedback.
2. Choose bias conditions that provide adequate output power.
3. Adjust the feedback to obtain appropriate negative resistance or conductance at a port.
4. Select a termination impedance, at that port, which satisfies the oscillation conditions.

These steps guarantee only that oscillation will begin at the desired frequency; they do not precisely establish the amplitude or frequency of the large-signal, steady-state oscillation. Without the use of nonlinear analysis, accurately estimating the output power can be especially difficult, because the factors that limit the output-voltage range are not easy to identify.

The transistor oscillators we examine in this section consist of a positive-feedback amplifier that has a resonator as an input termination. Selecting the oscillator circuit's structure primarily involves selecting the type of amplifier; the choice of amplifier depends strongly on the application of the oscillator. For example, a common-gate circuit is usually preferred for VCOs, but for fixed-frequency oscillators using dielectric resonators, a common-source configuration is often preferred. We shall examine this matter further in Section 12.1.5.1.

Bias conditions are chosen in a manner similar to that used for a class-A power amplifier: V_{dc} and I_{dc} are chosen to allow a wide enough variation of the RF voltage and current to provide acceptable output power. If output power is not an important consideration, any bias point in the transistor's saturation region that provides good transconductance is probably acceptable; V_{dc} is often made equal to the drain voltage that the device would have when used as an amplifier, and I_{dc} is often set to approximately half the maximum value, 0.5 I_{dss} for FETs.

12.1.5.1 Circuit Structure and Feedback

There exist a number of possible oscillator circuits and methods of obtaining feedback. Three of the most common are the following:

- A feedback configuration with a resonator providing the coupling;
- A transistor in common-gate or common-base configuration with an inductor in series with the gate or base;
- A transistor in common-source or common-emitter configuration with a capacitor in series with the source or emitter.

These configurations are shown in Figure 12.8. Although the figures show only FETs, the same configurations can be used with bipolar devices. Figure 12.8(a) shows a dielectric resonator, but a wide variety of resonant circuits are possible. Furthermore, although it is not shown explicitly, a dielectric resonator can be used with the configurations in Figures 12.8(b) and 12.8(c). Adding reactance in series with the FET's common terminal can introduce a negative real part into the input or output impedance. If the resistance is reasonably high (but not too high!), the designer has a large degree of freedom in selecting the load impedance and usually obtains well-behaved operation. It is also important to adjust the feedback and to design the load network so that oscillation can occur at only a single

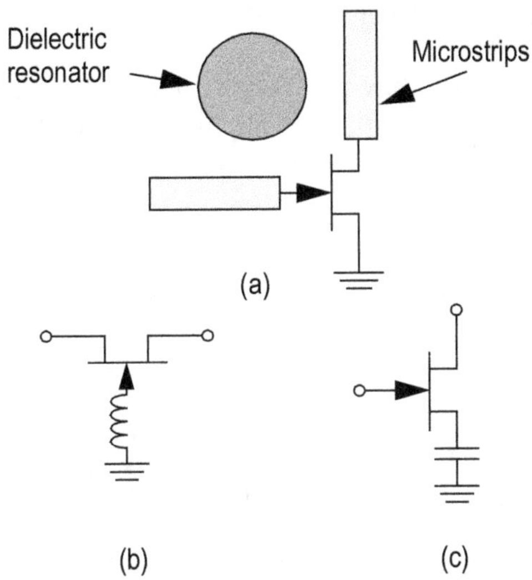

Figure 12.8 Three ways to obtain negative resistance: (a) coupling from the drain to the gate through a dielectric resonator; (b) series inductance in a common-gate circuit; (c) series capacitance in a common-emitter circuit.

frequency. If an additional resonance exists within the frequency range for which $\mathrm{Re}\{Y_s\} < 0$, the oscillator might oscillate at the frequency of that resonance instead of the desired one.

The choice of an oscillator configuration is rarely obvious. Because of resonator losses from weak coupling, feedback circuits, such as Figure 12.8(a), are practical only when the transistor's maximum available gain is high. The common-gate configuration in Figure 12.8(b) is probably the most practical; Figure 12.8(c) requires dc bypassing around the capacitor, usually an inductor, which could introduce a spurious resonance. Still, it is not usual to find that one of the configurations provides adequate negative resistance, while others do not. The superiority of one configuration or the other depends on frequency and characteristics of the particular device.

As with a small-signal amplifier, satisfying the oscillation and stability conditions at either the input or the output is enough to ensure oscillation.[2] Therefore, the load is usually chosen to provide adequate output power, and the input termination is chosen to satisfy the oscillation conditions. The design process is illustrated by the following example.

2. It is possible to show, in the linear case, that satisfying the oscillation conditions at one port of a feedback amplifier automatically satisfies them at the other. See [8.1].

12.1.5.2 Example: VCO Design

We design a 10-GHz VCO having approximately 10 dBm of output power. This output level is well within the capability of a small-signal Ku-band MESFET. Its small-signal S parameters at 10 GHz, in common-source configuration, are

$$S = \begin{bmatrix} 0.86\angle -102^\circ & 0.10\angle 48^\circ \\ 2.90\angle 104^\circ & 0.47\angle -48^\circ \end{bmatrix} \tag{12.14}$$

We also choose bias values of $V_{dc} = 3.0$V and $I_{dc} = 30$ mA, the conditions under which the S parameters were measured. The drain current is approximately 0.5 I_{dss}, the value that provides maximum gain in amplifier operation.

The MESFET is used in a common-gate configuration; an inductor in series with the gate provides feedback. The input (the MESFET's source terminal) is terminated by a varactor-tuned resonator, and the output is terminated by a load that ensures good output power. We choose the load impedance on the basis of the output-power requirement; we then design the resonator to satisfy the oscillation conditions.

We first estimate the load conductance. We find it by making a very rough estimate of the fundamental-frequency RF components of the drain voltage and current. If $I_{dc} = 30$ mA, the fundamental RF current must be less than 30 mA; we estimate it to be 25 mA. In Section 12.1.4 we saw that the fundamental RF drain voltage must be considerably less than V_{dc}. Although we derived this result by considering a common-source circuit, the same is approximately true for the common-gate circuit. Accordingly, we assume that the peak RF drain voltage is approximately 1V. The output power, P_{out}, is

$$P_{out} = \frac{1}{2} V_{d,1} I_{d,1} = 12.5 \text{ mW} \tag{12.15}$$

The real part of the load conductance, G_L, is

$$G_L = \frac{I_{d,1}}{V_{d,1}} = \frac{1}{40} \text{ S} \tag{12.16}$$

or a shunt resistance of 40Ω. Because the output impedance of a common-gate circuit is usually capacitive, the output load that provides maximum power is usually inductive. However, the susceptance found from small-signal S parameters may not be optimum for large-signal operation; furthermore, $S_{1,2}$ in the common-gate FET has a high value, approximately 1.8, indicating that the output admittance is sensitive to the input termination. Consequently, it is probably futile to predict the large-signal load susceptance from small-signal considerations. Furthermore, because of the complications introduced by the feedback inductance and the limited range of ΔV_d, the approach to output-load design used for the FET power amplifier is also invalid. For these reasons, in this first-order design we use a purely resistive load. The lack of a load susceptance may cost a decibel or two of output power; this power loss can be reclaimed by empirical tuning or by optimizing the design by nonlinear analysis.

We now must make certain that oscillation conditions are satisfied when this value of load resistance is used. To do so, we adjust the circuit to obtain negative resistance at the input port when the output termination is a 40Ω resistor. Immediately we are faced with a question: just how much negative resistance do we need? Again, we come to the profound conclusion that we want neither too much nor too little. If the negative resistance is low, any small series resistance, perhaps from circuit losses, may eliminate it; if it is too high, shunt conductivity may do the same thing. Empirically, we find a negative resistance around -40Ω to -100Ω to be about right. This value can be obtained easily by setting the port impedance to 100Ω and maximizing the magnitude of the input reflection coefficient, Γ_{in}.

We find that a feedback (gate) inductance of 1.24 nH maximizes $|\Gamma_{in}|$, giving $|\Gamma_{in}|^2 = 10.4$ dB, or $Z_{in} = -65 - j38$. To resonate this impedance, we need an inductive reactance of 38Ω at 10 GHz. To satisfy (12.8) strictly, we need a termination of impedance $65 + j38$; however, this would leave no room for the oscillation to grow, so we simply select $Z_L = j38$. We expect that, as the oscillation grows, the input reactance will change slightly, but the oscillation frequency will change to maintain resonance. The amount that the frequency must change depends inversely on the resonator's Q.

If a capacitive reactance were needed, we could simply connect a varactor directly to the FET's source terminal with, of course, necessary dc blocks and bias circuitry. Since an inductance is needed, however, we could use a varactor in series with an inductor or transmission line. A little experimentation shows, for example, that a 100Ω transmission line 55 degrees long, and a 0.55-pF varactor, provides the desired resonance. It is important to calculate the impedance of the combination over the range of frequencies where $|\Gamma_{in}| > 1$ (i.e., where $\text{Re}\{Z_{in}\} < 0$) to make sure that there are no other resonances at which oscillation could occur.

Better stability could be achieved by a resonator having a high Q and good frequency stability, such as a dielectric resonator. Because the phase of a resonator's reflection coefficient varies rapidly with frequency near resonance, the oscillator's frequency will remain close to the resonator's resonant frequency, even if the FET's S parameters drift enough, with temperature or dc bias, to change Z_{in}. A varactor can then be coupled to the resonator in order to vary its resonant frequency, and thus to vary the oscillator's frequency.

This circuit is a very popular one for realizing wideband VCOs: even simple oscillators of this form can achieve remarkably wide tuning bandwidths, often over an octave.

In designing the resonator and its tuning circuit, we find that there is a direct trade-off between tuning range and frequency stability; more generally, there is a trade-off in the design of any VCO between tuning range and phase noise. To achieve high stability (or low noise) one must use a stable, high-Q resonator (e.g., a dielectric resonator or a waveguide cavity) and couple the FET and varactor to it very weakly. This weak coupling limits the ability of the tuning varactor to vary the resonator's frequency, however, so the tuning range is narrow. Coupling the varactor more strongly to the circuit increases its effect on the circuit and provides a wider tuning range; unfortunately, such strong coupling also increases the effect of its poor thermal stability, and it may also reduce the Q of the resonator.

Another phenomenon worth noting is that Z_{in} is not constant with changes in the output load. As Z_L varies, so does Z_{in}; then, the frequency at which the oscillation conditions are satisfied must also vary, so the oscillator frequency must change. This phenomenon, called *pulling*, is also more serious when the Q of the resonator is low and the varactor is tightly coupled.

Although relatively crude estimates of the output conductance and power were necessary for the initial design, after the oscillator is fabricated one can use empirical techniques to obtain an improved estimate of the load impedance. One method that has been widely accepted is called a *device-line measurement* [12.3]. In this process, RF power is applied to the output terminals of the unmatched oscillator and the output-reflection coefficient Γ_{out} is measured under large-signal, nonoscillating conditions. Because the magnitude of the large-signal reflection coefficient is greater than unity, power is delivered by the oscillator to the measurement system during this test. That power, P_d, is

$$P_d = P_{av}(|\Gamma_{out}|^2 - 1) \tag{12.17}$$

where P_{av} is the available power of the excitation source. When the oscillator is delivering P_d to the measuring system, it has the large-signal output-reflection coefficient Γ_{out}. This is the same Γ_{out} it would have in normal operation, delivering P_d to a load Γ_L in which oscillation conditions were satisfied. Therefore, the load Γ_L must result in an output power P_d. The underpinnings of this argument are essentially the same as those of load-pull theory, which is applied to power amplifiers; the greatest limitation is that effect of the load impedance at harmonic frequencies is neglected.

12.2 NONLINEAR ANALYSIS OF TRANSISTOR OSCILLATORS

The oscillator design illustrated in Section 12.1.5.2 was adequate only to guarantee that the circuit would oscillate. Because it was based on small-signal, linear S parameters, it was not possible to estimate the output power or to find a load impedance that optimized the output power. Thus, we were forced, with much embarrassment, to use rather crude estimates in the design of the output network. We would prefer to use our knowledge of nonlinear analysis to define the load impedance and predict the output power more precisely. A second concern is the frequency of oscillation. The oscillation frequency determined through linear analysis is approximate. When a low-Q resonance is used, the frequency of oscillation can be significantly different from that predicted by the linear analysis. Finally, we might like to simulate certain characteristics of the oscillator, like phase noise, pulling, and dc-bias sensitivity, that are fundamentally nonlinear and therefore cannot be addressed by linear analysis.

Several problems arise when one attempts to analyze oscillators by harmonic-balance techniques: the first is that the saturation phenomena that limit the amplitude of the oscillation must be modeled very carefully. In particular, the channel current $I_d(V_g, V_d)$ as well as the gate-to-channel capacitances must be modeled accurately throughout both the linear and saturation regions. The second problem is more serious. Harmonic-balance analysis is used primarily when a nonlinear circuit is driven from an external source. It is assumed at the outset of a harmonic-balance analysis that the excitation frequency is known exactly. In an oscillator analysis, however, the frequency of oscillation is not known; it is one of the things that the analysis must determine. Clearly, the frequency of oscillation must be one of the independent variables of the harmonic-balance process.

Finally, the phase of the oscillation is indeterminate, so unless the phase is somehow constrained, it is impossible for a harmonic-balance process—however formulated—to converge.

One solution to the latter problem is to use time-domain analysis. In this case, which involves integrating the circuit's nonlinear differential equations numerically until a steady-state response is obtained. The process suffers from the standard limitations of time-domain analysis: transmission lines and lumped impedances often cannot be modeled adequately (a serious problem, in view of the fact that the resonators in microwave oscillators invariably use transmission lines), and long time constants may cause the settling time of the transient response to be long. Furthermore, the presence of a high-Q resonator in the oscillator's circuit may introduce numerical instability. A more subtle difficulty is that all oscillators have a valid "zero solution": nonoscillation invariably satisfies the circuit equations, and an analysis—either time-domain or harmonic-balance—can easily have this trivial result.

In view of the complexity of oscillator analysis, it is not surprising that many techniques have been developed. We examine a few of them in the following sections.

12.2.1 Numerical Device-Line Measurements

One possibility is to use harmonic-balance analysis to perform a numerical device-line measurement. This analysis is straightforward and can be performed on any harmonic-balance simulator. The process is as follows:

1. Design the oscillator as in the above example. Use a nonlinear model for the device; this design process is nonlinear, so S-parameter characterization cannot be used.
2. Connect a source and a power/impedance measuring device to the output port.
3. Excite the output port, and vary the power.
4. Find the point where output power is maximum; determine the port impedance at this power level.
5. Design an output network that provides this load impedance to the device.

Figure 12.9 illustrates the analysis. Figure 12.9(a) shows the oscillator, designed according to linear theory. Figure 12.9(b) shows the circuit for the power sweep; the block on the right represents the oscillator, which is treated as a subcircuit. The power-measuring element is a voltage/current sampler, which provides information to the circuit simulator for both power and impedance. Because negative resistances degrade matrix conditioning, harmonic-balance simulators often experience convergence difficulties in oscillator analysis. To avoid this problem, a 1,000Ω resistor is inserted in series with the port. The resistor makes high excitation power necessary, but this is of no consequence, as the excitation power is intrinsically meaningless. Note that we plot the real part of the output power, not the magnitude of the power.

The plot shows that maximum output power of 14.4 dBm occurs at an excitation level of 32 dBm. The impedance, at that same excitation level, is – 79 – j48. This impedance is the negative of the optimum load impedance; thus, the oscillator requires a load of 79 + j48. Synthesizing such a load is a minor, final step. After the output matching network is designed, it is wise to sweep the circuit over the entire range of frequencies where negative resistance exists, to make certain that no spurious resonances exist.

In the above example, the output impedance of the oscillator displayed a series resonance. In that case, a large series resistor was appropriate. Occasionally, however, an oscillator's output impedance is best modeled as a parallel resonance, and in that case a small, shunt resistance usually works better. In either case, however, the level of the excitation source is irrelevant [because the output power is determined by direct measurement in the voltage/current sampler, not by (12.17)], as is the value of the series or shunt resistor at the port; use whatever works best.

A disadvantage of this method is the poor treatment of harmonic terminations. In the analysis, the port is terminated, at all harmonics, in a high resistance. This situation is generally unrealistic. Another problem is that it cannot account for the effects of a buffer amplifier or other type of load; the method can be applied only to the oscillator stage.

12.2.2 Harmonic Balance: Method 1

It is clearly valuable to have a complete harmonic-balance analysis of oscillators, in which the user simply describes the circuit and the simulator provides output power, frequency, and all required voltage and current waveforms, much as is done in forced circuits. In this section, we describe two approaches to such an analysis. To do so, we must deal with the problems of (1) unknown frequency, (2) indeterminate phase, (3) spurious "zero solution," and (4) difficult convergence. Additionally, we should consider the need for synthesis as well as analysis: it may be more valuable to adjust some circuit parameter to achieve the desired frequency than to adjust the frequency to satisfy the circuit equations.

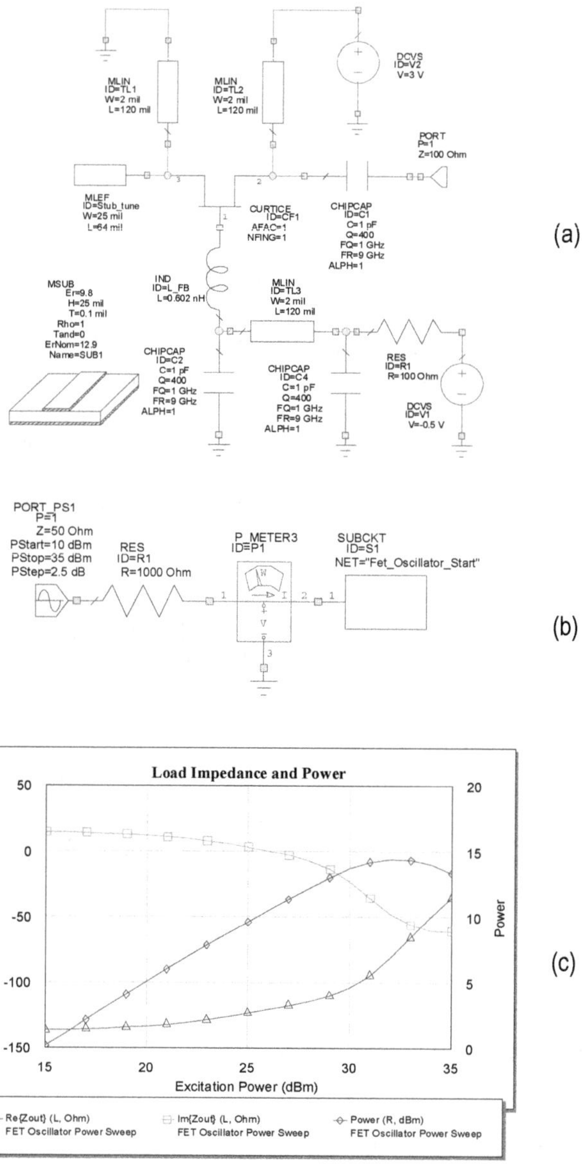

Figure 12.9 Oscillator design by a numerical device-line measurement: (a) the oscillator circuit; (b) circuit for performing the large-signal device-line simulation; (c) port impedance and output power.

One straightforward approach [12.4] is to use one or more circuit parameters as variables, along with the harmonic voltages across the nonlinear elements. The current-error function (3.1.20) becomes

$$\mathbf{F}(\mathbf{V}, \mathbf{P}) = \mathbf{I}_s(\mathbf{P}) + \mathbf{Y}(\mathbf{P})\mathbf{V} + j\Omega\mathbf{Q} + \mathbf{I}_G \tag{12.18}$$

where **P** is a set of circuit parameters. For oscillator analysis, **P** could include the frequency of oscillation; conversely, for synthesis, it could include a varactor capacitance that is adjusted to obtain the specified frequency of oscillation. The components of **P**, as well as the voltage variables, are adjusted by the appropriate harmonic-balance algorithm, and convergence is indicated, as usual, by minimal values of the components of **F**. The authors of [12.4] emphasize the fact that the variables **V** and **P** can be adjusted simultaneously in the solution algorithm; it is not necessary to solve the harmonic-balance equations to obtain **V** and then to optimize the circuit variables **P**. An advantage of the authors' approach is that, by defining the error function to include output power or other performance parameters, one can include performance optimization in conjunction with the standard harmonic-balance analysis. Additionally, for the analysis case, the Jacobian includes derivatives of the frequency with respect to the system voltages; this quantity can be very useful for phase-noise analysis.

12.2.3 Harmonic Balance: Method 2

Another clever method [12.5] involves inserting a probe into the circuit in some appropriate place. The probe consists of a voltage source and series impedance at the fundamental frequency, and an open circuit at harmonics. The magnitude and frequency of the excitation is adjusted, in an external optimization loop, until the harmonic-balance equations are satisfied. Oscillation conditions are satisfied when the fundamental-frequency current in the source is zero.

Figure 12.10 shows a feedback oscillator designed according to this method. The probe element is connected to the base of the HBT; because of its sensitivity, this is usually a good connection point. The frequency and voltage ranges over which the circuit simulator will search for a solution are parameters of the probe; it is set to search the range of 3 to 5 GHz in as many as 1,000 steps. Similarly, the voltage range is set to search over 250 steps. These parameters describe a coarse search used only to find the frequency and voltage region within which the oscillator operates; a fine search begins after the coarse one is completed. This method is very robust, and works well even when the resonator Q is quite high.

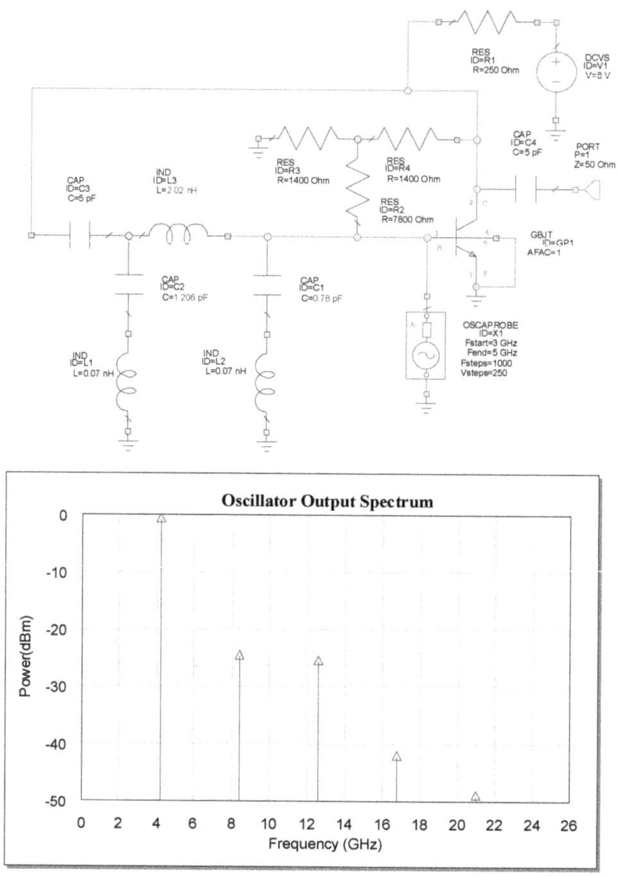

(a)

(b)

Figure 12.10 Oscillator designed according to [12.5]: (a) the oscillator circuit; (b) its output spectrum.

This method is similar, in many ways, to an automated version of the device-line measurement. However, it provides a more correct excitation source and is more versatile, allowing the probe to be used at any appropriate point in the circuit.

12.2.4 Eigenvalue Formulation

It is interesting to note that the oscillator problem can be formulated as a classical eigenvalue problem. This gives a rigorous method for determining

(1) oscillation frequency, (2) whether oscillation conditions are satisfied, (3) possibilities for satisfying oscillation conditions at multiple frequencies, and (4) the fundamental-frequency voltages at all nonlinear ports. It suggests, also, a large-signal method for analyzing an oscillator.

First, we examine the linear case. Imagine that the nonlinear circuit is linearized at the bias point. Let **Z** be the impedance matrix of the linear subcircuit and **Y** be the admittance matrix of the linearized, nonlinear subcircuit. Then,

$$-\mathbf{I} = \mathbf{YV} \tag{12.19}$$

where **I** is the vector of fundamental-frequency current components in the linear subcircuit. To have oscillations, this current must excite the linear subcircuit and generate a voltage equal to **V**:

$$\mathbf{V} = \mathbf{ZI} \tag{12.20}$$

This implies

$$-\mathbf{ZYV} = \mathbf{V} \tag{12.21}$$

In general, however, the voltage generated by the current is greater than **V**, and as oscillations build, the device saturates and (12.21) is satisfied. Therefore, at startup we have

$$-\mathbf{ZYV} = \lambda\mathbf{V} \tag{12.22}$$

where λ is real and $\lambda > 1$. This is a classical eigenvalue problem, and we can state that the oscillation conditions, for the linear circuit, are that the $-\mathbf{ZY}$ matrix must have an real eigenvalue equal to 1.0 (or, for the practical case, greater than 1.0). The corresponding eigenvector, **V**, gives the relative, but not absolute, port voltages. It is probably valid to claim that this applies in the Kurokawa sense; that is, if we define a large-signal admittance $\mathbf{Y}(\mathbf{V}_0)$ analogous to (12.10), oscillation conditions are satisfied when $\lambda = 1$.

In the strict harmonic-balance case, we must satisfy

$$-\mathbf{ZF}_\mathbf{I}(\mathbf{V}) = \mathbf{V} \tag{12.23}$$

where $\mathbf{F_I(V)}$ is the vector of fundamental-frequency currents at all the ports of the nonlinear subcircuit. The harmonic-balance problem could be solved by first using the linearized case as an initial estimate, then increasing the magnitude of the **V** eigenvector, in a continuation loop, correcting it with the harmonic-balance process as λ decreases. The process terminates when $\lambda = 1$.

12.3 PRACTICAL ASPECTS OF OSCILLATOR DESIGN

12.3.1 Multiple Resonances

Multiple resonances are a serious problem in oscillator design. If an oscillator has a resonance at some frequency other than the desired one, it can establish oscillations at that undesired resonant frequency. Often, the oscillator works properly at room temperature, with the expected bias voltages, but when the temperature changes or the bias voltage drifts, the oscillator suddenly jumps to an undesired frequency.

Undesired resonances can be introduced by matching and bias circuits. Especially at high frequencies, the additional parasitics of packaged devices can introduce unwanted resonances. Housings can also be sources of such resonances.

The fundamental cause of multiple resonances is complexity. If oscillators really could be made as simple as the idealized circuit in Figure 12.9, multiple resonances would rarely be a problem. To avoid such problems, the circuit should be made as simple as possible, and should be analyzed over the full range of frequencies where it exhibits negative resistance. Here, especially, Maas' First Law of Microwaves applies: the simplest circuit that works, works best.

12.3.2 Frequency Stability

Frequency stability is governed by the Q of the resonator and the sensitivity of its resonant frequency to temperature. Since the transistor is inevitably coupled to the resonator, its temperature-sensitive parasitic elements also affect the resonant frequency.

When the resonator's Q is high, it changes less when the transistor's characteristics drift with temperature. To illustrate this point, consider a series resonant circuit. Its Q can be written

$$Q = \frac{\omega_0}{2R}\frac{dX}{d\omega} \tag{12.24}$$

where ω_0 is the resonant frequency, R is the series resistance, and X is the total reactance. This can be rearranged to

$$d\omega = \frac{\omega_0}{2RQ}dX \tag{12.25}$$

showing that a high Q reduces the change in frequency $d\omega$ when the reactance, X, changes.

Of course, the resonant frequency of the resonator itself must be stable with temperature. In the past, temperature-stable metal alloys such as Invar were used for resonant cavities. Now, dielectric resonators are more likely to be used. Dielectric resonators can be formulated to have virtually no thermal drift, or even temperature coefficients that compensate for drift in other parts of the oscillator.

12.3.3 Dielectric Resonators

Many microwave systems require free-running sources that are highly stable. For these applications, fixed-frequency oscillators using dielectric resonators are ideal. Dielectric resonators are made from modern ceramic materials that have low thermal expansion coefficients and high dielectric constants, usually around 40. This makes them much smaller than waveguide or coaxial resonators. Because these materials have very low loss, the Qs of dielectric resonators are nearly as high as those of metal cavities. Furthermore, the temperature coefficients of the dielectrics can be adjusted by varying the composition of the ceramic. Dielectric resonators are usually solid cylinders, although occasionally they are realized as hollow cylinders or rectangular blocks.

When used in a microstrip circuit, a dielectric resonator is coupled magnetically to a microstrip line. The coupling coefficient is adjusted by varying the distance from the resonator to the edge of the microstrip, or by placing the resonator on top of a dielectric spacer. When coupled to a single line, the resonator open-circuits the line at its resonant frequency; when coupled to two microstrips, a dielectric resonator can be operated in a transmission mode.

Cylindrical dielectric resonators usually operate in their dominant mode, the $TE_{0,1\delta}$. This mode is shown qualitatively in Figure 12.11, and it

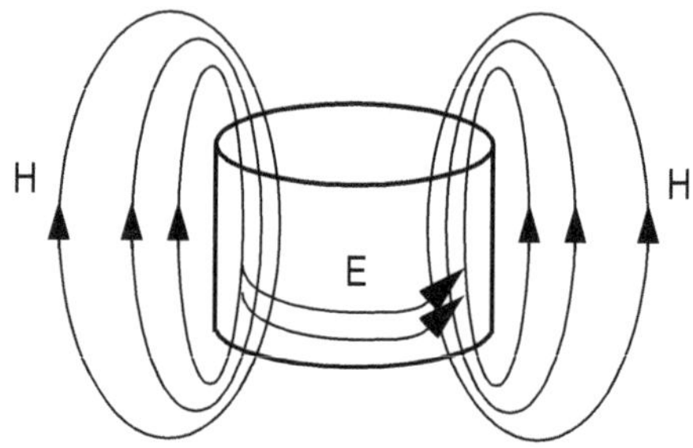

Figure 12.11 E and H fields of the $TE_{0,1\delta}$ mode in a solid cylindrical dielectric resonator.

should be clear, from its structure, that it couples nicely to the magnetic field around a microstrip transmission line. The "δ" in the mode indices indicates that the mode is not completely contained by the dielectric structure; the magnetic field, in particular, "leaks" from the dielectric, and unless measures are taken to prevent it, the radiative loss is substantial. A dielectric resonator therefore must be shielded, or radiative loss reduces its Q dramatically. Usually, the resonator is mounted, along with the rest of the oscillator circuit, in a metal cavity. The resonant frequency can be adjusted somewhat by a tuning screw located above the resonator.

Figure 12.12 shows a simple FET oscillator using a dielectric resonator. The design process is identical to the ones we have been discussing. For example, the design described in Section 12.1.5.2 can be converted to a dielectric-resonator oscillator. The resonating impedance of $+j38\Omega$ is realized by a piece of transmission line, and the dielectric resonator is coupled to the line at the point where the open circuit is required. Finally, a load resistor provides stability by preventing oscillation at other frequencies. The load has no effect at the frequency of oscillation, because it is decoupled from the circuit by the dielectric resonator.

The theory of dielectric resonators is a separate discipline; that theory, and further information on the use of dielectric resonators in filters as well as in oscillators, can be found in [12.6–12.10].

12.3.4 Hyperabrupt Varactors

We saw earlier that the capacitance of a *pn* or Schottky junction, $C(V)$, is given by

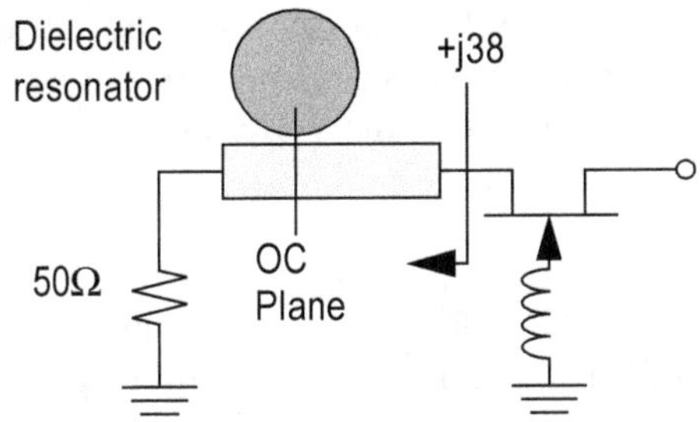

Figure 12.12 The oscillator of the design example, modified to use a dielectric-resonator.

$$C(V) = \frac{C_{j0}}{\left(1 - \frac{V}{\phi}\right)^{\gamma}} \tag{12.26}$$

where C_{j0} is the zero-voltage capacitance, V is the junction voltage, and ϕ is the built-in voltage. The parameter γ is usually close to 0.5 in Schottky junctions, but in *pn* junctions, it depends on the doping profile. A linearly graded junction has $\gamma = 0.33$, while an abrupt junction has $\gamma = 0.5$.

Suppose we wish to use a varactor diode as a tuning element. We might reasonably want a linear tuning characteristic, in which the resonant frequency is proportional to the control voltage, V. For an LC resonator, in which the entire capacitance comes from the diode, a little algebra shows that

$$\frac{df}{dV} = k\left(1 - \frac{V}{\phi}\right)^{\frac{\gamma}{2} - 1} \tag{12.27}$$

where k is a constant. We see that linear tuning requires $\gamma = 2.0$. This is a much stronger capacitive nonlinearity than is normally encountered in junction diodes. The use of a sharply graded doping profile in a *pn* junction, however, can approximate this condition over a modest range of junction voltages; such diodes are called *hyperabrupt varactors*. Even when a linear tuning characteristic is not possible, the wide capacitance range of such diodes is valuable for wide-range VCOs.

To achieve the wide capacitance variation, the doping density must be greater near the junction and decrease in the direction into the semiconductor. This results in a relatively high series resistance, and thus a lower Q. This lower Q is the price of the wide tuning range.

12.3.5 Phase Noise

Noise processes in semiconductor devices can modulate the phase of an oscillator and create noise sidebands in its output spectrum. This phase deviation is a serious problem in systems where a signal's phase carries information, especially in modern communication systems using phase or phase-amplitude modulation. Because high-frequency components of that noise are attenuated by the resonator, noise processes that generate low-frequency components are of most concern: $1/f$ noise is a particularly significant contributor to phase noise; accordingly, devices having low $1/f$ noise levels are usually preferred for use in oscillators. Bipolar transistors (both HBTs and conventional homojunction devices) have significantly lower $1/f$ noise levels than MESFETs or HEMTs, so they are often preferred for use in oscillators when possible.

Phase noise is characterized by the ratio of carrier noise spectral density or, equivalently, the spectral power in a 1-Hz bandwidth, at an offset f_m from the carrier. This quantity is designated $\mathscr{L}(f_m)$ and is illustrated in Figure 12.13.

Phase noise can be minimized not only by using a low-noise device, but also by using a resonator having a high loaded Q; in VCOs, a high-Q varactor is necessary. Because phase noise can also be caused by noise originating in the power supply or coupled to the dc bias circuits, power-supply filtering should not be overlooked as a means to minimize phase noise.

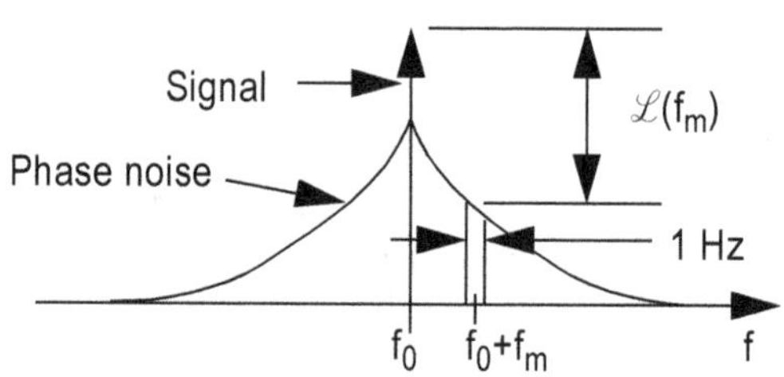

Figure 12.13 Signal and noise spectrum of an oscillator.

12.3.5.1 Phase-Noise Analysis

We can develop an understanding of phase noise by first viewing the noise as a sinusoidal phase perturbation. Suppose we have a signal, $v(t)$, with a sinusoidal phase perturbation of β radians at a radian frequency $\omega_m = 2\pi f_m$:

$$v(t) = V_s \cos(\omega_0 t + \Delta\phi \sin(\omega_m t)) \tag{12.28}$$

The frequency is the time derivative of phase. Differentiating the argument gives

$$\omega(t) = \omega_0 + \Delta\phi\omega_m \cos(\omega_m t) \tag{12.29}$$

showing that the frequency deviation $\Delta\omega = \Delta\phi\ \omega_m$, or

$$\Delta\phi = \frac{\Delta\omega}{\omega_m} \tag{12.30}$$

In frequency modulation (FM) theory, $\Delta\phi$ is sometimes called β, the *modulation index*. Clearly, the phase deviation must be small, so $\Delta\phi \ll 2\pi$. This condition corresponds to narrowband FM, so we can use the results of narrowband FM theory to obtain the spectrum. After consulting any good communications theory book, we obtain

$$\frac{V_{ssb}}{V_s} = \mathrm{J}_1(\Delta\phi) \approx \frac{\Delta\phi}{2} \tag{12.31}$$

where V_{ssb} is the level of a single sideband and J_1 is a Bessel function. The carrier-to-sideband ratio is then

$$\left(\frac{V_{ssb}}{V_s}\right)^2 = \frac{\Delta\phi^2}{4} \tag{12.32}$$

Since $\Delta\phi$ is small, this approximation is virtually exact; so, offending only the most anal-retentive mathematicians, we have replaced the approximation sign with an equality.

This is fine for sinusoidal phase deviations, but we are really interested in noise, not sinusoids. To convert the above results to noise, we equate the sinusoidal and noise cases on the basis of power. For a sinusoid, the RMS and peak values are related as

$$\sqrt{2}\Delta\phi_{RMS} = \Delta\phi_{\sin} \tag{12.33}$$

so the carrier to noise ratio $\mathcal{L}(f_m)$ becomes

$$L(f_m) = \left(\frac{V_{ssb}}{V_s}\right)^2 = \frac{1}{2}\Delta\phi^2{}_{RMS} \tag{12.34}$$

where $\Delta\phi^2{}_{RMS}$ represents the RMS value of either the sinusoid or the noise process.

Sinusoid and Noise

In many systems, noise is added to a carrier and the combination is limited in amplitude. The limiting removes the amplitude component of the noise, but not the phase component. The resulting phase noise can be found easily. Figure 12.14 shows the combined signal and noise phasors. From Figure 12.14, the phase deviation is

$$\Delta\phi(t) = \operatorname{acos}\left(\frac{n(t)}{v(t)}\right) \approx \frac{n(t)}{v(t)} \tag{12.35}$$

The mean square noise can be defined by a noise factor, F:

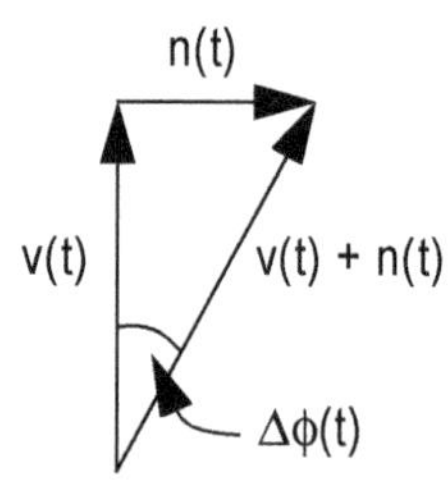

Figure 12.14 When a signal plus noise process is limited, amplitude variations are removed and only phase variations remain. Since the oscillator's transistor is driven into hard saturation, it acts as a limiter, removing most AM noise.

$$\overline{n^2(t)} = FKT_0R \tag{12.36}$$

where K is Boltzmann's constant, $1.37 \cdot 10^{-23}$ J/K; $T_0 = 290$K, by definition; and R is the load resistance at which $n(t)$, which has units of voltage, is measured. The signal power is

$$\overline{v^2(t)} = PR \tag{12.37}$$

where P is the power dissipated in R. Substituting (12.35) through (12.37) into (12.34) gives

$$L(f_m) = \frac{1}{2}\Delta\phi^2{}_{\text{RMS}} = \frac{1}{2}\frac{FKT_0}{P} \tag{12.38}$$

or, in dBC,

$$L(f_m) = -174 + F - P - 3 \tag{12.39}$$

Noise Spectrum and Leeson's Model

The previous relations assume that the noise is white. In reality, the dominant noise process is the upconversion of $1/f$ noise by the oscillator's nonlinearities. We can assume that this power spectrum is centered on the carrier and has the form

$$\overline{v_n^2(f_m)} = FKT_0\left[1 + \frac{f_c}{f_m}\right] \tag{12.40}$$

where f_c is the *corner frequency* of the noise and, as before, f_m is the deviation from the carrier in either a positive or negative direction. The power spectrum of the phase fluctuations, $S(f_m)$, is

$$S(f_m) = \Delta\phi^2{}_{\text{RMS}} = \overline{\Delta\phi^2} = \frac{FKT_0}{P}\left[1 + \frac{f_c}{f_m}\right] \tag{12.41}$$

Plotted on a logarithmic scale, the noise spectrum has the shape shown in Figure 12.15.

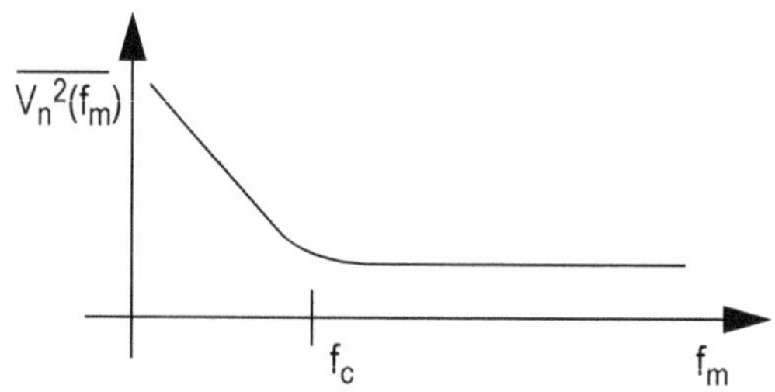

Figure 12.15 $1/f$ noise spectrum showing the corner frequency, f_c.

In 1966, Leeson [12.11] proposed a simple model of a noisy oscillator by treating it as a phase-feedback system with added noise. The added noise is a high-frequency noise spectrum, which consists of both broadband noise and upconverted $1/f$ noise. The model does not treat the upconversion process, so it is valuable only for its qualitative, not quantitative predictions. Even so, it provides considerable insight into oscillator operation.

The oscillator model is shown in Figure 12.16(a). It consists of an amplifier, a resonator, and feedback. Noise is added at the amplifier's input, and the oscillator's output port is the amplifier's output port. Leeson showed that this circuit can be represented as the baseband circuit in Figure 12.16(b), in which the variable quantity is the oscillator's phase. In Figure 12.16(b), the resonator becomes a low-pass filter and the "output" is the phase, not the signal itself.

We now can apply ordinary feedback theory to the circuit of Figure 12.16(b). The transfer function of the low-pass filter, $T(f_m)$, is

$$T(f_m) = \frac{1}{1 + j2Q_L\dfrac{f_m}{f_0}} \tag{12.42}$$

where Q_L is the loaded Q of the resonator and f_0 is the frequency of the oscillator. The transfer function between the phase of the noise and that of the oscillator's output is

$$\Delta\phi = \frac{1}{1 - T(f_m)}\Delta\theta \tag{12.43}$$

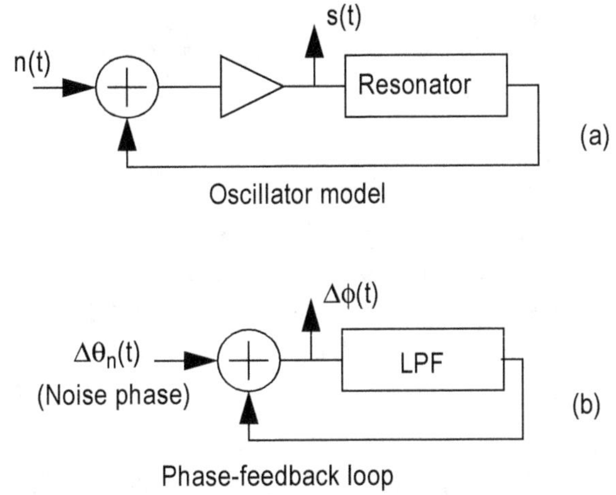

Figure 12.16 (a) Leeson's model of a noisy oscillator; (b) the equivalent circuit, in which phase is the variable.

Substituting (12.42) into (12.43) and using (12.41) and (12.38), we obtain

$$L(f_m) = \frac{1}{2}S_\phi(f_m) = \frac{1}{2}\frac{FKT_0}{P}\left[1 + \frac{f_c}{f_m}\right]\left[1 + \frac{f_0^2}{f_m^2}\frac{1}{4Q_L^2}\right] \tag{12.44}$$

The loop acts as a kind of filter on the phase noise. According to (12.44), there are two break points in the phase noise spectrum: one at the corner frequency, f_c, and another at $f_m = f_0/2Q_L$. At frequencies well below both break points, the phase-noise spectrum has a slope of 30 dB per decade; at higher frequencies, regions can exist where the slope is either 20 dB per decade or 10 dB per decade, depending upon the relative values of f_c and $f_0/2Q_L$. The possible spectra are shown in Figure 12.17. Note that these depend on the assumption that the dominant noise source has a $1/f$ spectrum; often the spectrum is not precisely $1/f$, so the phase-noise spectrum may deviate from this ideal case.

Other Sources of Phase Noise

It is important to recognize that phase noise can arise from sources other than the noise in the transistor. Some important sources are the following:

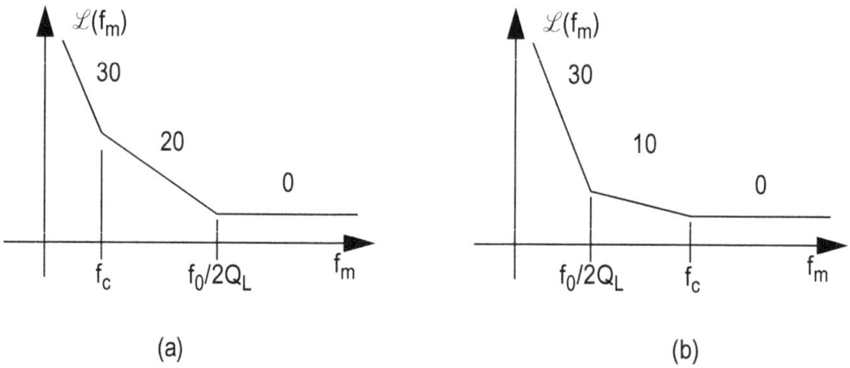

Figure 12.17 Phase noise spectra: (a) "low-Q" case, in which $f_0 / 2Q_L > f_c$; (b) "high-Q" case, in which $f_0 / 2Q_L < f_c$. The former corresponds to VCOs and oscillators having microstrip resonators; the latter, to DROs and oscillators using resonant cavities.

- Power supply noise can easily modulate the phase of an oscillator. The power supply must be well filtered to remove such noise. For measurements, a battery can be used to eliminate this noise source.
- Coupling from the ac line is invariably evident in measurements of the phase-noise spectrum of low-noise oscillators. Peaks at the ac line frequency, and its harmonics, are invariably present. If the peaks are not too great, they can simply be ignored; however, large peaks can degrade the accuracy of a phase noise measurement and make it difficult to interpret. In many cases, it may be necessary to shield the oscillator during the measurement.
- Mechanical vibration can generate phase fluctuations that appear as phase noise. Ambient mechanical vibration has frequency components from a few hertz to a few kilohertz; this is just the right range to corrupt most phase-noise measurements.
- The noise of a buffer amplifier, especially if it uses active biasing, can degrade the oscillator's phase noise.

12.3.5.2 Frequency Multiplication

Since frequency is the time derivative of phase, frequency multiplication is, in fact, phase multiplication. Multiplying the frequency by a factor, n, multiplies $\Delta\phi$ by n as well. From (12.38), we see that the phase noise increases by n^2 or, in decibels, 20 log(n).

12.3.6 Pushing and Pulling

In Section 12.1 we saw that changes in the load impedance could affect the oscillation frequency by changing the phase of Z_s. This phenomenon is called *pulling*. To some degree, pulling is inevitable; it occurs because the feedback necessary to make oscillation possible increases $S_{1,2}$, and thus increases the sensitivity of Z_s to the load impedance. Nevertheless, pulling can be minimized. Beyond the obvious solutions of using an output isolator or buffer amplifier, a high-Q resonator is effective in reducing pulling.

Similarly, changes in dc bias voltage can change the transistor's S parameters and Z_s, thus changing the oscillation frequency. This phenomenon is called *pushing*. The straightforward way to minimize pushing is to maintain adequate regulation in the oscillator's bias circuits. As with pulling, pushing is minimized by a high-Q resonator. Pushing is not always undesirable; it is sometimes used as a means to obtain voltage-tuning capability in a narrowband VCO.

12.3.7 Post-Tuning Drift

When the frequency of a VCO is changed, the RF current and voltage waveforms throughout the oscillator also change, as well as the dc bias current. As a result, the heat dissipated in the transistor and the tuning varactor, in blocking capacitors (which dissipate heat because of finite Q), and in coupling inductors all change as well. A small time interval is required before the circuit returns to thermal equilibrium and steady-state conditions. During this time the frequency may drift; this phenomenon is called *post-tuning drift.* It is most significant in fast-tuning, wide-range VCOs.

In a well-designed oscillator, the primary cause of post-tuning drift is heat dissipation in the varactor. Thus, careful thermal design of the varactor can reduce post-tuning drift significantly. If the varactor is mounted in a package, it should be mounted on a large metal surface; a beam-lead or chip device mounted on a substrate should be bonded to the substrate metallization over as large an area as possible.

12.3.8 Harmonics and Spurious Outputs

A well designed transistor oscillator should be free of spurious outputs that are not harmonically related to the frequency of oscillation. However, because the transistor is driven into saturation, most oscillators have significant harmonic outputs. Harmonic distortion can also occur in a buffer amplifier, which may be driven into saturation to level the output of a VCO. In most cases the designer has little control of the harmonic levels unless an output filter is used.

References

[12.1] R. W. Rhea, *Oscillator Design and Computer Simulation, Second Edition*, New York: McGraw-Hill, 1997.

[12.2] K. Kurokawa, "Some Basic Characteristics of Broadband Negative Resistance Oscillator Circuits," *Bell Sys. Tech. J.*, Vol. 48, 1969, p. 1937.

[12.3] W. Wagner, "Oscillator Design by Device Line Measurement," *Microwave J.*, Vol. 22, Feb. 1979, p. 43.

[12.4] V. Rizzoli, A. Lipparini, and E. Marazzi, "A General-Purpose Program for Nonlinear Microwave Circuit Design," *IEEE Trans. Microwave Theory Tech.*, Vol. MTT-31, 1983, p. 762.

[12.5] E. Ngoya et al., "Steady-State Analysis of Free or Forced Oscillators by Harmonic Balance and Stability Investigation of Periodic and Quasiperiodic Regimes," *Int. J. Microwave and Millimeter-Wave Computer Engineering*, Vol. 5, 1995, p. 210.

[12.6] D. Kajfez and P. Guillon (eds.), *Dielectric Resonators*, Norwood, MA: Artech House, 1986.

[12.7] S. J. Fiedziuszko, "Microwave Dielectric Resonators," *Microwave J.*, Vol. 29, Sept. 1986, p. 189.

[12.8] S. J. Fiedziuszko, "Dielectric Resonators Raise your High-*Q*," *IEEE Microwave Magazine*, Sept. 2001, p. 51.

[12.9] N. Elmi and M. Radmanesh, "Design of Low-Noise, Highly Stable GaAs Dielectric Resonator Oscillators," *Microwave J.*, Vol. 39, Nov. 1996, p. 104.

[12.10] M. Regis et al., "Design of a Low Phase Noise Ku-Band Oscillator Using a SiGe HBT," *Microwave J.*, Vol. 44, Oct. 2001, p.136.

[12.11] D. B. Leeson, "A Simple Model of Feedback Oscillator Noise Spectrum," *Proc. IEEE*, Vol. 54, Feb. 1966, p. 329.

About the Author

Steve Maas received BSEE and MSEE degrees in electrical engineering from the University of Pennsylvania in 1971 and 1972, respectively, and a Ph.D. in electrical engineering from UCLA in 1984. He joined the National Radio Astronomy Observatory in 1974, where he designed the low-noise receivers for the Very Large Array radio telescope. Subsequently, at Hughes Aircraft Company and TRW, he developed low-noise microwave and millimeter-wave systems and components, primarily FET amplifiers and diode and FET mixers, for space communication. He also has been employed as a research scientist at The Aerospace Corporation, where he worked on the optimization of nonlinear microwave circuits and the development of circuit-design software based on harmonic-balance, Volterra-series, and time-domain methods. He joined the UCLA Electrical Engineering Faculty in 1990 and left it in 1992. Since then, he has worked as an independent consultant and currently is chief scientist of Applied Wave Research, Inc.

Dr. Maas is the author of two other books, *Microwave Mixers* (1986 and 1992) and *The RF and Microwave Circuit Design Cookbook* (1998), both published by Artech House. From 1990 until 1992 he was the editor of the *IEEE Transactions on Microwave Theory and Techniques*, and from 1990 to 1993 he was an Adcom member and publications chairman of the IEEE MTT Society. He received the MTT Society's Microwave Prize in 1989 for his work on distortion in diode mixers and its Application Award in 2002 for his invention of the FET resistive mixer. He is a fellow of the IEEE.

Index

www.ingramcontent.com/pod-product-compliance
Lightning Source LLC
LaVergne TN
LVHW011855200926
848484LV00002B/457
* 9 7 8 1 5 8 0 5 3 4 8 4 0 *